ENCYCLOPEDIA OF

AQUARIUM
& POND FISH

ENCYCLOPEDIA OF

AQUARIUM & POND FISH

DAVID ALDERTON

PHOTOGRAPHY BY MAX GIBBS

THIRD EDITION

DK Delhi

SENIOR EDITOR Sreshtha Bhattacharya
ART EDITOR Mansi Agrawal
ASSISTANT EDITORS Ankona Das, Manan Kapoor
SENIOR PICTURE RESEARCHER Surya Sankash Sarangi
JACKET DESIGNER Priyanka Bansal
JACKETS EDITORIAL COORDINATOR Priyanka Sharma
DTP DESIGNERS Pawan Kumar, Rakesh Kumar, Syed Mohammad Farhan
PICTURE RESEARCH MANAGER Taiyaba Khatoon
MANAGING JACKETS EDITOR Saloni Singh
MANAGING EDITOR Kingshuk Ghoshal
MANAGING ART EDITOR Govind Mittal

DK London

SENIOR EDITOR Rob Houston
SENIOR ART EDITOR Duncan Turner
US EDITOR Jennette ElNaggar
US EXECUTIVE EDITOR Lori Cates Hand
JACKET EDITOR Emma Dawson
JACKET DESIGN DEVELOPMENT MANAGER Sophia MTT
PRODUCER, PREPRODUCTION Jacqueline Street-Elkayam
SENIOR PRODUCER Alex Bell
MANAGING EDITOR Angeles Gavira Guerrero
MANAGING ART EDITOR Michael Duffy
ASSOCIATE PUBLISHING DIRECTOR Liz Wheeler
ART DIRECTOR Karen Self
DESIGN DIRECTOR Philip Ormerod
PUBLISHING DIRECTOR Jonathan Metcalf

FIRST EDITION

Produced for Dorling Kindersley by

cobaltid

The Stables, Wood Farm, Deopham Road,
Attleborough, Norfolk NR17 1AJ
www.cobaltid.co.uk

ART EDITORS
Paul Reid, Darren Bland,
Pia Hietarinta, Lloyd Tilbury

EDITORS
Marek Walisiewicz, Kati Dye,
Maddy King, Steve Setford

PHOTOGRAPHY
PhotoMax/Max Gibbs and Craig Wardrop

EDITORIAL CONSULTANTS
Timothy Hovanec, PhD; Marshall Meyers; Robert Weintraub

For Dorling Kindersley

SENIOR ART EDITOR Joanne Doran
MANAGING ART EDITOR Lee Griffiths
DTP DESIGNER Louise Waller

SENIOR EDITOR Simon Tuite
MANAGING EDITOR Deirdre Headon
PRODUCTION CONTROLLER Kevin Ward

All information correct at press time. Readers should note that the care information in the book is general and not a substitute for professional advice. Neither the author nor the publishers can be liable or responsible for any loss, damage, or injury to health allegedly arising from any information or suggestion in this book. All fish shown in the pictures specially taken for this book were properly handled and not harmed in any way.

This American Edition, 2019
First American Edition, 2005
Published in the United States by DK Publishing, 1450 Broadway, Suite 801, New York, NY 10018, USA

A catalog record for this book is available from the Library of Congress.
ISBN 978-1-4654-8031-6

DK books are available at special discounts when purchased in bulk for sales promotions, premiums, fund-raising, or educational use. For details, contact: DK Publishing Special Markets, 1450 Broadway, Suite 801, New York, NY 10018, USA
SpecialSales@dk.com

Printed and bound in China

A WORLD OF IDEAS:
SEE ALL THERE IS TO KNOW

www.dk.com

Contents

Preface

Starting with a blank sheet of paper creates both challenges and possibilities, and planning this book proved no exception! The greatest headache when embarking on such a venture is choosing which species to include. Our selection is based on international trade data collated by the Pet Industry Joint Advisory Council (PIJAC) of the United States. This novel approach should ensure comprehensive coverage of the species most commonly sold by aquarium stores, while also allowing for the inclusion of some of the more unusual species that occasionally become available.

Having selected the fish, invertebrates, and plants, the next problem was to decide what to call them. This may seem a rather bizarre statement, but a single species may have eight or more scientific names, each of which has its own supporters, and a similar number of common names. As a result, we have incorporated a wide range of alternative names—both scientific and common— into the name index at the back of the encyclopedia. I hope that this index will form a useful reference resource in its own right and help to overcome the confusion caused by the widespread use of different names to describe the same species in books, in magazines, and on the Internet.

Another unique feature of this encyclopedia, achieved with the help of Max Gibbs's superb photography, is the behavioral studies that run through the directory sections. Understanding more about the lifestyles of fish will not only give you greater confidence as an aquarist but also help you to get the most out of this fascinating, rewarding, and immensely enjoyable hobby.

In this current edition, the latest accepted scientific names have been incorporated, along with other changes, in terms of distribution information and common names, too, relating to the fish, invertebrates, and plants, so the text is as up-to-date as possible. New innovations in the hobby, such as digital aquarium controls, have also been introduced.

David Alderton

David Alderton

How to use this book

This encyclopedia is divided into three main sections, covering freshwater aquariums, marine aquariums, and ponds. Each section includes practical advice on how to care for your fish, along with a directory of relevant species, organized into related groups.

Feature boxes in the directories of fish and invertebrates give fascinating insights into the lifestyles of fish and reveal amazing adaptations to different habitats.

CONFLICT RESOLUTION

Although many cichlids are aggressive, most disputes are resolved without actual physical conflict. The bright red of the Firemouth Cichlid (*Cichlasoma meeki*), shown below, warns other fish to steer clear. If this does not work as a deterrent, a Firemouth will inflate its throat and flare out its gill covers. This makes the fish appear larger and more intimidating and may persuade a would-be rival to back down and swim away. In the aquarium, however, conflict is more likely because the fish cannot avoid one another.

Each group of fish is introduced with an overview of the defining characteristics.

Specially commissioned photographs provide stunning close-ups of fish behavior and anatomical features.

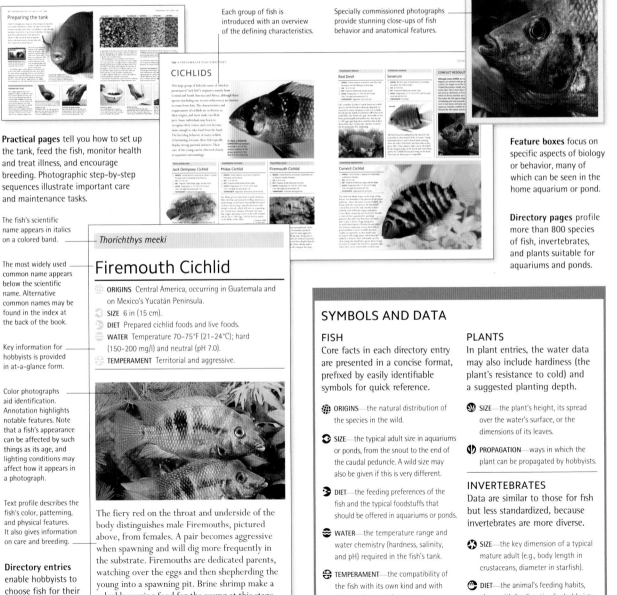

Practical pages tell you how to set up the tank, feed the fish, monitor health and treat illness, and encourage breeding. Photographic step-by-step sequences illustrate important care and maintenance tasks.

The fish's scientific name appears in italics on a colored band.

Feature boxes focus on specific aspects of biology or behavior, many of which can be seen in the home aquarium or pond.

Directory pages profile more than 800 species of fish, invertebrates, and plants suitable for aquariums and ponds.

The most widely used common name appears below the scientific name. Alternative common names may be found in the index at the back of the book.

Key information for hobbyists is provided in at-a-glance form.

Color photographs aid identification. Annotation highlights notable features. Note that a fish's appearance can be affected by such things as its age, and lighting conditions may affect how it appears in a photograph.

Text profile describes the fish's color, patterning, and physical features. It also gives information on care and breeding.

Directory entries enable hobbyists to choose fish for their tank and compare related species.

Thorichthys meeki

Firemouth Cichlid

- **ORIGINS** Central America, occurring in Guatemala and on Mexico's Yucatán Peninsula.
- **SIZE** 6 in (15 cm).
- **DIET** Prepared cichlid foods and live foods.
- **WATER** Temperature 70–75°F (21–24°C); hard (150–200 mg/l) and neutral (pH 7.0).
- **TEMPERAMENT** Territorial and aggressive.

The fiery red on the throat and underside of the body distinguishes male Firemouths, pictured above, from females. A pair becomes aggressive when spawning and will dig more frequently in the substrate. Firemouths are dedicated parents, watching over the eggs and then shepherding the young into a spawning pit. Brine shrimp make a valuable rearing food for the young at this stage.

SYMBOLS AND DATA

FISH
Core facts in each directory entry are presented in a concise format, prefixed by easily identifiable symbols for quick reference.

- **ORIGINS**—the natural distribution of the species in the wild.
- **SIZE**—the typical adult size in aquariums or ponds, from the snout to the end of the caudal peduncle. A wild size may also be given if this is very different.
- **DIET**—the feeding preferences of the fish and the typical foodstuffs that should be offered in aquariums or ponds.
- **WATER**—the temperature range and water chemistry (hardness, salinity, and pH) required in the fish's tank.
- **TEMPERAMENT**—the compatibility of the fish with its own kind and with other tank or pond occupants.

PLANTS
In plant entries, the water data may also include hardiness (the plant's resistance to cold) and a suggested planting depth.

- **SIZE**—the plant's height, its spread over the water's surface, or the dimensions of its leaves.
- **PROPAGATION**—ways in which the plant can be propagated by hobbyists.

INVERTEBRATES
Data are similar to those for fish but less standardized, because invertebrates are more diverse.

- **SIZE**—the key dimension of a typical mature adult (e.g., body length in crustaceans, diameter in starfish).
- **DIET**—the animal's feeding habits, along with feeding tips for hobbyists.

INTRODUCTION TO
FISHKEEPING

Diversity

Fish have colonized almost every aquatic environment on the planet. Blackfin Icefish inhabit the freezing depths of the Antarctic Ocean, while Desert Pupfish thrive at over 104°F (40°C) in pools in the Nevada Desert. The huge diversity of fish habitats worldwide has driven the evolution of the countless shapes, sizes, and colors that make fish so fascinating to keep.

The observed diversity in the appearance of fish is the product of millions of years of evolution. Imperceptibly, the forces of natural selection have shaped every fish's external form, internal anatomy, and behavior to deal with the challenges of its own very particular environment; in the process, they have created thousands of living species of fish in the world's seas and fresh waters. Most species are created by geographical separation; this occurs when one population of fish becomes fragmented into two or more smaller groups. Natural selection works on each group in slightly different ways and changes them so that if members of both groups meet again, they are too different to interbreed. By definition, a new species has been created.

Of the 25,000 living species of fish, about 60 percent are marine—a surprisingly low proportion given the extent of the world's oceans compared to its fresh waters. This is equivalent to one species for every 24,000 cubic miles (100,000

Lake Malawi is home to many species of cichlids. The species (of which a selection is shown below) have highly localized distributions in the lake, and many have color variants.

Maylandia zebra (Orange blotch morph)

M. zebra (Blue morph)

M. zebra (Cobalt blue morph)

M. lombardoi (Yellow morph)

M. lombardoi (Blue morph)

P. elongatus

Coral reefs are highly varied habitats —they have been called the tropical rain forests of the seas. They are areas of extremely high species diversity, for both fish and invertebrates.

Mangrove forests, where fresh- and saltwater habitats merge, are the natural home of a number of aquarium species, such as mudskippers and Archer Fish.

The Amazon is the world's largest river, measured by volume. Seasonal flooding of this vital fish habitat enriches the water with food and acts as a trigger for breeding.

Human activity has damaged or reduced many natural fish habitats. Some species, however, benefit from agriculture, spreading into drainage ditches and paddy fields.

km^3) of sea water, compared to one species for every 3½ cubic miles (15 km^3) of fresh water. The diversity of freshwater fish relative to the size of their habitat is due to the ease with which groups of fish can become separated and geographically isolated in rivers and pools, compared with the sea.

Geography and species formation

There is no better example of diversification and species formation than Lake Malawi in Africa's Great Rift Valley. Created about two million years ago by a geological fault, the 365-mile- (584-km-) long lake is today home to more than 1,600 species of cichlids—more than occur in all of the rest of Africa. It is thought that all these cichlids developed from just one or two ancestral species, which entered the lake at the time it was formed. The early lake cichlids adapted to the entire range of different habitats that they encountered in the lake. Some became predators; others plant eaters. Some became restricted to the shoreline; others occupied the depths of the lake. Some grew to large sizes; others diminished. They

New species are constantly being discovered. The Red Dwarf Pencilfish (*Nannostomus mortenthaleri*) was first collected as recently as 2000 from the Rio Nanay near the town of Albarenga, Peru.

also diversified in their breeding habits, some species scattering their eggs, others displaying a long period of parental care. By dividing up the biological "niches" available in the lake, the cichlids were able to explode in number, without directly competing with one another for limited resources.

BODY SHAPES

Fish occur in a wide range of different shapes, which usually relate to their lifestyle. Surface-dwelling fish, for example, have flattened backs and upturned mouths so that they can grab floating food. Body shape may, however, change with age. The discus, for example, has a compact body when young; it attains its flattened shape only at several months of age.

Tubular body shape

Tall, flattened body; barring helps to break up the fish's outline

Propulsive power comes from body, rather than fin, movements

Narrow, tubular, wormlike body lacks caudal, pelvic, and anal fins

Obviously flattened body; propulsion provided by the so-called "wings"

Tail assists movement

A Spotted Moray Eel (*Gymnothorax moringa*) hides in its lair. The body shape of this hunter means it can hide away in caves or under rocks, from where it can ambush prey.

A discus slips through dense weed to escape a predator. Its tall, narrow shape is typical of species that live in slow-moving waters. Fast currents would make swimming difficult for this fish.

A Pipefish (*Syngnathoides* sp.) drifts in a bed of sea grass, perfectly camouflaged by its shape and color. It even holds its body at a slight angle to accurately mimic the sea grass.

Flatfish spend their lives on or near the substrate. They have asymmetrical bodies, with both eyes on the same side of the head. They are able to burrow into the substrate, hiding most of the body.

What is a fish?

Defining a fish is harder than it seems. To most people, it is an animal that lives and breathes in water. Some fish, however, emerge onto land, breathe air, and use their fins like legs. And many other animals, including amphibians, mammals, and invertebrates, live in water. To add further confusion, some creatures called fish—starfish, jellyfish, and cuttlefish—are not really fish at all.

Fish are vertebrates, which means that—like humans, but unlike starfish, jellyfish, and cuttlefish—they possess a backbone. Most are cold-blooded (ectothermic), so they cannot raise their body temperature above that of their environment, unlike mammals and birds; this limits the effective range of fish to warmer waters (though there are many exceptions). All fish have gills, which they rely upon to varying degrees to obtain oxygen, and most species have two pairs of fins in place of arms and legs, as well as several other fins on the body. The majority of fish are covered in scales,

Eye, with outer iris and central dark pupil — Dorsal fin — Fin rays — Body scales — Mouth — Gill cover or operculum — Pectoral fin — Ventral or pelvic fin — Urogenital and anal openings — Anal fin — Caudal peduncle — Caudal or tail fin

Common Goldfish

The body shape of a fish is determined by its skeleton. Most fish have bony skeletons, but primitive species, notably sharks and rays, have skeletons of cartilage.

which are thin, overlapping outgrowths of the outer skin that protect the body. They secrete a slimy coating that protects them from parasites and bacteria and helps them slip through the water. The presence of scales helps to distinguish fish from amphibians—another group of water-dwelling vertebrates.

Body and fins

There are many alternative body plans for fish. A "typical" fish is designed to be streamlined so that it can cut through the water with the least effort. It has a spindle-shaped profile, though it is somewhat wider to the front of the midline, and its head joins the body without a neck. Its eyes are flush with the head, and only the fins extend beyond the body; even the fins can be pressed flat against the sides to minimize water resistance.

Fish rely on their fins for locomotion, though they may have more specialized uses in some species. Swimming through water, which is much more dense than air,

---

BODY COVERINGS

The skin of a fish is usually covered by protective scales or bony plates. A fish has the same number of scales throughout its life; if some are lost, they will be replaced, but new scales are not added. Several basic types of body covering are recognized.

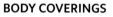

Ctenoid scales, such as those of a Queen Angelfish, have a comblike rear edge. Ctenoid and cycloid scales are found in the vast majority of bony fish.

Cycloid scales, such as those of a goldfish, have a smooth rear edge. Like ctenoid scales, they have a hard surface layer over a deeper fibrous layer made of collagen.

Bony plates, as seen in this catfish, offer better protection than scales but restrict mobility. They start as folds in the skin of fry; the folds harden and develop into bony plates.

Lacking scales or plates, Synodontis catfish rely on their thickened skin and plentiful mucus for protection. Many bottom-dwelling fish lack scales on their undersides.

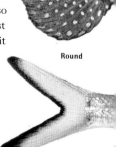

Crescent

Round

Deeply cleft

Selectively bred twintail

The shape of the caudal (tail) fin varies between species, and greatly affects swimming ability. Fish with deeply forked tails rank among the most powerful swimmers. In some cases, the tail has become enlarged naturally, or by selective breeding, into a more decorative feature.

The shape of the fins identifies the sexes in species like *Trichogaster* Gouramis *(see pp.112–113)*. Males have longer and more tapering dorsal and ventral fins than females.

Some cichlids, such as the Chessboard Cichlid *(see p.144)* pictured here, display lyre-tailed extensions on their caudal fins. The function of this elaborate tail is to attract and select mates.

The elaborate fins of the domesticated male Siamese Fighting Fish *(see pp.104–106)* are larger and more flamboyant than in its wild counterparts. Fish have been bred selectively for this characteristic.

requires considerable muscular effort, and the main thrust for swimming is provided by the caudal fin at the rear of the body; this starts the rippling movement that spreads through the fish's body. The dorsal fin, which runs down the center of the back, helps to keep the fish moving in a straight line and is counterbalanced on the underside of the body by the anal fin.

Paired ventral (or pelvic) fins are set on either side of the midline in front of the anal fin on the underside of the body. They act like stabilizers, keeping the fish upright, and in some species, such as corydoras catfish, they are used to hold the eggs during spawning. The pectoral fins, located farther forward, close to the gills on each side of the body, also help the fish to maneuver. In bottom-dwelling species, these fins may be adapted for use as props, or legs on which the fish can support themselves or even walk around. Sometimes the pectoral fins are equipped with spines for defense. In gouramis, the pectoral fins may be transformed into hairlike structures that help the fish locate food by detecting scents in the water.

Some fish, notably characoids, have an additional smaller fin behind the dorsal fin. This is known as the adipose fin, and as its name suggests, it acts as a store of adipose (fatty) tissue and has only a minor role in locomotion.

Fin variants

The shape and position of the fins vary between species and provide valuable clues to their lifestyle. For example, shoaling fish that live in areas of open water have forked caudal fins, which provide them with good propulsive power. The open V-shaped structure gives little drag but does not have the power to provide rapid propulsion from a stationary position. Sit-and-wait predators that catch their prey in a sudden burst of speed tend to have rounded, paddle-shaped caudal fins—ideal for fast acceleration. To move quickly from a standing start, some fish rapidly expel a stream of water from their gills in an aquatic form of jet propulsion.

The bladelike teeth of the piranha enable the fish to bite chunks out of its quarry. Fish teeth can be found in a variety of places. Some species have them on their jaw bones or on the bones of the roof of the mouth; others have patches of teeth on the tongue or pads of teeth on the gill arches in the throat.

MOUTH SHAPES

The shape, size, and position of a fish's mouth give a good insight into its feeding habits. Predators tend to have much larger mouths than omnivores. Some species have obvious canine-shaped teeth; in others, teeth are absent or less clearly visible.

Surface feeders, such as hatchetfish, have a relatively short upper jaw, which enables them to grab invertebrates at the water's surface easily.

Mid-water feeders, such as tetras, have slightly protrusive jaws so they can catch passing food particles. This is the most common mouth design.

Bottom-feeding fish, like the plecos, have suckerlike mouths for scraping food and algae from rocks, and scavenging from the bottom of the tank.

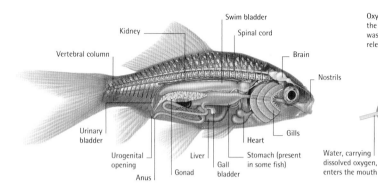

Kidney

Swim bladder

Spinal cord

Vertebral column

Brain

Nostrils

Urinary bladder

Gills

Heart

Urogenital opening

Liver

Stomach (present in some fish)

Anus

Gonad

Gall bladder

Oxygen is taken up over the gill surface, and waste carbon dioxide is released into the water

Water, carrying dissolved oxygen, enters the mouth

Deoxygenated water leaves the gill chamber via the gill flap (operculum)

Gills are highly efficient breathing organs, able to extract 80 percent of the oxygen dissolved in the water around them.

Fish have many organs—such as the brain, stomach, liver, and kidneys—in common with humans. Others, like the gills and swim bladder, are not present in our bodies.

Water position and buoyancy

All fish rely on their fins—especially the pectoral and ventral fins—to control their position in the water and prevent them from being swept away by currents. In fish that live in fast-flowing mountain streams, the fins can be fused together; the result is a suction cup that anchors the fish in place. This adaption is seen in the hillstream loaches of the family Balitoridae. Position in the water is also influenced by the swim bladder—an elongated gas-filled organ situated beneath the vertebral column. To achieve neutral buoyancy (when the fish neither rises nor sinks), the swim bladder must occupy about 8 percent of the fish's body volume. The amount of gas in the bladder can be adjusted in two ways: the fish can gulp down air, which enters the swim bladder via the foregut, or gas can be released into the bladder from blood vessels.

Digestion and respiration

The digestive system of a fish is typical of vertebrates; as with terrestrial species, herbivorous fish tend to have longer intestines than carnivores, because plant matter is tough, fibrous, and difficult to break down. The respiratory system, however, is unique to fish. Most fish extract oxygen from the water, rather than the air, using gills, which are located on the sides of the head behind the eyes, hidden under flaps known as opercula.

Gills are bony rods to which are attached fleshy filaments, rich in thin-walled blood capillaries. Water enters the fish's open mouth, which then closes. The water is forced over the filaments and out again through the opercula. Oxygen dissolved in the water is taken up into the bloodstream through the filaments, which usually have fine secondary flaps (or lamellae) to maximize the surface area available for gas exchange. Astonishingly, the total surface area of the gills can be more than 10 times the fish's outer body area. Within

Fish deter predators in a variety of ways. Some species use clever camouflage to break up their outline against the colorful reef background; others are armed with venomous spines or can inflate their bodies, making themselves too large to swallow.

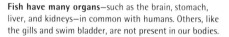

Porcupine fish

Spines on inflated fish deter attack

False eye confuses predators

Venomous spines

Threadfin Butterflyfish

Black bar masks body shape

Volitans Lionfish

the mouth, straining devices called gill rakers prevent food and debris from passing over and damaging the gills.

Some fish also gulp atmospheric oxygen using the swim bladder as a basic "lung." These species are usually the natural inhabitants of muddy pools, where dissolved oxygen may be in short supply.

Color and pattern

Almost all fish use color to aid camouflage or to attract mates. Some are colored with inconspicuous browns and greens to blend in with the background and escape the attention of predators; others—such as the flatfish—change their pattern to match their background. The brilliant colors displayed by many tropical species are also a form of camouflage; bold vertical stripes, for example, break up the outline of a body and make it hard to see. And dark stripes through the eyes often continue through the iris, making the eye almost invisible. Some fish have "false" eyespots (also called ocelli) on their tails; predators will attack what they believe to be the head, giving the prey a few moments to escape.

Fish that have a solid, dark color tend to have lots of pigment in their skin, while species that appear silvery have little skin pigment but rely more on the iridescence of the scales. This reflectiveness is caused by the presence of the chemical guanine—a waste product from the blood. Many fish have transparent bodies as fry and develop color only with age.

The coordinated movement of a shoal of fish increases the chances of survival of each individual. Many eyes are more effective than one at detecting danger, while swimming in close formation makes it harder for an individual to be targeted by a predator.

SENSES AND COMMUNICATION

Although fish brains are poorly developed compared to those of mammals, fish possess acute and often highly specialized senses and means of communication.

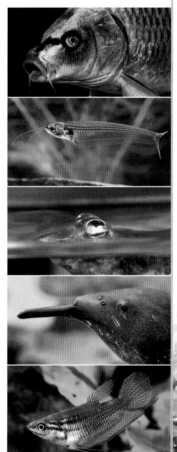

Barbels—structures on the lips that resemble elongated whiskers—are common in fish that live in water where visibility is poor. Barbels may contain touch and taste organs that help the fish navigate and find food.

The lateral line runs down each side of the fish's body. It comprises a row of pores opening into a channel that leads to the head. The channel is filled with a viscous jelly, which detects vibrations in the water.

Fish eyes are similar to those of other vertebrates and can see colors. Vision is particularly sharp in fish that live close to the surface. The Four-Eyes *(see pp.156–157)* can see in both air and water at the same time.

Electrical fields produced by mormyrids *(see p.186)*, like this Peter's Elephant-Nose, enable fish to sense their environment. Some experiments suggest that the electrical signals may also be used in communication.

Some species use sound to communicate with each other, such as the Croaking Gourami *(see p.110)*. Their "drumrolls" are produced by the action of muscles beating against the swim bladder.

Evolution and classification

Fish are the oldest of all vertebrates (animals with backbones), with an ancestry dating back more than 500 million years. However, the earliest fish to appear in the world's oceans were very different from those seen today, since they had no jaws or scales. They lacked specialized fins, so they relied solely on tail movements to propel them through the water. Internally, their spine was made of cartilage, rather than bone.

Lungfish, forming the family Ceratodiformes, have changed very little in appearance since they first evolved more than 400 million years ago.

The first scaly fish with movable jaws arose around 440 million years ago, their jaws having evolved from the front gill arches. These fish also possessed several pairs of spines along the lower sides of the body, from which paired fins later developed.

Fish did not colonize freshwater habitats until comparatively recently in their history, but by about 66 million years ago, there were recognizable forerunners of many of today's freshwater species, including *Hypsidoris*, a primitive catfish with sensory barbels and protective spines on its pectoral fins.

The fossil record shows that modern fish evolved from five ancient groupings. Two of these groups are now extinct; of the remaining three, the Osteichthyes—the bony fish—are the most numerous and diverse today.

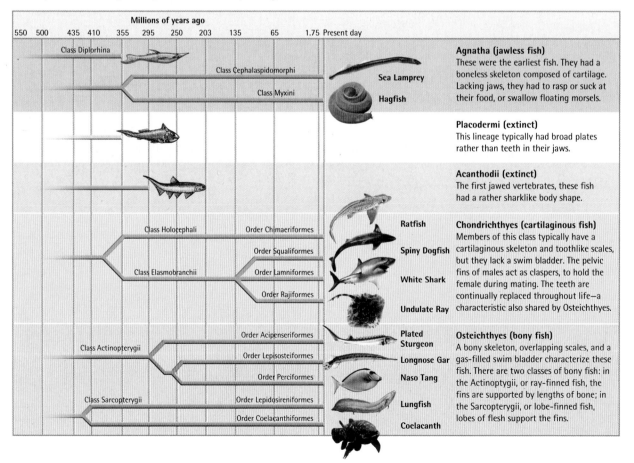

Millions of years ago

| 550 | 500 | 435 | 410 | 355 | 295 | 250 | 203 | 135 | 65 | 1.75 | Present day |

Agnatha (jawless fish)
These were the earliest fish. They had a boneless skeleton composed of cartilage. Lacking jaws, they had to rasp or suck at their food, or swallow floating morsels.

Class Diplorhina

Class Cephalaspidomorphi — Sea Lamprey

Class Myxini — Hagfish

Placodermi (extinct)
This lineage typically had broad plates rather than teeth in their jaws.

Acanthodii (extinct)
The first jawed vertebrates, these fish had a rather sharklike body shape.

Class Holocephali — Order Chimaeriformes — Ratfish

Chondrichthyes (cartilaginous fish)
Members of this class typically have a cartilaginous skeleton and toothlike scales, but they lack a swim bladder. The pelvic fins of males act as claspers, to hold the female during mating. The teeth are continually replaced throughout life—a characteristic also shared by Osteichthyes.

Order Squaliformes — Spiny Dogfish

Class Elasmobranchii — Order Lamniformes — White Shark

Order Rajiformes — Undulate Ray

Class Actinopterygii — Order Acipenseriformes — Plated Sturgeon

Osteichthyes (bony fish)
A bony skeleton, overlapping scales, and a gas-filled swim bladder characterize these fish. There are two classes of bony fish: in the Actinoptygii, or ray-finned fish, the fins are supported by lengths of bone; in the Sarcopterygii, or lobe-finned fish, lobes of flesh support the fins.

Order Lepisosteiformes — Longnose Gar

Order Perciformes — Naso Tang

Class Sarcopterygii — Order Lepidosireniformes — Lungfish

Order Coelacanthiformes — Coelacanth

With the exception of lampreys and hagfish, which are distant relatives of the early jawless fish, living fish fall into one of two groups. Sharks, rays, and their relatives are Chondrichthyes (cartilaginous fish), in which the skeleton is made of cartilage. The remainder, making up about 95 percent of all fish, are Osteichthyes (bony fish), which have a bony skeleton. Bony fish form the most diverse group of vertebrates on the planet, with about 23,500 different species.

How fish are classified

The classification of living things is called taxonomy. The basic unit of classification is the species—a collection of similar organisms that are capable of breeding together in the wild and producing fertile offspring. Related species are organized into groups called genera, which in turn are arranged into families. The grouping process continues, working upward through ever larger and more general groups known as orders, classes, phyla, and, lastly, kingdoms—the highest level in the hierarchy.

An integral part of classification is assigning unique scientific names to individual species. Although scientific names may seem unwieldy because they are usually in Latin, they are understood by scientists around the world and so are far more useful than common names, which often differ from country to country.

When a new fish is discovered, certain procedures have to be followed before it can be identified as a species in its own right. First, a specimen is lodged with a scientific institute, such as a museum. This is called the type specimen. Then a detailed description is published in a recognized scientific publication. Finally, the fish is ascribed a scientific name and placed in the genus containing those species to which it is most closely related. At present, classification relies primarily on comparing the anatomical features of the type specimen to those of other species, but DNA analysis is increasingly being used, since it gives a more accurate picture of the relationships between organisms.

The loricariid catfish family is a rapidly expanding group. To cope with the complexities of classifying these fish, the L-numbering system was devised. For example, this loricariid is known as L109.

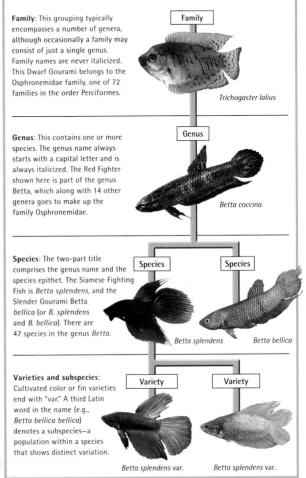

CLASSIFICATION AND SCIENTIFIC NAMES

Fish are members of the kingdom Animalia. Within this, they fall into the Chordata phylum, which contains all vertebrates. The farther down the hierarchy you go, the more the species in each group have in common. As far as hobbyists are concerned, it is the groupings from family downward that are most significant.

Family: This grouping typically encompasses a number of genera, although occasionally a family may consist of just a single genus. Family names are never italicized. This Dwarf Gourami belongs to the Osphronemidae family, one of 72 families in the order Perciformes.

Family

Trichogaster lalius

Genus: This contains one or more species. The genus name always starts with a capital letter and is always italicized. The Red Fighter shown here is part of the genus Betta, which along with 14 other genera goes to make up the family Osphronemidae.

Genus

Betta coccina

Species: The two-part title comprises the genus name and the species epithet. The Siamese Fighting Fish is *Betta splendens*, and the Slender Gourami Betta *bellica* (or *B. splendens* and *B. bellica*). There are 47 species in the genus Betta.

Species

Species

Betta splendens

Betta bellica

Varieties and subspecies: Cultivated color or fin varieties end with "var." A third Latin word in the name (e.g., *Betta bellica bellica*) denotes a subspecies—a population within a species that shows distinct variation.

Variety

Variety

Betta splendens var.

Betta splendens var.

The difficulties inherent in recognizing and differentiating both species and subspecies has led the aquarium trade to develop its own classification system, known as L-numbering, for loricariid catfish *(see pp.128–131)*— a little-studied family that shows remarkable diversity in appearance. This system uses numbers prefixed by the letter "L" to identify individual pattern and color forms that are not currently recognized by scientists. It sometimes happens that by the time a species is formally classified, it already has several L-numbers associated with it throughout its range.

The popularity of fishkeeping

It is unclear when people started keeping fish for their aesthetic qualities, rather than as a source of food, but the activity certainly dates back well over a thousand years. It probably began in China with the domestication of goldfish and koi and has spread worldwide with advances in aquarium technology and transportation. Today's aquatic industry produces a vast range of products to make fishkeeping more rewarding than ever before.

"Thence home and to see my Lady Pen, where my wife and I were shown a fine rarity: of fishes kept in a glass of water, that will live so for ever; and finely marked they are, being foreign."

This extract from the diary of Samuel Pepys, dated May 28, 1665, shows that ornamental goldfish were known in Restoration London. Even earlier records of fishkeeping in China have been found, and one of the oldest surviving essays on goldfish—*The Book of the Vermilion Fish*—dates back to 1596. While keeping coldwater fish seems to be an ancient pursuit, tropical fishkeeping is probably a more recent interest, although Siamese Fighting Fish have been bred selectively in Siam (now Thailand) for many centuries. Some of the hardier tropical fish—such as Paradise Fish *(see pp.108–109)*—were kept successfully in the late 19th century, decorating the Imperial Court of the Russian czar. The earliest tropical aquariums were equipped with slate bases, and heated from beneath using a naked flame, making a perilous life for the fish.

The design of public aquariums has changed radically over the last 100 years, but they continue to fascinate visitors of all ages.

The international trade in species for the aquarium is growing at a rate of 10–30 percent per year. Aquarium plants are today cultivated and imported on a massive scale.

Victorian enthusiasts

Interest in fishkeeping in the West blossomed in Victorian England, where the name "aquarium" was first coined, and keeping fish in the home became immensely fashionable. The first public aquarium opened at London Zoo in 1853, and scientific papers were published setting out the recipe for a "balanced aquarium." The public flocked to see amazing displays of native marine life, which were especially popular in the coastal resorts of Great Britain, where water could be pumped into the tanks from the sea.

In Japan, the keeping of ornamental fish—particularly colored carp—has a distinguished history that dates back more than 500 years. This Japanese woodcut from 1888 shows children enjoying the pastime.

Marine home aquariums, such as this centerpiece by Aquarium Design, have become extremely popular in recent years but are harder to establish than freshwater tanks.

Simple, compact, low-maintenance, acrylic tanks, such as this stylish BiOrb, make fishkeeping more accessible to the beginner.

CONSERVATION ISSUES

The vast majority of freshwater fish offered for sale to the aquarist are bred in captivity on a commercial scale. A small proportion of freshwater and many marine species, however, are wild-caught for the trade. While some people argue that the trade of live animals should be banned outright, conservationists increasingly agree that the sustainable harvesting of wild fish for the aquarium may benefit both fish populations and the wider environment. This is because the controlled collection of fish provides lucrative local employment and gives governments real incentives to monitor and safeguard precious habitats, such as tropical reefs and rain forests.

Transportation and technology

By the end of the 19th century, England and Germany dominated the aquarium pastime, exporting fish to the US and farther afield. However, tropical fishkeeping took off as a hobby only in the 20th century, when electricity supplies made lighting and water heating a reality. Commercial breeding of fish to meet growing demand began in Florida in 1926; the climate of the state and its proximity to the rivers of South America, where many popular aquarium fish originate, made it the ideal base for a fast-growing industry. Breeding of highly ornamental varieties in the US and elsewhere attracted more people to the hobby, while the expansion of air travel after World War II disseminated exotic varieties around the world with unprecedented speed.

Today, fishkeeping is big business. Constant improvements in aquarium technology and foods have made caring for fish in the home easier than ever. Furthermore, scientific research has confirmed what generations of fishkeepers have recognized: keeping fish brings measurable health benefits—lowering blood pressure and stress levels—and can promote a higher quality of life.

The growth of scuba diving has led to a greater interest in keeping marine fish. Recreating reef conditions within the aquarium has been made possible by specially formulated sea salts and efficient lighting.

INTRODUCTION TO
FRESHWATER FISH

What to consider

People keep fish for a huge range of reasons, from companionship to competition. Most of us, however, just enjoy the calming elegance and color that fish bring to the home. Choosing what fish to keep, and how to house them, is influenced by many subjective judgments, as well as practical considerations such as cost, available space, and ease of maintenance.

Adult Oscar, with mature coloration, shown at life size

Juvenile Oscars, as purchased in a store, shown at life size

Fish sold in aquarium stores are likely to be juveniles. They may change greatly in color, pattern, and size by the time they reach maturity. Always find out about a fish's development and requirements before making a purchase.

Fish for a busy life

The rise in popularity of fishkeeping can be attributed in part to the ready availability of spectacular species and advances in aquarium technology. But at least part of its appeal in the last few decades lies in our changing lifestyles. With leisure time diminishing, fish make ideal pets: they do not need walking or playing with; they can be kept in apartments; and they make no noise or mess. The effects on an electricity bill of running a single tank are minimal, and, after the initial investment in tank, equipment, and fish has been made, maintenance, food, and veterinary costs are very low. This is not to say that fish keep themselves. They must be fed regularly (though automatic feeders will reduce the time demanded here), and you will need to set aside a couple of hours every two weeks or so to carry out partial water changes in the aquarium, service the filter as required, and clean the sides of the tank.

Pets and show fish

Most people are attracted to fishkeeping by the idea of watching and nurturing a colorful collection of fish in the home. Some, however, prefer to keep just one or two fish that develop into real pets, capable of recognizing their owner and even feeding from the hand. Most fish in the latter category—including various cichlids and catfish—grow to a relatively large size and so need spacious accommodation. For this reason, "pet" fish usually require a higher investment, both at the outset and throughout their lives, in terms of lighting, heating, filtration, and feeding.

BUYING CHECKLIST

- Think about the size the fish will reach as adults. Do not exceed the aquarium's recommended stocking density.

- Consider the sociability of the fish; some species are highly aggressive, especially prior to breeding.

- Diet varies between species. Make sure the correct foodstuff is readily available.

- Fish life spans range from one year to more than a decade. If you are likely to become attached to your pet fish, choose long-lived species.

- Some species reproduce readily in home aquariums; others have never been captive-bred. Choose accordingly.

- Think carefully about tank size and location.

Feeding and water chemistry

Food and water requirements can impose real restraints on the plants and fish that can be used to stock an aquarium. Some predatory species, for example, can be difficult to wean off life foods, and may have to be fed small fish—not a practical option for a small home setup. Similarly, if you intend to establish lush, attractive vegetation in the tank, you should avoid species that are vegetarian, because they will nibble on the young plants.

Certain tropical fish are highly particular about water chemistry, while others are tolerant of varied water conditions. Discus, for example, need soft water; if you live in a hard-water area, you will need to invest in an ion-exchange water softener to keep these species successfully. However, your hard water will be ideal for keeping other species that enjoy these conditions, such as Rift Valley cichlids.

Tanks in the home

Some aquarists are primarily fascinated by the biology of the fish in their tank, or keep their fish for breeding, and are almost oblivious to the appearance of the aquarium. But for

EXTERNAL FISH HOUSE

As you become more serious about fishkeeping—and particularly if you want to breed fish—you may find that one tank is just not enough. Additional tanks may be needed to isolate young fish from adults or to treat sick fish. If space in the home is limited, a fish house may be the answer. This could be a well-insulated and/or heated outbuilding, such as a garage or shed, in which tanks can be supported on racks. Get a building inspector to make sure the building is strong enough to support the weight of tanks and water.

the majority, aesthetics are important—creating a harmonious design using the tank itself, the fish, plants, backgrounds, and tank furniture is part of the appeal of fishkeeping. Aquariums are available in all shapes and sizes. Some are designed to rest on existing furniture, while others can be supported on special stands or cabinets. Larger tanks can even be used architecturally—incorporated into the fabric of the home as room dividers, for example. In general, the tank should be visible at eye level, either when standing or when seated, but otherwise there are few rules about tank aesthetics.

Make sure that the stand can take the weight of the tank

Specially built cabinet supports the tank and conceals pumps filters, and other equipment

Tanks are available in almost any size and to fit almost any budget. Cabinet-mounted tanks (above) are pieces of furniture in themselves, while small "plug-and-go" tanks (right) are ideal for the novice (see p.31).

The Red-Tailed Shark is aggressive toward its own kind, or with other fish that display a similar coloration. Sociability is an important concern when selecting fish for the tank.

FISH MIXES

If you opt to keep large fish, one or two individuals of the same species will be enough to create a visual impact in the tank. Some smaller species—especially shoaling fish—also look their eye-catching best when kept in a single-species group (far right). Alternatively, fish may be mixed together in a community aquarium (right). This may be themed—perhaps a collection of fish from a particular part of the world or that share the same water chemistry requirements.

Breeding and longevity

Keeping fish is fun and brings great rewards, but breeding them in the home gives remarkable insights into their behavior and biology. Watching a fish build a nest from bubbles, for example, or a cichlid protecting its young is both fascinating and educational—especially for young children. Breeding fish for sale can also generate a little income to help support your hobby and offset some of the additional costs of breeding tanks and other necessary equipment.

If breeding fish is a priority, bear in mind that some species will reproduce in aquariums far more readily than others. In general, livebearers, such as guppies *(see pp. 165–167)*, are easier to breed than egg-laying species. With a little experience, more challenging species can be taken on—even those with a reputation for being reluctant to spawn in captivity or those in which reproduction is little documented.

Alongside the reproductive potential of a fish, it is worth considering its life span before buying. Adult guppies are notoriously short-lived, for example, and are likely to live only for a few months after purchase. Most tropical species live for about two to four years in the aquarium (although some catfish may live for well over a decade), which is longer than they would survive in the wild. As a general guide, larger fish tend to have a longer life span than smaller species.

SHOWING FISH

A number of tropical fish, such as guppies and discus (pictured below), have been selectively bred to accentuate their attractive characteristics, such as their color, patterning, and, in some cases, fin shape and size. Many breeders exhibit their fish, and judging standards have been set up for the most popular varieties, just as they have for breeds of dogs and cats.

The Chocolate Gourami is a fish with a reputation for being hard to keep, because it is susceptible to disease. It poses a worthy challenge to more experienced fishkeepers.

FRESHWATER FISH

SETTING UP
THE TANK

Choosing the tank

The tank is the most important piece of equipment you will buy, since it provides a home for your fish. Anyone starting out in fishkeeping faces a bewildering array of tanks to choose from, in a wide range of sizes and styles. Your budget and the space available in your home will influence your choice, but always make the welfare of the fish the prime consideration when buying a tank. Never select a tank simply because it looks good.

For fish, living space in an aquarium is at a premium, since the population density in the tank is much greater than in any natural aquatic habitat. Consequently, it is best to opt for the largest tank that you can afford and that space allows. Small tanks are initially cheaper to buy, but they are not necessarily any cheaper to run in the long term. What is more, you may find that your fish rapidly outgrow their accommodation.

Before purchasing your tank, it is worth thinking about the type and number of fish you want to keep and to find out their adult size. The key factor in assessing the correct stocking density of a tank is its surface area, because it is here, at the water–air interface, that gas exchange occurs. The greater the surface area, the more dissolved oxygen there will be in the water, and the more fish the tank will comfortably be able to support. It is usually recommended to allow about 12 sq. in (75 cm^2) of surface area per 1 in (2.5 cm) of adult fish body length (excluding the caudal fin).

In addition to the surface area, you also need to consider the volume of the tank, since the aquarium needs to provide adequate swimming space for the fish. Allow about 1 gallon of water per inch of adult fish body length (equivalent to about 2 liters per centimeter). When making your calculations, remember to deduct 10 percent of the total tank volume to take account of rockwork and other decor.

Fish need clean, well-oxygenated water and space to swim. If you overstock the tank, the fish's waste will pollute the water (see p.52).

BUYING AND TRANSPORTING A TANK

Staff at aquarium stores can advise you on the type of tank that will meet your needs. The tips below will enable you to do your own quality checks before purchasing and help you to get the tank home safely.

Spacer bars are set at intervals across the top of the tank

Larger tanks should be made from thicker glass. Tanks more than about 24 in (60 cm) long require spacer bars to reinforce the structure. These are broad glass struts held in place with silicone sealant. Make sure that the edges of the spacer bars are not rough; otherwise, you may cut yourself when servicing the tank. Covering the edges with plastic strips will prevent this.

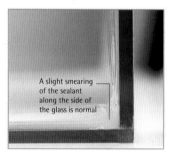
A slight smearing of the sealant along the side of the glass is normal

Check the joints between the glass panels to be sure that there is an even coverage of silicone sealant. If an area has been missed, the tank is likely to leak. Do not attempt to cut away any apparently surplus sealant, since this may seriously weaken the joints. Excess sealant may look ugly, but it will be inconspicuous against the substrate when the aquarium is complete.

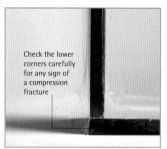
Check the lower corners carefully for any sign of a compression fracture

Examine the corners, which are potential weak spots on a glass tank. Larger, heavier tanks are especially vulnerable to damage if they have been tipped and supported on a corner first, rather than being lifted up horizontally. This can cause the glass to break into gritty fragments that may remain compressed in place but can still result in a serious weakness.

Protect all the corners with styrofoam

Be careful when taking your new purchase home, because tanks can be cumbersome and heavy. The store should tape styrofoam protectors over the vulnerable edges. When carrying the tank, always support it from beneath, regardless of its size. Lay the tank on a rug or similar soft material in the car to prevent it from being scratched, and make sure that it cannot slide around.

Tanks and stands

Colorful accessories like these lids and edging strips can instantly change the look of your tank.

A typical box-shaped tank is made from panels of glass held together by a special silicone sealant, free from chemicals that may harm fish. The silicone forms a strong, watertight bond and is also flexible, to prevent the panels from being pushed apart by the water pressure. Some aquariums have a protective frame of plastic or metal, although metal is best avoided, since it corrodes. Most tanks come with a hood that contains light fixtures and helps reduce evaporation from the water's surface.

Acrylic tanks are costlier than glass aquariums, but they are much stronger and lighter. Acrylic is also clearer than glass and a better insulator (so the tank loses less heat to the surroundings), but it scratches more easily and is harder to clean. "Plug-and-go" acrylic tanks can be bought with all the electrical equipment already in place, so you can simply add the substrate and decor, fill the tank, and turn on the power.

Whether you choose glass or acrylic, the finished tank is likely to be heavy—a 20-gallon (90-liter) tank, for example, can weigh 285 lb (130 kg) when full. Domestic furniture may not be able to bear such loads, so consider buying a stand or cabinet that is designed to take the weight of a full tank.

SITING THE TANK

- Choose a firm, level surface; use a level to make sure the site does not slope.
- Decide on a quiet position by an electric outlet where the risk of accidental knocks is low but with access for maintenance.
- Avoid drafty locations, such as in a hallway. Never put the tank by a window or anywhere else that receives direct sunlight.
- Place glass tanks on styrofoam or a specially made mat to absorb unevenness in the surface.

Second-hand tanks

A more economical way of starting off is to buy a second-hand tank. Always check glass tanks carefully for signs of leakage, and look for any scratches on the inner surface of the glass. Such scratches may seem innocuous at the time of purchase, but they will be unsightly if they later become colonized by algae, and the algae will be virtually impossible to remove. Acrylic tanks need to be inspected closely for scratches, discoloration, and cracks. Electrical equipment, such as a heater or fluorescent tube, is best replaced, and the wiring should be checked by a professional electrician.

Massive tower tank

TYPES OF TANKS

Today, there are many alternatives to the traditional rectangular design, from tall towers to hexagonal tanks and aquariums with curved surfaces. Often the tank's supporting structure or a cabinet hides all the electrical equipment from view so that all you see is the fish, plants, and substrate.

Small rectangular tank

Small corner tank

Small hexagonal tank

Large cabinet tank

Rounded glass tank

Double tank with connecting tunnels

Lighting and heating

To create a thriving aquarium, the natural requirements of its inhabitants for light and heat must be matched by artificial means. In tropical regions, water temperatures are around 79°F (26°C), changing little from one day to the next, while lighting conditions vary according to environment and time of day. Modern technology allows the optimum conditions to be created with relative ease.

Safety is paramount when using electricity near water. Rubber caps insulate contacts on fluorescent tubes.

The need for light

Almost all animals are dependent on light, and aquarium fish are no exception. Many species rely on their sense of sight to feed and communicate or use daylight to set their internal clocks, which govern many behavioral and biological processes. Light is vital for the healthy growth of living plants in the tank, and good illumination is necessary for showing fish at their best.

Warm white tube for optimum viewing — Full-spectrum tube for plant growth

PHOTOSYNTHESIS

Nearly all plants—terrestrial and aquatic—carry out photosynthesis. This process is vital to all life because it allows plants to manufacture sugars from carbon dioxide and water, which they use as food. The energy for photosynthesis comes from sunlight—but some wavelengths (colors) of light are more effective than others in energizing the process.

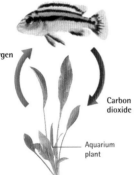

Oxygen

Carbon dioxide

Aquarium plant

Certain wavelengths of light are needed to promote photosynthesis. These wavelengths must be produced by any bulb or tube used to illuminate an aquarium.

Aquatic plants take up carbon dioxide produced by the fish and "fix" it into sugars. Oxygen given out by the plant as a by-product of photosynthesis is used by the fish in respiration.

Photosynthesis is most effective when plants are exposed to violet-blue and orange-red light

Relative rate of photosynthesis

Wavelength of light (nm)

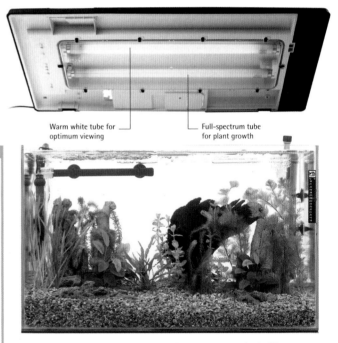

The reflective hood is often equipped with more than one kind of fluorescent tube. The tubes must be long enough to illuminate the entire tank. An electronic timer will automatically switch the light on and off in a daily cycle.

Ordinary tungsten bulbs, suspended over the water, can be used to illuminate a tank. However, they emit large amounts of heat (which causes rapid water evaporation), their light spectrum does not match that needed by plants, and they cast an unattractive yellow pallor over the tank. Today's aquarists are much more likely to use fluorescent tubes, which run cooler and can match the spectral qualities of natural light (see p.207). The tubes are typically housed in a reflector hood, placed on top of the tank. The light housing also plays a useful role as a lid, preventing the escape of heat, moisture, and sometimes even leaping fish, while keeping pets and children's hands out of the aquarium.

CHOOSING A HEATER

Thermostatic heaters are the most popular and reliable means of heating a tank. They are available in various lengths for aquariums of different depths and in different power (wattage) ratings. Allow about 100 watts for every 26 gallons (100 liters) of water, and choose a unit that can be fully submerged in your tank. Thermostatic heaters keep water temperature constant by switching the heating element on and off repeatedly. For this reason, they have a relatively short life span and should be replaced every two years or so. Although good-quality heaters are reliable, it is wise to put a separate thermometer in the tank to pick up any irregularities in temperature. This should not be positioned directly above the heater. Thermostatic heaters are suitable for use with most fish, but some aggressive species with razor-sharp teeth, such as piranhas *(see p.92)*, can cut through electrical cables. For these fish, undergravel heating units, which are not accessible, are preferred.

Thermostatic heaters can only add heat to the tank. In warm climates, overheated rooms, or tanks under bright lights, it may be necessary to install a chiller to reduce water temperature.

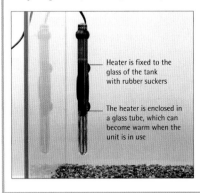

Heater is fixed to the glass of the tank with rubber suckers

The heater is enclosed in a glass tube, which can become warm when the unit is in use

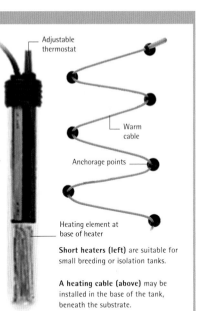

Adjustable thermostat

Warm cable

Anchorage points

Heating element at base of heater

Short heaters (left) are suitable for small breeding or isolation tanks.

A heating cable (above) may be installed in the base of the tank, beneath the substrate.

The amount of light needed in the tank depends largely on the plant species kept. In general, tanks with many substrate plants need stronger lighting than those with floating plants, and deeper tanks need more than shallow ones. For most tank setups, it is enough to leave the lights on for about 10 hours each day; too long a period of illumination will encourage the growth of algae—not only on the glass, but also on other tank plants, which may cause them to die. Fluorescent tubes should be replaced roughly every 12 months, even if they appear to be working. Light output falls and quality changes with the age of the tube, and although these changes may be imperceptible, they will stress the plants in the tank.

Heating the aquarium

A freshwater aquarium is usually heated to 76–79°F (24.5–26°C) using a thermostatic heater *(see box, above)*. This is a special waterproof electrical heating element that incorporates a thermostat; the thermostat measures water temperature and switches off the heater when the desired level has been reached. The most efficient units are those that can be fully submerged in the water.

A high-capacity tank may need two or even three heaters to maintain the target temperature throughout its whole volume, and more heaters may need to be added in the winter months. Even in a smaller tank, using two heaters is a sensible precaution; if one fails, the other will provide the necessary heat. The overall cost of heating the tank will remain the same as if one device were used.

There are two basic types of aquarium thermometers— the traditional alcohol-filled design (far left) and the LCD type (left), which fits onto the outside of the aquarium glass.

ELECTRICAL SAFETY

● Make sure the power supply is disconnected before placing your hands in the water.

● Allow the heater time to cool before lifting it out of the water.

● Avoid trailing cords and adaptors. Consider using a cable organizer instead.

Fish vary greatly in their temperature requirements. The White Cloud Mountain Minnow is adaptable, surviving happily in temperatures from 66°F (19°C) to 82°F (28°C).

Filtration and aeration

An effective aquarium filtration system not only removes waste products from the water by physical or chemical means but also mirrors the process of biological filtration—the nitrogen cycle—that occurs in the wild. Filtration goes hand in hand with aeration, in which water is circulated so that it can absorb oxygen from the air and lose unwanted carbon dioxide.

FILTRATION TIPS

● Add zeolite sachets to remove ammonia from the water and a starter seed culture of bacteria for the biological filter.

● Be careful not to overfeed the fish and burden the filter with decomposing food.

● Test the water quality regularly to check the filter's efficiency; frequent partial water changes (see pp.50-52) will ease the pressure on the filtration system.

In the confines of an aquarium, the waste produced by the fish can quickly build up to harmful levels without an effective filtration system. Filtration involves passing the water in the aquarium through one or several filtration media, which purify the water by biological, chemical, or mechanical means (see box, below). There are two basic methods of driving water through the media: using an electric pump, or relying on an airlift system, in which air bubbled into the tank through an airstone draws water up an airlift tube.

NITROGEN CYCLE

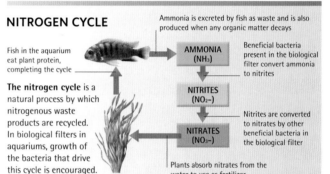

Ammonia is excreted by fish as waste and is also produced when any organic matter decays

Fish in the aquarium eat plant protein, completing the cycle

The nitrogen cycle is a natural process by which nitrogenous waste products are recycled. In biological filters in aquariums, growth of the bacteria that drive this cycle is encouraged.

AMMONIA (NH₃) — Beneficial bacteria present in the biological filter convert ammonia to nitrites

NITRITES (NO₂-)

NITRATES (NO₃-) — Nitrites are converted to nitrates by other beneficial bacteria in the biological filter

Plants absorb nitrates from the water to use as fertilizer

FILTRATION METHODS

There are three basic methods of filtration: mechanical, biological, and chemical (below). Aquarium filtration systems often involve more than one of these methods and may utilize all three. Many tanks have an undergravel filter, shown below, which is a simple biological filter. Internal power filters (right), also provide biological filtration but can include additional media for purifying water by mechanical and chemical means.

Mechanical filtration uses a filter medium, such as this filter wool, to sieve particulate waste from the water. The fibrous structure traps the waste, which can then be removed.

Biological filtration involves the breakdown of waste by beneficial bacteria that drive the nitrogen cycle. These multiply in media, such as this foam sponge, and in the substrate.

Chemical filtration relies mainly on activated carbon to eliminate dissolved waste from the water. Unfortunately, this process also neutralizes some medical treatments.

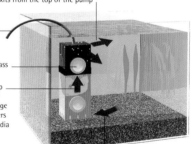

Clean water exits from the top of the pump

Power line

Suckers attach filter unit to glass

Water passes up the unit, either through a sponge or through layers of different media

Water is sucked into the unit at the base

Internal power filter An integral electric pump draws water through the filter unit, which contains one or several chambers housing filtration media inserts. After passing through the media, the clean water is returned to the main tank.

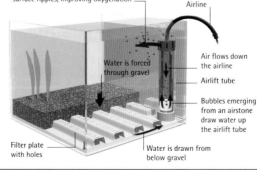

Water emerging from airlift creates surface ripples, improving oxygenation

Airline

Air flows down the airline

Airlift tube

Water is forced through gravel

Bubbles emerging from an airstone draw water up the airlift tube

Filter plate with holes

Water is drawn from below gravel

Undergravel filter Air pumped through an airstone within an airlift tube draws water from below the gravel substrate. This forces the water through the gravel, which contains beneficial bacteria that break down harmful waste products.

Undergravel filters—which are now used less commonly, compared with power filters—rely on a simple airlift system. A perforated corrugated or ridged plate is placed on the base of the tank and then covered by a substrate of gravel. The plate allows water to flow under the gravel, while the gravel particles—which should be at least ⅛ in (3 mm) in diameter to ensure good water movement—form the biological filter medium.

Power filters use an electric pump to drive water through the filtration media. There are two basic types: internal power filters, which sit inside the tank, and external filters, which are housed outside the aquarium and are generally used for larger aquariums. A range of different media are available for power filters: biological media, such as foam sponge and ceramic granules; mechanical media, such as filter wool; and chemical filtration media, such as carbon. The most efficient filters use layers of different media in combination. It is often

External power filters work on the same principle as other systems, but water is pumped out of the aquarium and passes through the filter unit, which contains one or several types of media, before being returned to the tank.

Cabinet hides unsightly equipment | External power filter

Powerhead mounted on airlift of undergravel filter

A powerhead, shown above, is a pump that can be added to the airlift tube of an undergravel filter to draw water more strongly through the system. Installing a powerhead also improves the aeration provided by an undergravel filter.

possible to add additional materials—for example, peat or coral sand—to this type of filter to alter the water chemistry.

Efficient aeration is vital for the maintenance of a healthy tank, providing a source of oxygen not only for the fish but also for the beneficial bacteria within the biological filter. Using an air pump to pump air through an airstone (see p.41) can help to aerate the tank: the bubbles produced cause surface ripples that increase the area exposed to the air where oxygen exchange can take place. However, if there is enough surface movement generated by the outflow of the filtration system, an additional air pump may not be needed.

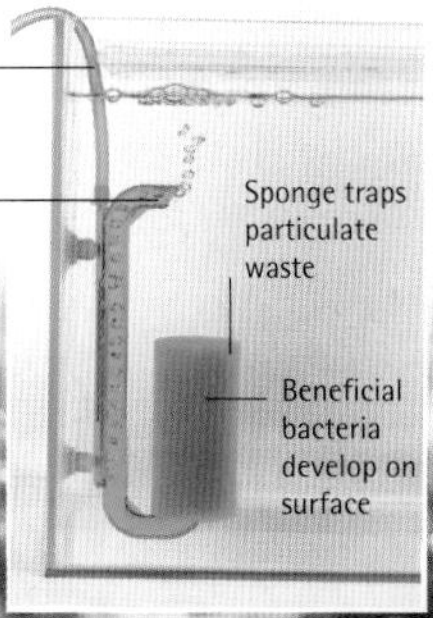

The small size of young fish, such as the Jaguar Cichlid (Parachromis managuensis) fry seen here with their mother, means that they are at risk of being sucked into a power filter. Consequently, breeding tanks require less vigorous filtration; a simple sponge filter, pictured left, is probably the best solution.

AIR PUMP TIPS

Airline with non-return valve

● Fit a non-return valve in the airline between the pump and the tank to keep water from entering the pump.

● If the pump gets noisier, have it serviced or replaced.

● Never cover an external pump; this creates a fire risk.

Airline

Filter outflow

Sponge traps particulate waste

Beneficial bacteria develop on surface

Choosing plants

Plants play a key role in the aquarium, offering shade and shelter so that fish feel secure and providing food for herbivorous species. Thriving freshwater plants will restrict algal growth and help to improve the water conditions by absorbing nitrate and giving off oxygen. With hundreds of different colors, leaf shapes, and sizes available, there is endless scope for you to exercise your creative powers in the design of your aquarium.

It is best to devise a rough planting plan before making any purchases. The standard approach is to have one or two eye-catching central plants toward the back of the tank, with taller plants flanking these and extending around the sides. Low-growing foreground plants will give your aquascape a more natural look while still allowing you to appreciate the fish and the background flora. Floating plants add another tier of interest and create attractive dappled shadows below.

When choosing plants, make sure they have requirements similar to those of the fish that will live in the tank. For example, not all plants will thrive in the hard water needed by Rift Valley cichlids, nor in brackish surroundings. The behavior of the fish should also influence your choice. Include bushy plants for nervous fish that like to hide in thick vegetation and broad-leaved plants for shade-loving bottom dwellers. Floating plants provide a refuge for fry and spawning sites for bubble nesters, such as gouramis. If the fish like to dig in the substrate, position substrate plants behind tank décor, where they are less likely to be uprooted.

Disguise pot with rockwork or gravel

Rooted substrate plants, such as this cryptocoryne, are often sold in open-mesh pots, which make them easy to transfer to the aquarium.

Roots trail freely in water

Floating plants, such as this Butterfly Fern (*Salvinia* species), are the easiest type of plant to maintain. They move around the tank on surface currents.

Healthy green color

Attaching a weight
to a cutting will anchor it in the substrate until it roots. Green Cabomba (*Cabomba caroliniana*), shown here, prefers bright light and soft-water conditions.

Weight

Leaves die back naturally about every nine months

Some plants have specific growing conditions. The tubers of this Wavy-Edged Swordplant (*Aponogeton crispus*) must rest in cool water at 50°F (10°C) for two months after their leaves die back. They will then sprout again.

Tuber

Establishing new plants

After buying plants, return them to water as soon as possible; allowing the leaves to dry out, even for a short time, may fatally damage the plant. Leaves also become bruised by careless handling, so hold the plant by its base or container. Check leaves for snail eggs, which are laid in jellylike masses, and trim off affected parts, along with dead or dying leaves.

Floating plants usually make a trouble-free transition to new surroundings, but do not allow condensation from the hood to drip on them—this will cause the plants to blacken and rot. Rooted substrate plants also establish themselves quickly, but you should constrain their root growth with pots; otherwise, the undergravel filtration will be impaired. Some substrate plants are sold as bunches of cuttings, which need to be separated and planted individually to give them space to grow. Other plants are available as swollen stems, called rhizomes and tubers, that you partially bury in the substrate, leaving any shoots uncovered. Rhizomes can be cut into slices, each of which will root. Tubers cannot be

It takes patience and care to establish an attractive, well-planted tank (left), just as it does with a garden. A planting tool (above) can be helpful for putting plants in place when the tank has been partially filled and for adding new plants without causing serious disruption.

LIVING OR PLASTIC PLANTS?

There are now highly realistic plastic substitutes for most of the popular varieties of aquarium plants. The obvious advantage of plastic plants is that they do not die back or spread out and dominate the aquarium. Plastic plants are equipped with a weight to anchor them in the substrate. Although there is no risk of their being destroyed by the fish, it is still possible that they may be dug up.

Living plant   Plastic plant

divided like this, since they have only one growing point. If an established tuberous plant is moved elsewhere, some of its leaves may die back, but provided the water and lighting conditions in the tank are favorable, it should soon recover.

Care and maintenance

Plants need nutrients for growth. They obtain many of these nutrients from the water, but adding aquarium fertilizer will ensure that they stay healthy. Rich in potassium, phosphorus, iron, and other elements, this fertilizer is available as pellets, which can be buried beside a plant, or as a liquid that is added to the tank water. Another way to boost the growth of plants is to place them in pots of enriched aquarium soil and then sink the pots into the gravel. Never use regular plant potting mix or garden compost, which will poison the fish.

Plants, like fish, benefit from an efficient filtration system, especially fine-leaved varieties, which can become clogged with suspended material. In addition, the gentle currents created by a power filter help to keep the foliage moving, to ensure that no part of the plant is permanently shaded. When the plants are established, you will periodically need to cut back excessive growth or prune straggly plants to encourage new shoots and a better overall shape.

BUYING TIPS

● Check plants for signs of damage, such as crushed stems or leaves.

● Avoid yellowing plants, which may be suffering from nutrient deficiencies.

● Plants that do not permanently grow under water are unsuitable, since they will soon begin to rot.

● Choose plants with lighting needs similar to the fish's.

Large herbivorous fish, such as this Silver Dollar *(see p.94)*, may destroy plants in a new aquarium before they have time to become fully established. Choose tough, quick-growing plants for this type of fish, and offer alternative foods, such as lettuce, to minimize damage to the plants in these vital early stages.

Preparing the tank

A little forethought goes a long way when setting up an aquarium. Give careful consideration to where you want to site the tank, because if you make a poor choice, you will have to empty the tank and strip it down before it can be moved elsewhere. It is also a good idea to plan the layout of the tank well in advance so that you end up with an aquarium that not only looks attractive but also allows the fish to display their natural behavior.

Visualize the finished tank. Knowing how you want your tank to look will make it much easier and quicker to assemble the different elements.

One of the most important elements in the aquarium will be the substrate—the material covering the floor of the tank. This not only forms an essential part of the habitat for the fish but also provides anchorage for the roots of aquarium plants and a surface on which beneficial bacteria can develop.

For fish that require a sandy substrate, use filtration sand or river sand, which are chemically inert (they do not affect the water) and non-compacting. However, most freshwater aquariums use gravel as the substrate, since the water passes through it more easily than sand, making the undergravel filter *(see p.34)* more effective. Aquarium gravel is available in various grain sizes, but make sure that the gravel you choose

Use gravel with rounded stones for bottom dwellers, substrate feeders, and fish that excavate the substrate for breeding purposes, such as this Black Belt Cichlid *(see p.138)*. Avoid rough or sharp-edged gravel, which may cut or scratch the fish.

PREPARATORY TASKS

Before adding components such as decor, equipment, and plants, you will need to wash the tank and the substrate with clean, warm water (do not use detergent) and install the undergravel filter. Handling a wet glass tank can be hazardous, so allow the outside of the tank to dry before placing it on its stand, which must be level and stable. As a general hygiene rule, keep equipment used for aquarium tasks, such as sponges and buckets, separate from items used for domestic chores. If possible, pour dirty water down outside drains, not into a kitchen or bathroom sink.

❶ Wash the tank
Add a little water to the tank and wipe the glass with a clean sponge. This will remove dust or tiny fragments of glass left over from the manufacturing process.

❷ Position on sponge matting
Glass tanks should stand on specially made sponge matting, to absorb any unevenness in the surface below, which could cause the joints between the glass panels to leak.

| COARSE GRAVEL | PEA GRAVEL | FINE-GRAINED GRIT | GLASS BEADS | YELLOW GRAVEL | BLUE GRAVEL |

is appropriate for the fish you intend to keep. The large stones of coarse-grained gravel, for example, will not be suitable for fish that habitually dig in the substrate, bury themselves in it, or sift the stones in search of food.

When buying gravel, bear in mind that a layer of about 2 in (5 cm) is needed to create the filter bed, so allow roughly 2 lb (1 kg) of gravel for every gallon (4 liters) of tank capacity. Although aquarium gravel is usually prewashed, this does not mean that it is necessarily clean enough for the tank. Rinse it thoroughly to remove all traces of sediment; otherwise, you may find that a muddy scum forms on the water when you fill the aquarium. It is also advisable to soak the gravel overnight in aquarium disinfectant, to reduce the likelihood of introducing disease or parasites into the tank.

If you intend to keep fish that require soft-water conditions, avoid gravel containing limestone (calcium carbonate), since

Pea gravel, the most commonly used type, has a grain size of $1/8$ in (3–4 mm); anything finer may reduce the efficiency of the undergravel filter. Coarser gravel tends to trap uneaten food in the gaps between the stones. Glass beads and dyed gravel provide attractive alternatives, but choose dyed gravel carefully, because strong colors may detract from the appearance of the fish.

this will dissolve slowly and increase the water's hardness. You can test gravel by adding vinegar to a small sample. If it contains limestone, the vinegar will fizz as it reacts with the calcium.

Rockwork and bogwood

With the gravel bed in place, you can begin to furnish the tank. Rockwork not only looks attractive but also provides egg-laying sites for a number of tropical species—especially cichlids, which favor slate. In addition, rockwork offers fish places to shelter, and more aggressive species may use it to mark out their territories.

❸ Clean the gravel
After soaking the gravel in aquarium disinfectant, pour it into a colander and rinse it under a running faucet, stirring the gravel occasionally with your hand.

❹ Place the filter
Lay an undergravel filter directly on the aquarium floor. Place the airlift at the back, or else a power filter can be put in place later, with the gravel carefully tipped in beforehand.

❺ Slope the gravel
Smooth out the gravel by hand, sloping it upward to the back of the tank. This will make it easier to spot any buildup of mulm (decayed organic matter) at the front.

Bogwood is popular with various catfish, which use it as a resting spot and rasp away at its surface with their teeth to obtain fiber. Bogwood may float if it has dried out previously, but once fully waterlogged, it should stay on the floor of the aquarium. Weigh it down with rocks if necessary.

As with gravel, the vinegar test will tell you whether the rocks you plan to use contain limestone. You can avoid this problem altogether if you opt instead for a chemically inert rock, such as granite or slate. Do not be put off by its relatively drab appearance out of water; the subtle hues will be much more obvious when the rock is submerged.

Before placing rocks in the aquarium, scrub them in a solution of aquarium disinfectant and give them a good rinse in clean water. Avoid positioning rocks in corners or other places where water cannot flow easily behind them, because mulm will accumulate. Embed each rock firmly in the substrate so that there is no risk of its toppling over or being undermined by the excavations of the fish. If you want to build rocks into towers, cement them together with a silicone sealant designed for aquarium use. Try not to clutter the tank

When choosing rocks for the aquarium, always wet the surface to get an idea of what the color will be like underwater. Only use rocks purchased from aquarium stores; never be tempted to collect your own, because they may leach toxins into the water or upset the water chemistry.

with an excessive amount of rock, because this will impair the efficiency of the undergravel filter by reducing the area of the substrate through which water can flow.

Bogwood—wood that has been either submerged in peaty water or buried in boggy ground for a long time—helps to give the aquarium a more natural feel. It also provides an attractive growing medium for plants such as Java Fern *(see p.194)*. It is especially suitable for aquariums where soft, acidic water conditions are required. Bogwood contains tannin, which will leach into the water and turn it brownish-yellow. The tannin is not harmful to fish, but it spoils the appearance of the tank. To prevent this, presoak bogwood in a bucket of water for several days. Change the water each day, and scrub the bogwood with a clean brush before adding it to the tank.

Catering to individual needs

Before finalizing your design for the tank, consider the fish you intend to keep, since they may influence the decor and the layout. For example, annual killifish *(see pp.170–172, 174–175)*

TANK BACKDROPS

To hide whatever is behind your aquarium, or simply to add another dimension to the tank decor, you can buy printed backdrops to stick on the outside of the glass. Scenes of rocks or plants can increase the naturalistic feel of the aquarium's design, while ancient ruins can add an air of fantasy. You can even achieve the surreal effect of displaying your fish against a desert landscape. Backdrops usually correspond to standard tank dimensions, although they can be cut to fit if necessary.

Sandstone has a rough-hewn appearance

Lava comes in various colors but often has sharp edges

Rose quartz can be stunning in an aquarium with a contemporary design

Gray Cumberland stone looks good alongside green plants

Marble is available in a wide range of colors and patterns

Rustic slate is ideal if you want tall, slender rock structures

Artificial rock is a lightweight alternative to the real thing

An airstone attached to the airline from a pump produces an attractive column of bubbles. The airstone can be disguised by placing a plant in front of it. The coarser the airstone used, the larger the bubbles.

FUN FURNISHINGS

There is an almost endless variety of fun objects that can add a sense of playfulness to your tank. Some are stand-alone ornaments, while others can be connected to an airline to serve as airstones. What to include is a matter of personal choice, but use only items designed for use in aquariums, because others may contain chemicals that are toxic to fish.

Ornate decorations can be hard to clean if algae develop on their surface

need open swimming space, just as they would have in the pools they inhabit in the wild. Conversely, doradid catfish *(see pp.118–119)*, which live in rocky streams, require plenty of hiding places in their aquarium, so you should arrange the rockwork into cavelike retreats. A few fish have very specific requirements. Some African cichlids, such as *Signatus (see p.155)*, spawn in snail shells, and you will have to include these in the tank if you want the fish to breed. The cleaned shells of large edible snails, as sold in gourmet shops, are ideal.

Assembling the components

Keep to your overall plan for the tank as you assemble the rest of the components. Position the heater unit toward the back of the tank so that it is both inconspicuous and accessible for maintenance. (Most tank hoods also have holes at the rear, through which you can run power cables to heating and filtration equipment.) Convection currents in the water distribute heat throughout the aquarium; warmed water rises toward

A clean terracotta flowerpot on its side forms a ready-made cave where shy fish can shelter. Some of the smaller catfish and cichlids, such as this Pearl Cichlid *(see p.139)*, may even spawn inside flowerpots.

the top of the tank, cools, and then sinks back down again. It is important that water can move freely around the heater; otherwise, this circulation will be disrupted and local hot spots will develop in the tank. In a large aquarium, it may be better to use two widely spaced thermostatic heaters, to ensure a more even distribution of warm water.

Place the thermometer (see p. 33) at the opposite end of the aquarium from the heater so that it records the minimum temperature in the tank. (If you are using a separate thermostat and heater, position the thermostat there, too.) An LCD thermometer attaches to the outside of the glass, but be careful not to touch the display with your fingers, since this will give a false reading. Keep it out of direct sunlight, too, which will have the same effect. An alcohol thermometer will not suffer from these problems because it attaches to the inside of the glass, but it may be dislodged by large, active fish.

Connect the airline from the air pump to the airlift tube of the undergravel filter. If you are also using an internal power filter, place this close to the surface, and well clear of the substrate, so that it can circulate water more effectively.

Final tasks

With all the decor and equipment in place, add the plants and water. When the tank is full, switch on the power. An indicator light on the heater will tell you whether it is operating, while bubbles produced by air pumps and power filters will show that they are functioning properly.

Before introducing the fish, add a commercial seed culture of beneficial bacteria to the gravel. This can also be useful in a tank that does not have an undergravel filter but still relies on a medium that carries out biological filtration (see p. 34) in some form, such as the sponge in an internal power filter.

ADDING WATER AND PLANTS

Never put tap water in the tank without first treating it with a water conditioner. This will neutralize chlorine and chloramine in the water, which can be fatal to fish. Some water conditioners contain aloe vera, which helps to heal any minor injuries that the fish sustain when they are caught and moved. When you add substrate plants, do not bury the crowns, or they will rot. If you want to include floating plants, add these when the tank is full, allowing at least 1 in (2.5 cm) of clear airspace between the surface and the hood.

❶ Add water conditioner
Using a watering can with a known volume of water, measure out and add the correct amount of water conditioner. Check the temperature before adding it to the tank.

❷ Partially fill the tank
Pour the water on to a small plate or saucer to avoid disturbing the gravel bed. Fill the tank until the water is at least as deep as the tallest plant you have bought for the tank.

❸ Add the plants
Set the plants in place using either your hands or a planting stick. Then continue filling the tank carefully, pouring the water slowly to avoid creating a strong current.

When the tank is finished and you are happy with the layout, put the hood on to prevent evaporation and to help prevent airborne pollutants from wafting on to the surface of the water.

Brackish water

The conditions encountered by fish living in estuaries and mangroves, where fresh water and saltwater meet, can be recreated in a brackish water aquarium. The equipment required is much the same as for freshwater aquariums, but the substrate is usually sandy, so power filters are more appropriate than undergravel systems.

Themed aquariums can be created, based on different aquatic habitats. The one shown here replicates the conditions in a shallow, slow-moving coastal stream in Central America.

The water in a brackish aquarium requires a salinity level, expressed in terms of specific gravity (SG), of 1.002–1.007 (marine aquariums have a salinity of at least SG 1.020). To achieve this, dissolve the appropriate amount of marine salt *(see pp.214–215)* in dechlorinated tap water, before filling the tank. The movement of water within the tank caused by the filter and heater will ensure an even concentration throughout the aquarium. Salinity increases as water evaporates from the tank, so when topping off to replace lost water, use dechlorinated tap water rather than saltwater to keep the salinity within the desired range. For significant partial water changes carried out during routine maintenance *(see pp.50–51)*, use salinated water.

It is important that the tank used for a brackish water aquarium is either all-glass or acrylic, with no metal surround. This is because saltwater rapidly corrodes metal, and any corrosion in the tank pollutes the water and can harm the fish.

CREATING A MANGROVE TANK

This fascinating habitat—part water, part land—requires young mangrove plants, which can be bought from aquatic nurseries. Pot the mangrove shoots to restrict their root growth. You may need to prune them when they start to spread across the tank. Use aquarium sand rather than gravel for the substrate, and keep the water level low to mimic the tidal shallows of a mangrove swamp. A thermostatic heater will not work with the low water level, so use a heating cable *(see p.33)* and a thermostat.

❶ Position the potted mangrove shoots
Lay a base of sand at least 1½ in (4 cm) deep. Place the mangrove pots toward the back of the tank. Sink their bases into the sand, but do not completely cover the pots.

❷ Add bogwood and stones
Conceal the mangrove pots behind a bank of tank decor, such as bogwood and large pebbles, so that the shoots project over the top. Make sure the structure is stable.

❸ Add water and shape the sand
The tank should be less than half full of water. Add more sand to the tank and shape it so that it resembles a beach. Some of the sand should be above the water level.

❹ The finished tank
The beachlike effect of the sandy slope can be enhanced by the ripples from the power filter. The mangrove plants will grow better under strong lighting.

Mudskippers *(see p.189)* are an ideal choice for a mangrove tank. They will emerge from the water on to the sandy "beach" for short periods.

The mudskipper uses its fins to move around on land

Choosing and introducing fish

Once the aquarium is fully prepared and you are certain that everything is functioning properly, you can start to introduce the fish. Although suppliers can ship stock to you, it is probably best to visit a local store, especially when starting out, so that you can see the fish firsthand. Ask experienced aquarists to recommend a store—ideally one that belongs to a recognized trade association. Such bodies run specialized training courses for staff, so you are more likely to receive genuine, professional advice.

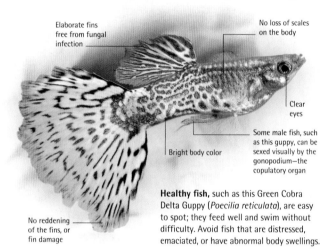

Elaborate fins free from fungal infection

No loss of scales on the body

Clear eyes

Some male fish, such as this guppy, can be sexed visually by the gonopodium—the copulatory organ

Bright body color

No reddening of the fins, or fin damage

Healthy fish, such as this Green Cobra Delta Guppy (*Poecilia reticulata*), are easy to spot; they feed well and swim without difficulty. Avoid fish that are distressed, emaciated, or have abnormal body swellings.

A good aquarium store will have helpful staff and clean, clearly labeled, well-stocked tanks (below). Disease spreads easily between fish in a tank, so when choosing fish, try not to focus on just one or two good specimens that catch your eye but also look at their tankmates, to make sure that they are all healthy (right).

Choosing freshwater fish from the vast array available can be a daunting task; a little advance research can help you to make the right choices. Select species that share similar water chemistry requirements, and avoid fish that will quickly outgrow the tank. Temperament and behavior should also be considered; lively, shoaling fish, for example, are not ideal tankmates for species that prefer calmer, less-populated surroundings. Do not mix active predators with smaller, placid fish that could become their prey. Aggressive fish may bully less bold companions and steal their food. Nervous fish, such as some dwarf cichlids, may benefit from the company of "dither" fish— more confident species whose calm presence helps the fish to feel more secure. Finally, try to obtain a mix of species that naturally inhabit different levels of the tank.

SOCIABILITY

The "traffic light" system is used in some aquatic stores to rank sociability.

● A red spot against a fish's name indicates that it must be kept on its own.

● Yellow means that the fish may have special requirements.

● Green indicates fish that can be kept in a community tank.

Checking for problems

When you have made your choice, ask to inspect the fish to be sure that they are healthy. Before visiting the store, familiarize yourself with the signs of disease in the chart on page 56 so that you know what to look for. It is important to view the fish from both sides, which is easiest to do when it has been caught and is in a plastic bag. It can be difficult to determine the condition of some catfish and other sedentary species, but if they are fairly plump and do not have a hollow-bellied appearance, the likelihood is that they are healthy.

Adding the fish to the tank

Before introducing the fish to the tank, check the water chemistry (*see p.46*) to make sure that it is suitable for your fish. It is advisable not to populate the tank to its maximum stocking density at first, in order to avoid putting a strain on the filtration system, which will not yet have a fully established colony of beneficial bacteria. Observe the fish closely in these early stages, to make sure that they are settling in peacefully.

INTRODUCING NEW FISH

Being moved can be traumatic for fish, and it takes several days for them to acclimate to their new home. Provide a vitamin C–rich diet at first, to boost the immune systems of the fish and help them avoid stress-related illness.

❶ Equalize water temperatures
Float the bag in the tank for about an hour, so that the temperature inside the bag gradually adjusts to match that in the tank.

❷ Catch the fish
Net the fish inside the bag, gripping the neck of the bag to stop water from escaping into the tank, which could introduce disease.

❸ Release the fish
Carefully allow the fish to swim out of the net. To minimize stress for the new arrivals, do not turn on the tank lights for a while.

CATCHING AND TRANSPORTING FISH

Most fish can be caught with a net. Patience is essential, since chasing the fish around the tank will simply cause them to panic, and the water resistance will make it even more difficult to catch them. Nets can transfer disease between tanks, so dip the net in a solution of aquarium disinfectant after use. Fish with spines should be steered into a bag, because they can become caught up in the mesh of a net. When buying fish, the supplier will catch the fish for you. Fish are usually transported in clear plastic bags, tied at the neck, with a ratio of about two-thirds air to one-third water. Put the plastic bag in a brown paper bag to make the journey less stressful for the fish.

Scoop up the fish from below when it is near the surface. As you lift it from the water, place your hand over the net (above) to stop the fish from escaping.

Neck securely tied during transit

The store may inflate the bag with oxygen

Checking the water

Maintaining water quality in the aquarium involves regular monitoring and adjusting. Prior to adding the fish, measure the water's pH (its acidity or alkalinity) and hardness (the level of mineral salts in solution). When the fish are established, test for toxins, such as ammonia and nitrite, which can be a problem in a new tank (see p.52). If you have a brackish water aquarium, you must also monitor the salinity.

The pH scale runs from 0 to 14, with pH7 being neutral. Water conditions giving a pH reading above 7 are described as alkaline; below this, they are acidic. Small changes in the pH value have very significant effects on water chemistry.

The pH level of the water can be raised by adding coral sand to the filter, while including aquarium peat acidifies the water. Hardness can be increased by placing limestone rocks or coral shells in the tank, which dissolve slowly. To soften the conditions, add tap water that has been passed through an ion-exchange column, which draws mineral salts from the water, or use a reverse osmosis (RO) system.

ELECTRONIC MONITORING SYSTEMS

A range of electronic monitoring devices are available, displaying various degrees of sophistication, typically reflected in their prices. They constantly monitor water quality and other parameters, such as temperature and power supply. Such units can alert you via smartphone or computer system if anything is seriously wrong, even if you are not at home, and so can prove to be a lifesaver for the tank occupants.

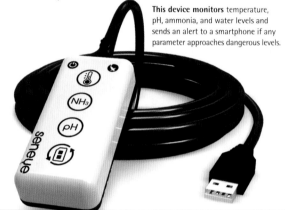

This device monitors temperature, pH, ammonia, and water levels and sends an alert to a smartphone if any parameter approaches dangerous levels.

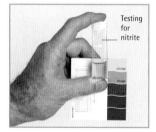

Testing for nitrite

Water test kits, which use reactive substances that change color when added to tank samples, can be used to check pH, hardness, and toxin levels.

Electronic meters provide a quick and easy way of testing hardness, pH, and salinity. Salinity can also be measured with a simple hydrometer.

It takes time for the plants to become established and for the filter to function with maximum efficiency, so frequent monitoring of the water conditions is essential in the first few weeks. Adding zeolite, a chemical that removes ammonia from the water, can help in these early stages. If you have snails that start damaging the plants, place a piece of cucumber under an upturned saucer, propped up so the snails can crawl underneath. Leave it overnight, and then collect them in the morning.

FRESHWATER FISH
MAINTENANCE

Feeding the fish

Feeding time offers the best opportunity to see your fish at close quarters and monitor their well-being. It is also an opportunity to tame your fish, especially if you feed them at the same times each day. Freshwater fish can be given a wide range of foods, but commercial products are the most convenient and also the safest option, since diseases may be introduced to the tank along with invertebrate live foods, such as tubifex worms.

Fish require the same key food groups as humans—proteins, carbohydrates, fats, vitamins, and minerals. Protein is vital for healthy growth, especially in young fish. Raising the level of protein in the diet—by providing more live foods, for example—helps to bring fish into breeding condition. Carbohydrates fuel the body's processes, and fat forms a protective cushion around organs and acts as an energy store. Vitamins and minerals are important for a fish's metabolism and overall health. Vitamin C, for example, helps protect against infection, while calcium and phosphorus are essential for sound skeletal structure.

SPECIALIST DIETS

Some prepared freshwater fish foods are specially formulated to cater to the dietary needs of particular groups of fish. Pellets for carnivorous catfish, for example, contain more oil than foodsticks for plant-eating cichlids. Aside from protein, oil (fat), and carbohydrates, prepared fish foods also contain fiber. In the wild, fish inadvertently consume a variety of indigestible items, ranging from plant matter to fish scales. This roughage helps to prevent blockages in the intestinal tract.

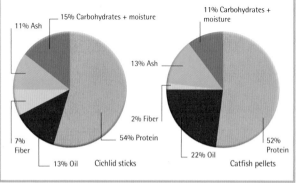

Cichlid sticks: 11% Ash, 15% Carbohydrates + moisture, 7% Fiber, 13% Oil, 54% Protein

Catfish pellets: 11% Carbohydrates + moisture, 13% Ash, 2% Fiber, 22% Oil, 52% Protein

Vegetables, like the pea pod being devoured by this Spotted Catfish, provide fiber as well as vitamins and other nutrients. It is better to use organic vegetables to avoid the risk of introducing harmful chemicals to the tank.

Special color foods in the form of flake may be used to enhance the intensity of coloration in red fish, such as this Red-Eyed Red Swordtail.

Quantity and frequency

Most fish-food packaging gives few details about the quantity that should be used, largely because this depends on the number and size of the fish in the tank. As a rule of thumb, feed fish little and often—offer food three or four times a day, providing no more than will be eaten within two or three minutes. In a mature tank, there will be edible items, such as algae, for the fish to browse on outside of their set feeding times.

Avoid overeating—unwanted food scraps will pollute the water, impairing the health of the fish. Providing an unbalanced diet also causes problems: an excess of fat, for example, leads to obesity, shortening a fish's life and possibly causing sterility. Guard against underfeeding, too. Fry, which need to eat more regularly than adults, may become stunted or deformed if they are underfed. Using dedicated fry foods will ensure that they get the nutrients they need. In a community aquarium, make sure nervous fish get a chance to feed; otherwise, they, too, may become malnourished. A number of catfish are night feeders, so drop food into their tank just before lights-out.

VACATION SOLUTIONS

Automatic battery-operated feeders of the type shown below can be set to dispense measured amounts of food at regular times, ensuring that the fish do not go hungry when you are away. Another long-term option is a food block placed on the aquarium floor, which will not pollute the water as other foods would if left uneaten.

TYPES OF FOOD

Commercially prepared foods sometimes need to be supplemented with fresh items and live foods. Freeze-dried live foods tend to be less palatable to fish than thawed live foods, because they have a lower water content. Some live foods can be cultured at home: a container of water left outdoors in summer will attract breeding midges and mosquitoes, and their larvae can be removed with a sieve. The way in which food is presented to the fish is also important. The means of delivery should take account of the fish's normal feeding habits. Bottom-feeding catfish, for example, are unlikely to be tempted by flake floating on the surface.

PREPARED FOODS

Pellets are good for large fish, such as pacus, which can swallow them whole. They may float for a time before sinking.

Granules are more dense than pellets and sink rapidly to the floor of the aquarium, making them ideal for bottom dwellers.

Flake floats, so it will attract fish to the surface to feed. Suitable for smaller fish, it can also be powdered and fed to fry.

Tablets are useful for group feeding. They are too large to be swallowed whole so give the fish time to take several bites.

LIVE FOODS

Brine shrimp in their larval form, called *nauplii*, are ideal for rearing fry. They can be hatched from eggs at home.

Daphnia ("water fleas") are a good conditioning food for smaller fish. These crustaceans can also be used to enhance color.

Chironomus worms, also known as bloodworms because of their color, are actually the larval form of a type of midge.

Mosquito larvae are eagerly eaten by many surface-dwelling fish. They may also help to stimulate breeding behavior.

FRESH FOODS

Carrot was used as a natural coloring agent before there were commercial color foods. It is high in fiber and may aid fertility.

Sliced cucumber and zucchini are a good source of vitamins and roughage for herbivorous catfish and some cichlids.

Shelled peas, both fresh and thawed, are a good dietary supplement for bottom dwellers; avoid using salted canned peas.

Beef heart, trimmed of fat, should be used sparingly, since it pollutes the water and may cause obesity in fish.

Routine tasks

A well-set-up tank that is functioning properly will require only a minimal amount of time spent on its maintenance. The aquarium should develop into a stable ecosystem, in which toxin levels are kept in check as part of the natural cycle. With regular checks and a few simple "housekeeping" tasks, such as partial water changes, your aquarium will look its best and your fish will stay healthy and content.

Regular partial water changes ensure that toxic chemicals do not build up in the tank and harm the fish. They should be carried out once a month or when indicated by water test results *(see p.46)*. Start by turning off the power to the tank. Fill a length of siphon tube *(see box, opposite)* with tap water, adding a drop of water conditioner to dechlorinate it, just in case any should escape into the tank. With a finger over each end of the tube, place one end in the tank and the other into a bucket. Release the tank finger first, followed by the finger in

Emperor tetras *(see p.100)* are very sensitive to accumulations of nitrogenous compounds in the tank water. Regular partial water changes are therefore especially important for the well-being of these fish.

the bucket; water should flow from the tank into the bucket. A gravel cleaner can be attached to the end of the siphon tube; as you move the cleaner over the substrate, the water flow stirs up the gravel and sucks out particulate waste. When you have removed enough water, simply lift the tube out of the

REGULAR MAINTENANCE TASKS

DAILY

● Check the thermometer; if the water temperature has changed, the heater or the thermostat may be faulty.

● When feeding the fish, watch out for any decline in appetite, since this is usually a sign of illness.

● Check the lights above the tank; replace a burned-out tube without delay.

● Make sure the filter is working effectively; if it is not, there may be a blockage in the system or even a power failure.

● Reposition any substrate plants that have become uprooted and floated to the surface.

WEEKLY

● Carry out water tests to monitor levels of nitrogenous waste. Keep a check on the pH as well, using either test kits or a meter.

● In a newly established aquarium, carry out a partial water change of up to 20 percent every week, since the filtration system will not yet be fully functional.

● Check for any change in the appearance or behavior of the fish that may indicate that they will soon be breeding.

● Siphon out any mulm accumulating on the substrate. This will reduce the burden on the filtration system.

MONTHLY

● Carry out a partial water change— approximately 25 percent of the functional tank volume—using a gravel cleaner as well.

● Trim dead stalks and leaves from plants. Add aquarium plant fertilizer to the water.

● Remove any buildup of algae in the tank by cleaning, and then adjust the period of light exposure within the aquarium.

● Where an internal power filter or an external filter are being used, strip down, check, and clean the filtration system.

Long-handled algal scrapers are invaluable for keeping the sides of the aquarium free of algal growth, enabling you to reach right down to the base of the tank. Magnetic scrapers use a short handle on the outside of the glass to control a scraper blade on the inside.

Algal overgrowth on aquarium plants is unsightly and can also prevent them from photosynthesizing, which will cause them to die back.

Some algal growth can be beneficial to fish, because it forms part of the natural diet of a number of species, including this Electric Blue.

tank. Before refilling the tank, add a suitable amount of water conditioner to the fresh water, and make sure that it is at the same temperature as that within the tank, or the fish will be stressed by the sudden change. Pour the water in slowly, taking care not to disturb the roots of the substrate plants.

If your tank has a biological filter, switch the power back on as soon as possible, because the aerobic bacteria that provide the basis for filtration will die if they do not receive oxygenated water for some time. Should you need to replace the filter sponge, be sure to add a seed culture of bacteria. Be prepared for an initial deterioration in water quality, since the biological filter will not work efficiently again until the bacteria have colonized the surface of the new filter sponge.

Problems with algae

Excessive algal growth may occur if the aquarium lights are left on for too long. It is especially likely if there are no plants in the tank, or if the plants in a new tank are not yet fully established, because plants naturally compete with algae for nitrates and other key growth compounds in the water. Without competition, algae spread more easily, not only growing on the glass but also covering rockwork and other

PARTIAL WATER CHANGE

Cleaning the gravel while siphoning water from the tank improves the filter's efficiency and ensures that the gravel bed does not become compacted. Save the tank water that you siphon into the bucket; you will need this to rinse the filter sponge.

Cleaning the filter sponge
Rinse out the sponge in water taken from the tank, to remove any debris that has collected.

❶ Siphon out the water
Place the bucket below the tank to ensure a good flow. Never suck water through the tube to start the flow—you could swallow harmful microbes.

❷ Clean the gravel
Take care not to uproot any substrate plants when using a gravel cleaner. The water flow will not be strong enough to suck gravel up the tube.

tank decor. Reducing the length of time that the aquarium lights are on will help to curb this problem, as will making regular partial water changes (which keep the nitrate levels low) and introducing fish that browse on algae.

Monitoring water quality

Test kits and meters *(see p.46)* should be used to make weekly checks on the water quality. Daily visual checks are also vital; any unexpected change in the appearance or behavior of your fish may indicate that there is a problem. For example, if the gills of a fish become brown instead of the normal pink, it is likely that there is too much nitrite in the water. This will interfere with the ability of red blood cells to carry oxygen. Regular partial water changes will usually cure nitrate excess. A rise in the level of nitrite or toxic ammonia may be due to "new tank syndrome" *(see right)*, caused by overfeeding on a regular basis *(see p.49)*, or a result of overstocking the tank, which places extra demands on the filtration system.

Fish differ in their susceptibilities to dissolved chemicals, which is often a reflection of their habitat in the wild. Those occurring in fast-flowing water where there is little opportunity for pollutants to accumulate, such as discus *(see pp.142–143)*, are much more vulnerable than those that naturally inhabit small ponds. Discus show obvious signs of

nitrite poisoning when this chemical is present in concentrations of just 0.5 mg/liter, whereas most other fish will be unaffected until the level rises to 10–20 mg/liter. Ammonia can be removed by the chemical zeolite, which is either added to the filter (if present) or simply dropped into the water.

NEW TANK SYNDROME

Water conditions in a new tank take time to stabilize. There is an initial rise in the level of ammonia, which the fish excrete as waste. As the biological filter starts to work, beneficial bacteria break down the ammonia into slightly less harmful nitrite, which is eventually converted into nitrate. Although this is called new tank syndrome, a similar situation can arise in a mature tank if the filter's efficiency is dramatically reduced. This could be caused by the use of antibiotics (which will kill the bacteria), a breakdown in the oxygenation of the filter bed, which is essential for the survival of these aerobic microbes, or simply the replacement of the filter sponge.

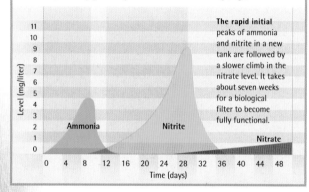

The rapid initial peaks of ammonia and nitrite in a new tank are followed by a slower climb in the nitrate level. It takes about seven weeks for a biological filter to become fully functional.

The stocking density of the tank has a direct impact on the efficiency of the filter. The larger the number of fish, the more difficult it will be for the filter to deal with their accumulated waste *(see p.30 for more on stocking densities)*.

VACATION ISSUES

● Arrange for a friend or neighbor to check the tank every day in case there is a power failure or any of the equipment malfunctions.

● If someone else is to feed the fish, show them exactly how much food they should give each time in order to prevent overfeeding.

● As an alternative, consider using an automatic feeder or a food block.

● Carry out a partial water change and check all the equipment before you leave.

● Leave a contact number in case of emergencies.

FRESHWATER FISH

ILLNESS AND TREATMENT

Health concerns

Fish are susceptible to a wide range of bacterial, viral, fungal, and parasitic diseases. The artificial conditions in the aquarium tend to concentrate the risk of disease and increase the speed at which illnesses spread, because fish health is strongly influenced by the quality of the environment. Early intervention can head off many of the most common conditions, but, as with humans, prevention is always better than cure.

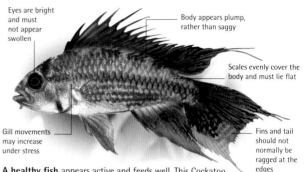

A healthy fish appears active and feeds well. This Cockatoo Dwarf Cichlid (*Apistogramma cacatuoides*) demonstrates what to look for in a well-kept specimen.

Eyes are bright and must not appear swollen

Body appears plump, rather than saggy

Scales evenly cover the body and must lie flat

Gill movements may increase under stress

Fins and tail should not normally be ragged at the edges

Minor injuries, such as the rip in the dorsal fin of this corydoras catfish, can provide an entry point for life-threatening infections.

Fungal infections are opportunistic, lurking in the tank and invading at sites of injury. If untreated, they spread fast and may kill the fish.

Get to know your fish

Most common diseases of aquarium fish are treatable, provided they are identified in their early stages. Later in the course of a disease, obvious symptoms, such as major color changes or large parasites, will become apparent, but it may be too late for successful intervention. The best strategy for disease prevention is therefore close observation; knowing how your fish look, feed, swim, and interact with others allows you to spot subtle changes in appearance and behavior that may signal stress or the early stages of disease.

If you suspect that a fish may be ill, check the tables on pages 56–58 and attempt a diagnosis. If in doubt, consult an expert at your local aquarium store. Move the affected fish out of the main tank into a hospital tank *(see opposite)* for further observation and treatment. This will isolate the fish from its tankmates and prevent cross-infection. Always check the quality of water in the tank; dirty or unbalanced water may cause illness directly or put the fish under stress, thus predisposing it to attack by pathogens that may already be in the tank. Most healthy fish carry a natural burden of parasites, but environmental stress may lower the immune resistance of the fish and allow the parasites to multiply.

Risk factors

A fish housed alone is less likely to succumb to illness than one in a community aquarium because it is not subject to bullying by tankmates. This often causes minor injuries, such as damaged fins and scales, which may then become infected. Aggression usually increases at the beginning of the spawning period, so a close watch should be kept on fish at this time.

The greatest risks to health occur when fish are introduced to the aquarium. New acquisitions should be held in a simple isolation tank for two weeks before transfer to the main tank, by which time any serious health problems should have emerged. With new fish, it is often helpful to remove all the occupants from the tank, rearrange the tank decor, then introduce new fish together with the previous occupants. This strategy significantly reduces outbreaks of bullying.

WHITE SPOT

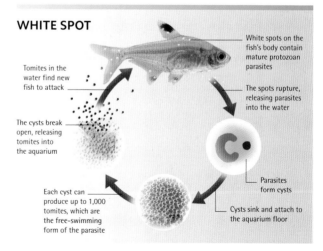

White spots on the fish's body contain mature protozoan parasites

The spots rupture, releasing parasites into the water

Parasites form cysts

Cysts sink and attach to the aquarium floor

Each cyst can produce up to 1,000 tomites, which are the free-swimming form of the parasite

The cysts break open, releasing tomites into the aquarium

Tomites in the water find new fish to attack

Protozoan parasitic infections, such as white spot (also known as "ich"), spread fast within the aquarium. The life cycles of the parasites are completed rapidly, and thousands more can be produced in a single cycle.

Effective treatments

The effective commercial remedies now available treat the great majority of tropical fish conditions; a knowledgeable retailer should be able to guide you to the most suitable one. The treatment may be given in the home tank, in a stripped-down hospital tank, or in a small bath. Carefully follow the instructions given on the medication. Overdosing is likely to be harmful, especially for a fish already weakened by illness; similarly, do not stop the treatment earlier than advised if the fish seems to be recovering. Remember to dechlorinate any water used to make up treatment solutions and make sure it is at the same temperature as that in the main tank, in order to lessen the stress on the fish.

If an outbreak of unspecified disease occurs in an established group of fish (rather than in one individual that can be isolated from the group), regular partial water changes will always help, because they lessen the concentration of disease-causing organisms present.

Some illnesses simply defy treatment. If you need to kill a fish painlessly, carefully cut through the spinal cord behind the head with a sharp knife or take it to your veterinarian. Never be tempted to flush a fish—living or dead—down the drain.

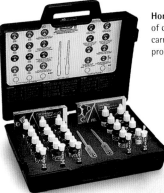

Home test kits include a wide variety of chemical analyses that can be carried out to identify almost any problem associated with water quality.

The design of the aquarium itself is influential in keeping the fish healthy. A well-planned tank provides retreats for shy species and lessens the risk of bullying in more territorial species.

ISOLATION AND MEDICATION

In some cases, it is desirable or essential to treat a fish in a separate hospital tank—a relatively stress-free environment where sick fish can recuperate. This should be a small tank (about 12 gallons or 50 liters) containing no living plants, but with a simple filtration system, and perhaps a flowerpot or plastic plant as a refuge.

Using a hospital tank to quarantine new fish before transferring them to the main tank is a sensible way of reducing the risk that you will introduce diseases to your other fish.

Simple filter in hospital tank

Main tank

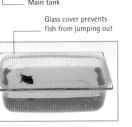

Glass cover prevents fish from jumping out

Eyedropper

A small glass dish can be used to give concentrated medication—for example, to remove external bacteria, parasites, and other disease-causing organisms.

Methylene blue is a traditional remedy used for a range of ailments. It will, however, stain silicone sealant and should never be added to the main aquarium.

HUMAN HEALTH

● Switch off the power before placing your hands in the water.

● Always wear a pair of rubber gloves when coming in contact with tank water.

● Never try to create a flow of water through a siphon by sucking it.

● Don't dump dirty water down the kitchen sink; use an outside drain if possible.

DIAGNOSIS OF COMMON PROBLEMS

It is not always easy to distinguish between the different ailments that can affect tropical fish, especially in their early stages. Diagnosis may be further complicated by the fact that more than one infective agent is often responsible for the visible symptoms. For example, when a fish develops white spot *(see p.54)*, bacteria and fungi may gain access to the body at the site where the parasites initially attacked, confusing the diagnostic picture.

The tables on the following pages will help you diagnose the most common conditions affecting freshwater species and point you toward appropriate treatments. First, in the table below, identify the part of the fish's body that appears to be affected. Next, find the signs of illness that most closely match those displayed by the fish. Note the number(s) associated with the relevant signs and refer to pages 57 and 58 for a fuller description of the possible conditions and their treatment.

SIGNS OF DISEASE

EYES

- One or both eyes appear to be bulging abnormally from their sockets ⑤ ⑨ ⑫
- Bulging eye or eyes, accompanied by loss of appetite and color change ①
- One or both eyes have a cloudy appearance in the central pupillary area ①

- One or both eyes missing from their sockets, with no other obvious physical symptoms or behavioral abnormalities ① ㉔
- Area around the eyes becomes abnormally white ⑨
- Eyes develop a glazed appearance, and the fish has difficulty coordinating its movements ㉓

SKIN

- Skin loses its color, becoming paler than normal, and the fish appears less active than usual ⑮ ⑲
- Growth resembling cotton fluff appears on an area of the body or on a fin ③ ⑦
- White, pus-filled pores develop near the head, becoming ulcerated. Whitish feces may hang from the body ⑭

- Skin has a more slimy appearance than normal ⑬ ⑯ ⑰ ⑱ ⑲ ㉓
- Straggly pieces resembling cotton appear to trail down from the sides of the body ⑰
- Strangely shaped blemishes become evident on the sides of the body ⑥ ⑰

FINS

- Fins develop an abnormally ragged appearance, especially evident in the caudal fin ②
- Red streaks start to develop in the fins, which appear to be congested with blood ② ⑤
- Areas missing from the fins; especially evident in fish with unusually long fins ㉔

- White spots appear over the surface of the fins as well as on the body ⑬
- Fins are clamped down against the body, making it difficult for the fish to swim normally ㉑ ㉓
- Splits develop in the fins, vertically in the dorsal fin and more horizontally in the caudal fin ㉔

COLOR

- Fish becomes paler in color and loses its appetite ⑮ ⑲
- Reddish areas become evident on various parts of the body, including the mouth, anal region, and bases of the fins ⑤ ⑥
- Body develops a yellowish-gray cast, which extends to the fins as well as the body ⑱

- Body color alters, with markings appearing less defined than normal ① ⑬ ⑯ ⑰ ⑱ ⑲ ㉓
- Pale areas develop on the head ⑭
- Eggs develop an abnormal cloudy, milky appearance, especially if infertile ⑧

SHAPE

- Abdominal area swells significantly, to the extent that the fish has difficulty swimming ① ⑫ ㉓
- One or more unusual swellings appear anywhere on the surface of the fish's body ⑨ ⑯
- Fish undergoes a progressive loss of weight, in spite of the fact that it appears to have a healthy appetite ①

- Surface of the body erodes, forming ulcers ⑤ ⑥ ⑬ ⑭ ⑰ ⑱
- Fish's profile alters, with its belly area starting to bulge noticeably upward, indicating emaciation ① ⑭

BEHAVIOR

- Fish starts to spend longer than normal at the surface of the tank, sometimes hanging there at an abnormal angle ⑯ ⑳ ㉒
- Gill movements become very apparent, causing the fish to appear as if it is gasping ⑱ ⑳ ㉒ ㉓
- Fish starts to rub itself against rockwork and other objects in the aquarium ⑬ ⑯ ⑰ ⑱ ⑲

- Loss of appetite, with the fish becoming less active, and lying on the floor of the aquarium ⑤ ⑥ ⑨ ⑩ ⑪ ㉓ ㉔
- One individual starts to be picked on by other fish in the tank and starts to hide away ① ⑮ ㉔
- Fish begins to swim at a strange angle in the tank, sometimes even upside down ④ ⑲ ㉑ ㉓

BACTERIAL DISEASES

CONDITION	AT RISK	SYMPTOMS	TREATMENT
① Piscine tuberculosis	All fish	Weight loss, especially evident on the underparts, with loss of color and often bulging eyes (exophthalmia). Loss of appetite. May try to hide.	No effective treatment for the causal mycobacterium. Entire aquarium needs to be stripped down and disinfected after an outbreak before any restocking.
② Fin rot	All fish	Erosion of the edges of the fins, often with some reddening suggesting inflammation. May follow fin-nipping by tankmates.	Check and improve water quality as necessary. Identify and remove fin-nipper. Feed Vitamin C–enriched food. Treat with antifungal remedy.
③ Mouth fungus	All fish, but especially livebearers	Cottony substance on the jaws. Fish loses appetite and often displays shimmying movements. Caused by *Flexibacter* bacteria.	Often a sign of deteriorating water conditions. Use an antibiotic or other commercial remedy.
④ Swim bladder disorder	All fish, especially Balloon Mollies (see p.169)	Fish has consistent difficulty maintaining its balance in the water, listing or even floating on its back. May be caused by an infection, chilling, or anatomical problems linked with a more corpulent body shape.	Try to identify the cause. Check water temperature to be sure heater is working. If an infection is suspected, a medicated bath may help. Lowering the water level may aid recovery.
⑤ Vibriosis	All fish	Lethargy, skin discoloration, exophthalmia, reddish staining of the fins, abdominal swelling. Spreads fast and can cause rapid mortality. Caused by *Vibrio* bacteria.	Medicated antibiotic food may help to contain an outbreak. Good quarantine practice and water quality management will help prevent outbreaks.
⑥ Pseudomoniasis	All fish	Results from infection by *Pseudomonas* bacteria. May result in hemorrhages in the mouth and ulceration on the sides of the body. Small hemorrhages can occur internally, affecting the liver and kidneys.	Treatment can be difficult, because many strains are resistant to commonly used antibiotics. Treat with medicated foods or possibly direct injection.

FUNGAL AND VIRAL DISEASES

CONDITION	AT RISK	SYMPTOMS	TREATMENT
⑦ Saprolegnia fungus	All fish, especially brackish water species kept in freshwater	May follow an injury, resulting in whitish, furlike areas on the affected part of the body. Spores are invariably present in aquarium water.	Treat using a commercial remedy in a medicated bath. Give food rich in Vitamin C.
⑧ Egg fungus	Most species, especially Bumblebee Gobies (see p.189)	Furry growth on eggs. The fungus may initially attack dead eggs but will rapidly spread to healthy neighbors if unchecked, compromising hatching rates.	Remove eggs, and add dye such as methylene blue to the water in the hatching tank.
⑨ Lymphocystis	All fish	The most common viral disease, resulting in cauliflower-like growths over the body surface and white areas around the eyes. May sometimes develop internally.	Not highly infectious but can spread through direct contact. Vaccination may sometimes help to provoke immune response and overcome the virus.
⑩ Iridovirus	Many fish, especially gouramis, angelfish, and Ramirez Dwarf Cichlids	Weakness, loss of appetite, and darkening in color. The abdomen may be visibly distended, indicating an enlarged spleen. Some types of iridovirus are responsible for lymphocystis (see above).	High mortality. No effective treatment.
⑪ Singapore angel disease (SAD)	Angelfish (see pp.140–141)	Fish become inactive, lose appetite, and die rapidly. Epidemics of this disease, first recorded in Singapore, have caused great damage to the trade in angelfish in the past.	No treatment possible. Quarantine affords best hope of containing an outbreak.
⑫ Malawi bloat	Cichlids from Lake Malawi and neighboring lakes	This disease (suspected to be viral) causes symptoms resembling dropsy. Fish have swollen abdomens and sometimes pop-eyes. Especially affects vegetarian species.	No effective treatment. Offering a high-fiber diet may provide some protection.

PARASITES

CONDITION	AT RISK	SYMPTOMS	TREATMENT
⑬ White spot (*Ichthyophthirius multifiliis*)	Many species, but especially mollies (*see pp.168–169*)	Small white spots over the body, which ulcerate and are likely to become infected. Increased mucus production indicates skin irritation.	Raise water temperature slightly to shorten the life span of the free-swimming tomites. Treat water with commercial remedy to kill tomites.
⑭ Hole-in-the-head (*Hexamita* species)	Discus (*see pp.142–143*) and other cichlids	Pale areas develop on the head, becoming ulcerated. Secondary bacterial infection often sets in, and signs of emaciation develop.	Treat rapidly with micronazole or similar medication to prevent lifelong scarring and promote healing. Improve diet, using food containing Vitamin C.
⑮ Neon Tetra disease	Neon Tetras (*see p.99*) and related species	Loss of color and white areas under the skin. Caused by a microsporan parasite.	Treatment not possible. Remove affected fish at once, as well as any that die, since cannibalism is likely to spread this protozoal infection.
⑯ Skin and gill flukes (*Gyrodactylus* and *Dactylogyrus*)	All fish	Gill flukes cause labored breathing; fish commonly hang below the surface, reluctant to move. Skin flukes cause irritation; the fish scrapes itself against objects, gills pump fast and may appear red.	Specific commercial medications will kill these trematodes. Watch for reinfestation, because the parasites' eggs are hard to destroy.
⑰ Anchor worm (*Lernaea* species)	All fish	Elongated parasites up to 3/4 in (20 mm) in length attach to the sides of the body, resulting in ulceration and irritation. Patches of inflammation may become infected.	Remove visible parasites with forceps and treat affected fish in a medicated bath. Add insecticide to the tank to kill the free-swimming young parasites.
⑱ Velvet disease (*Oodinium* species)	Many species, especially anabantoids (*see pp.104–115*) and danios (*see pp.76–77*)	A yellow-gray coating develops on the fins and skin, resembling gold dust. The fish try to relieve irritation by rubbing; the skin may peel away in strips. Breathing may become labored, and the fish may be lethargic.	Raise water temperature slightly. Use a proprietary treatment to kill the free-swimming stage in the life cycle of these protozoa.

ENVIRONMENTAL CONDITIONS AND INJURIES

CONDITION	AT RISK	SYMPTOMS	TREATMENT
⑲ Chlorine poisoning	All fish	Fish appear pale and covered in mucus; some show redness on parts of the body. Often seem stressed, rubbing against rockwork and swimming erratically.	Immediately remove fish to a tank containing chlorine-free water. Alternatively, add appropriate chemicals to remove the chlorine from solution.
⑳ Hypoxia (lack of oxygen dissolved in the water)	All fish	Fish spend much of their time at the surface of the water, where the concentration of dissolved oxygen is higher.	Carry out a partial water change without delay. Check the system—including pump and filter—and monitor water conditions. Make sure tank is not overcrowded.
㉑ Chilling	All fish	Loss of activity and balance. Symptoms appear suddenly, and all of the fish in a tank are likely to be affected.	Check the water temperature—the power may be off or the heater broken. Do not suddenly add hot water, but allow the temperature to rise gradually.
㉒ Nitrogen toxicity	All fish	Protruding eyes. Fish gasp for breath at the water surface or hang near water outlets. Gills appear brown and move more rapidly than usual.	Change water immediately and remove fish to an aquarium or other container with clean water. Monitor nitrogen waste levels regularly; if they remain high, perform partial water changes daily.
㉓ Other poisoning	All fish	Symptoms are variable, ranging from mild distress to sudden death; often, they begin with the fish swimming in circles, with clamped fins.	Remove source of contamination. Move fish out of the environment, and run a carbon filter to deactivate the chemical. Carry out a major water change.
㉔ Attack by tankmate	All fish	Injured fish floating on the surface. Fish are often noticed in this condition first thing in the morning, because many attacks occur under cover of darkness.	Move individual to a different tank to recover. Use aloe vera water conditioner, which may assist healing. Treat with commercial antibacterial and antifungal treatments. Do not reintroduce to existing aquarium.

FRESHWATER FISH

BREEDING

Reproductive behavior

Fish can be divided into two broad categories on the basis of their breeding behavior. Egg-layers, which form the majority, practice external fertilization, while in livebearers the eggs are fertilized in the female's body and emerge as fully formed young. Some species show little regard for their offspring and cannibalize their own eggs or fry. Others display surprisingly high levels of parental care.

When fish mate, it is known as spawning. In egg-layers, this involves the male's releasing sperm into the water at roughly the same time that the female expels her eggs nearby so that fertilization occurs outside the female's body. This is a rather

The male Siamese Fighting Fish uses his elaborate fins for display purposes, not just to entice potential female mates, but also to intimidate rival males.

This temporarily flooded forest in the Amazon region of South America provides a spawning ground for egg-laying fish such as discus. Breeding behavior in the wild is often seasonal and influenced by rainfall patterns.

haphazard method of reproduction, and, not surprisingly, a significant proportion of the eggs fail to become fertilized. In order to maximize the chances of fertilization occurring, some egg-layers attempt to maneuver their genital openings as close to each other as possible before spawning occurs, and there is occasionally contact between the fish, such as embraces with the fins.

Reproduction in livebearers

The likelihood of successful fertilization is greatly enhanced in livebearing species, which practice internal fertilization. The male livebearer introduces his sperm directly into the body of the female using his anal fin, which is typically modified into a tubelike projection called a gonopodium. The eggs are fertilized inside the female and then develop in the relative safety of her body.

The male's gonopodium carries sensory organs that help him to guide it into the female's genital pore. The shape of the gonopodium varies between different livebearing species, and its appearance is used by zoologists as a means of distinguishing between them. Males with a long gonopodium are able to mate easily. Spawning is more intimate for males with a short gonopodium, and courtship plays a greater role in such species, helping the female to feel comfortable with the male being so close. During mating, sperm are transferred into the female's body in the form of a sperm packet. This dissolves within about 15 minutes, freeing the sperm. Any sperm that fail to

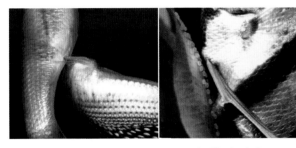

When livebearers, such as these mollies, mate, the gonopodium of the male enters the female's genital pore and transfers a tiny sperm packet that contains up to 3,000 individual sperm.

External fertilization is the most typical method of fish reproduction. Here, a male angelfish releases his sperm over eggs that a female has just laid on a leaf. It is unlikely that all the eggs will be fertilized.

fertilize eggs immediately do not die off, as happens in the mammalian reproductive tract, but instead remain viable for the life of the female, probably nourished by an output of sugars from the ovaries. This is why a female livebearer housed on her own can still give birth to successive broods of fry, using stored sperm from past matings to fertilize the eggs. As a result, a number of males may be responsible for the young born in a single brood.

In some livebearers, including guppies *(see pp.165–167),* the embryos are nourished by their yolk sac as the eggs develop in their mother's body. As she gives birth, the fry hatch from their eggs. In the One-Sided Livebearer (*Jenynsia lineata*), placentalike links form between the body of the female and the embryos. The mother provides nutrients through cords that develop between the ovarian wall and either the anus or mouth and gills of the young fish.

Breeding strategies in egg-layers

Since livebearers nurture their young through the most vulnerable early stages of life, they do not need to produce as many offspring as egg-layers. An egg-laying female may lay hundreds or thousands of eggs, of which only a small proportion survive and hatch. Although the incubation period is brief—often about 36 hours—the eggs are at great risk during this time. Eggs are eagerly devoured by fish (sometimes even the parents) and other aquatic animals, since they provide a rich source of protein. Some egg-layers deposit their eggs in relatively inaccessible places where they are more likely to escape the attentions of predators, such as in caves, on the underside of leaves, or even, in the case of *Lamprologus* cichlids *(see p.155),* in empty snail shells. Others guard their eggs until they hatch.

To protect their eggs from the dangers of incubating in the open, a number of species exhibit a behavior called mouth-brooding. After spawning, one or the other of the pair carries the eggs in the mouth for up to three weeks until they hatch,

Haplochromid cichlids are mouth-brooders. After spawning, the female nips at "egg spots," shown here, on the male's anal fin. As she does so, she swallows his sperm, ensuring that the eggs in her mouth become fertilized.

Heavily pregnant female is less agile when swimming

Abdomen becomes swollen

The bulging body of this Platy is a clear sign that she is in breeding condition. When a female swells with eggs or developing young, she is said to be "gravid."

during which time the adult fish does not feed. The fry emerge from their parent's mouth when they are able to swim freely. Mouth-brooding is best documented in members of the cichlid family, but it also occurs in some bettas and catfish. These fish produce far fewer eggs than other egg-layers, but their eggs stand a better chance of survival.

Survival of the fry

Fry that hatch in the open initially lie largely immobile, close to where they hatched, until they have digested the remains of their yolk sac—a process that may last several days. Many species that guard their eggs also show parental care toward their fry. Some cichlids, for example, herd their offspring into spawning pits that they excavate in the substrate, where they can keep watch over the brood. Several such spawning pits may be dug and used during the early days after the young have hatched. The combined tail movements of the fry set up tiny currents above the pit, increasing the flow of oxygen-rich water over the developing young. Even when the fry are free-swimming, the adults will chase after and catch individuals that separate from the school and carry

EGG-LAYING BEHAVIORS

Fish display a wide range of egg-laying behavior, reflecting the way in which they have adapted to a variety of environments. Even related species may have very different egg-laying habits. For example, many anabantoids inhabit calm waters and build bubble nests for their eggs. Some anabantoids, however, have changed to mouth-brooding in order to exploit more turbulent waters that would destroy bubble nests.

BEHAVIOR	DESCRIPTION	TYPICAL EXAMPLES
Egg-scattering	The female releases a large number of eggs at random, some of which stick to plant leaves, while others become lodged in the substrate. The fish become increasingly active prior to spawning, with the males pursuing the females vigorously. The fish show no parental care and may eat their own eggs and any fry that hatch.	Egg-scatterers include cyprinids, such as barbs, danios, and rasboras, as well as tetras and many other characoids.
Egg-depositing	Egg-depositors often form pair bonds. A pair will carefully choose a safe, clean spawning site where they can lay their eggs. This may be a rock, the underside of a leaf, or even a cave. After spawning, the fish stay in the vicinity of their chosen site, driving off other fish that come too close to the eggs. They also jealously guard their fry.	Discus, angelfish, and various other cichlids. A number of catfish, such as bristlenoses and loricariids, also reproduce in this way.
Egg-burying	This behavior is shown by species dwelling in muddy pools that dry up each year. After the water has evaporated, the eggs lie dormant in the dried mud and hatch when the rains return. The fish must develop rapidly and spawn again before the dry season begins.	Annual killifish, such as the nothobranch group, and Argentine Pearl Fish.
Mouth-brooding	One of the parent fish takes the eggs into its mouth and retains them there until they hatch. This behavior keeps the eggs out of sight of predators in open habitats and also helps to prevent them from being swept away by strong currents.	Various cichlids, including many African Rift Valley species, as well as anabantoids that occur in fast-flowing water, and some catfish.
Bubble-nesting	The nest is made of mucus bubbles blown at the surface by the male and may be anchored to plants. It holds the eggs, and later the fry, together in one place where they can be closely guarded. The male often drives the female away after spawning and keeps watch alone.	Some anabantoids, including Siamese Fighting Fish and Pearl Gouramis. Hoplos catfish also exhibit bubble-nesting behavior.

The best-known bubble nester is the Siamese Fighting Fish. The eggs hatch about 48 hours after being laid. The male guards the fry until they are free-swimming.

Remaining close to its chosen spawning site, this male Ram (*Mikrogeophagus ramirezi*), a cichlid, guards his partner's eggs, ready to defend them against predators.

Eggs laid on rock

These tiny discus fry are feeding on a nutrient-rich secretion, called discus milk, produced by one of their parents. Nourishing the fry like this may help to keep them close to the adults. Despite this parental care, many still fall victim to predators in the wild.

them back to the group in their mouths. They warn the young of possible danger by distinctive movements of their brightly colored pelvic fins. This behavior, known as jolting, sends out both a visual signal and a pressure wave that alert the fry to an approaching threat. Some cichlids, including discus, even produce food for their young in the form of secretions that the fry nibble from the flanks of the adults.

Cichlids are not the only species that display parental care. A number of catfish also guard their eggs, as do various anabantoids, including the popular Siamese Fighting Fish *(see pp.104–106).* The male of this species constructs a special nest for the eggs by blowing air bubbles and then guards the nest and watches over its fry during the immediate post-hatching period, when they are at their most vulnerable.

The more developed fry are when they emerge into their surroundings, the better their chances of survival. Mouth-brooding helps to give the fry of some egg-layers a head start in life. The female Mosquito Fish *(see p.157),* a livebearer, uses a different process, called superfetation, to improve the odds for her young. She does not produce a single brood but instead gives birth to a few offspring at a time every two days or so. Using sperm stored in her body, she regularly fertilizes a small number of eggs so that her reproductive tract contains young at varying stages of growth. As a result, Least Killifish fry are proportionately larger and better developed than those of similar livebearers when they are born.

Large oral cavity to house young fish

Fry stay close to the parent fish

Parental care in mouth-brooders, such as this Redhump Eartheater (*Geophagus steindachneri*), does not end when the fry are free-swimming. Should danger threaten, the young fish will dart back into the security of their parent's mouth.

A pair of Ram (the male is on the right) guard their four-day-old fry. Many aquarium strains show less parental care than their wild relatives, because they have been tank-bred for generations in the absence of predators.

Breeding in aquariums

One of the most challenging yet rewarding aspects of fishkeeping is to breed your own fish. Think carefully about which fish you should breed and where you will house the spawning fish, and decide how many fry you can comfortably cope with and how you will feed them. Make a record of your successes and failures so that future breeding attempts go more smoothly.

Many fish spawn in pairs, but some species form small breeding groups—such as a male and two females, or vice versa—while others spawn communally in shoals. You will need to sex your stock in order to be sure that you have fish of each sex. Sexing is easy with species that display sexual dimorphism (see p.61). However, some fish, such as barbs and tetras, are visually alike and hence impossible to sex outside the breeding period. Starting with at least six individuals of such species should guarantee that there is at least one pair in the group.

Compatibility—the ability of a pair to interact favorably and spawn successfully—is unlikely to be a problem with barbs, tetras, and other species in which the sexes come together only briefly to mate. However, it is more of an issue

A male Siamese Fighting Fish, at the left of the picture, courts a female, adopting a special posture and displaying his long, flowing fins. Always monitor any pairings that you make; separate the fish if there is serious aggression.

in species that display a degree of shared parental care, such as angelfish (see pp.140–141) or discus (see pp.142–143), since they necessarily spend more time with their partners. A pair of discus, for example, will simply refuse to breed if they are incompatible. Changing their partners or, better still, rearing the fish in a group so that they can choose their own mates as they mature should overcome this difficulty. By watching the behavior of the fish, you should be able to see signs of compatibility, because naturally matched pairs will swim

Courtship takes many forms in tropical fish. Here, the smaller, darker male Auratus Cichlid nudges the flanks of a female near her vent in order to stimulate her into breeding. This is a polygamous species, which means that a single male will mate with several females. There is also a blue form of this cichlid (see p.151).

BREEDING TIPS

● Thoroughly research the breeding habits of your fish.

● When choosing breeding stock, select young, healthy adults, with good markings, fin shape, and coloring.

● With fish that spawn in small groups or shoals, rather than in pairs, make sure that you have a large enough group, with the right gender mix, to ensure breeding success.

● Successful breeding may result in more fry than you can adequately care for, so find alternative homes for surplus fry in advance.

● Avoid allowing different species to breed together (hybridize). Fellow breeders will have little interest in the resulting offspring.

SPAWNING TANKS FOR EGG-LAYING SPECIES

The spawning tank must reflect the breeding habits of the fish it is to house. Include plants, among which the female can seek refuge if the male becomes aggressive. If the adults are likely to eat their eggs or fry, they should be transferred back to the main aquarium after spawning, leaving the eggs to hatch on their own. Filtration in this type of tank is gentle, so it is best not to feed the adults while they are in the tank, to avoid polluting the water unnecessarily.

Cave-spawners

These fish require an artificial cave, such as a clean, partially buried clay flowerpot or a section of coconut shell. Alternatively, you can build a cave out of rocks. Cave-spawners may also breed successfully in the main aquarium.

Substrate can be aquarium peat or sphagnum moss

Flowerpot provides shelter

Eggs fall into the gaps between the marbles

Fine-leaved plants, such as Java Moss, provide cover for females

Egg-scatterers

A layer of marbles on the tank floor will help to prevent the adult fish from eating the eggs, which will fall between the marbles and out of reach. Alternatively, a mesh net across the tank can be used to let the eggs fall safely through.

Egg-buriers

A soft peaty substrate is essential for fish that bury their eggs. After spawning, the peat, complete with eggs, can be removed and stored in a warm, dark place. Immersing the peat in tank water again will cause the eggs to hatch.

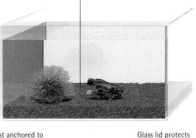

Spawning mop consists of synthetic yarn attached to a float

Egg-depositors

Fish that deposit eggs on plant leaves can be persuaded to lay them on an artificial spawning mop instead. With fish that spawn over several days, the mop can be replaced regularly and the eggs hatched safely in a separate tank.

Bubble-nest anchored to large aquarium plant

Glass lid protects the nest

Bubble-nest builders

A glass lid will prevent drafts from damaging the nest or chilling the eggs and keep the air over the water warm and humid. Provide tall plants to which the nest can be attached. These fish may also breed successfully in the main aquarium.

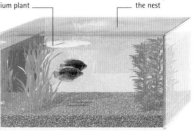

together and remain in relatively close contact. Alternatively, you can avoid these problems altogether by buying a proven spawning pair from an aquarium store.

The males of some larger species, such as a number of the Central American cichlids, will become aggressive at spawning time. If the female is harassed by her intended partner, she may lose her breeding condition, and it may be necessary to remove the male for a period of time to enable her to recover. Incompatibility is not something to be taken lightly—in extreme cases, a male may bully a female to death.

As spawning time approaches, a female egg-layer will swell with eggs. She will also start to attract the attention of mature males. After the first successful spawning, she is likely to spawn again within weeks.

Non-spawning female Rosy Barb

Female Rosy Barb swollen with eggs

Conditioning your fish

Having chosen your breeding stock, the next task is to condition them for spawning. This involves feeding them well, especially with protein-rich foods, so that they produce plenty of healthy sperm and eggs. For those egg-laying species that spawn seasonally in nature, it may also involve altering conditions in the tank to mimic the environmental changes that occur in the wild. With Amazonian species, which breed when heavy rains raise the water level in their habitat,

Breeding traps protect newborn livebearers from their mother and other tank occupants. In the trap above, the mother rests in the top section, while the fry slip through a slit in the floor into the lower chamber. The trap on the right has a double birthing chamber, to house two pregnant females.

The lower chamber protects the fry from being eaten

Air trapped under each end provides buoyancy

Slits in the floor of the birthing chamber allow fry to pass through

to 90°F (32°C). The disadvantage of speeding up gestation is that higher temperatures also quicken the rest of the life cycle of the fish, thus shortening their life span.

Although most livebearers produce offspring readily, even in a community tank, few of the fry survive to adulthood because they are eagerly devoured by other tank occupants—often including their own mother. To avoid this unnecessary carnage, you can transfer the pregnant female to a breeding trap, or spawning box. The breeding trap, which hooks over the side of the tank or floats freely in the water, has a birthing chamber that houses the female. Small holes or slits in the floor or walls of the chamber allow newborn fry to escape into a secondary chamber, where they can develop in safety, out of reach of the female and separate from the main tank. Once they have grown too large to be eaten, the fry can be released into the main aquarium.

Avoid buying small breeding traps, since gravid females will become distressed if they are confined in too small a space. Do not wait until immediately before the birth before transferring the female, since this is likely to cause her to abort her brood. The ideal time to move her is about a week before the brood is due.

spawning can be triggered by making a partial water change and dropping the water temperature slightly. Feeding extra live foods will also help, since invertebrates naturally become more numerous in the floodwaters at this time. In contrast, annual killifish can be brought into breeding condition by lowering the water level and slightly increasing the temperature, since in the wild they spawn when the sun begins to dry up the pools in which they naturally live.

It is usually possible to tell from the changing appearance and behavior of the fish that spawning is imminent. The males may take on a more intense breeding coloration and show aggression toward one another. They will actively pursue the females, whose body becomes swollen with developing eggs. In some species, you may also notice that the fish perform courtship rituals or carefully clean spawning sites, such as rocks or leaves. This is the time to transfer the fish to a spawning tank (see p.65). In territorial species that spawn in pairs or small groups, always move the females first to allow them to settle in the tank before the males are introduced. Communal spawners can be introduced as a shoal.

Breeding livebearers

Livebearers need little encouragement to breed in aquariums. Like egg-layers, they benefit from protein-rich conditioning foods, but many originate from relatively stable habitats and are not seasonal spawners, so they do not need environmental changes to trigger spawning. The water temperature does, however, influence the gestation period. For example, at 77°F (25°C), gestation in guppies takes 28 days, but this figure falls to just 19 days when the temperature is raised

NEWLY HATCHED FRY

After hatching, a young fish is initially sustained by nutrients in the remains of its yolk sac, which attaches to the underside of the fish's body. In this picture of Arawana fry (see p.182), the yolk sacs are the reddish-orange "bags" dangling beneath the fish. Only when the yolk sac has been fully absorbed will a fish start to swim around the aquarium actively seeking food. Until that time, it rests on the floor of the tank or elsewhere out of sight. The fry that emerge from the eggs are usually tiny replicas of the adult fish. In a few species, such as discus, the young have a body shape very different from their parents' but come to resemble them as they increase in size.

FRY SIZE AND REARING

Mouth-brooders, such as the Pearl of Likoma cichlid (upper image), produce fewer but proportionately larger offspring than egg-laying species, such as the Firemouth Cichlid (lower image). Being larger, young mouth-brooders are easier to feed, and since there are fewer of them, there is less risk of overcrowding the tank or polluting the water. You may periodically have to remove the largest, fastest-growing members of a brood so that they do not cannibalize smaller siblings.

The female Red Devil cichlid (*Amphilophus labiatus*) keeps a protective watch over her free-swimming fry. In the wild, this helps to ensure that more of her offspring survive the critical early days, when they are at most risk of predation. Sometimes the fry of another Red Devil will join her brood, and she will show similar care toward these adopted young.

Rearing tanks

Most breeders prefer to use a completely separate tank for rearing the fry of livebearers. The female can give birth there in a breeding trap, before being moved back to the main aquarium after she has recovered. A special V-shaped partition can be inserted into the tank as an alternative to a breeding trap. This has a narrow gap at the apex of the V through which the fry can slip. Alternatively, a net with a wide mesh can be used to separate the female from her offspring. When the female has been removed, the fry can be reared in the tank on their own.

A power filter cannot be used safely in any tank that is to house young fish, whether livebearers or egg-layers, because small fry are likely to be sucked into the filter. Filtration must be gentle, so use a simple sponge filter instead, possibly in combination with an undergravel filter. Prime the filter in advance with a culture of beneficial bacteria so that it has time to become active before any fish are introduced. The water must be well aerated, and the heater should have a special protective cover so that the fry do not burn themselves.

Feeding the fry

An essential part of establishing a successful breeding regimen is making sure that you have sufficient stocks of the correct foods to nourish the young fish. Tiny fry will initially need to be given a specially formulated liquid fry food or microscopic aquatic creatures called infusoria. You can culture infusoria

In most species, rivalry between males is nonlethal; however, male Siamese Fighting Fish are likely to injure one another so must be kept apart.

yourself by placing a glass jar containing chopped lettuce and water in bright light, perhaps on a windowsill. After a few days, the water will turn pinkish as it becomes colonized by infusoria. Small amounts of this water can then be sprayed on the surface of the rearing tank. As the fry grow, they can progress to newly hatched brine shrimp. Larger fry can be given brine shrimp as a first food and subsequently small *Daphnia* and ground flakes.

Young fish need to be fed two, three, or even four times a day. They are not particularly mobile at this stage, so it is vital that food is evenly distributed throughout the tank and within easy reach; otherwise their growth will be checked. The sponge filter, which should be mature by this stage, will have tiny edible particles on it that the young fish can nibble. The gentle currents that such a filter creates, running off an airline, will help to waft floating food scraps toward the fry.

BRINE SHRIMP

Most young fish, even those that are vegetarian in later life, need animal protein during the first week or so after they become free-swimming. The most popular rearing food for fry in home aquariums is the larvae of brine shrimp (*Artemia* species), which are also known as *nauplii*. It is important to set up your brine shrimp hatchery in advance so that you can be sure of having enough food for the arrival of the young fish.

Brine shrimp eggs are sold in airtight containers. The eggs absorb atmospheric moisture readily, so avoid exposing them to the air before you need to use them. Very few will hatch if they become too moist.

Hatch the eggs in a breeding bottle. Add saltwater (made with marine salt) and oxygenate it via an airstone and airline. The bottle can be attached to the side of the tank with suckers if necessary.

Hatching takes about a day at 77°F (25°C). Sieve the nauplii from their empty shells. Before giving the *nauplii* to the fry, dip them briefly in dechlorinated freshwater to wash off salty residues.

Dangers of overpopulation

Successful breeding can leave you with a large number of fry to care for. Regular partial water changes, perhaps as often as once a day, will be vital to make sure that the water quality does not deteriorate as a result of accumulated waste and uneaten food. As the fry increase in size, a more efficient filtration system can be incorporated into their tank.

Eventually, the fry will need to be either moved to a much larger aquarium or divided between several different tanks, to give them sufficient growing space. Overcrowding the fish may stunt their growth and induce stress-related illnesses, as well as making it more likely that there may be a sudden, and potentially fatal, decline in water quality.

Specialist requirements

Some fry have special rearing needs. A rearing tank for the fry of gouramis (see pp.109–113) and related species needs to be kept covered so that the air immediately above the surface is at approximately the same temperature as the water. This is because the fish have anabantoid organs, which allow them to breathe air directly. If the air above the

Prepared fry foods in liquid form are invaluable for feeding tiny fry, as well as larger young, such as those of various anabantoids, whose mouths are too small to take brine shrimp.

GROWTH AND DEVELOPMENT OF A PLATY

Young livebearers, such as platies, are free-swimming at birth and have a remarkably consistent growth rate, averaging up to $1/100$ in (0.3 mm) per day. Initially, both sexes develop at the same pace, but the females have a growth spurt after maturity and eventually outgrow the males.

Three weeks after birth, traces of color are starting to show on this young fish, but it is still quite inconspicuous in the tank. It is now large enough to be safe from being eaten by the other tank occupants.

At five weeks old, the coloration of the fish is becoming clearer. The fins are proportionately larger and more elaborate. The body is less streamlined, with a more angular back and a bulkier abdomen.

By nine weeks of age, the patterning and richness of coloration are fully apparent. Now sexually mature, the fish can be identified as a male by its gonopodium, which is just visible behind the pelvic fin.

Many freshwater species are now bred commercially for the aquarium trade. This breeding farm is in Singapore, where the climate allows fry to be reared in outdoor ponds. The fish eat naturally occurring foods plus dietary supplements.

Problems and solutions

Difficulties can crop up at any stage during the breeding cycle, starting with the failure of the fish to spawn at all. It may be that you do not have a pair, or that the fish are not yet mature enough to breed. If the fish spawn but the eggs prove to be infertile, the male of the pair could either be too old or, if he is a livebearer, have a damaged gonopodium. Swap the males of different pairs around to see if this has any effect. With egg-layers such as barbs, add an extra male to the spawning tank to increase the likelihood of success. If eggs are attacked by fungus, you can treat the water with a little methylene blue to control the problem, although the presence of fungus may also be an indication that the water temperature is too low.

If the young fail to thrive, or even die off, reexamine their feeding regimen. Study the abdomens of fry with a hand lens; their bodies should be transparent enough to see if there is food in the gut. If there is not, the food you are giving the fry may be too large for them to eat. The fry may also perish if the water quality deteriorates, so monitor this closely during the rearing period. There will inevitably be a few deformed fry in any brood, and these should be humanely culled.

water is too cold when their anabantoid organs start to function at about three weeks of age, the fry could become fatally chilled when they try to breathe at the surface.

With territorial species, separate young males before they start to become aggressive toward the rest of the brood. Male Siamese Fighting Fish *(see pp.104–106)*, for example, must be kept apart by the time they are three months old. Young livebearers should also be separated as soon as you can sex them, to prevent littermates from interbreeding. A female livebearer can store sperm in her body throughout her life, so any unplanned matings like this will endanger your breeding program, since you will not know for sure which male sired the fry. Avoid housing together the young of related species, such as swordtails *(see pp.160–161)* and platies *(see pp.162–164)*, which will readily crossbreed, or hybridize. It is difficult to predict the appearance of hybrids, and they tend to be less attractive and less fertile than the species from which they originate.

When the fry become sufficiently mature, they can gradually be switched to their adult diet. Make sure that you provide enough to meet their rapid growth rate, while taking care not to overfeed them.

Fish genetics

As fish have been spawned in aquariums over generations, there has been an understandable tendency to select the most colorful individuals for breeding purposes. For this reason, many of the most popular tropical aquarium fish kept today are more brightly colored than their wild relatives. Modern techniques have allowed the development of varieties to be taken to new extremes, with the first genetically modified fish now available.

The science of genetics is concerned with the way in which characteristics pass from one generation to the next. The characteristics of all living organisms, including fish, are contained in the genes, which are carried on rodlike structures called chromosomes, present within the nucleus of every living cell in the body. Mutations (or unexpected changes) in the genetic structure can occur and sometimes result in a change to the fish's physical appearance, such as its color, patterning, or body and fin shape. When mutations occur in certain cells, such as egg or sperm cells, the change is passed to the next generation. Most mutations make only subtle changes, which can be enhanced by selective breeding *(see opposite)* over many generations. On occasions, however, a mutation that creates a radical change can occur. The golden coloration of the Butterball Angelfish (*Pterophyllum* sp. var.), for example, resulted from a mutation that caused the loss of a dark pigment that normally masks this yellow shade. Mutations can also be harmful; this is clearly illustrated in very large groups of fry, which may contain a few deformed individuals.

GENETIC MODIFICATION OF FISH

This controversial laboratory method for developing fish with unusual characteristics originated in scientific research. By introducing the luminescence gene from a jellyfish into the genetic makeup of fish (a standard technique used to "visualize" genes), researchers created fish in which every cell in the body emitted a green glow. Rice Fish that carry this jellyfish gene, pictured below, were developed in Taiwan during 2001 as part of a medical research program, but their commercial potential was quickly realized, and they became the first genetically modified aquarium fish available to hobbyists. Subsequently, Zebra Danios *(see p.76)* that carry the jellyfish luminescence gene have been developed at the National University of Singapore as part of a scientific study into detecting environmental pollution. Trademarked as "Glofish," they also went on sale in pet stores in some parts of the US late in 2003.

A tremendous range of color forms of the Blue Discus *(see pp.142–143)* now exists. Naturally occurring color variants of the wild Blue Discus have been enhanced and improved by selective breeding to create the domesticated strains of today.

Wild Discus

Relatively dull body coloration

Vertical stripes provide camouflage

Ghost Discus

Vivid blue on the sides of the body

Tangerine Orange Discus

Selective breeding

When a mutation that confers a desirable modification—like a color change—occurs, it must be established in future generations by a careful breeding program. This not only retains the favored characteristic but also ensures that the new variety does not carry any genetic problems. A common way to do this is outcrossing, which involves pairing the fish with an unrelated individual of the same species that displays normal coloring and patterning. The resulting offspring will be normal in color, but they will carry the mutation for the color change. After further pairings with the progeny of similar matings, a number of the new-colored fish will be present among the offspring. These can be separated out and allowed to mate with each other. Such a breeding program improves the genetic base, ensuring that the resulting individuals are less closely related. Mating together two fish from the same brood—or inbreeding—can result in offspring that produce reduced numbers of fry or are infertile, especially when this is carried out over several generations.

Hybridization

Under normal circumstances, members of a single species breed only with other individuals from that species. It is sometimes possible, however, to persuade individuals from two different species to pair, creating offspring with novel characteristics. This method for creating new aquarium varieties is controversial, because these fish would never exist in the wild. Hybridization has been used very widely to create variant forms of the popular livebearers, including the well-known Black Molly (see p.168), strains of the more colorful gouramis, and the spectacular Parrot Cichlid (see above).

The Parrot Cichlid, developed in the late 20th century in Asia, is perhaps the most controversial example of hybridization. Its ancestry includes Firemouth and Midas Cichlids (see p.136).

Once a successful variety, such as this Black Variegated Delta Guppy (*Poecilia reticulata*), has been established, the best parents are chosen for each future pairing to ensure that the strain is maintained and improved.

Brown-Red Alancea Discus

Pure blue markings

Red-Spotted Turquoise Discus

Streaks and spots over the body

Turquoise Discus

Extensive turquoise-blue coloration over the whole body

DIRECTORY OF
FRESHWATER FISH

CYPRINIDS

The cyprinid family is the largest single grouping of freshwater fish. It includes not only popular tropical aquarium fish but also their coldwater counterparts, notably goldfish (*Carassius auratus*) and koi (*Cyprinus carpio*). The majority of those kept in tropical aquariums have a strong shoaling instinct and are best kept in groups. Since a number of the most widely kept species are bred commercially on a large scale, color and fin variants occur, some of which have been developed into distinct ornamental strains.

The classification of cyprinids is still changing. The **Narrow-Wedge Harlequin** used to be called *Rasbora espei*, but in 1999 its scientific name was changed to *Trigonostigma espei*.

Balantiocheilos melanopterus	*Crossocheilus oblongus*	*Epalzeorhynchos kalopterus*
## Tri-Color Shark	## Siamese Flying Fox	## Flying Fox

Tri-Color Shark

- 🌐 **ORIGINS** Southeast Asia, occurring in Thailand, the Malay Peninsula, Sumatra, and Borneo.
- 💧 **SIZE** 12 in (30 cm).
- 🍴 **DIET** Live foods and flake.
- 🌊 **WATER** Temperature 72–77°F (22–25°C); soft (50 mg/l) and acidic (pH 6.0–6.5).
- 😊 **TEMPERAMENT** Relatively peaceful.

A silvery body and a deeply forked yellowish caudal fin edged with black help to identify Tri-Color Sharks. These active fish are fast swimmers and good jumpers, so they need to be caught with care. As they grow, the females can be recognized by the rounder shape of their underparts. Breeding requires more space than is available in the typical home aquarium. Large Tri-Color Sharks may prey on smaller companions.

Siamese Flying Fox

- 🌐 **ORIGINS** Southeast Asia, occurring in Thailand and the Malay Peninsula.
- 💧 **SIZE** 5½ in (14 cm).
- 🍴 **DIET** Algae and live foods.
- 🌊 **WATER** Temperature 72–77°F (22–25°C); soft (50 mg/l) and acidic (pH 6.0–6.5).
- 😊 **TEMPERAMENT** Relatively peaceful.

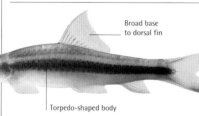

Broad base to dorsal fin

Torpedo-shaped body

The Siamese Flying Fox has a broad, dark line along its body. It can be distinguished from the Flying Fox *(see right)* by its clear fins and the fact that it has just one set of barbels. There is no obvious way of sexing individuals. These fish are highly valued because they keep algal growth under control but do not damage aquarium plants. They need well-oxygenated water and often rest close to the bottom.

Flying Fox

- 🌐 **ORIGINS** From northern India and Thailand to parts of Indonesia, including Sumatra and Borneo.
- 💧 **SIZE** 6 in (15 cm).
- 🍴 **DIET** Algae and live foods.
- 🌊 **WATER** Temperature 72–77°F (22–25°C); soft (50 mg/l) and acidic (pH 6.0–6.5).
- 😊 **TEMPERAMENT** Relatively peaceful.

The Flying Fox has a body pattern very similar to its Siamese relative. It, too, feeds on algae, nibbling algal growth from rocks with its low-slung mouth. Flying Foxes require a range of other foods to supplement the algae in their diet, with worms being especially favored. The sleek body shape reflects the fact that these fish are strong swimmers that naturally inhabit fast-flowing water. They often become more territorial with age.

Epalzeorhynchos bicolor

Red-Tailed Shark

- **ORIGINS** Southeast Asia, occurring in Thailand, notably in the central area of the country.
- **SIZE** 6 in (15 cm).
- **DIET** Flake and small live foods.
- **WATER** Temperature 72–79°F (22–26°C); soft to hard (50–150 mg/l) and acidic to neutral (pH 6.0–7.0).
- **TEMPERAMENT** Intolerant of its own kind.

Forked caudal fin

The velvet-black body of these cyprinids is offset by the bright red coloration of the tail. Females are larger than males, and their caudal fin is not as brightly colored. Breeding in the typical home aquarium is unrealistic because of their aggressive nature. A single fish in a community is unlikely to be disruptive, but avoid tankmates with a similar coloration, since they may be attacked. Young Red-Tailed Sharks are silvery at first, gaining their red caudal fin when about seven weeks old.

ON THE ALERT

Fish rely on a range of senses to locate food, including smell, sight, and touch. There are significant differences in the sensory equipment and capabilities of different groups, and even of individual species within groups. Most cyprinids, such as the Red-Fin Shark pictured here, have paired sensory barbels. These help the fish find edible items in the substrate, keeping the eyes free to detect approaching predators. The barbels may not be conspicuous in some cyprinids because they keep them folded back along the side of the face, giving the fish a more streamlined shape when swimming.

Epalzeorhynchos frenatus

Red-Fin Shark

- **ORIGINS** Southeast Asia, being restricted to northern parts of Thailand.
- **SIZE** 6 in (15 cm).
- **DIET** Flake and small live foods.
- **WATER** Temperature 72–77°F (22–25°C); soft to hard (50–150 mg/l) and acidic to neutral (pH 6.0–7.0).
- **TEMPERAMENT** Relatively peaceful.

One of the more colorful cyprinids, this fish can be distinguished at a glance from its red-tailed relative *(see left)* because all its fins are reddish in color. Red-Fin Sharks require an aquarium incorporating suitable retreats, such as a clay flowerpot set in the substrate, and some floating plants at the surface to create dappled lighting. These fish are more

The white form of the Red-Fin Shark is now widely available. It retains the reddish fin markings, but the body is mostly whitish or transparent.

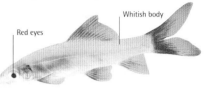

Whitish body

Red eyes

tolerant of their own kind than Red-Tails, so they can be bred more easily. The females swell with eggs, laying up to 4,000 at a single spawning. The resulting young are soon large enough to feed on brine shrimp.

Osteochilus vittatus

Golden Bony-Lipped Barb

- **ORIGINS** Southeast Asia, from Thailand across the Malay Peninsula to parts of Indonesia and Borneo.
- **SIZE** 12½ in (32 cm).
- **DIET** Flake and small live foods.
- **WATER** Temperature 72–77°F (22–25°C); soft (50 mg/l) and acidic (pH 6.0–6.5).
- **TEMPERAMENT** Peaceful.

Golden markings are apparent on the bodies of these barbs, which grow to a large size and need spacious accommodation. There is no visual difference between the sexes. Golden Bony-Lipped Barbs thrive in groups. They use their strong lips to dig around in the substrate in search of food and will uproot plants that are not set in pots. Should they start to nibble the plants, you may be able to deter them by offering foods such as spinach.

Danio rerio

Zebra Danio

- **ORIGINS** Asia, occurring in eastern India, where it ranges from Kolkata to Masulipatam.
- **SIZE** 2¹/₂ in (6 cm).
- **DIET** Flake and live foods.
- **WATER** Temperature 68–75°F (20–24°C); soft (50 mg/l) and acidic (pH 6.0–6.5).
- **TEMPERAMENT** Lively, shoaling fish.

These small danios show well in shoals as part of a community aquarium. The basic body patterning of horizontal blue and cream stripes varies between individuals, with the blue stripes sometimes being broken into streaks and spots. Few fish are more devoted to their mates than Zebras. Whiteworm is a good conditioning food for spawning purposes, with males becoming more brightly colored at this time and the females swelling with eggs. The water in a spawning tank for danios must be about 6 in (15 cm) deep. Placing a grid on the bottom will prevent the pair from eating their spawn. Egg-laying occurs in the morning, triggered partly by sunlight falling on the tank. The adults should then be removed. The young Zebras emerge two days later and are free-swimming in another five days.

Golden Zebra Danio This is one of the most popular color variants now established, having a more yellowish appearance than the normal variety.

Striped patterning still evident

Slim body — *Patterning across the tail*

Longfin Zebra Danio This is another extremely popular domesticated variant, in which the fins are larger than those of the naturally occurring species.

Elongated dorsal fin

Individual fin patterning

Metallic Longfin Zebra Danio Commercially bred variants—especially longfins such as this one, which has a metallic sheen—are more delicate than the wild form.

Danio rerio var. *frankei*

Leopard Danio

- **ORIGINS** Asia, probably a domestic variant, although its precise origins are not documented.
- **SIZE** 2¹/₂ in (6 cm).
- **DIET** Flake and live foods.
- **WATER** Temperature 72–75°F (22–24°C); soft (50 mg/l) and acidic (pH 6.0–6.5).
- **TEMPERAMENT** Lively, shoaling fish.

First described in 1953, the Leopard was initially thought to be a species in its own right, but recent genetic studies have revealed that it is a variant of the Zebra Danio. The Leopard can be identified by its predominantly spotted rather than lined body pattern, with the lines being more obvious close to the caudal fin. Its care and breeding requirements are identical to those of the Zebra. A long-finned form of the Leopard Danio has also been bred.

Dorsal fin set well back — *Disrupted striped markings*

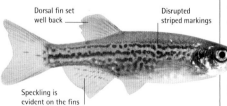

Speckling is evident on the fins

Danio albolineatus

Pearl Danio

- **ORIGINS** Southeast Asia, from Myanmar (Burma) to Thailand, the Malay Peninsula, and Sumatra.
- **SIZE** 2¹/₂ in (6 cm).
- **DIET** Flake and live foods.
- **WATER** Temperature 72–75°F (22–24°C); soft (50 mg/l) and acidic to neutral (pH 6.5–7.0).
- **TEMPERAMENT** Lively, active shoalers.

The Pearl Danio's violet-blue coloration is most evident over the rear of its body, where it is broken by a thin, yellowish-red stripe along the midline. In the picture above, the slimmer-bodied fish to the right is a male. Occupying the upper levels of the tank, this very active danio needs plenty of clear space for swimming. Include fine-leaved plants in the spawning tank. Feed young Pearl Danios fry food at first then brine shrimp.

ASYMMETRICAL MARKINGS

One of the features of many danio species is the way that patterning can differ quite significantly between fish, enabling individuals to be recognized at close quarters. Furthermore, there is generally no symmetry in the patterning, so each side of a danio's body may show different markings. These markings will be consistent throughout the fish's life. Collectively, in a tight shoal, this variance in patterning may help to protect the fish. Any predator attempting to track its prey by visually locking on to a particular pattern may lose its target in the throng when the shoal turns and its intended victim suddenly reveals a different pattern.

Danio nigrofasciatus

Spotted Danio

- **ORIGINS** Southern Asia, occurring only within the boundaries of Myanmar (Burma).
- **SIZE** 2 in (5 cm).
- **DIET** Flake and live foods.
- **WATER** Temperature 72–75°F (22–24°C); soft (50 mg/l) and acidic to neutral (pH 6.5–7.0).
- **TEMPERAMENT** Lively and social.

A strong, dark stripe, becoming blue above the anal fin, helps to distinguish this danio. There is also an irregular pattern of spots occurring below this stripe on the lower half of the body. Males are smaller than females and have a light brown edge to the anal fin. Increasing the water temperature slightly will help to trigger spawning behavior. These danios maintain a strong pair bond, even though they live in shoals. They scatter their eggs in batches among fine-leaved plants.

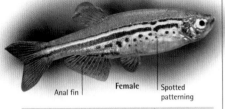

Anal fin | **Female** | Spotted patterning

Danio kerri

Blue Danio

- **ORIGINS** Southeast Asia, being restricted to the Koh Yao islands of Thailand.
- **SIZE** 2 in (5 cm).
- **DIET** Flake and live foods.
- **WATER** Temperature 72–77°F (22–25°C); soft (50 mg/l) and acidic to neutral (pH 6.5–7.0).
- **TEMPERAMENT** Lively, active, shoaling fish.

A pair of parallel, yellowish stripes along the side of the bluish body is the key feature of this species. Blue Danios will breed readily in the home aquarium, with the female simply dispersing her eggs over the base of the spawning tank. It takes about four days for the eggs to hatch. These fish, like all danios, are quite capable of leaping out of their tank, so make sure it is always covered.

Devario aequipinnatus

Great Danio

- **ORIGINS** Asia, occurring in western India as well as on the neighboring island of Sri Lanka.
- **SIZE** 6 in (15 cm).
- **DIET** Flake and live foods.
- **WATER** Temperature 72–75°F (22–24°C); soft (50 mg/l) and acidic to neutral (pH 6.5–7.0).
- **TEMPERAMENT** Peaceful and social.

A series of alternating narrow yellow and broad blue horizontal lines typify the body coloration of the Great Danio. In spite of its name, this fish rarely attains maximum size in home aquariums, and is quite suitable for inclusion in a community tank. In common with many other danios, an increase in water temperature and live foods will help to encourage breeding behavior. A typical spawning will comprise 300 to 400 eggs.

Straight blue stripe indicates a male

Males have thinner bodies than females

Devario devario

Bengal Danio

- **ORIGINS** Asia, occurring in northern and eastern parts of India, as well as in Pakistan and Bangladesh.
- **SIZE** 6 in (15 cm).
- **DIET** Flake and live foods.
- **WATER** Temperature 72–75°F (22–24°C); soft (50 mg/l) and acidic to neutral (pH 6.5–7.0).
- **TEMPERAMENT** Active and social.

The Bengal Danio's attractive patterning of yellow markings on a bluish background is seen to best effect in relatively dark surroundings where floating plants diffuse the lighting. Keep Bengals in shoals to maintain their appearance, since single individuals may feel nervous and become paler as a result. There is usually some slight variance between the sexes, with the female (the upper fish in the pair shown here) having a broader body than the male. The eggs will be strewn around the spawning tank and should hatch within two days. Young Bengal Danios require fry food or a suitable substitute once they are free-swimming. They can then be reared on a variety of other foods, ranging from powdered egg yolk and flake to small live foods. Carry out regular partial water changes to maintain good water quality as the young danios grow, and divide them into separate groups to prevent overcrowding.

Boraras maculatus

Spotted Rasbora

- **ORIGINS** Southeast Asia, extending from western Malaysia southward to western Sumatra.
- **SIZE** 1 in (2.5 cm).
- **DIET** Flake and live foods.
- **WATER** Temperature 75–79°F (24–26°C); soft (50 mg/l) and acidic (pH 6.0–6.5).
- **TEMPERAMENT** Peaceful and social.

This rasbora is the smallest member not only of its group but also of the entire cyprinid family. The Spotted Rasbora has a long, narrow caudal peduncle with three dark spots along the side of the body. The pattern of dark markings differs between individuals. The males are recognizable by their brighter coloration and flat underparts; females have a slightly curved lower outline. Spotted Rasboras, which look best in shoals, can be mixed with suitable companions of a similar size, including other small rasboras. A pair will spawn in a breeding tank that is well planted with Sagittaria and cryptocorynes, but they will eat their eggs if left there. About 50 eggs are produced, and they hatch in about 36 hours. Very fine fry food is essential for the young fish at first.

Boraras urophthalmoides

Glass Rasbora

- **ORIGINS** Southeast Asia, where its distribution extends from Vietnam south to Sumatra.
- **SIZE** 1½ in (3.5 cm).
- **DIET** Flake and live foods.
- **WATER** Temperature 73–77°F (23–25°C); soft (50 mg/l) and acidic (pH 6.0–6.5).
- **TEMPERAMENT** Peaceful and social.

These small, attractive rasboras have a red stripe running along each side of the body. There is a broader, iridescent area beneath this stripe that gradually tapers to a point along the caudal peduncle. Male Glass Rasboras have a very evident white spot at the base of the dorsal fin and a larger adjacent black area, both of which the females lack. Relatively dark surroundings, with bogwood decor and suitable plants, should help to overcome the natural nervousness of these fish. It has proved possible to breed Glass Rasboras in a planted spawning tank, provided that the water temperature is kept slightly higher than that in the main aquarium, at up to 82°F (28°C). About 50 eggs are laid on the underside of the leaves of cryptocoryne plants, after which the adult pair should be removed to protect the eggs. They hatch after about two days.

Sundadamo axelrodi

Axelrod's Rasbora

- **ORIGINS** Southeast Asia, where this species is restricted to the Indonesian island of Sumatra.
- **SIZE** 1¼ in (3 cm).
- **DIET** Flake and live foods.
- **WATER** Temperature 73–79°F (23–26°C); soft (50 mg/l) and acidic (pH 6.0–6.5).
- **TEMPERAMENT** Peaceful and social.

These beautiful fish are instantly recognizable, since their body is fluorescent green on top and reddish below, with red coloration also evident on the ventral fin. Axelrod's Rasboras appear to lack the sensory lateral line along the sides of the body, which is perhaps compensated for by their large eyes. Discovered in 1976, this species was named after Dr. Herbert Axelrod, a leading American ichthyologist and fishkeeper.

GROUP DYNAMICS

The majority of rasboras are shoaling fish, and if they are kept singly in the aquarium, they become shy and nervous. There appears to be some recognition between members of a shoal, since new individuals added to an aquarium containing an established group are unlikely to be accepted readily by the rest of the shoal. As a result, the newcomers may form a separate group, but there is rarely any conflict between the shoals. Changing the decor of the tank just before the newcomers are introduced seems to improve the chances that the fish will integrate.

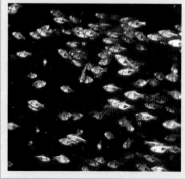

Rasboroides vaterifloris

Orange-Finned Rasbora

- **ORIGINS** Southern Asia, restricted to the island of Sri Lanka.
- **SIZE** 1½ in (4 cm).
- **DIET** Live foods, plus flake.
- **WATER** Temperature 77–84°F (25–29°C); soft (50 mg/l) and acidic (pH 6.0–6.5).
- **TEMPERAMENT** Peaceful.

The appearance of the Orange-Finned Rasbora differs across its range: individuals found between Valallavites and Meegahatenne are predominantly red, while specimens from around Gilimale on the Kaluganga River are greenish. The depth of the orange coloration around the eye socket also varies. Males are more colorful and slimmer-bodied than females. In addition to their regular diet, Orange-Finned Rasboras will sometimes take small amounts of pollen, available in health-food stores. Rich in protein, pollen can help to encourage spawning, as can live foods. Orange-Finned Rasboras are often the only fish that occur in parts of their natural habitat, but they can still be mixed with other nonaggressive fish in community aquariums. Their isolation in the wild may also explain why they have a shorter lateral line than other rasboras: being less exposed to predators, they may not have such a great need for this sensory warning system.

Rasbora kalochroma

Big-Spot Rasbora

- **ORIGINS** Southeast Asia, where it is found in parts of Malaysia, Sumatra, and Borneo.
- **SIZE** 4 in (10 cm).
- **DIET** Flake and live foods.
- **WATER** Temperature 77–82°F (25–28°C); soft (50 mg/l) and acidic (pH 6.0–6.5).
- **TEMPERAMENT** Peaceful and social.

In spite of its relatively large size, the Big-Spot Rasbora is rather sensitive to water quality, and its true beauty will be apparent only if conditions are optimal. A dark base in the aquarium and subdued lighting are also needed for this fish to look its best. Big-Spot Rasboras should be kept in groups, but their shoaling instincts are not as strong as in other rasboras. Spawning behavior in home aquariums has not been documented, but it is probably similar to that shown by other rasboras.

Bluish-black blotches

Pinkish-orange body color

Rasbora einthovenii

Long-Band Rasbora

- **ORIGINS** Southeast Asia, occurring in Malaysia, Sumatra, and Borneo.
- **SIZE** 4 in (10 cm).
- **DIET** Flake and live foods.
- **WATER** Temperature 72–77°F (22–25°C); soft (50 mg/l) and acidic (pH 6.0–6.5).
- **TEMPERAMENT** Peaceful and social.

Males have slimmer bodies than females

Broad stripe

The dark stripe along the midline of this rasbora extends from the mouth all the way through to the caudal fin. It is broader in males, which have a purplish cast to the body and a hint of red on the dorsal fin; this coloration is most obvious when they are in top condition. Females, in contrast, are a more greenish shade. A spawning tank for these fish needs a concentration of fine-leaved plants at one end in which eggs can be laid.

Rasbora trilineata

Scissortail Rasbora

- **ORIGINS** Southeast Asia, occurring in parts of Malaysia, Sumatra, and Borneo.
- **SIZE** 6 in (15 cm).
- **DIET** Flake and live foods.
- **WATER** Temperature 73–77°F (23–25°C); soft (50 mg/l) and acidic (pH 6.0–6.5).
- **TEMPERAMENT** Peaceful and social.

The striped pattern on the sides of the Scissortail Rasbora's silvery body becomes more evident toward the caudal fin. This rasbora's common name derives from the way the deep fork of the caudal fin tends to close up, like scissor blades, as the fish moves through the water. The size and shoaling behavior of Scissortail Rasboras means that they will benefit from being housed in a relatively long aquarium, where planting is restricted to the sides of the tank.

Dark spot on each lobe of caudal fin

Rasbora borapetensis

False Magnificent Rasbora

- ⊕ **ORIGINS** Southeast Asia, found in parts of Thailand and western Malaysia.
- ⟳ **SIZE** 2 in (5 cm).
- ⟩ **DIET** Flake and live foods.
- ≋ **WATER** Temperature 72–79°F (22–26°C); soft (50 mg/l) and acidic (pH 6.0–6.5).
- ⊜ **TEMPERAMENT** Peaceful and social.

False Magnificent Rasboras have a dark red area at the base of the caudal fin. They also display a black line that runs from behind the gills along the midline of the body. If their surroundings are less than ideal, the red area on the tail becomes much paler. Floating plants on the surface will ensure that the lighting is not too bright, while adding aquarium peat to the filtration system will help to maintain the water quality.

Rasbora elegans

Elegant Rasbora

- ⊕ **ORIGINS** Southeast Asia, on the Malay Peninsula, and southward to Sumatra and Borneo.
- ⟳ **SIZE** 8 in (20 cm).
- ⟩ **DIET** Flake and live foods.
- ≋ **WATER** Temperature 72–77°F (22–25°C); soft (50 mg/l) and acidic (pH 6.0–6.5).
- ⊜ **TEMPERAMENT** Peaceful and social.

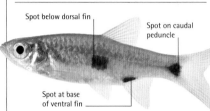

Spot below dorsal fin

Spot on caudal peduncle

Spot at base of ventral fin

Elegant Rasboras display three black areas on each side of the body. The body itself is greenish-brown above and silvery below. Females can be recognized when spawning approaches, because they become paler than the males and their bodies swell with eggs. They spawn in typical rasbora fashion, with the young requiring special fry food or infusoria once they are free-swimming. Elegant Rasboras will rarely reach their maximum size in home aquariums.

RASBORA JAW STRUCTURE

Many cyprinids have downward-pointing mouthparts adapted for feeding on the substrate. Rasboras are not substrate feeders, however, but quick, agile, sharp-eyed predators that seek their food in the middle and upper layers of the water. As a result, their lower jaw curves up, rather than down, as seen here on this Narrow-Wedge Harlequin (*see opposite*). This jaw structure makes it easier for the rasboras to snatch aquatic invertebrates swimming just above them in the water and also to grab any flying insects that may settle on the surface.

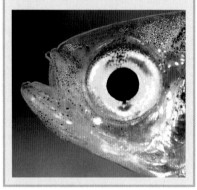

Rasbora caudimaculata

Spot-Tailed Rasbora

- ⊕ **ORIGINS** Southeast Asia, occurring on the Malay Peninsula, Sumatra, and Borneo.
- ⟳ **SIZE** 4¾ in (12 cm).
- ⟩ **DIET** Flake and live foods.
- ≋ **WATER** Temperature 68–77°F (20–25°C); soft (50 mg/l) and acidic (pH 6.0–6.5).
- ⊜ **TEMPERAMENT** Peaceful and social.

Dark band

Slightly pinkish dorsal fin

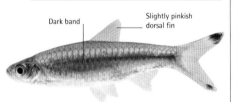

These rasboras are generally smaller than Scissortail Rasboras (*see p.79*), and their silvery body color also terminates in a deeply forked caudal fin that resembles a pair of scissors. Each lobe of the caudal fin has a colored area and ends in a dark tip. The slim body shape of these fish reveals their active nature. They can also jump effectively, which means that the aquarium or spawning tank should always be kept covered.

Rasbora daniconius

Slender Rasbora

- ⊕ **ORIGINS** Asia, from India and Sri Lanka, via Myanmar (Burma) and Thailand, to the Greater Sunda Islands.
- ⟳ **SIZE** 4 in (10 cm).
- ⟩ **DIET** Flake and live foods.
- ≋ **WATER** Temperature 75–79°F (24–26°C); soft (50 mg/l) and acidic (pH 6.0–6.5).
- ⊜ **TEMPERAMENT** Peaceful and social.

The appearance of this species is not dissimilar to that of the Long-Band Rasbora (*see p.79*), but its dorsal fin is set slightly farther back and its body color may be a little darker. Males have a yellow tinge to their fins and are slimmer than females, which swell noticeably with eggs. Spawning is reasonably easy to accomplish under favorable water conditions; this may entail dropping the pH slightly and raising the water temperature.

Largely transparent fins

Bluish-black stripe extends to caudal fin

Trigonopoma pauciperforatum

Redline Rasbora

- ⊕ **ORIGINS** Southeast Asia, present on the Malay Peninsula and as far south as Sumatra.
- ⟳ **SIZE** 3 in (7 cm).
- ⟩ **DIET** Flake and live foods.
- ≋ **WATER** Temperature 73–77°F (23–25°C); soft (50 mg/l) and acidic (pH 6.0–6.5).
- ⊜ **TEMPERAMENT** Peaceful and social.

Red stripe runs through the eye

The brilliant red stripe along each side of the body helps to sex these rasboras. When females swell with spawn, the stripe has a more curved appearance toward the rear of the body, but in the slimmer males it is always relatively straight. Compatibility between individuals can be an obstacle to successful breeding, but in a shoal the sexes will be able to pair naturally; paired fish remain in close contact with one another.

Trigonostigma espei

Narrow-Wedge Harlequin

- **ORIGINS** Southeast Asia, where it is restricted to parts of Thailand.
- **SIZE** 1³/₄ in (4.5 cm).
- **DIET** Flake and live foods.
- **WATER** Temperature 73–82°F (23–28°C); soft (50 mg/l) and acidic (pH 6.0–6.5).
- **TEMPERAMENT** Peaceful and social.

Once regarded as a subspecies of the Harlequin Rasbora (*see below*), this fish is now considered to be a separate species. It is more colorful than the Harlequin, with males displaying an especially rich red coloration. Although the two fish have similar markings, the Narrow-Wedge has an extra black line at the back of the gills, while its black flank markings are less prominent. For breeding, house a single female and two males in the spawning tank, which should include broad-leaved plants such as cryptocorynes. Transfer the adults back to the main aquarium to rejoin the shoal as soon as spawning is finished.

Trigonostigma heteromorpha

Harlequin Rasbora

- **ORIGINS** Southeast Asia, occurring in Thailand and Malaysia, and southward to the island of Sumatra.
- **SIZE** 2 in (5 cm).
- **DIET** Flake and live foods.
- **WATER** Temperature 73–77°F (23–25°C); soft (50 mg/l) and acidic (pH 6.0–6.5).
- **TEMPERAMENT** Peaceful and social.

This long-standing aquarium favorite has been popular with hobbyists for nearly a century, which is why it is sometimes simply referred to as "the Rasbora." The dark, triangular-shaped marking on the rear part of the body, which tapers down the caudal peduncle to the base of the caudal fin, makes the Harlequin easy to recognize. Males are more colorful than females.

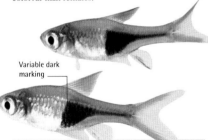

Variable dark marking

Brevibora dorsiocellata

Hi-Spot Rasbora

- **ORIGINS** Southeast Asia, occurring throughout the Malay Peninsula and southward to Sumatra.
- **SIZE** 2¹/₂ in (6.5 cm).
- **DIET** Flake and live foods.
- **WATER** Temperature 68–77°F (20–25°C); soft (50 mg/l) and acidic (pH 6.0–6.5).
- **TEMPERAMENT** Peaceful.

Dorsal eyespot

This golden-bodied fish has a black eyespot on its dorsal fin. A red, rather than yellow, tinge to the caudal fin indicates a male. The spawning tank should contain only about 6 in (15 cm) of water, and there should be marbles on the base. This will protect the eggs after spawning, because the fish will have less time to eat them before the eggs fall out of reach. If the female is placed in the tank a few days in advance, spawning should occur quite rapidly. The eggs hatch in a day.

Tanichthys albonubes

White Cloud Mountain Minnow

- **ORIGINS** Asia, occurring in the White Cloud Mountain region around Canton, China.
- **SIZE** 1¹/₄ in (4.5 cm).
- **DIET** Flake and live foods.
- **WATER** Temperature 64–77°F (18–25°C); soft (50 mg/l) and acidic (pH 6.0–6.5).
- **TEMPERAMENT** Peaceful and social.

The original form of the White Cloud Mountain Minnow has pale yellow edges on its fins, but a strain found in Hong Kong, which developed from aquarium escapees, displays red fin-edging. Females are not as brightly colored as males. These rasboras can be kept in much cooler waters than most, down to 40°F (4.5°C), but for breeding the optimum temperature is about 70°F (21°C). They spawn among vegetation, with the eggs hatching just over a day after being laid.

Deep red area on tail

Pethia conchonius

Rosy Barb

- **ORIGINS** Asia, where it occurs in northern India, in areas including Assam and Bengal.
- **SIZE** 6 in (15 cm).
- **DIET** Prepared foods and live foods.
- **WATER** Temperature 64–77°F (18–25°C); hard (100–150 mg/l) and around neutral (pH 7.0).
- **TEMPERAMENT** Peaceful and social.

The characteristic rosy coloration of these barbs is seen at its brightest in males that are in spawning condition. Only male fish have pinkish fins, while those of females are clear. Rosy Barbs are relatively undemanding in terms of their water chemistry needs, but they will benefit from being kept in a tank with subdued lighting, beneath floating plants. This is one of the easier egg-laying species to breed, with live foods being valuable for conditioning purposes. Prior to spawning, transfer pairs to a separate aquarium with fine-leaved plants such as *Myriophyllum* and a reasonably low water level. The eggs, which are scattered among the plants, hatch in about

a day; the fry become free-swimming after a further two or three days. Rosy Barbs can be reared on commercial foods formulated for egg-layers. Their relative hardiness means that they can be housed in well-planted outdoor ponds in subtropical and tropical parts of the world. There is likely to be enough natural food in the pond to nourish the young fry at first, and they are soon able to take food provided for the adult fish.

Longfin variant The longfin characteristic has been combined both with the native form seen here and also with recently developed colors, like the coppery variant.

Rosy Barb Selective breeding has tended to enhance the natural coloration of these barbs, with some strains now having a distinctive coppery-red appearance, as seen here.

Desmopuntius pentazona

Banded Barb

- **ORIGINS** Southeast Asia, occurring throughout the Malay Peninsula; also present in Borneo.
- **SIZE** 2 in (5 cm).
- **DIET** Prepared foods and live foods.
- **WATER** Temperature 68–75°F (20–24°C); hard (100–150 mg/l) and around neutral (pH 7.0).
- **TEMPERAMENT** Peaceful and social.

Dark bands may be broken in some cases

The first of the five bluish-black bands encircling the body of this fish runs through the eyes; the broadest lies at the front of the dorsal fin. Males are more brightly colored than females, with a richer shade of coppery-red on the back and paler underparts. These barbs are a good choice for community tanks, although they need to be transferred to a spawning tank for breeding. The young can eat brine shrimp at about 10 days old.

Pethia nigrofasciata

Black Ruby Barb

- **ORIGINS** Asia, found in sluggish streams in the mountains of Sri Lanka.
- **SIZE** 2½ in (6.5 cm).
- **DIET** Prepared foods and live foods.
- **WATER** Temperature 72–79°F (22–26°C); soft (50–100 mg/l) and acidic (pH 6.0).
- **TEMPERAMENT** Peaceful and social.

The male Black Ruby Barb, shown below, is larger and more colorful than the silvery female. A tank for these barbs needs subdued lighting, to prevent the fish from becoming nervous, and it must also include retreats. In the wild, seasonal changes in the water temperature trigger breeding; raising the temperature in the tank to the upper end of the specified range will achieve the same result. Spawning is likely soon after sunrise.

Dawkinsia filamentosa

Black-Spot Filament Barb

- **ORIGINS** Asia, occurring in mountain streams in south and southwestern India, as well as in Sri Lanka.
- **SIZE** 6 in (15 cm).
- **DIET** Prepared foods and live foods.
- **WATER** Temperature 68–75°F (20–24°C); soft (50–100 mg/l) and acidic (pH 6.0).
- **TEMPERAMENT** Peaceful and social.

Juvenile

The scientific name of this barb derives from the long filaments that develop on the dorsal rays of the mature male. In spawning condition, the male develops small white spots on its face. These barbs should be kept as a shoal, so they require a relatively large aquarium. As with other barbs, Black-Spot Filament Barbs are more likely to spawn successfully if the sexes are separated beforehand. Each female will produce as many as 1,000 eggs.

Pethia ticto

Tic-Tac-Toe Barb

- ⊕ **ORIGINS** Asia, occurring from southern India to the Himalayas; also in Sri Lanka.
- ◌ **SIZE** 4 in (10 cm).
- ◌ **DIET** Prepared foods and live foods.
- ◌ **WATER** Temperature 57–79°F (14–26°C); hard (100–150 mg/l) and around neutral (pH 7.0).
- ⊕ **TEMPERAMENT** Peaceful and social.

The more colorful male Tic-Tac-Toe usually has black markings on its dorsal fin, which are missing in the female. Keep these fish in cool water for a time, to mimic winter conditions in the wild. A gradual increase in temperature triggers spawning behavior. Place two males in a spawning tank with a single female. The Odessa Barb is a variant of this species created by breeders in Moscow.

Bronze body

Black spots

Male

Puntigrus tetrazona

Tiger Barb

- ⊕ **ORIGINS** Asia, in Sumatra and elsewhere in Indonesia; also in Borneo and possibly Thailand.
- ◌ **SIZE** 2¾ in (7 cm).
- ◌ **DIET** Prepared foods and live foods.
- ◌ **WATER** Temperature 68–79°F (20–26°C); soft (50–100 mg/l) and slightly acidic (pH 6.5).
- ⊕ **TEMPERAMENT** Social, sometimes aggressive.

The dark banding of these barbs may vary between individuals, while the red areas on the body are usually brighter in males. These barbs are best kept in large groups to lessen the risk of bullying. Tiger Barbs are active by nature and need adequate open swimming areas in their tank. Their breeding requirements are similar to those of related species. When purchasing stock, check carefully for any signs of white spot (*see p.58*), to which Tiger Barbs are especially vulnerable.

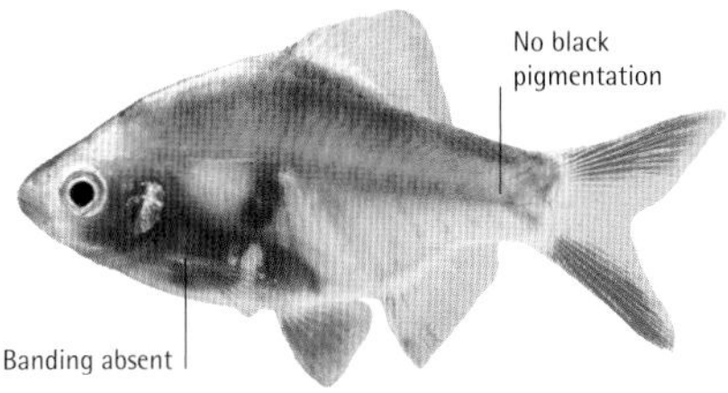

Green Tiger Barb The extensive, greenish-black coloration has led to this variant also being called the Moss-Banded Barb.

Male

No black pigmentation

Banding absent

Red Tiger Barb (above) Tiger Barbs are available in several color variants. This form has a reddish appearance, with no dark markings or banding.

Red markings on the fins

FIN-NIPPING BEHAVIOR

Although barbs are generally placid, certain species can be troublesome because they nip at the fins of other fish sharing their tank. This applies especially to Tiger Barbs *(see above, right)*. Fish with trailing fins are most at risk, including young angelfish (*Pterophyllum* species) and various gouramis, as well as male Siamese Fighting Fish (*Betta splendens*). Long-finned variants of other species, such as danios and guppies, are also vulnerable to being harried in this way. The picture shows a Tiger Barb nipping the caudal fin of a Yellow Veiltail Guppy.

The reason for this fin-nipping behavior is unclear. Unfortunately, there is little that can be done to prevent it, so take great care when choosing tankmates for Tiger Barbs and other fin-nippers. Minor fin damage usually heals over the course of several weeks, but repeated assaults and constant harassment by a group of Tiger Barbs can prove fatal. Tiger Barbs will even occasionally show aggression toward each other, to reinforce the order of dominance within a shoal. Any Tiger Barbs added later to an existing group are likely to be bullied.

Striuntius lineatus

Striped Barb

- ⬡ **ORIGINS** Southeast Asia, being present on the Malay Peninsula and extending to parts of Indonesia.
- ⬡ **SIZE** 4³/₄ in (12 cm).
- ⬡ **DIET** Flake and small live foods, some plant matter.
- ⬡ **WATER** Temperature 70–75°F (21–24°C); soft (50 mg/l) and acidic (pH 6.0–6.5).
- ⬡ **TEMPERAMENT** Peaceful and social.

The beautiful yellow-and-black horizontal striping on the sides of these barbs will only be apparent if they are housed in an aquarium where the lighting is subdued. Otherwise, they will look paler and much less attractive. Sexing is possible since females have a broader body and a more steeply curved back than males. Striped Barbs are rather nervous by nature, so their aquarium needs to incorporate plenty of retreats in the form of bogwood and aquatic vegetation. These fish should be housed in groups, and they will mix well with other nonaggressive species that require similar conditions, such as danios. Feeding is fairly straightforward, since Striped Barbs will eat a wide variety of foods. For spawning purposes, raise the proportion of live foods in their diet.

CANNIBALIZING THE YOUNG

Many barbs are opportunistic feeders, eating whatever they can swallow easily. They have relatively large mouths, as displayed by this Tiger Barb (*Puntigrus tetrazona*). A barb's appetite will even extend to its own eggs and fry. Barbs display no parental instincts at all, although once the young are large enough not to be regarded as food by the adults, they are accepted as part of the shoal. Although barbs typically produce hundreds of eggs at a single spawning, only a tiny number of the resulting fry survive to adulthood in the wild.

Barbodes everetti

Clown Barb

- ⬡ **ORIGINS** Southeast Asia, occurring in Singapore, Borneo, and the Bunguran Islands.
- ⬡ **SIZE** 4 in (10 cm).
- ⬡ **DIET** Flake and small live foods, some plant matter.
- ⬡ **WATER** Temperature 75–86°F (24–30°C); soft (50 mg/l) and acidic (pH 6.0–6.5).
- ⬡ **TEMPERAMENT** Relatively peaceful.

Large, irregular, dark blotches on a yellowish background typify the Clown Barb. The male fish becomes brighter as it starts to mature at about 18 months. Clown Barbs may damage aquarium plants, especially if their diet is deficient in vegetable matter. These are difficult barbs to spawn, although raising the water temperature, separating the sexes for several weeks, and giving them more live food will all help.

Blotched patterning is not symmetrical

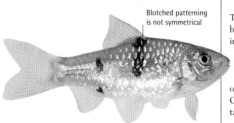

Puntius titteya

Cherry Barb

- ⬡ **ORIGINS** Asia, being restricted to the streams and rivers of southwestern Sri Lanka.
- ⬡ **SIZE** 2 in (5 cm).
- ⬡ **DIET** Flake and small live foods, some plant matter.
- ⬡ **WATER** Temperature 72–75°F (22–24°C); soft (50 mg/l) and acidic (pH 6.0–6.5).
- ⬡ **TEMPERAMENT** Placid and somewhat timid.

Dark coloration with evident reddish suffusion

The cherry-red coloration associated with these barbs is most evident in fully mature males that are in spawning condition. Females, in contrast, are significantly duller, with the reddish coloration being largely restricted to the area around the gills. Being relatively shy fish, they display less of a shoaling instinct than many related barbs. Cherry Barbs need a densely planted spawning tank, since their eggs stick to aquatic plants.

Barbodes semifasciolatus

Gold Barb

- ⬡ **ORIGINS** Eastern Asia, occurring in southeastern China, from Hong Kong to Hainan Island.
- ⬡ **SIZE** 4 in (10 cm).
- ⬡ **DIET** Flake and small live foods, some plant matter.
- ⬡ **WATER** Temperature 64–75°F (18–24°C); soft (50 mg/l) and acidic (pH 6.0–6.5).
- ⬡ **TEMPERAMENT** Peaceful.

The rich orange-yellow coloration of male Gold Barbs has been emphasized by selective breeding. There is also a relatively inconspicuous darker line running along the side of the body. Originating from farther north than many barbs, the Gold Barb does not require such a high water temperature, but raising it slightly can help to trigger spawning. As with related species, a separate spawning tank will be required, because otherwise the eggs are likely to be eaten by the adults soon after they have been laid.

Faint dark line along body

Barbonymus schwanenfeldii

Tinfoil Barb

- ⊕ **ORIGINS** Southeast Asia, from Thailand and the Malay Peninsula to Sumatra and Borneo.
- ⟳ **SIZE** 14 in (35 cm).
- ⟐ **DIET** Flake and live foods, some plant matter.
- ⊜ **WATER** Temperature 72–77°F (22–25°C); soft (50 mg/l) and acidic (pH 6.0–6.5).
- ⊜ **TEMPERAMENT** Agreeable with fish of a similar size.

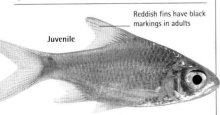

Reddish fins have black markings in adults

Juvenile

Tinfoil Barbs are not suitable for a community aquarium, since they grow rapidly and will soon dwarf the other occupants and possibly even prey on them. However, it is usually possible to house them in a large tank with compatible cichlids of a similar size. Tinfoil Barbs will dig in the substrate and are likely to uproot any plants that are not in pots. Breeding is unlikely in home aquaria. Aside from the natural silver form, an attractive gold-bodied variant has also been established.

Dawkinsia arulius

Arulius Barb

- ⊕ **ORIGINS** Asia, restricted to southern and southeastern parts of India.
- ⟳ **SIZE** 4¼ in (12 cm).
- ⟐ **DIET** Flake and live foods, some plant matter.
- ⊜ **WATER** Temperature 66–77°F (19–25°C); soft (50 mg/l) and acidic (pH 6.0–6.5).
- ⊜ **TEMPERAMENT** Peaceful and social.

In Arulius Barbs, dark banding running down the body from the dorsal fin is offset against pale underparts. The male, shown below, has extended rays on the dorsal fin that trail back and almost reach the tail. When in breeding condition—which is reached only after the fish are about 18 months old—the male develops white spots around the mouth and some iridescence. Females typically produce fewer than 100 eggs at a time.

Pethia gelius

Golden Barb

- ⊕ **ORIGINS** Asia, where it has been recorded in the central, eastern, and northeastern parts of India.
- ⟳ **SIZE** 1½ in (4 cm).
- ⟐ **DIET** Flake and small live foods, some plant matter.
- ⊜ **WATER** Temperature 64–72°F (18–22°C); soft (50 mg/l) and acidic to neutral (pH 6.5–7.0).
- ⊜ **TEMPERAMENT** Peaceful.

Tall dorsal fin

These barbs can be recognized easily by their small size and mottled body, with darker markings often extending onto the leading edge of the dorsal fin. They are highly social fish and should be kept in groups, with the males identifiable by their brighter coloration. The breeding habits are unusual, because females deposit their eggs on the undersides of the leaves of aquarium plants, with *Ludwigia* being favored for this purpose.

Puntius oligolepis

Checkered Barb

- ⊕ **ORIGINS** Asia, where it is widely distributed across the many islands that comprise Indonesia.
- ⟳ **SIZE** 5¾ in (15 cm).
- ⟐ **DIET** Flake and live foods, some plant matter.
- ⊜ **WATER** Temperature 68–75°F (20–24°C); soft (50 mg/l) and acidic (pH 6.0–6.5).
- ⊜ **TEMPERAMENT** Placid.

A dark, rather metallic sheen to the upperparts of the body helps to characterize the Checkered Barb. Males can be recognized by their deeper coloration, with the edge of their fins being black. Young Checkered Barbs grow rapidly and may be sexually mature by just four months of age. Spawning occurs near the surface on water plants such as *Myriophyllum*. A pair may produce up to 300 eggs in the spawning tank, after which they should be moved back to the main tank. The fry hatch within two days and can be reared initially on fry foods, and then on live foods, such as brine shrimp. Algae will later be beneficial to improve their coloration. Regular partial water changes are vital to maintain the water quality. Overcrowding the young fish may stunt their growth and will increase the risk of environmental diseases.

Barbodes lateristriga

T-Barb

- ⚙ **ORIGINS** Southeast Asia, from Thailand across the Malay Peninsula to Sumatra, Java, and Borneo.
- 🔄 **SIZE** 7 in (18 cm).
- 🍴 **DIET** Prepared foods and live foods.
- 〰 **WATER** Temperature 72–79°F (22–26°C); soft (50–100 mg/l) and acidic to neutral (pH 6.5–7.0).
- ☺ **TEMPERAMENT** Peaceful and social.

T-shaped black marking

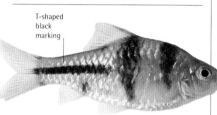

Set against the pale, golden-yellow body color, the black patterning of this barb is reminiscent of the silhouette of a monkey wrench. The patterning fades, however, as these relatively large barbs mature. T-Barbs have proved harder to breed than related species, but when they do spawn, the females can produce several hundred eggs in a well-planted tank. The adults should be removed immediately afterward, to prevent them from devouring their own eggs.

Desmopuntius rhomboocellatus

Round-Banded Barb

- ⚙ **ORIGINS** Asia, where it is thought to be restricted to the island of Borneo.
- 🔄 **SIZE** 2 in (5 cm).
- 🍴 **DIET** Prepared foods and live foods.
- 〰 **WATER** Temperature 73–82°F (23–28°C); soft (50–100 mg/l) and acidic (pH 6.5).
- ☺ **TEMPERAMENT** Peaceful and social.

This is one of the rarer Asiatic barbs. Under good water conditions and subdued lighting, the body coloration of Round-Banded Barbs is a faint pink. It will turn more silvery if the light is too bright, with the broad, irregular black markings also fading. Frequent partial water changes are vital to keep nitrate levels low, and aquarium peat should be added to the filter (*see p.46*).

Variable dark patterning

Hypsibarbus wetmorei

Pale Barb

- ⚙ **ORIGINS** Southeast Asia, where its distribution is restricted to Thailand.
- 🔄 **SIZE** 7 in (18 cm).
- 🍴 **DIET** Prepared foods and live foods.
- 〰 **WATER** Temperature 72–79°F (22–26°C); hard (100–150 mg/l) and neutral (pH 7.0).
- ☺ **TEMPERAMENT** Peaceful and social.

Dark edging to the scales   Humped back

The characteristic lemony-orange fin coloration is evident only on the lower fins of this barb; the dorsal fin is relatively transparent. A silvery body color and large scales are also typical. Unfortunately, these fish grow to a large size, so they require a suitably spacious aquarium to provide adequate swimming space. Pale Barbs can be sexed easily only during the spawning period, when females swell with eggs.

Enteromius fasciolatus

African Barb

- ⚙ **ORIGINS** Southern Africa, occurring in southern Zaire, Angola, Zambia, and Zimbabwe.
- 🔄 **SIZE** 2 in (5 cm).
- 🍴 **DIET** Prepared foods and live foods.
- 〰 **WATER** Temperature 73–79°F (23–26°C); soft (50–100 mg/l) and acidic (pH 6.0–6.5).
- ☺ **TEMPERAMENT** Peaceful and social.

An elongated body and striking orange coloration make African Barbs unmistakable. Although the distinctive black striping varies between individuals, male African Barbs can be distinguished by their slimmer shape. These lively fish should be kept in small groups containing at least six individuals. They need a darkened tank, with an open area for swimming and a well-planted perimeter. Include bogwood to provide them with hiding places; otherwise, they will swim around nervously. African Barbs spawn in clumps of vegetation, but are likely to eat their eggs and fry if left in the tank with them. Give the young fry food once they are free-swimming, and then progress to brine shrimp.

Puntius bimaculatus

Red-Striped Barb

- **ORIGINS** Southeast Asia, occurring in Sri Lanka, as well as in Mysore, southern India.
- **SIZE** 2¾ in (7 cm).
- **DIET** Prepared foods and live foods.
- **WATER** Temperature 72–79°F (22–26°C); soft (50–100 mg/l) and acidic (pH 6.5).
- **TEMPERAMENT** Peaceful and social.

The intensity of the red stripe on this Asiatic barb varies significantly, and in some individuals it is so faint that it is barely visible. All Red-Striped Barbs, however, have a black spot in the middle of the base of the dorsal fin and another black spot on the caudal peduncle. Males are more slender in appearance than females. These fish are quite nervous, even when kept in a shoal, and they prefer to occupy the lower areas of a tank. Increasing the water temperature toward the upper end of the stated range and adhering to the water chemistry criteria given here should encourage spawning behavior. Red-Striped Barbs can be prolific; the females are capable of breeding several times during the year and laying as many as 400 eggs on each occasion.

BARB BREEDING BEHAVIOR

The shoaling instincts of barbs, such as these Red Tiger Barbs *(see p.83)*, even extend to their spawning behavior. Several males take part in this process in the wild, so it is best to house a single female in the breeding tank with at least two males. Females release hundreds of eggs at each spawning, so the risk of a large number not being fertilized is greater if there is only one male present. The age of the male barbs can also be a factor in successful spawning, as some species may not be fully mature until they are 18 months old.

Pethia cumingii

Cuming's Barb

- **ORIGINS** Asia, where it is restricted to the southwestern part of Sri Lanka.
- **SIZE** 2 in (5 cm).
- **DIET** Prepared foods and live foods.
- **WATER** Temperature 72–81°F (22–27°C); soft (50–100 mg/l) and neutral (pH 7.0).
- **TEMPERAMENT** Peaceful and social.

On its native island of Sri Lanka, the red-finned form of Cuming's Barb, pictured here, occurs from the Kelani River northward, while a yellow-finned variant extends as far north as the Kalu River. Like most barbs, Cuming's is suitable for a community tank with nonaggressive fish that need similar water conditions. Breeding is rare in home aquariums, usually because of a lack of compatibility.

Fins have a reddish tinge

Two black marks on bronze body

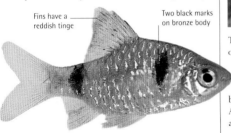

Enteromius trispilos

Black-Spotted Gold Barb

- **ORIGINS** Western Africa, with distribution extending from the western part of the Ivory Coast to Guinea.
- **SIZE** 3½ in (9 cm).
- **DIET** Prepared foods and live foods.
- **WATER** Temperature 72–77°F (22–25°C); soft (50–100 mg/l) and acidic (pH 6.0).
- **TEMPERAMENT** Peaceful and social.

This barb may have as many as five black spots on each side of its bronze body. Ensure that the tank is well planted, but allow space at the front for swimming. The male Black-Spotted Gold Barb, shown here, is slimmer and more brightly colored than the female. Like other African barbs, this species is less commonly available than many of its Asiatic relatives.

Barbus roloffi

Roloff's Barb

- **ORIGINS** Southeast Asia, where its distribution centers on Thailand.
- **SIZE** 2 in (5 cm).
- **DIET** Prepared foods and live foods
- **WATER** Temperature 72–79°F (22–26°C); soft (50–100 mg/l) and acidic to neutral (pH 6.5–7.0).
- **TEMPERAMENT** Placid and social.

This small barb has a deeply forked caudal fin and a streamlined, silvery body, with a prominent black marking on its dorsal fin. Roloff's Barb is an agile, fast-swimming species that should be kept in groups. Provide well-planted areas in the tank to give the fish a sense of security. These barbs eat a variety of foods, including flake, and benefit from the regular addition of small live foods to their diet.

Black blotch on large dorsal fin

CHARACOIDS

It is surprisingly difficult to describe exactly what defines a characoid, in spite of the fact that many of them are popular aquarium fish. While they usually have teeth, they are not the only group of fish with this characteristic. Most characoids have a small, remote fin located on the back, just in front of the caudal fin. This fin, which lacks any rays, is called an adipose fin, but then again, not every member of the group displays this feature. Although the majority of characoid species are of New World origin, the group is also well represented in Africa.

Body shape is no guide as to what constitutes a member of this group of fish. While pencilfish, such as these **Dwarf Pencilfish** (*Nannostomus marginatus*), have thin, elongated bodies, other characoids, including hatchetfish, have very deep profiles.

Nannostomus beckfordi

Golden Pencilfish

- **ORIGINS** Northern South America, in the central Amazon region, the Rio Negro, and Guyana.
- **SIZE** 3 in (7.5 cm).
- **DIET** Prepared diets, plant matter, and live foods.
- **WATER** Temperature 75–82°F (24–28°C); hard (150–200 mg/l) and acidic to neutral (pH 6.0–7.0).
- **TEMPERAMENT** Social, peaceful, and timid.

Sky-blue tips to the clear ventral fins are a feature of these fish, which also have a broad dark stripe along the body, with a red line above, and perhaps a hint of gold in between. *Nannostomus* pencilfish have small mouths, and they may choke or simply go hungry if their food is too large. Flake and small live foods such as daphnia are ideal. Sex can be determined by the shape of the anal fin, which is more rounded in males.

Chilodus punctatus

Spotted Headstander

- **ORIGINS** Found in northeastern South America, including parts of Guyana.
- **SIZE** 4³⁄₄ in (12 cm).
- **DIET** Live foods preferred.
- **WATER** Temperature 75–82°F (24–28°C); soft (50–100 mg/l) and acidic to neutral (pH 6.5–7.0).
- **TEMPERAMENT** Social, peaceful, and often retiring.

The Spotted Headstander displays rows of dark, similar-sized spots, with a more definite line running through the eye to the tail. Some individuals have a more golden background color than others. The spotted patterning is replaced by black blotches behind each eye when the fish come into spawning condition.

Spotted pattern on the body

Like other headstanders, these fish often hang in the water at an angle of approximately 45°, which explains the common name. Filtration through peat is recommended *(see p.46)*. Spotted headstanders can be bred successfully in the home aquarium.

Abramites hypselonotus

High-Backed Headstander

- **ORIGINS** South America, occurring in the Amazon, Orinoco, and Paraguay river basins.
- **SIZE** 5 in (12.5 cm).
- **DIET** Vegetable matter essential.
- **WATER** Temperature 75–82°F (24–28°C); hard (100–150 mg/l) and acidic to neutral (pH 6.0–7.0).
- **TEMPERAMENT** Intolerant of its own kind.

The body of this fish is marked by broad, blackish, vertical bands. The silvery background becomes more yellowish on the upperparts. The young do not display the high back, which is limited to the adults. Decorate the aquarium with rockwork and bogwood. Choose plants carefully because they are likely to be eaten, or use plastic plants instead. Cover the tank, because these headstanders will jump readily. Their intolerance makes pairing difficult.

Banded patterning

Silver Hatchetfish

Gasteropelecus sternicla

- ⊕ **ORIGINS** South America, in Surinam and Guyana, plus southern tributaries of the Amazon.
- ⟳ **SIZE** 2¹/₂ in (6.5 cm).
- ⊳ **DIET** Live foods preferred, plus flake.
- ≋ **WATER** Temperature 73–81°F (23–27°C); hard 100–150 mg/l) and acidic to neutral (pH 6.0–7.0).
- ⊕ **TEMPERAMENT** Peaceful and social, but nervous.

Transparent fins

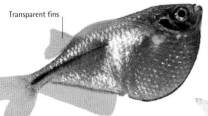

The Silver Hatchetfish has a distinguishing black stripe on its rear half that extends to the caudal peduncle. Males look slimmer than females when viewed from above These fish prey on invertebrates that congregate near the surface, such as mosquito larvae, and feed eagerly on wingless fruit flies or baby crickets dropped onto the water. Make sure the water surface is not blocked by floating plants.

Marbled Hatchetfish

Carnegiella strigata

- ⊕ **ORIGINS** Northern South America, occurring in Guyana and the middle Amazon.
- ⟳ **SIZE** 2¹/₂ in (6 cm).
- ⊳ **DIET** Small live foods and flake.
- ≋ **WATER** Temperature 75–82°F (24–28°C); soft (50–100 mg/l) and acidic (pH 6.0).
- ⊕ **TEMPERAMENT** Social and inoffensive.

This hatchetfish has silver-and-black marbling and a yellowish line from the eye to the caudal peduncle. It is easy to breed if regularly given live foods. Use blackwater extract to help create suitable water conditions. The eggs, laid among the roots of floating plants, fall to the base of the tank. Give the fry infusoria or suitable fry food once they are free-swimming at about five days old and brine shrimp after a week or more.

Dwarf Hatchetfish

Carnegiella myersi

- ⊕ **ORIGINS** South America, occurring in the Rio Ucayali, Peru, and also in Bolivia.
- ⟳ **SIZE** 1 in (2.5 cm).
- ⊳ **DIET** Small live foods and flake.
- ≋ **WATER** Temperature 73–79°F (23–26°C); soft (50–100 mg/l) and acidic (pH 6.0).
- ⊕ **TEMPERAMENT** Social and inoffensive.

The internal organs of this hatchetfish are visible through its semitransparent body. There is a black stripe running from the eye to the tail. The Dwarf is the smallest of all the hatchetfish, but it has a lifestyle similar to its larger relatives, living and feeding near the surface. It is relatively trustworthy with the fry of other fish and can sometimes even be housed in rearing tanks.

FLYING HATCHETS

The aerodynamic shape of hatchetfish, with the deep, boatlike keel, means that these fish are sufficiently streamlined to be able to leap out of the water without difficulty, as shown in this picture. Their flat topline reflects the fact that hatchetfish normally lurk just below the surface, grabbing surface-swimming invertebrates and others that touch down momentarily on the water. The upturned mouth is another adaptation to surface feeding. When hatchetfish are viewed from above, their flat body shape makes them difficult to spot, even near the surface, because so little of the body is visible. They have sharp eyesight to help them catch prey and avoid predators. On occasion, usually to escape would-be predators approaching them in the water, these fish will take to the air. The flapping movements of their pectoral fins are powerful enough to keep them airborne, enabling them to cover distances of up to 4 ft (1.2 m) before reentering the water. While in the air, they use their caudal fin to provide them with some directional guidance. It is essential that an aquarium housing hatchetfish is kept covered, even when it is being serviced, because otherwise they are likely to leap out into the room.

Copella nigrofasciata

Black-Banded Pyrrhulina

🌐 **ORIGINS** Eastern South America, found in Brazil in the vicinity of Rio de Janeiro.

💧 **SIZE** 2½ in (6 cm).

🍽 **DIET** Flake and small live foods.

💧 **WATER** Temperature 70–77°F (21–25°C); soft (50–100 mg/l) and acidic (pH 6.0).

😊 **TEMPERAMENT** Peaceful.

Subdued lighting emphasizes the color of these pyrrhulinas, in particular the pale blue stripes on the sides of the body. A narrow black line runs through the eye. Sexing is straightforward, since the fins are more pointed and brightly colored in the male, shown above. A pair will spawn on a leaf, which is first carefully cleaned by the male. The eggs hatch after about a day, and the fry are free-swimming by the time they are 5 days old.

Corynopoma riisei

Swordtail Characin

🌐 **ORIGINS** Found in northern South America, where it occurs in Colombia's Rio Meta.

💧 **SIZE** 2½ in (6.5 cm).

🍽 **DIET** Prepared foods and small live foods.

💧 **WATER** Temperature 72–82°F (22–28°C); soft (50–100 mg/l) and neutral (pH 7.0).

😊 **TEMPERAMENT** Peaceful and social.

Like the female, pictured below, the male is silvery in color but has a swordlike extension on the lower lobe of the caudal fin and a filament on each gill cover, which is probably used to steer the female into position for mating. Fertilization is internal, and the eggs are laid in the male's absence. Hatching occurs up to 36 hours later, and the young Swordtails can be reared on fry food, followed by brine shrimp.

Copella arnoldi

Splashing Tetra

🌐 **ORIGINS** South America, occurring in Guyana, in the lower part of the Amazon basin.

💧 **SIZE** 3¼ in (8 cm).

🍽 **DIET** Prepared foods and small live foods.

💧 **WATER** Temperature 77–84°F (25–29°C); soft (50–100 mg/l) and acidic (pH 6.0).

😊 **TEMPERAMENT** Peaceful and social.

Black spot on dorsal fin

Pale underparts

Male Splashing Tetras are larger than females, with red markings on the caudal fin. Because of the athletic behavior of these fish, keep the aquarium covered and the water level reasonably low. A spawning pair will jump up and deposit as many as 15 small batches of eggs on a broad leaf protruding above the water line, or even on the glass of the tank. The male keeps the eggs moist by splashing them with its tail every 30 seconds or so, continuing until they hatch about three days later.

Copeina guttata

Red-Spotted Copeina

🌐 **ORIGINS** Northern South America, occurring widely throughout the Amazon basin.

💧 **SIZE** 6 in (15 cm).

🍽 **DIET** Prepared foods and small live foods.

💧 **WATER** Temperature 73–84°F (23–29°C); soft (50–100 mg/l) and acidic (pH 6.0).

😊 **TEMPERAMENT** May be aggressive when spawning.

The pattern of red spots is most apparent in the male, shown here; the male also displays an enlarged upper lobe on the caudal fin and has a

stronger body color. These fish can be disruptive in the tank, especially at the start of the spawning period, when the male may chase its intended partner relentlessly. The eggs are laid in batches in a scrape in the substrate. They are guarded by the male, who also oxygenates them by fanning them with his fins. The young, which emerge after about two days, need a fry food once they are free-swimming. For breeding, ensure that there is a soft, sandy substrate that the male can excavate. Include a clear digging area so that plants are not uprooted by the male's actions. Raising the water temperature is a proven spawning trigger.

Thayeria boehlkei

Hockey Sticks

🌐 **ORIGINS** Northern South America, in the Peruvian part of the Amazon basin, and in Brazil's Rio Araguaia.

💧 **SIZE** 3 in (7.5 cm).

🍽 **DIET** Prepared foods and small live foods.

💧 **WATER** Temperature 73–84°F (23–29°C); soft (50–100 mg/l) and acidic (pH 6.0).

😊 **TEMPERAMENT** Peaceful and social.

Deeply forked caudal fin

Extensive black stripe

Hockey Sticks have a black stripe that extends from behind the gills back to the caudal peduncle and then diverts down across the lower lobe of the caudal fin. A faint golden line running beneath this stripe is also apparent. In the Penguin Fish (*T. obliqua*), the black stripe is shorter and reaches only as far as the base of the dorsal fin. It is essential to make a partial water change every two weeks in order to prevent any harmful nitrate buildup.

Brycinus longipinnis

Long-Finned African Tetra

- **ORIGINS** Western Africa, present in the Niger delta, Ghana, Sierra Leone, and Togo.
- **SIZE** 5 in (13 cm).
- **DIET** Prepared foods and small live foods.
- **WATER** Temperature 73–79°F (23–26°C); soft (50–100 mg/l) and acidic (pH 6.5).
- **TEMPERAMENT** Peaceful and social.

A prominent black stripe along the lower part of the caudal peduncle helps to identify this characoid. The remainder of its body has a silvery-green hue, which is most apparent in low-level lighting. This nervous yet active fish should be housed in small groups, with spacious surroundings to provide plenty of swimming space and floating plants on the surface to provide some cover. Only a mature male—the upper fish of the pair pictured here—has extended dorsal fin rays. Good water quality serves to improve the males' coloration. Avoid nitrate accumulation by carrying out partial water changes every two weeks or so. Long-Finned African Tetras require a separate spawning tank, with vegetation among which they can scatter their eggs (an artificial spawning mop may be used instead of plants). Hatching can take as long as six days.

HANGING AROUND

Penguin fish have such a strong shoaling instinct that they even rest together as a group, hanging at an angle in the water with all the fish facing the same direction, as illustrated by the group of Hockey Sticks below. In this position, the mass of stripes makes the shoal resemble underwater vegetation when viewed from a distance, camouflaging the fish in reedy stretches of water where many shoots grow up toward the surface. The name "penguin fish" alludes to the way in which penguins move on land, with their bodies leaning slightly forward.

Chalceus erythrurus

Watermelon Fish

- **ORIGINS** Northern South America, present in the Amazon basin and extending north to the Guianas.
- **SIZE** 10 in (25 cm).
- **DIET** Prepared foods and live foods.
- **WATER** Temperature 73–82°F (23–28°C); soft (50–100 mg/l) and acidic (pH 6.0).
- **TEMPERAMENT** An active predator.

The slim, silvery body and reddish-pink caudal fin are characteristic features of the Watermelon Fish. In some populations, the other fins are yellowish. This is one of the larger and more predatory characoid species, so any tankmates should be chosen carefully. It may be necessary to wean this fish off live foods; thawed items are likely to prove more palatable than freeze-dried alternatives. Nothing is known about the breeding habits of Watermelon Fish.

Anostomus anostomus

Striped Headstander

- **ORIGINS** Northern South America, from Colombia down the Amazon to Manaus; also found in the Orinoco.
- **SIZE** 7 in (18 cm).
- **DIET** Prepared foods, live foods, and vegetable matter.
- **WATER** Temperature 73–82°F (23–28°C); soft (50–100 mg/l) and slightly acidic (pH 6.5).
- **TEMPERAMENT** Can be aggressive.

The attractive patterning of these headstanders (so-called because they often rest with the head pointing downward) consists of alternating stripes of black and yellow. Striped Headstanders need powerful filtration and good lighting to encourage algal growth, which makes up part of their natural diet. Retreats in the form of rocks and bogwood are also important. Aggression levels will be lower if these fish are kept in shoals.

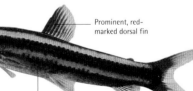

Prominent, red-marked dorsal fin

Sexes look identical

DREADED HUNTERS

Few fish inspire greater fear than piranhas. While many predatory fish live solitary lives, piranhas are like packs of wolves, living and hunting together in groups and occasionally taking prey as large as horses or even people. Indeed, keeping these notorious fish is outlawed in some parts of the world because of fears over their likely impact if they were to become established in the wild outside their normal range. Piranhas have a sharp sense of smell, which enables them to detect blood in the water. Repeated splashing, indicating a creature out of its depth, is equally likely to attract the attention of a hungry shoal. They then go into a feeding frenzy, ripping repeatedly into their victim, using their sharp teeth to pull off mouthfuls of flesh. The short upper jaw helps the fish to gain a firm anchorage, while the powerful lower jaw provides strength, enabling the small, triangular, interlocking teeth to shear through tissue with ease. Being instinctive bullies, piranhas will turn on weak or injured members of their own kind, especially if just two or three are housed in a tank. A larger group, from five upward, will reduce the likelihood of this, since there will be less of an established hierarchy within the shoal.

Serrasalmus nattereri

Red-Bellied Piranha

- **ORIGINS** South America, in both the Orinoco and Amazon River basins, from Guyana south to La Plata.
- **SIZE** 12 in (30 cm).
- **DIET** Meat-based foods and live foods.
- **WATER** Temperature 77–81°F (25–27°C); soft (50–100 mg/l) and acidic (pH 6.5).
- **TEMPERAMENT** Aggressive and territorial.

Red-Bellied Piranhas require a spacious setup, decorated with some bogwood to provide retreats, as well as subdued lighting and a good filter. They are not active fish by nature, except when feeding. Adults have a silvery, speckled appearance with red on the underparts. Juveniles also have black spots on their bodies. Piranhas can inflict a painful bite. Large individuals can bite through cabling, so an undergravel heating unit is advisable.

Ichthyborus ornatus

Ornate Fin-Nipper

- **ORIGINS** Occurs in the tropical area of western Africa, where it is found in the Zaire basin.
- **SIZE** 8 in (20 cm).
- **DIET** Larger aquatic invertebrates and fish-based foods.
- **WATER** Temperature 73–82°F (23–28°C); hard (100–150 mg/l) and neutral (pH 7.0).
- **TEMPERAMENT** Aggressive yet shy.

The silvery underparts and the brown on the upper body help to set this highly predatory species apart from related fin-nippers, including the African Pike Characin (*P. loricatus*). The term "fin-nipper" comes from their habit of biting pieces from the fins of fish that are too large to swallow whole. Keep them separate, therefore, in a relatively dark yet well-planted tank. It must offer adequate retreats, since all fin-nippers are nervous.

Exodon paradoxus

Bucktooth Tetra

- **ORIGINS** South America, occurring in various localities in the Amazon basin from Guyana to Brazil.
- **SIZE** 6 in (15 cm).
- **DIET** Thawed, freeze-dried, and fresh live foods.
- **WATER** Temperature 73–82°F (23–28°C); hard (up to 200 mg/l) and neutral (pH 7.0).
- **TEMPERAMENT** Not suitable for a community aquarium.

These large tetras have a distinctive black spot just in front of and below the dorsal fin, with another on the caudal peduncle. Beautiful they may be, but they are brutal, too, simply slicing chunks out of prey that are above swallowable size. Keep Bucktooth Tetras in shoals of a dozen or so to lessen their aggressive tendencies; even the young will cannibalize one another. Adding peat to the filter (*see p.46*) will create good water conditions.

Ctenolucius hujeta

Gar Characin

- 🌐 **ORIGINS** South America, ranging from Panama southward to Colombia and Venezuela.
- 🌀 **SIZE** 12 in (30 cm).
- 🍃 **DIET** Invertebrates and fish-based diets.
- 🌊 **WATER** Temperature 72–77°F (22–25°C); hard (up to 200 mg/l) and neutral (pH 7.0).
- 😐 **TEMPERAMENT** Aggressive but nervous.

The upper jaw of these fish is slightly shorter than the lower jaw. They are predominantly silvery in color, but there are slight differences between the three recognized subspecies. A tank for Gar Characins needs a clear area for swimming, with a good water current created by the filter. A dense surface covering of aquatic plants is also needed. When mating, the male uses his anal fin to hold the female's abdomen. Anywhere from 1,000 to 3,000 eggs may be laid during a single spawning, and the fry hatch within a day.

Acestrorhynchus altus

Torpedo Pike Characin

- 🌐 **ORIGINS** South America, occurring in both the Amazon basin and the more southerly Paraguay basin.
- 🌀 **SIZE** 14 in (35 cm).
- 🍃 **DIET** Larger invertebrates and prepared foods.
- 🌊 **WATER** Temperature 72–79°F (22–26°C); hard (up to 200 mg/l) and neutral (pH 7.0).
- 😐 **TEMPERAMENT** Predatory, best kept alone.

This is a more colorful form of the Amazon Cachorro (*A. falcirostris*), displaying attractive reddish coloration on the caudal and anal fins. There is also a large black spot at the base of the caudal fin and others farther up the body near the gills. When designing an aquarium for these fish, be sure to incorporate sufficient retreats so that they will not injure themselves by swimming into the glass. Floating plants will make them feel more secure.

Chalceus macrolepidotus

Pink-Tailed Characin

- 🌐 **ORIGINS** Northern South America, particularly the Guianas and the Amazon basin.
- 🌀 **SIZE** 10 in (25 cm).
- 🍃 **DIET** Thawed, fresh, and freeze-dried live foods.
- 🌊 **WATER** Temperature 73–82°F (23–28°C); hard (100–150 mg/l) and neutral (pH 7.0).
- 😐 **TEMPERAMENT** Aggressive and predatory.

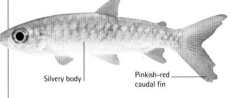

Silvery body   Pinkish-red caudal fin

Pink-Tailed Characins have a slim body and a relatively large and deeply forked tail fin. These fish need spacious surroundings but are quite adaptable as far as water conditions are concerned. The young may be persuaded to eat flake food and subsequently encouraged to take tablets. The species must never be mixed with smaller companions, which are likely to be eaten. These characoids can be long-lived, with a life expectancy in aquariums of nearly 20 years.

Hoplias malabaricus

Wolf Fish

- 🌐 **ORIGINS** Occurs throughout much of Central America and northern parts of South America.
- 🌀 **SIZE** 20 in (50 cm).
- 🍃 **DIET** Live foods and fish-based foods.
- 🌊 **WATER** Temperature 72–81°F (22–27°C); hard (100–150 mg/l) and neutral (pH 7.0).
- 😐 **TEMPERAMENT** Aggressive and predatory.

Attractive patterning and a relatively tall, banded dorsal fin are characteristic features of Wolf Fish. Males are slimmer, while females have a slight upward curvature to the lower surface of their body when viewed in profile. One difficulty with keeping Wolf Fish and similar predatory characoids is persuading them to take an inert diet rather than smaller fish, which are their natural prey. This is more easily achieved by starting out with juveniles, which are more adaptable in their feeding habits.

Hepsetus odoe

Kafue Pike Characoid

- 🌐 **ORIGINS** Widely distributed throughout much of tropical Africa, although absent from the Nile basin.
- 🌀 **SIZE** 28 in (70 cm).
- 🍃 **DIET** Thawed and fresh meat-based foods.
- 🌊 **WATER** Temperature 79–82°F (26–28°C); hard (100–150 mg/l) and neutral (pH 7.0).
- 😐 **TEMPERAMENT** A powerful predator.

Although young Kafue Pike Characoids are occasionally available, bear in mind when thinking of buying one that it will grow fast and eventually require a very large aquarium. The predatory habits of these fish mean that they are best housed alone. The tank must afford plenty of cover, since they are instinctively nervous and like to hide away in vegetation or caves, which can be created with bogwood as well as rockwork. Take particular care with a newly acquired individual that has not yet settled in the aquarium, because it is likely to swim around wildly and could injure itself while you are attending to its needs. The eggs, which the female guards, hatch in a bubble nest, suggesting an affinity with anabantoids (see pp. 104–115). Successful breeding in aquariums is highly unlikely.

Metynnis argenteus

Silver Dollar

- **ORIGINS** South America, occurring in the Amazon region east of the Rio Negro and into Guyana.
- **SIZE** 5 in (14 cm).
- **DIET** Vegetable flakes, pellets, and fresh foods.
- **WATER** Temperature 73–81°F (23–27°C); soft (50–100 mg/l) and acidic to neutral (pH 5.0–7.0).
- **TEMPERAMENT** Peaceful, shoaling fish.

Shaped rather like a coin, with a silvery body color, the Silver Dollar makes an impressive tank occupant, especially when a shoal of these fish are housed together. They can be sexed quite easily, since the anal fin is longer in males and has a reddish tinge at the front. These fish sometimes display small dark spots on the sides of their bodies, which may look like a sign of disease but are actually quite normal. An aquarium for Silver Dollars needs to be relatively large, with open areas for swimming and caves for the fish to use as retreats. Like other large vegetarian characoids, Silver Dollars are likely to damage the plants in their tank. It is advisable to include plastic plants to augment living specimens, which will be at risk. Living Java moss (*Vesicularia dubyana*) is useful, however, because it is attractive yet will not be eaten by most fish. It will also provide fish with potential spawning sites.

Mylossoma duriventre

Hard-Bellied Silver Dollar

- **ORIGINS** South America, with a distribution from the southern Amazon basin down as far as Argentina.
- **SIZE** 9 in (23 cm).
- **DIET** Vegetable-based foods of all types.
- **WATER** Temperature 72–81°F (22–27°C); soft (50–100 mg/l) and acidic to neutral (pH 5.0–7.0).
- **TEMPERAMENT** Peaceful and social.

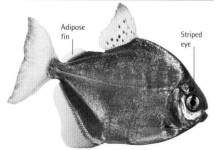

Adipose fin

Striped eye

The silvery body color may vary in its depth, but the red edging to the caudal fin is a consistent feature of this characoid. The description "hard-bellied" refers to the narrow, serrated area on the underside of the body between the pelvic fins and the anal fin. The care needs of this fish match those of the Silver Dollar itself.

Hemiodus gracilis

Slender Hemiodus

- **ORIGINS** South America, occurring in the Amazon region and into Guyana.
- **SIZE** 6 in (15 cm).
- **DIET** Vegetable foods plus small invertebrates.
- **WATER** Temperature 73–81°F (23–27°C); soft (50–100 mg/l) and acidic to neutral (pH 6.5–7.0).
- **TEMPERAMENT** Peaceful and social.

The Slender Hemiodus has a thin, silvery body with a black stripe in the midline behind the dorsal fin. This stripe curves down the lower lobe of the caudal fin, where the area beneath it is red. This characoid needs well-oxygenated water, and the addition of peat to the filtration system is recommended, especially at first. Silver Hemiodus will eat algae as well as plants in their aquarium.

Colossoma macropomum

Black-Fin Pacu

- **ORIGINS** South America, widely distributed throughout most of the Amazon region.
- **SIZE** 16 in (40 cm).
- **DIET** Vegetarian diets and fresh foods.
- **WATER** Temperature 73–81°F (23–27°C); soft (50–100 mg/l) and acidic (pH 6.5).
- **TEMPERAMENT** Peaceful, even when larger.

The dark edging to the fins, especially the caudal fin, helps to identify the Black-Fin Pacu. The body is a metallic silver, sometimes with a pattern of darker spots, and there is a variable orange-red area around the throat. These sizable fish need a large aquarium, which should be provided from the start because the young grow rapidly. Retreats can be provided by rockwork or bogwood.

Citharinus citharus

Lined Citharoid

🔹 ORIGINS West Africa, where it ranges from Senegal through to the Nile basin.
🔹 SIZE 20 in (50 cm).
🔹 DIET Vegetable matter and live foods.
🔹 WATER Temperature 73–81°F (23–27°C); soft to hard (50–150 mg/l) and neutral to alkaline (pH 7.0–8.0).
🔹 TEMPERAMENT Peaceful shoaler.

This fine-scaled characoid is silvery in color, with a steeply curved topline. The dorsal fin is located just below the highest point on the back. Young Lined Citharoids may sometimes display horizontal black stripes on the body. The Lined Citharoid is a large species and is usually kept in groups. Although it typically grows to a much smaller size in aquariums than it does in the wild, it still needs a sizable tank.

Semaprochilodus taeniurus

Silver Prochilodus

🔹 ORIGINS South America, where it is found in Colombia and in the Rio Purus in western Brazil.
🔹 SIZE 12 in (30 cm).
🔹 DIET Mainly vegetable matter.
🔹 WATER Temperature 73–81°F (23–27°C); soft to hard (50–200 mg/l) and acidic to alkaline (pH 6.0–7.8).
🔹 TEMPERAMENT Peaceful shoaler.

The Silver Prochilodus has random spots on the rear of its body, and striped markings on its tail; the pelvic fins are reddish. The stomach consists of two parts; one of these parts contains little more than mud scooped from the substrate and is thought to help in digestion. These fish are able jumpers. They put their ability to good use in the wild, leaping over obstacles as they migrate upstream to spawn.

Distichodus sexfasciatus

Six-Striped Distichodus

🔹 ORIGINS Africa, occurring in the Zaire basin and also in Angola.
🔹 SIZE 10 in (25 cm).
🔹 DIET Vegetable-based foods.
🔹 WATER Temperature 73–79°F (23–26°C); soft to hard (50–150 mg/l) and acidic to neutral (pH 6.5–7.0).
🔹 TEMPERAMENT Nonaggressive.

Orange background color

Tail turns grayer with age

Juvenile

Unfortunately, only young Six-Striped Distichodus have the attractive striped patterning and red fins, with the adults being much plainer. Initially, the young fish may be confused with the Long-Nosed Distichodus (*D. lusosso*), although their snouts are rounded rather than pointed. They tend to be destructive toward plants growing in the aquarium, but Java Fern (*Microsorum pteropus*) is usually ignored.

Distichodus affinis

Silver Distichodus

🔹 ORIGINS Africa, occurring in the lower part of the Zaire basin.
🔹 SIZE 5 in (13 cm).
🔹 DIET Vegetable-based foods.
🔹 WATER Temperature 73–81°F (23–27°C); soft to hard (50–150 mg/l) and acidic to neutral (pH 6.5–7.0).
🔹 TEMPERAMENT Nonaggressive.

The relatively small size of the Silver Distichodus means that this fish is suitable for a community aquarium. There are three very similar *Distichodus* species, all of which have red fins and a black spot at the front of the dorsal fin. This particular *Distichodus* can be distinguished by the fact that its dorsal fin is shorter than its anal fin. All *Distichodus* have similar care requirements. Their breeding habits have yet to be documented in any detail.

A VEGETARIAN DIET

Browsing on plants calls for strong jaws that can cut through leaves and young stems without difficulty. A relatively wide gape is also helpful when it comes to swallowing plant matter. In the case of the Silver Dollar (*Metynnis argenteus*) seen here, the bottom jaw acts as the main cutting blade, slicing upward against the more rigid top jaw. Internally, the digestive system has to be adapted to processing large quantities of food with low nutritional value. In order to obtain enough nutrients, vegetarian species must eat more frequently than their carnivorous counterparts.

Hyphessobrycon megalopterus

Black Phantom Tetra

- **ORIGINS** South America, where it occurs in areas of central Brazil and Bolivia.
- **SIZE** 2 in (5 cm).
- **DIET** Prepared foods and small live foods.
- **WATER** Temperature 73–82°F (23–28°C); soft (50–100 mg/l) and acidic (pH 6.5).
- **TEMPERAMENT** Social and nonaggressive.

A black blotch behind the gills is a characteristic of these tetras, as is the relatively transparent body—hence the "phantom" of their common name. They look elegant in a shoal, thanks in part to the tall dorsal fin, which is more prominent in males. Include floating plants in their tank to mimic their natural habitat. Commercial breeding has made Black Phantom Tetras more tolerant in terms of their water chemistry needs, but soft, acidic conditions will give the best chance of spawning success. The eggs are scattered at random, after which the adults must be removed from the spawning tank so that they do not devour them. It takes about five days before the young are free-swimming, at which stage they should be offered a suitable fry food.

Hyphessobrycon sweglesi

Swegles's Tetra

- **ORIGINS** Northern South America, occurring in the upper Amazon and Orinoco basins.
- **SIZE** 1½ in (4 cm).
- **DIET** Prepared foods and small live foods.
- **WATER** Temperature 68–73°F (20–23°C); soft (50–100 mg/l) and acidic (pH 6.5).
- **TEMPERAMENT** Peaceful and social.

Swegles's Tetras are very sensitive to water chemistry, and their impressive red coloration will be seen clearly only under ideal conditions. They need frequent water changes and a relatively low water temperature if they are to thrive and spawn in the aquarium. A dark substrate in the spawning tank is necessary to protect their eggs from light. The female can be distinguished from the male by the white tip on the dorsal fin.

Hemigrammus pulcher

Garnet Tetra

- **ORIGINS** Northern South America, found in the Peruvian part of the Amazon River.
- **SIZE** 2 in (5 cm).
- **DIET** Prepared foods and small live foods.
- **WATER** Temperature 73–81°F (23–27°C); soft (50–100 mg/l) and acidic (pH 6.0).
- **TEMPERAMENT** Peaceful and social.

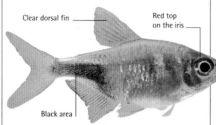

Clear dorsal fin — Red top on the iris

Black area

The pale-lemon background color of the Garnet Tetra contrasts with the black area on the flanks, which extends to the caudal peduncle. The body appears iridescent under subdued lighting. This fish can be sexed by the swim bladder, which is visible through the semitransparent body: it is pointed in males and rounded in females. Compatibility can be a problem when attempting to breed these fish.

Hemigrammus ocellifer

Head-and-Taillight Tetra

- **ORIGINS** Northern South America, occurring in the Amazon region, eastward as far as French Guiana.
- **SIZE** 2 in (5 cm).
- **DIET** Prepared foods and small live foods.
- **WATER** Temperature 75–82°F (24–28°C); soft (50–100 mg/l) and acidic (pH 6.0).
- **TEMPERAMENT** Peaceful and social.

This tetra's rather plain body is enhanced by a red, beaconlike area on the upper eye. There is also a yellow area with a golden spot and a black blotch around the caudal peduncle. The subspecies shown here (*H. O. falsus*) has a black line on the tail, which is missing in other forms. A female in spawning condition lays around 300 eggs, among fine-leaved aquatic vegetation, two days after a pair is moved to the breeding tank.

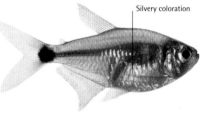

Silvery coloration

Hyphessobrycon peruvianus

Peruvian Tetra

- ⊕ **ORIGINS** Northern South America, in the Peruvian part of the Amazon river, in the vicinity of Iquitos.
- ◯ **SIZE** 2 in (5 cm).
- ◔ **DIET** Prepared foods and small live foods.
- ≋ **WATER** Temperature 73–82°F (23–28°C); soft (50–100 mg/l) and acidic (pH 6.0).
- ⊜ **TEMPERAMENT** Peaceful and social.

The patterning of these tetras is variable; the black stripe that runs along the lower body is more extensive in some individuals than others. The area adjacent to this stripe is usually pale blue. Differences in coloration do not help when sexing these fish, although males tend to be more slender than females. A typical Amazonian tank is ideal for these tetras; water should be filtered through aquarium peat (see p.46).

Hemigrammus erthyrozonus

Glowlight Tetra

- ⊕ **ORIGINS** South America, where it occurs in the Essequibo River basin region in Guyana.
- ◯ **SIZE** 1¹/₂ in (4 cm).
- ◔ **DIET** Prepared foods and small live foods.
- ≋ **WATER** Temperature 73–82°F (23–28°C); soft (50–100 mg/l) and acidic (pH 6.0).
- ⊜ **TEMPERAMENT** Peaceful and social.

Red marking on dorsal fin

Reddish-gold line

The reddish-gold stripe of the Glowlight Tetra runs the entire length of its body. Males are more brightly colored than females, with a slimmer profile. During spawning, the male grips the female with its fins. If the plants in the spawning tank are too dense to allow the mating pair to swim through them with ease, many of the resulting eggs will be infertile. Up to 200 eggs are laid in total, often sinking to the substrate.

Hyphessobrycon anisitsi

Buenos Aires Tetra

- ⊕ **ORIGINS** South America, in the Plate River basin of Argentina, extending to Brazil and Paraguay.
- ◯ **SIZE** 4 in (10 cm).
- ◔ **DIET** Prepared foods and small live foods.
- ≋ **WATER** Temperature 64–82°F (18–28°C); soft (50–100 mg/l) and acidic (pH 6.0).
- ⊜ **TEMPERAMENT** Can be aggressive.

The Buenos Aires Tetra has a black stripe on its caudal fin that joins a bluish line along the side of its body. Males are more brightly colored at spawning time, while females often become aggressive toward their intended mates. Because it naturally occurs farther south than most tetras, this species can be kept at a lower water temperature in the aquarium. It will eat the leaves of aquatic plants, so avoid including delicate varieties in its tank.

BLACKWATER DWELLERS

Many wild tetras live in water that appears blackish because it contains chemicals called tannins, which result from the breakdown of leaves and other organic matter. Tannins acidify the water, and if this water chemistry is not replicated in the aquarium, the tetras may not show their vibrant coloration, and breeding will

be unlikely. Adding manufactured "blackwater extract" helps to recreate the conditions under which tetras live in the wild. Peat has a similar effect on the water chemistry, as does bogwood (wood extracted from peaty surroundings). However, blackwater conditions reduce light penetration and may thus restrict plant growth.

Hemigrammus rhodostomus

Banded Rummy-Nose

- ⊕ **ORIGINS** South America, in the lower reaches of the Brazilian Amazon, near Belem.
- ◯ **SIZE** 2 in (5 cm).
- ◔ **DIET** Prepared foods and small live foods.
- ≋ **WATER** Temperature 73–82°F (23–28°C); soft (50–100 mg/l) and acidic (pH 6.0).
- ⊜ **TEMPERAMENT** Peaceful and social.

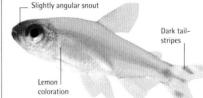

Slightly angular snout

Dark tail-stripes

Lemon coloration

This species can be differentiated from the Brilliant Rummy-Nose Tetra (*H. bleheri*) by the patch of red on its head, which is smaller than in its close relative and does not extend to the body. Females tend to be slightly larger and more rounded than males. Aquarium breeding is possible. As with other tetras, mosquito larvae make an excellent conditioning food before the fish are transferred to the breeding tank.

Hyphessobrycon erythrostigma

Bleeding Heart Tetra

- **ORIGINS** Northern parts of South America, especially in the upper Amazon region.
- **SIZE** 4 in (10 cm).
- **DIET** Prepared foods and small live foods.
- **WATER** Temperature 75–82°F (23–28°C); soft (50–100 mg/l) and acidic (pH 5.6–6.9).
- **TEMPERAMENT** Placid, but does not breed readily.

The Bleeding Heart, so called because of the red patch on the side of its body, is one of the larger tetras. This species is not aggressive and can be kept in groups in a community aquarium, along with other placid species. Bleeding Hearts can be susceptible to fungus when introduced to the tank. To combat this, keep them at the upper end of the temperature range at first and offer food enriched with vitamin C. Over the longer term, the addition of peat to the filter will help to maintain suitable water conditions for these attractive tetras. Once established, they may live for as long as four years. Unfortunately, females are frequently reluctant to respond to the courtship of males, which are recognizable by their prominent black dorsal fin and longer anal fin.

Hemigrammus ulreyi

Ulreyi Tetra

- **ORIGINS** South America, occurring in the Paraguay River basin, south of the Amazon.
- **SIZE** 2 in (5 cm).
- **DIET** Eats both prepared foods and live foods.
- **WATER** Temperature 75–82°F (24–28°C); soft (50–100 mg/l) and acidic to neutral (pH 6.5–7.0).
- **TEMPERAMENT** Nonaggressive, shoaling fish.

This tetra is often confused with the Flag Tetra (*H. heterorhabdus*), from the lower Amazon basin in Brazil. The key difference is the narrow line of color above the black stripe along the body. This is orange in the Ulreyi Tetra but red in the Flag Tetra. The scales on the caudal fin are a less obvious distinguishing feature; these are missing in the Flag Tetra. This fish is named after Albert B. Ulrey, the biologist who first described it in 1895. Ulrey also discovered the X-Ray Tetra (see p.101).

Hyphessobrycon bentosi

Rosy Tetra

- **ORIGINS** South America, occurring in Guyana and throughout the lower Amazon basin.
- **SIZE** 1¾ in (4.5 cm).
- **DIET** Both prepared foods and small live foods.
- **WATER** Temperature 75–82°F (24–28°C); soft (50–100 mg/l) and acidic to neutral (pH 6.5–7.0).
- **TEMPERAMENT** Social and nonaggressive.

Rosy Tetras can be sexed without difficulty, since males have a longer dorsal fin that curls over at the tip, while females have a red tip on this fin. The anal fin is also longer and more concave in the male. Raising the water temperature, typically to 80°F (27°C), preceded by an increase in the live foods offered, usually stimulates spawning behavior. The fry hatch after about three days.

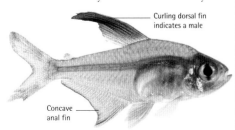

Curling dorsal fin indicates a male

Concave anal fin

Hyphessobrycon pulchripinnis

Lemon Tetra

- **ORIGINS** The Amazon basin in central Brazil, South America, occurring in slow-flowing, narrow streams.
- **SIZE** 2 in (5 cm).
- **DIET** Prepared diets and live foods.
- **WATER** Temperature 73–82°F (23–28°C); soft (50–100 mg/l) and acidic (pH 6.0).
- **TEMPERAMENT** Placid and social.

The beautiful lemon coloration of these tetras shines through only once they are well established and the water conditions become ideal. Until then, they may look rather drab. Subdued lighting, a dark substrate, and the occasional use of a color food may help to improve their appearance.

Hyphessobrycon herbertaxelrodi

Black Neon Tetra

- **ORIGINS** Distribution restricted to the Taquari River in the Mato Grosso region of Brazil.
- **SIZE** 2 in (5 cm).
- **DIET** Eats both prepared foods and small live foods.
- **WATER** Temperature 73–81°F (23–27°C); soft (50–100 mg/l) and acidic (pH 6.5).
- **TEMPERAMENT** Thrives in a shoal.

In spite of its name, this species is not related to the Neon Tetra *(see opposite)*. Females tend to be slightly larger than males, with a more rounded body shape. Breeding will be most likely if you transfer a number of these tetras as a shoal to a spawning tank, but make sure the water is softer than in the main aquarium. Live foods will help to bring the fish into breeding condition. After the eggs have been scattered among fine-leaved plants, the adult fish will need to be removed from the spawning tank.

Raised dorsal fin

TETRA BIOLOGY

Tetras are lively, active fish that instinctively stay close together as a shoal. A torpedolike body shape makes them suitable for the middle and upper layers of the tank. Their streamlining enables them to swim fast, which is important in the wild because their small size leaves them vulnerable to attack. They have good eyesight, however, which makes it difficult for predators to approach a shoal without being detected. Tetras can also hear underwater sounds, thanks to a connection between their swim bladder and inner ear.

Hyphessobrycon bifasciatus

Yellow Tetra

- **ORIGINS** South America, specifically the southeastern region of Brazil, in the vicinity of Rio de Janeiro.
- **SIZE** 2½ in (6.5 cm).
- **DIET** Prepared foods and small live foods.
- **WATER** Temperature 68–77°F (20–25°C); hard (100–150 mg/l) and neutral (pH 7.0).
- **TEMPERAMENT** Social and placid.

Not to be confused with the Lemon Tetra (*H. pulchripinnis*), the Yellow Tetra has two dark bands just behind the gills. These are most apparent in males, which are more brightly colored overall. The males also have a more concave anal fin than the females. Successful spawning is not difficult to achieve if the fish are in good condition. Fry foods intended for egg-laying species can be used for the young tetras once they are free-swimming.

Yellow coloration along upper body

Paracheirodon axelrodi

Cardinal Tetra

- **ORIGINS** Tributaries of the Rio Negro and Orinoco in the northwestern region of South America.
- **SIZE** Up to 2 in (4.5 cm).
- **DIET** Prepared foods, such as flake, and small live foods.
- **WATER** Temperature 73–79°F (23–26°C); soft (0–50 mg/l) and acidic (pH 5.8).
- **TEMPERAMENT** Social; should be kept in shoals.

One of the most beautiful of all freshwater fish, the Cardinal Tetra can have its coloration enhanced by good water conditions. It is not an easy species to breed, but the use of blackwater extract may help to encourage reproductive behavior. Cardinal Tetras need to be transferred to a separate tank for spawning purposes. The eggs hatch in about a day, with the fry becoming free-swimming about five days later.

Paracheirodon innesi

Neon Tetra

- **ORIGINS** Occurs only in Rio Putumayo in eastern Peru, South America.
- **SIZE** 1½ in (4 cm).
- **DIET** Prepared foods and small live foods.
- **WATER** Temperature 68–79°F (20–26°C); soft (50–100 mg/l) and acidic (pH 6.0).
- **TEMPERAMENT** Highly social.

These tetras are so beautiful that when they first became available to aquarists in the 1930s, they sold for the equivalent of more than a month's wages. Today, they are likely to be found in almost every tropical fish store, being widely bred commercially. There are now even variants—including a yellow strain, christened "Mon Cheri," and a long-finned strain—although none rivals the natural form in popularity. Neons can be easily distinguished at a glance from Cardinal Tetras *(see top right)*, because the red stripe on their bodies is restricted to the tail region, rather than extending all along the lower body. Females tend to have plumper bodies than males, sometimes causing their stripes to look more curved. As with most other tetras, they are shoaling fish and will thrive in groups, tending to occupy the midwater region of the aquarium.

Nematobrycon palmeri

Emperor Tetra

- **ORIGINS** Northern South America, especially western parts of Colombia.
- **SIZE** 3 in (7.5 cm).
- **DIET** Prepared foods and live foods.
- **WATER** Temperature 73–81°F (23–27°C); soft (50–100 mg/l) and acidic (pH 6.5).
- **TEMPERAMENT** Placid. Avoid boisterous companions.

The true purple hue of these stunning tetras will be most apparent in an aquarium with relatively subdued lighting and floating plants at the surface to diffuse the light. If the aquarium is too bright, their yellow coloration will be more dominant. Male Emperor Tetras are usually larger and more colorful than females. Provide cover on the floor of the spawning tank; otherwise, the eggs are likely to be eaten as they are laid.

Curved dorsal fin indicates a male

LIFE SPANS IN AQUARIUMS

Most aquarium fish live longer than their wild counterparts, provided they are correctly housed and fed. Modern diets have reduced the risks of nutritional diseases that could reduce their life span, and aquarium fish will not be at risk from predators if their tankmates are carefully chosen. This enables small fish such as tetras to come close to attaining their maximum life span. Members of the *Nematobrycon* genus, such as these Emperor Tetras, are typically the longest-lived tetras, living for up to six years in aquariums.

Phenacogrammus interruptus

Congo Tetra

- **ORIGINS** Occurs in western Africa, in parts of the upper Congo basin.
- **SIZE** 3 in (7.5 cm).
- **DIET** Prepared foods and small live foods.
- **WATER** Temperature 73–82°F (23–28°C); soft (50–100 mg/l) and acidic (pH 6.0).
- **TEMPERAMENT** Highly social but nervous.

The iridescence of Congo Tetras means that their appearance can change noticeably, depending on the lighting conditions. Subdued lighting suits

these fish best, helping to emphasize rather than fade their body colors. Males are larger than females and can have longer fins, which may be ragged on occasions. Do not mix these fish with fin-nipping species, since this can lead to serious problems. It is possible to persuade pairs of Congo Tetras to spawn, but you will have more chance of success if you have a small shoal, since these fish normally reproduce in a group. Adding blackwater extract to the water to raise the tannic acid level will also help. The female may produce as many as 300 eggs, typically quite early in the morning, which will then take about five days to hatch.

Ladigesia roloffi

Jelly Bean Tetra

- **ORIGINS** The Yung River in West Africa, occurring in Sierra Leone, Liberia, Ghana, and Ivory Coast.
- **SIZE** 1½ in (4 cm).
- **DIET** Eats prepared foods and small live foods.
- **WATER** Temperature 72–79°F (22–26°C); soft (50–100 mg/l) and acidic (pH 6.0).
- **TEMPERAMENT** Placid shoaler.

Female has a short, straight anal fin

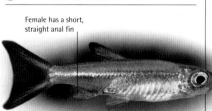

Jelly Bean Tetras are so called because of their size and shape. One of the less commonly encountered tetra species, they are not difficult to keep, as long as the aquarium's water chemistry is very similar to that which they encounter in the wild. Fading of the red from their fins is a sign that conditions are less than ideal. The tank must be covered, because Jelly Bean Tetras can jump well, and it should include floating plants to give shade. Pairs will spawn on a layer of aquarium peat.

Crenuchus spilurus

Sailfin Tetra

- **ORIGINS** Northern South America, occurring in Guyana and the middle Amazon.
- **SIZE** 2½ in (6 cm).
- **DIET** Fresh and thawed live foods preferred.
- **WATER** Temperature 75–82°F (24–28°C); soft (50–100 mg/l) and acidic (pH 6.0).
- **TEMPERAMENT** Keep singly or in pairs.

Sexing these tetras is simple; males are larger and have more prominent dorsal fins than females. Sailfins are predators. They usually eat invertebrates, but they consume any smaller fish sharing their tank, especially fry. Pairs spawn seasonally on flat rocks in the wild, guarding the eggs until they hatch and wafting water over them with their ventral fins.

Raised dorsal fin of male

Arnoldichthys spilopterus

African Red-Eyed Characin

- **ORIGINS** West Africa, ranging from Lagos, Nigeria, to the Niger Delta.
- **SIZE** 4 in (10 cm).
- **DIET** Eats prepared foods and live foods.
- **WATER** Temperature 73–82°F (23–28°C); hard (150–200 mg/l) and neutral (pH 7.0).
- **TEMPERAMENT** Placid shoaler.

Black marking on the dorsal fin

Both sexes show the characteristic red eye marking, but they can usually be sexed on the basis of the anal fin. In the males, this is convex and more colorful, displaying red, yellow, and black bands. The extremely active nature of African Red-Eyed Characins precludes dense planting in their tank. Pairs can be prolific, with females producing more than 1,000 eggs at a single spawning. A soft base in the spawning tank is vital, since the fry will dive to the bottom if scared. They grow fast, reaching almost 2 in (5 cm) in length within seven weeks.

Inpaichthys kerri

Blue Emperor

- **ORIGINS** South America. A relatively recent discovery from the Rio Aripuana in Amazonia.
- **SIZE** 2 in (5 cm).
- **DIET** Eats both prepared diets and small live foods.
- **WATER** Temperature 75–81°F (24–27°C); soft (50–100 mg/l) and neutral (pH 7.0).
- **TEMPERAMENT** Placid.

The stunning blue of the adult male, seen below, is maintained only in good water conditions. The female is smaller and duller, with brownish-yellow upperparts and a broad black band passing through the eyes and along the body. These fish need a well-planted tank with an open area for swimming. They are best kept in groups and are perfect for a community aquarium, making ideal companions for nonaggressive species such as corydoras catfish.

Pristella maxillaris

X-Ray Tetra

- **ORIGINS** South America, in Venezuela, Guyana, and the lower Amazon in Brazil.
- **SIZE** 1¾ in (4.5 cm).
- **DIET** Prepared foods and live foods.
- **WATER** Temperature 73–82°F (23–28°C); soft (50–100 mg/l) and acidic to neutral (pH 6.0–7.0).
- **TEMPERAMENT** Nonaggressive.

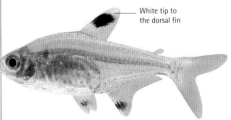

White tip to the dorsal fin

The transparent appearance of X-Ray Tetras allows them to be sexed on the basis of the shape of their swim bladder, which is more pointed in the males. Despite their transparency, they are not colorless, with the golden, red, white, and black areas on their bodies giving rise to their alternative name of Water Goldfinch. The coloration of X-Ray Tetras can be emphasized if they are kept in an aquarium with a dark base and subdued lighting.

Moenkhausia sanctaefilomenae

Yellow-Banded Tetra

- **ORIGINS** South America, occurring in eastern Peru and Bolivia, Paraguay, and western Brazil.
- **SIZE** 3 in (7.5 cm).
- **DIET** Eats both prepared foods and live foods.
- **WATER** Temperature 73–82°F (23–28°C); hard (100–150 mg/l) and neutral (pH 7.0).
- **TEMPERAMENT** Has a strong shoaling instinct.

Attention is immediately drawn to the head of this fish by the red on the upper part of the iris. The yellow band on the caudal peduncle distinguishes it from the larger and more aggressive Glass Tetra (*M. oligolepis*). Yellow-Banded Tetras are active fish, so their tank should have planted areas around the sides and to the rear for use as retreats and a clear area at the front for swimming. These fish are quite adaptable in terms of their water chemistry requirements. Sexing is difficult unless the fish are in breeding condition, when females appear swollen with spawn. The spawning tank must be densely planted to deter the fish from eating their own eggs. The female scatters several hundred eggs in vegetation, often among the roots of floating plants. The adults will then need to be removed. The young emerge after about two days; by a week old, they should be large enough to be weaned off fry foods and onto brine shrimp.

Nannaethiops unitaeniatus

One-Lined African Tetra

- **ORIGINS** Equatorial Africa, occurring in the Niger, Congo, and Nile Rivers.
- **SIZE** 3 in (7.5 cm).
- **DIET** Prepared foods, live foods, and vegetable matter.
- **WATER** Temperature 73–79°F (23–26°C); soft (50–100 mg/l) and neutral (pH 7.0).
- **TEMPERAMENT** Placid and social.

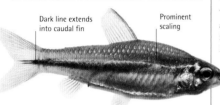

Dark line extends into caudal fin

Prominent scaling

In this species, the sexes may be distinguished only when the fish are in spawning condition; at this time, males develop red areas on the upper part of the caudal fin and on the square-shaped dorsal fin. The odd, jerky swimming motion of these tetras is perfectly normal. Provide fine-leaved plants in the spawning tank, among which females will lay up to 500 eggs. Hatching takes two days, and the fry are free-swimming five days later.

Mimagoniates microlepis

Croaking Tetra

- **ORIGINS** Southern South America, distributed throughout southeastern Brazil.
- **SIZE** 2¼ in (5.5 cm).
- **DIET** Flake and small live foods.
- **WATER** Temperature 66–77°F (19–25°C); soft (50–100 mg/l) and acidic (pH 6.5).
- **TEMPERAMENT** Placid and social.

The soft blue coloration of these tetras is especially apparent in males, which are slightly larger than females and have longer fins when full-grown. Males are unusual in that they attract females by releasing chemical messengers called pheromones from a gland located at the base of the caudal fin. Live foods such as mosquito larvae are a valuable conditioning food; the tetras subsequently spawn among vegetation. Croaking Tetras can make croaking sounds via their air bladder.

Aphyocharax anisitsi

Argentine Bloodfin

- **ORIGINS** Southern South America, where it is restricted to the Rio Parana in Argentina.
- **SIZE** 2 in (5 cm).
- **DIET** Flake and small live foods.
- **WATER** Temperature 64–82°F (18–28°C); soft (50–100 mg/l) and around neutral (pH 7.0).
- **TEMPERAMENT** Placid and social.

The blood-red coloration on the fins of this tetra is not a consistent feature and is more pronounced in some individuals than others. Male fish have a hook on their anal fin. Pairs spawn quite readily among vegetation, but the adults will rapidly eat the eggs, numbering several hundred, if they get the chance. Argentine Bloodfins are easy to keep and have a life expectancy of more than 10 years.

Neolebias ansorgii

Ansorge's Neolebias

- **ORIGINS** Central equatorial Africa, where it is found in shallow pools in Angola and Cameroon.
- **SIZE** 1½ in (3.5 cm).
- **DIET** Flake and small live foods.
- **WATER** Temperature 75–82°F (24–28°C); soft (50–100 mg/l) and acidic (pH 6.5).
- **TEMPERAMENT** Placid and social.

Although this is the most colorful fish in its genus, it has never been very popular with aquarists. Unusually for a characoid, Ansorge's Neolebias lacks an adipose fin. The more brightly colored male, pictured below, has a reddish-orange body, with a darker stripe along the side and a black spot at the base of the tail. This coloration is improved by subdued lighting conditions and a dark base in the aquarium. Ansorge's Neolebias is best kept in a single-species tank. It naturally inhabits shallow pools, so make sure the water is no more than 8 in (20 cm) deep. The breeding tank should have a peat-covered base, on which the fish will spawn. Hatching occurs within a day, and the young will need a fry food at first. The new generation will be ready to breed when the fish are seven months old.

Astyanax mexicanus

Blind Cave Tetra

- **ORIGINS** Central America, restricted to the San Luis Potosi region of Mexico.
- **SIZE** 3½ in (9 cm).
- **DIET** Prepared foods and small live foods.
- **WATER** Temperature 64–82°F (18–28°C); hard (100–150 mg/l) and alkaline (pH 7.5).
- **TEMPERAMENT** Placid and social.

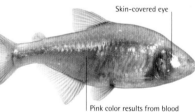

Skin-covered eye

Pink color results from blood flowing around the body

Blind because they live in darkness in the wild, these fish should be kept in an aquarium decorated with slate, devoid of plants, and lit by a nighttime fluorescent tube. A single-species setup is preferable, and some vegetable matter should be provided in the diet. Eggs will be scattered over a suitable spawning mop, after which the adults must be removed to protect the eggs.

Bryconaethiops microstoma

Small-Mouth Featherfin Tetra

- **ORIGINS** Central equatorial Africa, occurring in Stanley Pool, lower Zaire, in fast-flowing water.
- **SIZE** 6 in (15 cm).
- **DIET** Flake, small live foods, and algae.
- **WATER** Temperature 75–82°F (24–28°C); soft (50–100 mg/l) and acidic (pH 6.5).
- **TEMPERAMENT** Placid, but does not shoal.

The long trailing filaments on the dorsal fin are seen only in the male Small-Mouth Featherfin, shown here; in the female, the dorsal fin is shorter and more rounded. Young Small-Mouth Featherfins have a silvery body color, which darkens with age. The water chemistry requirements of Small-Mouth Featherfins are quite specific; the water must be well oxygenated, with good movement to replicate the fast-flowing stretches of water that they inhabit in the wild. Being rather nervous fish, they need a well-planted tank to provide plenty of hiding places. The tank should also be covered, because they are excellent jumpers. Although algae form an important part of their natural diet, they will often ignore alternative plant foods offered to them. Small-Mouth Featherfins have an unusually small mouth compared to other tetras and so should only be offered live foods of an appropriate size.

ADAPTED TO LIFE UNDERGROUND

The Blind Cave Tetra *(see opposite)* occurs only in a single Mexican cave system. Outside, in nearby rivers, the ancestral form of this fish can still be found, complete with natural pigmentation and fully functional eyes. Reflecting their origins, Blind Cave Tetras can see when they hatch but lose vision as skin grows over the eyes. This happens even when they are reared in a well-lit environment. The principal sensory input of these fish comes from the lateral line system, which runs along each side of the body. This detects changes in water movements. Blind Cave Tetras probably also have a good sense of smell, because they have no problem finding enough food when competing with fully sighted species in aquarium surroundings.

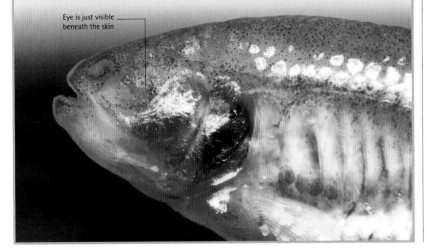

Eye is just visible beneath the skin

Prionobrama filigera

Glass Bloodfin

- **ORIGINS** South America, occurring in central and southern Brazil, Paraguay, and Argentina.
- **SIZE** 2½ in (6 cm).
- **DIET** Prepared foods and small live foods.
- **WATER** Temperature 72–86°F (22–30°C); soft (50–100 mg/l) and acidic (pH 6.5).
- **TEMPERAMENT** Placid and social.

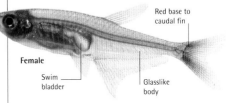

Red base to caudal fin

Female

Swim bladder

Glasslike body

When backlit, the body of the Glass Bloodfin is completely transparent. Sexing is straightforward, because only mature males display a white extension of the outer edge of the anal fin. An aquarium for Glass Bloodfins should be shaded and incorporate floating plants and retreats around the sides. These active fish must also have plenty of space in which to swim. When spawning, Glass Bloodfins scatter their eggs among clumps of vegetation. The eggs hatch about three days later.

Hasemania nana

False Silver-Tipped Tetra

- **ORIGINS** South America, in the Rio Purus tributaries in western Brazil and the San Francisco basin in the east.
- **SIZE** 2 in (5 cm).
- **DIET** Prepared foods and small live foods.
- **WATER** Temperature 72–82°F (22–28°C); soft (50–100 mg/l) and acidic (pH 6.5).
- **TEMPERAMENT** Placid and social.

A coppery body and contrasting silver tips on the fins, including both lobes of the caudal fin, typify the males of this small, shoaling species. Females have a paler, silvery-gold coloration, as well as a more rounded body shape, which is most obvious when they are in spawning condition. A separate spawning tank is needed to prevent the eggs from being devoured. Hatching takes place in a day; the free-swimming young require fry foods at first.

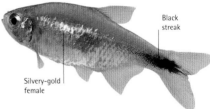

Black streak

Silvery-gold female

ANABANTOIDS

The defining characteristic of this group of fish is a special structure called the labyrinth organ. Located close to the gills, it enables the fish to breathe atmospheric air directly. Anabantoid species, which include the popular bettas and gouramis, are found in parts of southern Asia and in Africa. Virtually all anabantoids display a degree of parental care. In most cases, the male constructs a bubble nest for the eggs, although some species are mouth-brooders. While males can be kept in community aquariums, they are frequently too aggressive to make suitable companions for each other.

Siamese Fighting Fish (*Betta splendens*) are undoubtedly the best-known anabantoids among hobbyists, with more than 100 different strains available.

Betta splendens

Siamese Fighting Fish

- **ORIGINS** Southeast Asia, occurring in Thailand, although its exact range is uncertain.
- **SIZE** 2¼ in (6 cm).
- **DIET** Prepared foods and live foods.
- **WATER** Temperature 75–82°F (24–28°C); soft (50–100 mg/l) and acidic (pH 6.0–6.5).
- **TEMPERAMENT** Males are aggressive toward each other.

The Siamese Fighting Fish is also referred to as the Betta, particularly in North America. This species was widely kept in Thailand for more than 200 years before it became known in the West in the late 1800s. In their homeland, where they frequent the canals, or *klongs*, that flow through many Thai cities, Siamese Fighting Fish were selectively bred not only for their color but also for their fighting ability, with significant amounts of money being bet on the outcome of contests between the more aggressive males.

Different strains evolved from cross-breeding fish obtained from various parts of Thailand. As a result, it is now practically impossible to be certain of the original distribution of these fish, or of their natural coloration, even though alleged "wild type" specimens are occasionally offered for sale. Current thinking is that wild forms were originally

dark red, probably with bluish streaking on their fins and a pair of vertical lines on the side of the head behind the eyes. Certainly, the wild ancestors of today's Siamese Fighting Fish had simpler fins than those seen in modern strains.

It is likely that interest in keeping these fish for fighting purposes began, not in Thailand, but in neighboring Cambodia (Kampuchea). In fact, the Thai name for these fish is *pla kat khmer*, which translates as "fin-biter in Khmer" (Khmer is a former name of Cambodia). Since being introduced to the West, however,

Dorsal fin far back on the body

Rounded caudal fin is similar in shape to that of a wild Betta

Half-Moon Betta female Sexing is quite straightforward with the Betta, since the fins of females are generally far less elaborate in appearance than those of males.

breeders have concentrated on establishing a wide range of color forms, ranging from white through yellow to purple. Selective breeding has also been used to modify the fins, which are always more elaborate in the males. With the exception of the female Half-Moon Betta, pictured below left, all the specimens shown here are male.

Color becomes blue-black at spawning time

Depth of black coloration is variable

Half-Moon Betta male Black is not a popular color for bettas, since it is linked with a lethal gene. If two black fish mate, some of the offspring will die before hatching.

Betta splendens (continued)

In Thailand, it is traditional to house Siamese Fighting Fish in small jars, but these provide little swimming space and make it difficult to maintain the water quality. The natural grace and elegance of the fish will be more apparent in an aquarium. A single male can be kept in a tank with several females, or even as part of a community aquarium. However, avoid mixing these fish with fin-nipping species, which will attack the flowing fins, or with fish of a similar coloration, which may themselves be attacked by the Siamese Fighting Fish.

These fish are easy to care for, but they are not especially long-lived, with an average life span of about two years. Pairs of seven or eight months old are best for breeding. They need a relatively shallow spawning tank, about 8 in (20 cm) deep. It must be covered and include floating plants, among which the male will build a bubble nest. Thai breeders often add the leaves of the Ketapang or Indian almond tree (*Terminalia catappa*) to assist with the conditioning of the water. These leaves are available in the West through specialist suppliers.

Raising the water temperature can trigger spawning, as can increasing the amount of live food in the diet. Check that the female is in breeding condition, because otherwise the male may harass her. Aside from her slightly swollen belly, one of the surest indicators of the female's readiness to spawn is when she develops yellowish stripes on her body. She will actively seek out the male at this stage, rather than trying to avoid him.

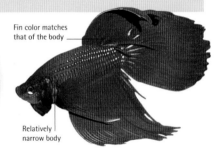

Fin color matches that of the body

Relatively narrow body

Red Betta Ranked as the most colorful Betta variety, this fish comes in several different shades, from the very bright shade shown above through to rich crimson.

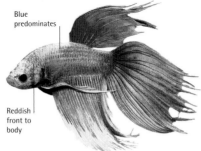

Upturned jaw

Ventral fin reaches back to the caudal

Steel-Blue Betta Blue coloration is common in the Betta. This specimen has excellent fins that are unblemished and show no signs of damage.

Blue predominates

Reddish front to body

Violet Betta This color form is largely blue with violet undertones. However, the patterning and intensity varies greatly, allowing individuals to be easily distinguished.

Salmonlike hue

White on fins

Cambodian Betta This white color variant was thought to be a separate species when it first became known in the West. It is sometimes referred to as the Plakat.

Crown Betta Selective breeding has given the Crown Betta rough-edged fins. Like other variants, the Crown Betta uses its fins to display to rivals and potential mates.

Uneven edging to the fins

LOOKING FOR A FIGHT

Siamese Fighting Fish have good color vision, which enables them to spot rivals with relative ease. Even when faced with its own reflection in a mirror, as here, the male responds by adopting an aggressive posture. Aggression is one of the traits developed in this species over the centuries, and most modern strains are far more belligerent than their wild ancestors. However, actual conflict is still a last resort, and the fish initially try to deter one another with ritualized displays, notably raising the fins and flaring the gill covers to make themselves look bigger.

Betta splendens (continued)

It is the responsibility of the male Siamese Fighting Fish to construct a bubble nest. Spawning occurs nearby, with the pair wrapping around one another. The female will then float upside down, as though stunned, while the male collects the 15 or so eggs in its mouth and carries them to the bubble nest. Mating resumes once he has gathered all the eggs. This sequence is repeated until some 500 eggs have been produced, with the entire process lasting about two hours. It is then best to remove the female while the male guards the nest; otherwise, he may attack her. If the tank is very large and well planted, however, it may be safe for her to stay put.

Hatching occurs 48 hours after mating, and the young fry are free-swimming within a further four days. Rear them on fry foods at first, and gently circulate the water with an airstone to convey food particles to them. Powdered flake and brine shrimp can be provided as they grow. The large number of fry means that gentle filtration is needed to maintain water quality, and partial water changes are required every three days. Once the males in the brood can be identified, usually at about two months of age, they should be moved to individual accommodation to prevent fighting. Prior to this, keep the aquarium covered to keep the young fish from becoming chilled, since this will impair the development of their labyrinth organs *(see p.110)*.

Body is paler than the fins

Long ventral fin

Split-Tailed Betta In good specimens, the divided caudal fin is symmetrical, with both branches of the tail being equal in size and shape.

Round-Tailed Cambodian Betta Some individuals are paler than others, being, in effect, albinos. The dark-eyed form retains some color pigment.

Round-Tailed Betta The caudal fin of this fish is relatively small and rounded, rather than long and trailing as in most strains. The colorful rays add to its appeal, as does the dark edging on the fins.

Rounded dorsal fin

Prominent pelvic fins

Betta smaragdina

Emerald-Green Betta

- **ORIGINS** Southeast Asia, present in Cambodia (Kampuchea), Laos, and eastern parts of Thailand.
- **SIZE** 2¼ in (7 cm).
- **DIET** Prepared foods and live foods.
- **WATER** Temperature 75–82°F (24–28°C); soft (50–100 mg/l) and acidic (pH 6.0–6.5).
- **TEMPERAMENT** Males may quarrel.

The body of the male Emerald-Green Betta displays areas of red, blue, and green. The female is plainer but will develop stripes as the time

for spawning approaches. Males can be mixed with unrelated fish, but if they are housed together, they will fight, with potentially fatal consequences. In the wild, Emerald-Green Bettas inhabit shallow areas of water that flood during the wet season, and this flooding marks the onset of the breeding period. A significant partial water change in their tank may thus trigger spawning. The male builds a bubble nest among vegetation. After spawning, remove the female in case the male becomes aggressive toward her, and leave the male to guard the nest, which contains up to 100 eggs. Give the young fry food at first and then brine shrimp.

Mixed coloration of male

Betta pugnax

Mouth-Brooding Betta

- **ORIGINS** Southeast Asia; occurs widely on the Malay Peninsula.
- **SIZE** 4 in (10 cm).
- **DIET** Prepared foods and live foods.
- **WATER** Temperature 73–77°F (23–25°C); soft (50–100 mg/l) and acidic (pH 6.0–6.5).
- **TEMPERAMENT** Males likely to be pugnacious.

Faint banding on body

Green edge on gill covers

The appearance of these fish is highly variable, depending partly on their origins and partly on their overall condition. They tend to have a reddish-brown background color, with green spots evident on the individual scales. Cooler water conditions suit this mouth-brooder, which often occurs in flowing waters. The young are sexually mature by about six months.

Betta bellica

Slim Betta

- **ORIGINS** Southeast Asia, where its distribution is centered on the Malay Peninsula.
- **SIZE** 4½ in (11 cm).
- **DIET** Prepared foods and live foods.
- **WATER** Temperature 75–82°F (24–28°C); soft (50–100 mg/l) and acidic (pH 6.0–6.5).
- **TEMPERAMENT** Males only aggressive when breeding.

The Slim Betta shows greenish-blue markings set against a brownish background. Although introduced to Europe as far back as 1905, this species has not become as popular as the Siamese Fighting Fish *(see pp. 104–106)*. Males can be identified by their brighter coloration and longer fins. Slim Bettas are now regarded as being the same species as the Striped Betta—a darker blue variant from Sumatra that was previously known as *Betta fasciata*. The male Slim Betta constructs a bubble nest at the surface among vegetation, so make sure that floating plants are included in the tank. These bettas occasionally emerge to rest on the plants, slipping back into the water if disturbed. The young, which are free-swimming within three days of hatching, require a fry food at first. After a similar interval, introduce brine shrimp to their diet.

JAW SHAPE AND BREEDING

Head structure is a good guide to breeding habits in *Betta* species. The jaws of mouth-brooders are modified to provide more space in the oral cavity for the developing brood. As a result, the head is taller and the jawline less rounded, as seen in this Krabi Mouth-Brooder *(Betta simplex)*. Bubble-nesting species have smaller heads, since they simply have to pick up the eggs and transfer them to the nest. The enlargement of the oral cavity in mouth-brooders develops only when the fish become sexually mature, and it is only seen in the sex that is responsible for brooding the young.

Betta coccina

Wine-Red Betta

- **ORIGINS** Southeast Asia, present on Sumatra and the southern tip of the Malay Peninsula.
- **SIZE** 2½ in (6 cm).
- **DIET** Prepared foods and live foods.
- **WATER** Temperature 75–82°F (24–28°C); soft (50 mg/l) and acidic (pH 5.0–6.0).
- **TEMPERAMENT** Males are aggressive toward each other.

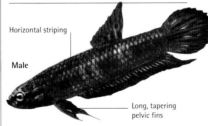

Horizontal striping

Male

Long, tapering pelvic fins

The Wine-Red Betta first became available to hobbyists in the 1980s. The body color that gives the fish its name is restricted to the male; the female is decidedly brownish. Males must be kept individually, to prevent fighting. Water quality is especially important for the health of this rather delicate species, and peat filtration is advisable. Up to 60 eggs form the typical brood, and both parents can be safely left with their young.

Betta imbellis

Crescent Betta

- **ORIGINS** Southeast Asia, on the Malay Peninsula and nearby islands, notably Phuket and Pinang.
- **SIZE** 2 in (5 cm).
- **DIET** Prepared foods and live foods.
- **WATER** Temperature 75–82°F (24–28°C); soft (50–100 mg/l) and acidic (pH 6.0–6.5).
- **TEMPERAMENT** Males may fight each other.

Male Crescent Bettas are more colorful than females, their brownish bodies displaying blue hues. There are distinct regional variations, with fish from Pinang Island being yellowish in color. Male Crescent Bettas are far from docile, especially when in breeding condition. If well fed, a pair of these bubble-nesting fish will spawn several times in quick succession, producing as many as 150 eggs per batch.

Male coloration

Long anal fin

Macropodus opercularis

Paradise Fish

- **ORIGINS** Eastern Asia, occurring in China, Korea, Vietnam, and islands including Taiwan and Hainan.
- **SIZE** 4½ in (11 cm).
- **DIET** Prepared foods and live foods.
- **WATER** Temperature 61–82°F (16–28°C); soft (50 mg/l) and acidic (pH 6.0–6.5).
- **TEMPERAMENT** May be aggressive.

The blue body color of the Paradise Fish is broken by vertical coppery-orange bands, and black blotches are evident on its head. This species was probably the first tropical fish to be kept in Europe, its attractive patterning and hardiness making it very popular during the late 1800s. Males, which can be recognized by their longer fins, grow to a slightly larger size than females. Include retreats in the tank, and add areas of dense planting. Floating plants are important for spawning, because this is where the male will construct the bubble nest. As many as 500 eggs are laid, with the fry emerging and becoming free-swimming within about four days. The young, which grow rapidly, will initially require a fry food or infusoria.

Albino Paradise Fish Although known as the Albino, this variant is not pure white, since it retains the coppery bands along the sides of its body.

Male Paradise Fish
The males are sometimes very aggressive toward their intended mates at spawning time.

Macropodus spechti

Black Paradise Fish

- **ORIGINS** Asia, where its distribution is restricted to Vietnam and southern China.
- **SIZE** 4 in (12 cm).
- **DIET** Prepared foods and live foods.
- **WATER** Temperature 68–79°F (20–26°C); soft (50 mg/l) and acidic (pH 6.0–6.5).
- **TEMPERAMENT** Breeding males quarrel.

With a brownish body and dark-edged scales, this species is not as brightly colored as the Paradise Fish itself *(see above)*. However, Black Paradise Fish are better suited to mixed aquariums, because they are generally tolerant of other fish. The males may be more aggressive toward each other when they are in breeding condition, which is indicated by a blackening of their coloration. Red ventral fins enable males to be distinguished from females.

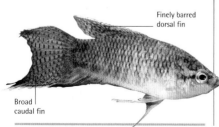

Finely barred dorsal fin

Broad caudal fin

Macropodus ocellatus

Chinese Paradise Fish

- **ORIGINS** Asia, where it occurs in Korea, eastern China, and Vietnam.
- **SIZE** 3 in (7.5 cm).
- **DIET** Prepared foods and live foods.
- **WATER** Temperature 61–82°F (16–28°C); soft (50 mg/l) and acidic (pH 6.0–6.5).
- **TEMPERAMENT** Can be belligerent.

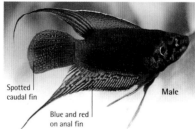

Spotted caudal fin

Blue and red on anal fin

Male

Chinese Paradise Fish have blackish bodies, and males are easily distinguished by their long, flowing fins. They should not be allowed to hybridize with related species, because the female offspring may be infertile. Spawning can be encouraged by allowing the water temperature to fall over the winter and then raising it gradually in the spring.

Pseudosphromenus cupanus

Spike-Tailed Paradise Fish

- **ORIGINS** Asia, being found in southeastern parts of India, and also on the island of Sri Lanka.
- **SIZE** 2½ in (6 cm).
- **DIET** Prepared foods and live foods.
- **WATER** Temperature 75–82°F (24–28°C); soft (50 mg/l) and neutral (pH 7.0).
- **TEMPERAMENT** Nonaggressive.

A coppery-brown color typifies these bubble-nesting Paradise Fish. The sexes are similar, but the males have longer tips on their dorsal and anal fins, and they become more colorful at spawning time, with red more evident on their underparts. These fish need a well-planted, partially filled tank with subdued lighting, some algal growth, and plenty of retreats.

Long ventral fin extends along undersides

Pseudosphromenus dayi

Day's Paradise Fish

- **ORIGINS** Asia, in southwestern India; some reports suggest eastern India and Malaysia as well.
- **SIZE** 3 in (7.5 cm).
- **DIET** Prepared foods and live foods.
- **WATER** Temperature 75–82°F (24–28°C); soft (50 mg/l) and neutral (pH 7.0).
- **TEMPERAMENT** Avoid mixing with aggressive fish.

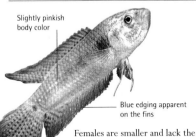

Slightly pinkish body color

Blue edging apparent on the fins

Females are smaller and lack the characteristic "spike" formed by the central rays of the male's caudal fin. Day's Paradise Fish is a docile species; males do not harass their partners, even at spawning time. Subdued lighting conditions and a well-planted aquarium will help to emphasize the coloration of these fish. Raising the water temperature should help to trigger breeding behavior.

Parosphromenus nagyi

Nagy's Licorice Gourami

- **ORIGINS** Southeast Asia, restricted to the vicinity of Kuantan in eastern Malaysia.
- **SIZE** 1½ in (4 cm).
- **DIET** Prepared foods and live foods.
- **WATER** Temperature 68–75°F (20–24°C); soft (50 mg/l) and acidic (pH 5.0).
- **TEMPERAMENT** Quite social.

The normally dull coloration of dark brown with lighter stripes alters significantly when these gouramis are in spawning condition. Males then become blackish with vibrant blue fin markings, while females lose their striped markings and turn a pale brownish-yellow. Keep these fish in a single-species setup, and provide each male with its own cave. Up to 40 eggs are laid, and the fry become free-swimming about a week later.

Parosphromenus deissneri

Licorice Gourami

- **ORIGINS** Southeast Asia, on the Malay Peninsula and the islands of Sumatra and nearby Banka.
- **SIZE** 1½ in (4 cm).
- **DIET** Prepared foods and live foods.
- **WATER** Temperature 75–82°F (24–28°C); soft (50 mg/l) and acidic (pH 6.0–6.5).
- **TEMPERAMENT** Nonaggressive.

Males have more elaborate fins than females

Male

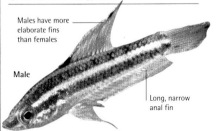

Long, narrow anal fin

The two yellowish stripes running along the body of this species are separated by licorice-colored bands. Licorice Gouramis have a rather undeserved reputation for being delicate; in fact, their care is quite straightforward when they have good water conditions. Filtration over peat *(see p.46)* to acidify the water can help in this regard. These bubble-nesting cave-spawners need a well-planted aquarium, since they are rather nervous by nature.

BLOWING BUBBLES

In many anabantoid species, the male builds a bubble nest for the eggs out of mucus and air, sometimes including plant matter in its construction as well. The eggs naturally float because of their oil content, and the bubble nest, which is often anchored to aquatic vegetation, traps the eggs and keeps them all safely together in one place. The male Paradise Fish, pictured here, then guards the bubble nest until the fry hatch. Bubble nests vary in size, with those built by the Giant Gourami (*Osphronemus goramy*) measuring up to 20 in (50 cm) in diameter and 10 in (25 cm) in height. Some smaller members of the anabantoid group prefer to make their bubble nests in underwater caves, where they will be less conspicuous.

The other method of reproduction often associated with anabantoids is mouth-brooding, but even in mouth-brooding species it is not uncommon for the males to display rudimentary bubble-nesting behavior. This suggests that the switch from bubble-nesting to mouth-brooding is a comparatively recent development. It may have arisen as a way of adapting to faster-flowing stretches of water, where the current would break up bubble nests and sweep the eggs away.

Trichopsis vittata

Croaking Gourami

- **ORIGINS** Southeast Asia, from Vietnam, Thailand, and the Malay Peninsula to Sumatra, Java, and Borneo.
- **SIZE** 3 in (7.5 cm).
- **DIET** Prepared foods and live foods.
- **WATER** Temperature 77–82°F (25–28°C); soft (50 mg/l) and acidic (pH 6.5).
- **TEMPERAMENT** Rather shy.

Chocolate striping along the side of the body helps to identify this species. Raising the temperature of the water and lowering the level to about 4 in (10 cm) should help to trigger spawning behavior, when the croaking calls uttered by these fish are most likely to be heard. The male may use floating plants as anchorage points for the bubble nest, which will trap up to 200 eggs laid by the female.

Coloration varies between individuals

Relatively slender body shape

Trichopsis pumila

Pygmy Gourami

- **ORIGINS** Southeast Asia, in Cambodia (Kampuchea), Vietnam, Thailand, and the Malay Peninsula.
- **SIZE** 1½ in (4 cm).
- **DIET** Prepared foods and live foods.
- **WATER** Temperature 77–82°F (25–28°C); soft (50 mg/l) and acidic (pH 6.5).
- **TEMPERAMENT** Only aggressive when spawning.

This small gourami has semitransparent fins patterned with red and blue, while the body displays variable brownish markings that form two lines along the body. The male builds a small, fragile bubble nest under the leaf of a large plant, such as a suitable cryptocoryne. Nearly 200 eggs are laid; the fry hatch after two days and become free-swimming after a similar period.

Ctenops nobilis

Frail Gourami

- **ORIGINS** Asia, where it occurs in eastern parts of India and in Bangladesh.
- **SIZE** 4 in (10 cm).
- **DIET** Prepared foods and live foods.
- **WATER** Temperature 68–75°F (20–24°C); hard (100–150 mg/l) and neutral (pH 7.0).
- **TEMPERAMENT** Aggressive.

Brown coloration and variable lighter markings are characteristic of the Frail Gourami. The pale stripes along the sides of young fish fade with age. Males, which have red edging on their caudal and anal fins, must not be kept together because they will fight. A tank for Frail Gouramis should be well planted, with calm water conditions. This is a mouth-brooding gourami, and the young leave the female's mouth about two weeks after spawning.

Long, pointed head

Dark speckling on the caudal fin

SURVIVING IN THE MURKY SHALLOWS

In the wild, anabantoids are often found in shallow, muddy waters. The temperature in these surroundings can rise rapidly during the day under the heat of the tropical sun, with the result that less oxygen is available to the fish. To help them survive in such a poorly oxygenated environment, anabantoids have developed labyrinth organs. Located on either side of the head, close to the gills, these structures enable the fish to breathe atmospheric air directly, supplementing the meager amounts of oxygen that they can extract from the water with their gills.

Another anabantoid adaptation to living in muddy water is the way in which the pelvic fins have moved right to the front of the body and developed into hairlike feelers. These feelers not only enable anabantoids to sense their surroundings by touch, but they also help the fish to find mates. The male uses his feelers to detect chemicals called pheromones, which females release into the water when they are ready to spawn.

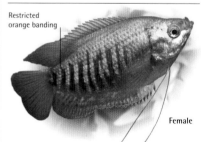

Trichogaster labiosa

Thick-Lipped Gourami

- **ORIGINS** Asia, being present in parts of northern India and Myanmar (Burma).
- **SIZE** 4 in (10 cm).
- **DIET** Prepared foods, algae, and live foods.
- **WATER** Temperature 72–82°F (22–28°C); soft (50 mg/l) and acidic (pH 6.0–6.5).
- **TEMPERAMENT** Suitable for a community tank.

Restricted orange banding

Female

Pointed tips on the dorsal fin and a brighter coloration make males of this species easy to spot. They construct a large, relatively fragile bubble nest at the surface, often continuing to increase its size even after egg-laying. As in other cases, the filtration in the tank must be gentle in order to prevent damage to the bubble nest.

Trichogaster lalius

Dwarf Gourami

- **ORIGINS** Asia, occurring in the drainage basins of the Brahmaputra, Ganges, and Indus Rivers in northern India.
- **SIZE** 2¹/₄ in (5.5 cm).
- **DIET** Prepared foods and live foods.
- **WATER** Temperature 72–82°F (22–28°C); soft (50 mg/l) and acidic (pH 6.0–6.5).
- **TEMPERAMENT** Ideal for community tanks.

Extensive orange striping across the body is set against a sky-blue background in males of this species; females, in contrast, are predominantly silvery-gray. A pair of Dwarf Gouramis will form a strong bond and stay relatively close to each other. They are rather nervous fish and need to be kept in a well planted tank. The bubble nest is built using pieces of vegetation. It is often located amongst floating plants and can be up to 1 in (2.5 cm) deep. Females produce as many as 600 eggs at a single spawning, which are then guarded by the male until they hatch about a day later. The young have particularly tiny mouths, even for anabantoid fry, and must have infusoria or a suitable substitute as their first food. Keep the tank covered, especially at four weeks, when their labyrinth organ *(see opposite)* is developing, because chilling at this stage can be fatal.

Blue Dwarf Gouramis Color variants of this species have been bred since the 1980s. The Blue Dwarf has solid blue over the whole body.

Dwarf Gourami The male, seen below, is very colorful in appearance. Red and blue variants have been created, both of which have highly individual patterning.

Reddish iris

Even blue coloration

Trichogaster fasciata

Little Giant Gourami

- **ORIGINS** Asia, present throughout the Indian peninsula, with the exception of the far south and southwest.
- **SIZE** 4 in (10 cm).
- **DIET** Prepared foods and live foods.
- **WATER** Temperature 72–82°F (22–28°C); soft (50 mg/l) and acidic (pH 6.5).
- **TEMPERAMENT** Territorial when breeding.

This species grows to just a fraction of the size of the Giant Gourami, *Osphronemus goramy (see p.113).* Once Little Giant Gourami are about 2 in (5 cm) long, they can be sexed by the male's brighter color and the pointed tips of its dorsal and anal fins. After spawning, remove the female; otherwise, she may be attacked by her partner. Give the developing young plenty of space.

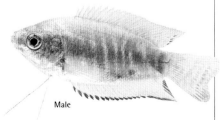

Male

Trichogaster chuna

Honey Gourami

- **ORIGINS** Asia, in northeastern India and Bangladesh, notably in the Brahmaputra River and the lower Ganges.
- **SIZE** 2 in (5 cm).
- **DIET** Prepared foods and live foods.
- **WATER** Temperature 72–82°F (22–28°C); soft (50 mg/l) and acidic (pH 6.5).
- **TEMPERAMENT** Territorial when spawning.

Honey Gouramis are brownish in color, with a bluish hue around the hairlike pelvic fins. When the male, seen here, is in breeding condition, it turns a rich amber, displaying dark markings on the underparts and along part of the anal fin. The female, in turn, develops a dark stripe along the side of its body. Outside the breeding period, the longer tips of the male's caudal and dorsal fins distinguish the sexes. For breeding purposes, a pair needs their own spawning tank. The male constructs a bubble nest among floating plants, into which the female deposits up to 200 eggs. This species is particularly susceptible to Velvet Disease *(see p.58).*

Belontia signata

Combtail

- **ORIGINS** Southern Asia, restricted to the island of Sri Lanka.
- **SIZE** 5 in (12.5 cm).
- **DIET** Prepared foods and live foods.
- **WATER** Temperature 75–82°F (24–28°C); soft (50 mg/l) and acidic (pH 6.0–6.5).
- **TEMPERAMENT** Lively by nature.

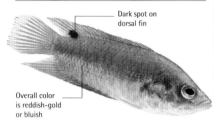

Dark spot on dorsal fin

Overall color is reddish-gold or bluish

The extended caudal-fin rays of this fish resemble the teeth of a comb. The coloration is naturally variable, ranging from red through yellowish-orange to shades of brown. Combtails are not suitable for a community tank, because they tend to be disruptive. They can also be rather nervous; house them in a tank with plants and bogwood for cover. Females must be removed after spawning, because males then become very aggressive.

Belontia hasselti

Java Combtail

- **ORIGINS** Southeast Asia, found on the Malay Peninsula, Java, Sumatra, Borneo, and other islands.
- **SIZE** 8 in (20 cm).
- **DIET** Prepared foods and live foods.
- **WATER** Temperature 77–86°F (25–30°C); hard (100–150 mg/l) and neutral (pH 7.0).
- **TEMPERAMENT** Aggressive when spawning.

Java Combtails from different islands can vary markedly in appearance. The characteristic honeycombed patterning is most prominent in males in breeding condition. Males do not build a bubble nest—the eggs simply float to the surface—but if danger threatens, they take the eggs into their mouth and move them elsewhere. Hatching occurs within a day, and the fry are free-swimming after a further three days. Unlike many anabantoid fry, the young can be reared immediately on brine shrimp.

Sphaerichthys osphromenoides

Chocolate Gourami

- **ORIGINS** Southeast Asia, occurring on the Malay Peninsula, as well as on Sumatra and Borneo.
- **SIZE** 2½ in (6 cm).
- **DIET** Prepared foods and live foods.
- **WATER** Temperature 61–82°F (16–28°C); soft (50 mg/l) and acidic (pH 6.0–6.5).
- **TEMPERAMENT** Rather shy and nervous.

A variable chocolate-brown coloration and paler vertical bands characterize this mouth-brooder. The male, pictured above, has a more pointed dorsal fin than the female. Chocolate Gouramis from Borneo have red markings on their fins. Peat and regular water changes are needed to maintain ideal water conditions.

Trichopodus leerii

Pearl Gourami

- **ORIGINS** Southeast Asia, occurring on the Malay Peninsula, as well as on Sumatra and Borneo.
- **SIZE** 4 in (10 cm).
- **DIET** Prepared foods, vegetables, and live foods.
- **WATER** Temperature 75–82°F (24–28°C); soft (50 mg/l) and neutral (pH 7.0).
- **TEMPERAMENT** Males may quarrel.

These gouramis display a delicate patterning of whitish, pearl-like spots. Males are easily recognizable from the age of about seven months by the reddish coloration on their underparts. They can be kept in community aquariums, but as with all gouramis and some other anabantoids, avoid mixing them with fin-nipping fish such as barbs. A breeding tank for a pair of Pearl Gouramis should have a relatively low water level—about 4 in (10 cm) deep—and must be spacious. The male builds a large bubble nest among plants, with the female laying up to 1,000 eggs. The fry will hatch within a day or so and be free-swimming within a further three days. They should be given infusoria or a commercial substitute as their first food.

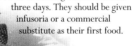

Trichopodus pectoralis

Snakeskin Gourami

- **ORIGINS** Southeast Asia, occurring in Cambodia (Kampuchea), Thailand, and the Malay Peninsula.
- **SIZE** 8 in (20 cm).
- **DIET** Prepared foods and live foods.
- **WATER** Temperature 73–82°F (23–28°C); soft (50 mg/l) and acidic (pH 6.0–6.5).
- **TEMPERAMENT** Nonaggressive.

The mottled, almost vertical markings of these fish bear some resemblance to snakeskin. Males can be recognized by their taller dorsal fins, while females have yellow rather than orange edging on their fins. Snakeskin Gouramis live and breed in shallow waters, often in rice paddies. Including an airstone in the rearing tank will help to circulate the microscopic food needed by newly hatched fry.

Darker horizontal line runs along body

Trichopodus microlepis

Moonlight Gourami

- **ORIGINS** Southeast Asia, occurring mainly in Thailand, but also found in Cambodia (Kampuchea).
- **SIZE** 6 in (15 cm).
- **DIET** Prepared foods, vegetables, and live foods.
- **WATER** Temperature 79–86°F (26–30°C); soft (50 mg/l) and acidic (pH 6.0–6.5).
- **TEMPERAMENT** Nonaggressive.

Tiny, silvery scales give Moonlight Gouramis a very shiny appearance, typically with a greenish tinge. The long, trailing pelvic fins help to distinguish the sexes, being red in males and yellow in females. Floating plants are important to provide cover in the main aquarium and anchorage for the large bubble nest in the breeding tank. The nest itself may extend up to 2 in (5 cm) above the surface.

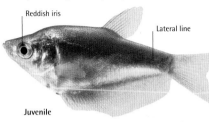

Reddish iris
Lateral line
Juvenile

Trichopodus trichopterus

Three-Spot Gourami

- **ORIGINS** Southeast Asia, where it occurs in Myanmar (Burma), Vietnam, Thailand, Malaysia, and Indonesia.
- **SIZE** 6 in (15 cm).
- **DIET** Prepared foods and live foods.
- **WATER** Temperature 72–82°F (22–28°C); soft (50 mg/l) and acidic (pH 6.0–6.5).
- **TEMPERAMENT** Males are aggressive toward each other.

Contrary to their name, the spotted patterning of Three-Spot Gouramis is variable, as is their coloration. These fish can be sexed by the dorsal fin, which has a more elongated tip in males. Three-Spots are easy to keep, because they are hardy and unfussy eaters. The male creates a bubble nest up to 10 in (25 cm) across at the surface among floating plants. After spawning, remove the female to protect her from the male. The fry should initially be reared on infusoria. Good water quality needs to be maintained in the tank as the fry develop.

Blue Gourami This subspecies (*T. t. sumatranus*) from Sumatra has been used to breed ornamental strains. Two black spots characterize the wild form.

Traces of dark vertical bands
Pelvic "feelers"
Golden Opaline Gourami These fish are of a rich yellow shade, with spotted patterning on the fins.

Osphronemus goramy

Giant Gourami

- **ORIGINS** Southern Asia; original range is unknown, because it has been captive-reared for centuries.
- **SIZE** 28 in (70 cm).
- **DIET** Prepared foods and live foods.
- **WATER** Temperature 68–86°F (20–30°C); soft (50 mg/l) and acidic (pH 6.0–6.5).
- **TEMPERAMENT** May be aggressive and predatory.

Older adult
Lacelike body pattern

Young Giant Gouramis are different in appearance from adults, having a narrow, pointed head and vertical body stripes. As they mature, they develop a more rounded head shape and a grayish body. As the largest anabantoids, these fish need a spacious, well-filtered tank, if not a pond. They are prolific spawners, with females laying up to 20,000 eggs.

Helostoma temminckii

Kissing Gourami

- **ORIGINS** Southeast Asia, from Thailand and the Malay Peninsula to Sumatra and Borneo.
- **SIZE** 12 in (30 cm).
- **DIET** Prepared foods, vegetables, and live foods.
- **WATER** Temperature 72–82°F (22–28°C); soft (50 mg/l) and neutral (pH 7.0).
- **TEMPERAMENT** Not normally disruptive.

There are two forms of Kissing Gourami; the wild-type fish is silvery in appearance, while the domesticated strain has a pinkish body color. Visual sexing is difficult, although females swell with spawn when in breeding condition. In the aquarium, these gouramis dig for food in the substrate, and also browse on algae growing on rocks or the glass of the tank. They may even nibble at plants, especially if any of the more delicate species are growing in their tank.

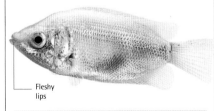

Fleshy lips

KISSING CONTESTS

Part of the appeal of Kissing Gouramis comes from the way in which individuals lock their fleshy lips together, as if kissing. In reality, however, this lip-locking is a test of strength between the individuals, the gouramis remaining attached by the lips until the weaker fish breaks away. Although this "kissing" behavior is primarily a way of settling disputes, it can also have a breeding function. Gouramis of opposite sexes may behave in this way—sometimes locking on to the body instead of the lips—in order to find a compatible partner of similar strength.

Badis badis

Indian Chameleon Fish

- **ORIGINS** The Ganges region of India, Bangladesh, and Nepal. Also recorded from Pakistan and Bhutan.
- **SIZE** 3 in (7.5 cm).
- **DIET** Prepared foods and live foods.
- **WATER** Temperature 73–82°F (23–28°C); soft (50 mg/l) and acidic (pH 6.0–6.5).
- **TEMPERAMENT** Usually quite peaceful.

Male

Female

These fish are sometimes called Chameleon Fish due to their variable coloration, which is influenced by their surroundings. Males are slightly larger and more brightly colored than females. Variations in appearance between chameleon fish may be slight. The Indian species shown here has bluer tones than its Burmese relative (*Badis ruber*), which is more reddish. These fish need a densely planted tank with a sandy base. Include a flowerpot laid on its side to act as a spawning cave. The female will lay up to 100 eggs in the cave, and the male will watch over them until they hatch about three days later. Use brine shrimp as a rearing food for the free-swimming fry. Raising the water temperature slightly can trigger spawning behavior. Males become increasingly territorial at this stage, so breeding pairs are best given their own tank.

Anabas testudineus

Climbing Perch

- **ORIGINS** Asia, widely distributed from India to southern China and parts of Indonesia.
- **SIZE** 10 in (25 cm).
- **DIET** Prepared foods and live foods.
- **WATER** Temperature 61–82°F (16–28°C); soft (50 mg/l) and acidic (pH 6.0–6.5).
- **TEMPERAMENT** Occasionally aggressive.

Males have longer anal fins

Brownish in color, the Climbing Perch is a hardy species. If its habitat dries up, it can use its pectoral fins to drag itself short distances over land to find a new stretch of water. When on land, it relies on its labyrinth organs (*see p.110*) to meet its oxygen needs. Climbing Perch need to be housed in a covered tank to prevent them from climbing out. Their eggs may simply be left to float in the water, since the males often do not construct bubble nests.

Microctenopoma fasciolatum

Banded Bushfish

- **ORIGINS** Central Africa, in the Congo River, Zaire, from close to Mosembe down to Boma at the river's mouth.
- **SIZE** 3½ in (9 cm).
- **DIET** Prepared foods and live foods.
- **WATER** Temperature 75–82°F (24–28°C); soft (50 mg/l) and neutral (pH 7.0).
- **TEMPERAMENT** Aggressive.

Irregular dark bands run vertically across the body of this fish. The dorsal and anal fins are more pointed in males. The coloration is quite variable, with some individuals having blue in their fins. The banding is less obvious in juveniles, which tend to be grayer overall. The Banded Bushfish is a bubble-nesting species, and up to 1,000 eggs may be produced at a single spawning. The fry are not usually harmed by their parents.

Ctenopoma acutirostre

Leopard Ctenopoma

- **ORIGINS** Central Africa, occurring in central and lower parts of the Congo River in Zaire.
- **SIZE** 8 in (20 cm).
- **DIET** Prepared foods and live foods.
- **WATER** Temperature 73–82°F (23–28°C); soft (50 mg/l) and acidic (pH 6.5).
- **TEMPERAMENT** Relatively placid but predatory.

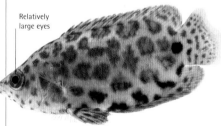

Relatively large eyes

This ctenopoma has a leopardlike pattern of spots, with a much darker spot at the base of the caudal fin. Leopard Ctenopomas tend to be more active after dark. They are nervous by nature, but if their tank provides plenty of cover, the fish may eventually become tame. In the wild, they prey on small fish and other aquatic creatures. Males have small patches of defensive spines on their bodies.

Microctenopoma ansorgii

Ornate Ctenopoma

- ⬡ **ORIGINS** Central Africa, in southern Cameroon and the Congo basin, including Stanley Pool in Zaire.
- ⬡ **SIZE** 3 in (7.5 cm).
- ⬡ **DIET** Prepared foods and live foods.
- ⬡ **WATER** Temperature 75–82°F (24–28°C); soft (50 mg/l) and acidic (pH 6.5).
- ⬡ **TEMPERAMENT** Relatively peaceful.

The Ornate Ctenopoma is one of the most colorful members of its group. Its contrasting patterning looks best when viewed under subdued lighting, when the six dark bands on the body will be clearly visible. The male, which is more brightly colored than the female, builds a relatively small bubble nest at the surface of the tank. Young Ornate Ctenopomas develop their striped pattern when they are about four weeks old.

Long, serrated dorsal fin

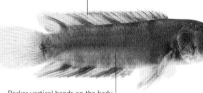

Darker vertical bands on the body

Nandus nandus

Nandus

- ⬡ **ORIGINS** Asia, ranging from India eastward through Myanmar (Burma) to Thailand.
- ⬡ **SIZE** 8 in (20 cm).
- ⬡ **DIET** Live foods.
- ⬡ **WATER** Temperature 72–79°F (22–26°C); hard (100–150 mg/l) and alkaline (pH 7.5).
- ⬡ **TEMPERAMENT** Predatory.

These mottled fish are more active than leaf fish, their close relatives. They must not be mixed with small companions because of their predatory nature. However, they can be kept alongside either of the Asiatic chromide species *(see p.146)*, which require similar water conditions, including the addition of some marine salt to create a slightly brackish environment. Unlike leaf fish, Nandus do not guard their brood.

Polycentrus schomburgkii

Schomburgk's Leaf Fish

- ⬡ **ORIGINS** Northern South America, found in Venezuela and Guyana, as well as on Trinidad.
- ⬡ **SIZE** 4 in (10 cm).
- ⬡ **DIET** Live foods.
- ⬡ **WATER** Temperature 72–79°F (22–26°C); hard (150–200 mg/l) and neutral (pH 7.0).
- ⬡ **TEMPERAMENT** Predatory and shy.

Female Schomburgk's Leaf Fish differ significantly in color from males, being a much lighter shade of brown. Males turn virtually black when in spawning condition. These leaf fish should be mixed only with similar-sized, nonaggressive species. They hide away for much of the day, typically becoming more active at dusk. Spawning occurs in a cave, where the female lays up to 600 eggs. The male guards the eggs until they hatch.

Ragged edge to dorsal fin

Monocirrhus polyacanthus

Barbeled Leaf Fish

- ⬡ **ORIGINS** Western South America, occurring in sluggish streams in the Amazonian region of Peru.
- ⬡ **SIZE** 4 in (10 cm).
- ⬡ **DIET** Live foods.
- ⬡ **WATER** Temperature 72–77°F (22–25°C); soft (50 mg/l) and acidic (pH 6.0–6.5).
- ⬡ **TEMPERAMENT** Predatory.

A narrow, yellowish body and markings similar to those of a dead leaf characterize this unusual-looking species. The aquarium must include dense vegetation, among which the fish can hide. Sexing is difficult, although females are often larger than males. Spawning occurs on rocks or leaves. The male guards the eggs, which number up to 300. Hatching occurs about four days later.

PREDATORY DRIFTER

Everything about the Barbeled Leaf Fish reflects its highly specialized predatory lifestyle. Not only does it look like a leaf, but it also mimics the movement of a drifting leaf under water. Its hunting technique is not to swiftly pursue its victims but to ambush them. Hanging at an angle in the water, it is carried along by the current until it comes within range of a smaller fish. It then seizes the unsuspecting prey with a lightning-fast snap of the jaws and gulps it down head first. Often the Leaf Fish does not need to bother making a lunge for its quarry, because at close range the rapid opening of its large jaws creates a pressure difference in the water that sucks the unfortunate individual into its mouth. The predatory instincts of the Barbeled Leaf Fish are so strong that the fish can prove problematic in aquarium surroundings, making it virtually impossible to persuade it to eat inert foodstuffs. Fortunately, this species will take live invertebrates in addition to fish.

CATFISH

Catfish represent one of the largest groups of aquarium fish and also one of the most diverse in terms of appearance and lifestyle. There are more than 2,000 species in approximately 30 families. Some of these fish are sedentary and suitable for a community tank, while others are active predators that grow to a large size. Identification is not always easy, partly because members of the same species often show differences in patterning and partly because new species are constantly being discovered. The complex and expanding nature of one catfish family, the Loricariids, has led to the development of a classification system based on "L" numbers *(see p. 21).*

Glass Catfish (*Kryptopterus bicirrhis*) rely on camouflage for protection, but other catfish can defend themselves with spines, poison, and even electric shocks.

Auchenipterichthys thoracatus

Midnight Catfish

- **ORIGINS** South America, in the upper reaches of the Amazon and its southern tributaries in Peru.
- **SIZE** 4½ in (11 cm).
- **DIET** Flake food, tablets, and small live foods.
- **WATER** Temperature 68–75°F (20–24°C); hard (100–150 mg/l) and neutral (pH 7.0).
- **TEMPERAMENT** Social.

The patterning of light spots on a dark background resembles stars in the midnight sky. As their name suggests, these catfish are nocturnal in their habits. They like to hide themselves away in holes in bogwood, anchoring themselves in place using the rays of their dorsal and pelvic fins, which are more prominent in males. Midnight Catfish will come up to the surface to feed and can be housed with nonaggressive companions of similar size.

Raised dorsal fin

Dark eye

Kryptopterus bicirrhis

Glass Catfish

- **ORIGINS** Southern Asia, from eastern India, Thailand, and Malaysia to Java, Sumatra, and Borneo.
- **SIZE** 4 in (10 cm).
- **DIET** Flake and freeze-dried foods, and small live foods.
- **WATER** Temperature 68–79°F (20–26°C); soft (50–100 mg/l) and acidic (pH 6.5).
- **TEMPERAMENT** Placid.

The long body of this catfish is almost completely transparent, allowing it to assume the color of its surroundings. The only color on the fish itself is a small reddish-violet patch behind the gills. The barbels are long, as is the anal fin, which runs virtually the length of the underparts. In contrast, there is just a hint of a dorsal fin, with the pelvic fins also being rudimentary. These active catfish naturally inhabit fast-flowing streams.

Chaca bankanensis

Chocolate Frogmouth

- **ORIGINS** Southeast Asia, occurring in parts of Thailand and Malaysia, extending to Indonesia.
- **SIZE** 8 in (20 cm).
- **DIET** Prefers live foods, but will take tablets.
- **WATER** Temperature 72–75°F (22–24°C); soft to hard (50–200 mg/l) and acidic to alkaline (pH 6.0–8.0).
- **TEMPERAMENT** Predatory.

The Chocolate Frogmouth is distinguished from the Squarehead Frogmouth (*C. chaca*) by its richer shade of brown, but both species share similar habits. Chocolate Frogmouths hide under or behind rockwork during the day, becoming active and feeding at night. They are likely to hunt down smaller fish, so it is unwise to house them with such companions. A third species, *C. burmensis,* was discovered in 1988, but is little documented.

PUTTING OUT FEELERS

The different shapes and sizes of the sensory barbels surrounding the mouths of catfish provide an insight into their differing lifestyles. Often likened to feelers, barbels give catfish additional sensory information about their environment. Active predatory catfish hunt by sweeping long barbels from side to side as they swim, which helps them to detect and home in on potential prey. The longest barbels are seen in nocturnal species, or species that hunt in turbid waters where visibility is restricted. In contrast, more sedentary or vegetarian catfish have very much shorter barbels, which act as a sensory supplement to the eyes.

Bunocephalus coracoideus

Banjo Catfish

- **ORIGINS** South America, occurring in the Amazon basin, extending to the vicinity of La Plata.
- **SIZE** 6 in (15 cm).
- **DIET** Pellets, tablets, and sinking live food.
- **WATER** Temperature 70–79°F (21–26°C); soft to hard (50–200 mg/l) and acidic to alkaline (pH 6.0–7.8).
- **TEMPERAMENT** Placid and sedentary.

This catfish rarely strays far from the substrate and will often try to conceal itself by burrowing in the gravel in the base of the aquarium, just as it would on riverbeds in the wild. The inclusion of shriveled oak leaves that have been previously soaked will provide extra camouflage. Pairs of these fish can breed in aquaria. The female lays up to 5,000 eggs in a pit; the male guards the eggs until they hatch.

Pareutropius debauwi

African Glass Catfish

- **ORIGINS** West Africa, occurring in the Congo and also in Gabon.
- **SIZE** 3 in (7.5 cm).
- **DIET** Small live foods, fresh or freeze-dried, plus flake.
- **WATER** Temperature 72–79°F (22–26°C); soft to hard (50–150 mg/l) and acidic to near neutral (pH 6.5–7.2).
- **TEMPERAMENT** Very placid and social.

These fish should be kept in small groups to ensure the presence of at least one pair, because it is difficult to distinguish the sexes until females swell with eggs. Ideally, two males should be housed in the breeding tank with a single female. Typically, she will lay up to 100 white eggs in the morning, after which the fish should be returned to the main aquarium. The catfish fry will hatch within three days, and are easy to rear on brine shrimp. Adults will take insect larvae very readily, especially bloodworm.

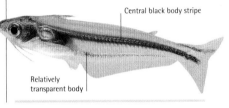

Central black body stripe

Relatively transparent body

Ariopsis seemanni

Black-Fin Shark Catfish

- **ORIGINS** Occurs in rivers and estuaries on the Pacific coast, from North America to Colombia and Peru.
- **SIZE** 13 in (33 cm).
- **DIET** Thawed and fresh live foods, plus prepared foods.
- **WATER** Temperature 72–82°F (22–28°C); soft to hard (50–300 mg/l) and neutral (pH 7.0).
- **TEMPERAMENT** Not to be trusted with smaller tankmates.

The combination of the streamlined body shape and the large dorsal fin explains why this species is known as the Black-Fin Shark Catfish.

The fins on the underside of the body are edged with white. Although primarily a freshwater species, the Black-Fin Shark sometimes enters brackish waters in river estuaries. In an aquarium setting, especially at first, the addition of a small quantity of sea salt (about a level teaspoon per gallon/ 4.5 liters) to the water is recommended. As the fish grow older, their characteristic coloring fades. Plenty of open space for swimming, plus a strong filter current, are essential features of their aquarium. These catfish are mouth-brooders, with the male carrying out this task.

Farlowella acus

Whiptail Catfish

- **ORIGINS** South America, occurring in southern tributaries of the Amazon, in the vicinity of La Plata.
- **SIZE** 6 in (15 cm).
- **DIET** Mainly vegetarian, including vegetable flake.
- **WATER** Temperature 75–79°F (24–26°C); soft (50–100 mg/l) and acidic to neutral (pH 6.0–7.0).
- **TEMPERAMENT** Very inactive.

The slender, brown body of this catfish makes it easy to confuse with a twig. Mature males can be recognized by bristles on their snouts. About a day before spawning occurs, the female develops an egg-laying tube called an ovipositor. She lays a clutch of up to 60 adhesive eggs, typically on rocks. The eggs are guarded, usually by the male, until the young emerge.

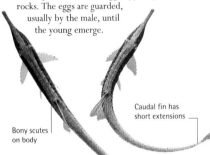

Bony scutes on body

Caudal fin has short extensions

Platydoras costatus

Chocolate Doradid

- **ORIGINS** South America, ranging from Peru eastward through the Amazonas region of Brazil.
- **SIZE** 8½ in (22 cm).
- **DIET** Catfish pellets, vegetable matter, and live foods.
- **WATER** Temperature 75–86°F (24–30°C); soft (50–100 mg/l) and acidic to neutral (pH 6.0–7.0).
- **TEMPERAMENT** Territorial toward its own kind.

The coloration of this species consists of distinct brownish-black and whitish-cream stripes. The "*costatus*" part of the scientific name refers to the riblike projections along the sides of the body.

These projections will become stuck in the material of a net, so Chocolate Doradids need to be caught with great care. This species occupies the lower levels in an aquarium and will burrow down into the substrate, which must be fine-grained and kept clean to reduce the risk of infection. Bogwood and partially buried clean flowerpots should also be included in the tank. Chocolate Doradids are aggressive toward their own kind, so they cannot be housed together. They can, however, be kept with other species, especially those that swim at higher levels in the tank. There appear to be no records of this species being bred successfully in the home aquarium.

Relatively large eyes

Gagata cenia

Indian Clown Catfish

- **ORIGINS** Southern Asia, occurring in northern India, Nepal, Pakistan, Bangladesh, and Myanmar (Burma).
- **SIZE** 6 in (15 cm).
- **DIET** Floating foods and live foods.
- **WATER** Temperature 68–75°F (20–24°C); hard (100–150 mg/l) and acidic to neutral (pH 6.0–7.0).
- **TEMPERAMENT** Nonaggressive.

This species is one of the Asian hillstream catfish, so it requires well-filtered and aerated water to mimic the conditions in a relatively fast-flowing stream. These fish will not harm vegetation, since they naturally feed on insects at the water's surface. Consequently, it is better to give live foods (including freeze-dried items) and flake foods rather than sinking pellets. Indian Clown Catfish are lively, active fish and will thrive if kept in small groups.

Platydoras hancockii

Hancock's Doradid

- **ORIGINS** Northern South America, ranging from Colombia eastward as far as Guyana.
- **SIZE** 6 in (15 cm).
- **DIET** Catfish pellets and live foods.
- **WATER** Temperature 77–84°F (25–29°C); soft (50–150 mg/l) and acidic to neutral (pH 6.0–7.0).
- **TEMPERAMENT** Nonaggressive.

Tall dorsal fin

Variable dark markings

Pectoral fins

Dark, mottled-brown shades predominate in Hancock's Doradid. These catfish belong to a group often described as talking catfish, because of their ability to make sounds by moving the spines of their pectoral fins. The resulting sounds are then amplified by the swim bladder. Hancock's is a bubble-nesting species. It builds its nest among floating plants at the surface; the water level in the tank should be kept low to assist this process.

Hara hara

Butterfly Catfish

- **ORIGINS** Asia, where it occurs in northern India, Bangladesh, and Nepal.
- **SIZE** 1½ in (4 cm).
- **DIET** Small live foods preferred.
- **WATER** Temperature 64–82°F (18–28°C); soft (50–150 mg/l) and acidic to neutral (pH 6.0–7.0).
- **TEMPERAMENT** Nonaggressive.

Though not often available, these small catfish make interesting aquarium occupants—they are social by nature and can be kept in groups. They are attractively and individually marked in various shades of brown, with some cream-colored areas on their bodies, too. The only drawback is that they are nocturnal and rather shy. Consequently, an aquarium for Butterfly Catfish should incorporate suitable retreats. Feed these catfish at dusk.

Oxydoras niger

Black Doradid

- **ORIGINS** South America, in the Amazon region, where it extends from Peru to parts of Brazil.
- **SIZE** 30 in (80 cm).
- **DIET** Catfish pellets, tablets, and live foods.
- **WATER** Temperature 70–75°F (21–24°C); soft (50–100 mg/l) and acidic to neutral (pH 6.0–7.0).
- **TEMPERAMENT** Peaceful, despite its large size.

This is not a species to acquire without careful thought about its future accommodation, since it will attain a very large size. This catfish also has a big appetite, especially when growing, so maintaining adequate filtration in the aquarium is essential for the health of both the catfish and any companions. Black Doradids are gentle, bottom-dwelling fish that will not trouble smaller species, although when frightened they may lash out powerfully with their tail. Their breeding habits are not documented.

Lighter spines along the sides of the dark body

THE SOUND OF SCALES

Members of the doradid group are well protected by the rows of protective spines along the body, which make these catfish difficult for predators to swallow. The spines on the pectoral fins are also used for communication—they produce an audible, scratchy sound when rubbed together. Exactly why doradids have evolved this method of communication is unclear, but since they are largely nocturnal, it may be that the sounds help them to locate each other for spawning purposes in the dark surroundings. It is possible that their well-developed sense of smell also assists them in their search for mates.

Agamyxis pectinifrons

White-Spotted Doradid

- **ORIGINS** South America, occurring in parts of Ecuador and Peru.
- **SIZE** 6 in (15 cm).
- **DIET** Catfish pellets and live foods.
- **WATER** Temperature 68–79°F (20–26°C); soft (50–100 mg/l) and acidic to neutral (pH 6.0–7.0).
- **TEMPERAMENT** Nonaggressive.

Sensory barbels

Banding on the tail

Whitish spots and blotches on a blackish background help to identify White-Spotted Doradids. Their tank should not be brightly lit and must include plenty of retreats where the catfish can hide. They will become more active when it is dark, emerging to hunt for food over the substrate. Like Hancock's Doradid *(see opposite)*, they can make sounds via the swim bladder. These catfish are probably substrate spawners, but breeding is unknown in aquariums.

Megalodoras uranoscopus

Irwin's Soldier Cat

- **ORIGINS** South America, where it ranges through much of the Brazilian Amazon and into the Guianas.
- **SIZE** 24 in (60 cm).
- **DIET** Catfish pellets and meat-based foods.
- **WATER** Temperature 72–79°F (22–26°C); soft (50–100 mg/l) and acidic to neutral (pH 6.0–7.0).
- **TEMPERAMENT** Peaceful.

Not to be confused with the Black Doradid *(see opposite)*, this snail-eating species does, however, attain a similarly large size, although it tends to grow more slowly. Its body is covered with very distinctive brown plates, overlaid in part with rows of spines, while darker shading helps to break up its outline. Snails make up the bulk of this catfish's diet in the wild, augmented by palm fruits and other items that fall into the water. Fortunately, in the home aquarium it is possible to wean Irwin's Soldier Cats onto artificial diets without much difficulty. It may be necessary to offer foods such as thawed shrimp at first, but before long, they should be eating tablets and similar prepared foods. These catfish need a large tank with secure decor that cannot be dislodged easily. The sharp projections on the body mean that it is unwise to transport this species in plastic bags, especially larger specimens.

Dianema longibarbus

Porthole Catfish

- **ORIGINS** Northwestern South America, in the Ambylac and Pacaya Rivers in Peru.
- **SIZE** 4 in (10 cm).
- **DIET** Prepared foods and live foods.
- **WATER** Temperature 72–79°F (22–26°C); soft (50–100 mg/l) and acidic (pH 6.5).
- **TEMPERAMENT** Social and nonaggressive.

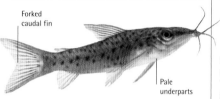

Forked caudal fin

Pale underparts

The pattern of dark speckling on the body varies between individual Porthole Catfish, but it never extends to the fins. Distinguishing the sexes is difficult, but pairs will breed in aquarium surroundings, where the male creates a bubble nest for the eggs. Lowering the water level and raising its temperature to 82°F (28°C) may help to trigger spawning. When swimming, the longer pair of barbels are held horizontally, while the shorter pair points downward.

Callichthys callichthys

Slender Armored Catfish

- **ORIGINS** South America, from Peru eastward to Guyana, eastern Brazil, Paraguay, and Bolivia.
- **SIZE** 8 in (20 cm).
- **DIET** Prepared foods and live foods.
- **WATER** Temperature 64–82°F (18–28°C); soft (50–100 mg/l) and acidic (pH 6.5).
- **TEMPERAMENT** Social and nonaggressive.

The steeply sloping head of this catfish helps to distinguish it from corydoras, which have a similar reflective sheen on their bodies. Slender Armored Catfish are also significantly larger than corydoras and may prey on smaller companions in the aquarium. These catfish prefer naturally shady surroundings in which they can hide among bogwood and rocks during the day. They are primarily nocturnal in their habits and are most active toward dusk.

Megalechis thoracata

Thorocatum Catfish

- **ORIGINS** Northern South America, reaching as far south as Paraguay; also offshore islands, including Trinidad.
- **SIZE** 8 in (20 cm).
- **DIET** Prepared foods and live foods.
- **WATER** Temperature 64–82°F (18–28°C); soft (50–100 mg/l) and acidic (pH 6.5).
- **TEMPERAMENT** Relatively peaceful.

Barbels of different lengths

Coloration is variable

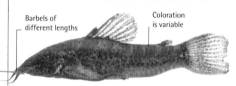

This catfish is usually bluish-black or reddish-brown, with some speckling. Thorocatum Catfish are easy to sex when in breeding condition, since the rays of the male's pectoral fins turn orange, while the underparts suffuse with a bluish-violet coloration. Spawning is very vigorous, with the male often pursuing the female aggressively. The eggs are deposited in a bubble nest at the surface among vegetation and will be guarded until they hatch. In the wild, this species inhabits muddy shallows that are thick with aquatic plant growth.

Aspidoras lakoi

Spotline Aspidoras

- **ORIGINS** South America, where it has been recorded in the state of Minas Gerais, Brazil.
- **SIZE** 1½ in (4 cm).
- **DIET** Prepared foods and small live foods.
- **WATER** Temperature 72–79°F (22–26°C); soft (50–100 mg/l) and acidic (pH 6.5).
- **TEMPERAMENT** Social and nonaggressive.

These small aspidoras catfish are closely related to the corydoras group and require similar care, although they are not as commonly available. In the Spotline, the silvery-bronze body is decorated with dark brown or black mottling, which in places forms rough horizontal lines. Dark lines also arc across the caudal fin. The key feature that sets aspidoras apart from corydoras is the presence of two pores in the skull (although these will obviously not be apparent in a live specimen). Spotline Aspidoras prefer to remain close to the substrate, which should be fine-grained so that they can dig into it. These fish will also appreciate the addition of bogwood and rockwork to their tank, especially small pieces of slate laid horizontally to create platforms on which the catfish can rest.

Aspidoras pauciradiatus

Sixray Corydoras

- **ORIGINS** South America, occurring in the Araguaia River, and the Rio Negro in Brazil.
- **SIZE** 1½ in (4 cm).
- **DIET** Prepared foods and small live foods.
- **WATER** Temperature 73–82°F (23–28°C); soft (50–100 mg/l) and acidic (pH 6.5).
- **TEMPERAMENT** Social and nonaggressive.

A large blotched area on the dorsal fin identifies this mottled catfish, which also has a strongly marked caudal fin. Sixray Corydoras are not as well known as true corydoras; however, they are equally attractive, and it can sometimes be difficult to distinguish between the two species. The tank needs well-oxygenated water as well as a special sandy substrate in which the fish will instinctively dig for small worms and other edible items.

Brochis multiradiatus

Long-Finned Brochis

- **ORIGINS** South America, occurring in the upper Napo River in Ecuador.
- **SIZE** 4 in (10 cm).
- **DIET** Prepared foods and small live foods.
- **WATER** Temperature 70–75°F (21–24°C); soft (50–100 mg/l) and acidic (pH 6.5).
- **TEMPERAMENT** Social and nonaggressive.

This species looks similar to the Greenhump Catfish, but a longer head profile helps to set it apart from its close relative. The Long-Finned Brochis also lacks the bony plate that covers the underside of the jaw in the Greenhump. A standard diet of catfish pellets needs to be supplemented with other foods, such as pieces of shrimp. This is likely to encourage spawning.

Corydoras splendens

Common Brochis

- **ORIGINS** South America, in the upper Amazon region; recorded in parts of Ecuador, Peru, and Brazil.
- **SIZE** 3¼ in (8 cm).
- **DIET** Prepared foods and small live foods.
- **WATER** Temperature 70–82°F (21–28°C); soft (50–100 mg/l) and acidic (pH 6.5).
- **TEMPERAMENT** Social and nonaggressive.

Scientific name changes mean that not all catfish now included in the genus *Corydoras* have this description incorporated into their common name, as typified by the case of the Brochis catfish. Green iridescence is particularly prominent in this Brochis species, covering the head as well as much of the body. The underparts have a pinkish suffusion, which is most evident in males. For breeding purposes, it is recommended to keep a small group of three males and two females. The eggs are scattered among aquatic vegetation, and the catfish must be removed before they eat them.

Emerald-green sheen   Sloping head

BONY CASING

The body covering of the so-called armored catfish, which form the Callichthyidae family, differs significantly from that of other fish. They are covered not with scales but with two rows of bony plates, sometimes described as scutes, which meet in the midline on each side of the body. The scutes form a more rigid barrier than scales and thus provide better protection against would-be predators. However, this body casing may make it harder for the fish to breathe via their gill flaps, so armored catfish should always be kept in well-oxygenated surroundings.

Corydoras britskii

Greenhump Catfish

- **ORIGINS** South America, in the Mato Grosso, Brazil, and also in the Rio Paraguay, Paraguay.
- **SIZE** 5 in (13 cm).
- **DIET** Prepared foods and small live foods.
- **WATER** Temperature 73–77°F (23–25°C); soft (50–100 mg/l) and acidic (pH 6.5).
- **TEMPERAMENT** Social and nonaggressive.

Greenhump Catfish are orange beneath and iridescent green on top. The intensity of their coloration depends on lighting conditions. The long mouth ends in prominent barbels on the jaws. Greenhump Catfish resemble corydoras, although they can be distinguished at a glance by their more compressed body shape and longer dorsal fin. They do not grow to a particularly large size, and they are generally suitable for a community aquarium. Young individuals often swim in the mid-water zone, while adults spend more time close to the floor of the tank, seeking food. They dig in the substrate with their mouthparts, so make sure that this is sandy. If kept on sharp gravel, Greenhump Catfish are likely to develop sore mouths, which can in turn become infected.

To maintain good water quality, change about 30 percent of the water every two weeks or so. The male and female are not easy to tell apart, but the male may be slightly smaller overall, with a more colorful appearance. Increasing the amount of live food in the diet should encourage the fish to spawn.

Corydoras schwartzi

Schwartz's Corydoras

- ⚙ **ORIGINS** South America, where it is restricted to a tributary of the Rio Purus in eastern Brazil.
- ⟳ **SIZE** 2½ in (6 cm).
- ⬤ **DIET** Live foods and prepared catfish foods.
- ≋ **WATER** Temperature 72–79°F (22–26°C); soft to hard (50–150 mg/l) and acidic to neutral (pH 6.0–7.0).
- ⊕ **TEMPERAMENT** Peaceful and social.

Schwartz's Corydoras is distinguished by dark marks that run through each eye and meet in front of the dorsal fin on the top of the head. There are about 150 different species of corydoras catfish, distributed in the more southerly parts of Central America and also through northern South America. Corydoras are small, attractive catfish. They have proved very popular, being ideal for a mixed community aquarium. Conspicuous and active during the day, corydoras tend to occupy the lower part of the tank. Low rockwork, such as small pieces of slate, will serve as vantage points for these catfish. Schwartz's Corydoras, like many species, has a limited distribution in the wild but is quite adaptable in the aquarium. Regular water changes are important, however, to keep nitrate levels low.

Corydoras trilineatus

Three-Striped Corydoras

- ⚙ **ORIGINS** South America, occurring in the Rio Ucayali, Rio Ampiyacu, and the Yarina Cocha River in Peru.
- ⟳ **SIZE** 2 in (5 cm).
- ⬤ **DIET** Live foods and prepared catfish foods.
- ≋ **WATER** Temperature 68–79°F (0–26°C); soft to hard (50–150 mg/l) and acidic to neutral (pH 6.0–7.0).
- ⊕ **TEMPERAMENT** Peaceful and social.

A variable but vibrant body patterning, including a large black mark on the dorsal fin, characterizes these corydoras. Females are generally slightly paler in color than males, with a smaller patch on the dorsal fin. The "three stripes" in the common name of this species refer to the central stripe running along the side of the body and the lighter lines above and below.

Corydoras aeneus

Green Corydoras

- ⚙ **ORIGINS** South America, widely distributed from Venezuela and Trinidad down to La Plata, Argentina.
- ⟳ **SIZE** 2¾ in (7 cm).
- ⬤ **DIET** Live foods and prepared catfish foods.
- ≋ **WATER** Temperature 68–79°F (20–26°C); soft to hard (50–150 mg/l) and acidic to neutral (pH 6.5–7.0).
- ⊕ **TEMPERAMENT** Peaceful and social.

Clear, slightly grayish fins

This popular corydoras lacks dark markings on its yellowish-brown body. Instead, there is a strong iridescence on the face and along the upper part of the back. This iridescence ranges from green to copper, depending on the light. Several color variants exist, including an albino. The larger, often more rotund, female actively initiates spawning. The eggs are laid in small batches among aquatic vegetation, and hatching takes about five days. The young initially need fry foods.

Corydoras panda

Panda Corydoras

- ⚙ **ORIGINS** South America, where it is confined to the Ucayali River system in Peru.
- ⟳ **SIZE** 1¾ in (4 cm).
- ⬤ **DIET** Live foods, algae, and prepared catfish foods.
- ≋ **WATER** Temperature 68–77°F (20–25°C); soft to hard (50–150 mg/l) and acidic to neutral (pH 6.0–7.0).
- ⊕ **TEMPERAMENT** Peaceful and social.

These tiny corydoras are named for their pattern of black markings on a pale background, reminiscent of a panda's coloration. They will scavenge any uneaten food on the floor of the tank before it starts to decompose and reduce the water quality. Nevertheless, all corydoras should be given their own food. Partial water changes every three weeks will aid the water quality, and an effective filter will help oxygenation.

Corydoras melanistius

Black Sail Corydoras

- **ORIGINS** South America, where it is restricted to the Orinoco River in Venezuela.
- **SIZE** 2½ in (6 cm).
- **DIET** Live foods and prepared catfish foods.
- **WATER** Temperature 68–75°F (20–24°C); soft to hard (50–150 mg/l) and acidic to neutral (pH 6.0–7.0).
- **TEMPERAMENT** Peaceful and social.

A striking silvery background is the backdrop for the irregular rows of spotted markings extending over the body of these corydoras. There is a black band through each eye and a dark area at the front of the dorsal fin. It is not easy to distinguish the sexes, but females can be identified at spawning time as they swell with eggs. Black Sail Corydoras have been bred successfully in aquariums.

Black at the front of the sail-like dorsal fin

Corydoras paleatus

Peppered Corydoras

- **ORIGINS** South America, occurring in southeastern parts of Brazil and the La Plata basin, Argentina.
- **SIZE** 2¾ in (7 cm).
- **DIET** Live foods and prepared catfish foods.
- **WATER** Temperature 66–79°F (19–26°C); soft to hard (50–150 mg/l) and acidic to neutral (pH 6.0–7.0).
- **TEMPERAMENT** Peaceful and social.

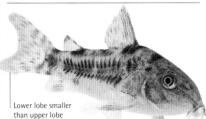

Lower lobe smaller than upper lobe

The Peppered Corydoras, one of the most widely kept species in the group, has black spots on the flanks that may sometimes fuse together to create a banded appearance. Iridescence may also be evident over this part of the body. This was one of the first tropical fish to be bred in Europe, having been spawned successfully in Paris in 1878, and it will reproduce readily in aquariums. Peppered Corydoras are relatively hardy fish.

Corydoras pygmaeus

Pygmy Catfish

- **ORIGINS** South America, occurring in Brazil in the Rio Madeira and its tributaries.
- **SIZE** 1 in (2.5 cm).
- **DIET** Live foods, algae, and prepared catfish foods.
- **WATER** Temperature 72–79°F (22–26°C); soft to hard (50–150 mg/l) and acidic to neutral (pH 6.0–7.0).
- **TEMPERAMENT** Peaceful and social.

These tiny corydoras are the smallest members of the group, as their name suggests. Aside from their size, however, they can be distinguished by their pattern of horizontal black stripes, with the central stripe broadening near the dorsal fin. They tend to swim throughout the tank more than most corydoras, which usually prefer the lower reaches. A small shoal of Pygmy Catfish will make ideal companions for other small, nonaggressive fish.

FEEDING BY TOUCH

One of the most distinctive features of corydoras catfish is the presence of six barbels arranged in pairs around their extendible mouthparts. Two pairs are located above the upper jaw, with the longest pair occasionally extending as far back as the gill openings. The third pair of barbels, which are very short and relatively inconspicuous, are on the chin itself. The mouth, situated on the underside of the body, enables these fish to feed by swimming just above the substrate and combing the surface with the barbels for edible items. Corydoras are also very efficient excavators, digging into the substrate with their mouthparts and using the barbels, which are covered in touch-sensitive cells, to direct them toward small worms and other concealed creatures. Sand rather than gravel is sometimes used as a substrate in an aquarium for corydoras, since it allows the fish to dig more easily. The drawback to sand, however, is that it is more likely to become compacted, so it needs to be turned over with a spoon at regular intervals. Since corydoras will rest on the substrate, any buildup of dirt here is likely to damage their barbels. To guard against this, use a gravel cleaner when a partial water change is undertaken.

Corydoras rabauti

Rabaut's Corydoras

🌐 **ORIGINS** South America, close to the confluence of the
Rio Negro and the Amazon, near Tabatinga, Brazil.

💧 **SIZE** 2½ in (6 cm).

🐟 **DIET** Live foods and prepared catfish foods.

🌊 **WATER** Temperature 72–79°F (22–26°C); soft to hard
(50–150 mg/l) and acidic to neutral (pH 6.0–7.0).

☺ **TEMPERAMENT** Peaceful and social.

Rabaut's Corydoras is one of the most striking
members of the group, thanks to the contrast
between its orange background color and its black
stripe. It is now usually considered to be synonymous
with Myers' Corydoras, which occurs in tributaries
of the upper Amazon. Females may lay their eggs at
or even just above the aquarium waterline on the
glass or utilize floating plants for this purpose.

Corydoras haraldschultzi

Harald Schultz's Corydoras

🌐 **ORIGINS** South America, occurring in Rio Tocantins and
Rio Araguaia in central Brazil.

💧 **SIZE** 3 in (7.5 cm).

🐟 **DIET** Live foods and prepared catfish foods.

🌊 **WATER** Temperature 75–82°F (24–28°C); soft to hard
(50–150 mg/l) and acidic to neutral (pH 6.0–7.0).

☺ **TEMPERAMENT** Peaceful and social.

The underparts of this corydoras are pinkish, while
the spots along the sides of the body are so close
together that they merge in places to form irregular
stripes. The name of this corydoras commemorates
the late Brazilian fish exporter Harald Schultz, who
brought a number of today's most popular fish,
including the Blue Discus *(see pp.142–143)*, to the
attention of aquarists around the world.

Spotted patterning
extends to the fins

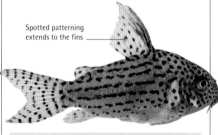

Corydoras agassizii

Agassiz's Corydoras

🌐 **ORIGINS** Western South America, where it is found in
the vicinity of Iquitos, Peru.

💧 **SIZE** 3½ in (9 cm).

🐟 **DIET** Live foods and prepared catfish foods.

🌊 **WATER** Temperature 72–79°F (22–26°C); soft to hard
(50–150 mg/l) and acidic to neutral (pH 6.0–7.0).

☺ **TEMPERAMENT** Peaceful and social.

This attractive corydoras has a patterning of dark
spots against a reddish background. Sexing is
difficult, although just prior to spawning, females
have a more rotund appearance. If you obtain
a small group of these catfish, however, the
likelihood is that you will have at least one pair
among them. Soft-water conditions are most
likely to encourage breeding behavior.

Corydoras robineae

Robina's Corydoras

🌐 **ORIGINS** South America, where it occurs in the upper
parts of the Rio Negro system, notably the Rio Aiuana.

💧 **SIZE** 3 in (7.5 cm).

🐟 **DIET** Live foods and prepared catfish foods.

🌊 **WATER** Temperature 72–79°F (22–26°C); soft to hard
(50–150 mg/l) and acidic to neutral (pH 6.0–7.0).

☺ **TEMPERAMENT** Peaceful and social.

The striped patterning of this species is most
evident toward the rear of the body, where black
horizontal stripes extend out across the caudal
fin. Robina's and other corydoras prefer subdued
lighting, which also helps to make their patterning
more apparent. In some lights, there may be
iridescence behind the gills and also down the
flanks. Like various members of the group, these
fish should be kept in shoals, rather than singly.

COMING UP FOR AIR

Corydoras originate from rivers and streams
with a relatively low oxygen content. Their
gills cannot extract enough oxygen from the
water to sustain them, so these small catfish
have also developed a means of breathing
air directly. Every now and again they will
suddenly swim upward, almost vertically
from the bottom, and break the water's
surface. They gulp down a mouthful of air
and dart back down again very quickly to the
substrate. Back on the bottom, the corydoras
supplement their blood-oxygen levels by
absorbing oxygen from the gulp of air,
which they store in their hind gut.

Corydoras adolfoi

Adolfo's Corydoras

- **ORIGINS** South America, occurring in Brazil, where it is found in the Rio Negro and the Rio Uapes.
- **SIZE** 2½ in (6 cm).
- **DIET** Live foods and prepared catfish foods.
- **WATER** Temperature 72–79°F (22–26°C); soft to hard (50–150 mg/l) and acidic to neutral (pH 6.0–7.0).
- **TEMPERAMENT** Peaceful and social.

Adolfo's Corydoras can be distinguished from the similarly patterned Imitator Corydoras (*C. imitator*) by its shorter snout and the red coloration in front of its dorsal fin. As with other corydoras, a well-planted breeding tank with a relatively low water level is recommended for spawning purposes. It should house a single female and two males. The eggs, about 30 of which are laid in groups on plant leaves or even on the sides of the aquarium, hatch after four days. The young will congregate on the substrate until they have digested the remains of their yolk sacs and become free-swimming. At this stage, they will need to be reared on fry foods, before taking flake when a week old. If the adults are well fed, they are likely to breed again a couple of weeks later.

Corydoras arcuatus

Skunk Corydoras

- **ORIGINS** South America, where it is found in parts of Ecuador, Peru, and Brazil.
- **SIZE** 3 in (7.5 cm).
- **DIET** Live foods and prepared catfish foods.
- **WATER** Temperature 72–79°F (22–26°C); soft to hard (50–150 mg/l) and acidic to neutral (pH 6.0–7.0).
- **TEMPERAMENT** Peaceful and social.

This pink-bodied species has a black stripe that curves along its upper body from the mouth to the base of the tail. Some individuals display slight black markings on the sides of the body as well. It is often confused with the Black-Top Corydoras, which has a very similar scientific name: *Corydoras acutus*. Corydoras can be kept at a slightly lower temperature than most tropical fish.

Arching black stripe below the dorsal fin

Corydoras metae

Masked Corydoras

- **ORIGINS** South America, occurring in Colombia's Rio Meta and its tributaries.
- **SIZE** 2 in (5 cm).
- **DIET** Prefers fresh and thawed live foods.
- **WATER** Temperature 72–79°F (22–26°C); soft to hard (50–150 mg/l) and acidic to neutral (pH 6.0–7.0).
- **TEMPERAMENT** Peaceful and social.

Prominent triangular dorsal fin

The black "mask" over the eyes of this corydoras explains its common name, although this feature is also seen in some other species. The body is plain, aside from a black area on the dorsal fin that extends down to the caudal peduncle. These catfish seek their food on the tank floor, so offer pellets that sink rapidly. Live foods such as small worms are a useful conditioning food for corydoras.

Scleromystax barbatus

Bearded Corydoras

- **ORIGINS** South America, occurring in the vicinity of both Rio de Janeiro and Sao Paulo, Brazil.
- **SIZE** 5 in (12.5 cm).
- **DIET** Live foods and prepared catfish foods.
- **WATER** Temperature 68–77°F (20–25°C); soft to hard (50–150 mg/l) and acidic to neutral (pH 6.0–7.0).
- **TEMPERAMENT** Peaceful and social.

This species, one of the largest known corydoras, exists as two distinct variants. Bearded Corydoras from around Sao Paulo are lighter in color and have less yellow on their bodies than those from the vicinity of Rio de Janeiro. These fish stay close to the bottoms of streams and rivers, hiding in aquatic vegetation when danger threatens. Include rocks and plants in their tank to provide hiding places.

Synodontis schoutedeni

Vermiculated Synodontis

- **ORIGINS** Africa, restricted to the vicinity of central Congo.
- **SIZE** 6 in (15 cm).
- **DIET** Prepared foods and live foods.
- **WATER** Temperature 72–79°F (22–26°C); soft to hard (50–150 mg/l) and acidic to alkaline (pH 6.0–7.5).
- **TEMPERAMENT** Placid, but may bully their own kind.

Synodontids are a popular group of catfish for the aquarium, partly because they do not grow to a large size, but also because they are active and attractively patterned in many cases. The marbling of the Vermiculated Synodontis makes this species easy to identify, although the precise patterning does differ slightly between individuals. During the daytime, a Vermiculated Synodontis may seek to conceal its presence in the aquarium by lying camouflaged and motionless on a rock, perhaps using its mouthparts to anchor itself in place. Vermiculated Synodontis tend to be quite solitary, and it is not wise to try to keep them in groups, because bullying is likely to occur. This will be not only distressing for the victims but also disruptive for other tank occupants.

Synodontis flavitaeniatus

Striped Synodontis

- **ORIGINS** Africa, occurring in Zaire, in the vicinity of the Stanley Pool and the Chiloango River.
- **SIZE** 8 in (20 cm).
- **DIET** Prefers vegetable matter and small live foods.
- **WATER** Temperature 73–79°F (23–26°C); soft to hard (50–200 mg/l) and acidic to alkaline (pH 6.5–8.0).
- **TEMPERAMENT** Nonaggressive.

This boldly patterned catfish can be distinguished by its horizontal yellowish stripes and extensive patches of blackish-brown coloration. The Striped Synodontis, which is not as widely available as most other members of this group, tends to burrow less in the substrate than related species. There need to be suitable retreats in its aquarium, in the form of pieces of bogwood or rocks. The Striped Synodontis is easy to cater to, since it will eat a wide range of foods, even browsing on algae.

Synodontis angelicus

Polkadot African Catfish

- **ORIGINS** Africa, occurring in Zaire around Mousembe and Stanley Pool, and also in Cameroon.
- **SIZE** 6½ in (18 cm).
- **DIET** Smaller live foods, algae, and prepared foods.
- **WATER** Temperature 72–82°F (22–28°C); soft to hard (50–150 mg/l) and acidic to alkaline (pH 6.5–7.5).
- **TEMPERAMENT** Placid.

A variable patterning of creamy-white spots on a dark chocolate-brown background characterizes these catfish. The spots become less pronounced with age, and the background color also becomes grayer. Polkadot African Catfish are largely nocturnal in their habits, emerging from their hiding places under cover of darkness. Like others of their kind, these catfish will excavate the substrate in search of edible items, and they may nibble aquatic plants.

The spots are of a consistent size

Synodontis greshoffi

Congo Synodontis

- **ORIGINS** Africa, occurring in the Congo basin, although absent from the lower Congo and the Luapula River.
- **SIZE** 8 in (20 cm).
- **DIET** Prefers live foods, but will take tablets.
- **WATER** Temperature 73–81°F (23–27°C); soft to hard (50–200 mg/l) and acidic to neutral (pH 6.5–7.0).
- **TEMPERAMENT** Placid.

The Congo Synodontis has relatively broad, creamy stripes set against a brown background color. Some authorities consider it to be the same species as *S. afrofischeri*, whose distribution extends to the Nile basin, where it occurs in Lake Victoria and other localities. The Congo Synodontis has care needs similar to other species in the group.

Synodontis alberti

Alberti Catfish

- **ORIGINS** Africa, occurring in various localities in Zaire, including Stanley Pool, Lukulu River, and near Kinshasa.
- **SIZE** 8 in (20 cm).
- **DIET** Prefers live foods, but will take tablets.
- **WATER** Temperature 72–81°F (22–27°C); soft to hard (50–200 mg/l) and acidic to alkaline (pH 6.0–8.0).
- **TEMPERAMENT** Nonaggressive.

A brown-blotched pattern helps to distinguish this species from its catfish relatives. The Alberti Catfish has special teeth in its lower jaw, which it uses to rasp algal growths off rocks; long barbels help the fish to orient itself and to search for food. The fish should be provided with algae (or a suitable substitute) to eat. It is best to grow algae for food in a separate tank, not the fish's home tank. This is because the algae require good illumination for productive growth, but the catfish prefers subdued lighting.

Erect dorsal fin

Forked caudal fin

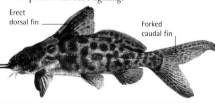

Synodontis notatus

Notatus

- **ORIGINS** West Africa, occurring in parts of Senegal, Gambia, Volta, Chad, and Niger.
- **SIZE** 8 in (20 cm).
- **DIET** Prefers live foods, but will take tablets.
- **WATER** Temperature 72–75°F (22–24°C); soft to hard (50–200 mg/l) and acidic to alkaline (pH 6.0–8.0).
- **TEMPERAMENT** Lively but nonaggressive.

Unfortunately, sexing the Notatus is visually impossible. In common with other synodontids, it is very reluctant to spawn in aquarium surroundings, and very little has been recorded about its breeding behavior. From what is known, the eggs take up to a week to hatch. The young can be reared on brine shrimp after about four days, when they are free-swimming.

Patterning is variable

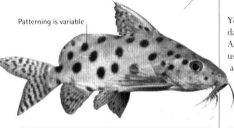

Synodontis eupterus

Featherfin Synodontis

- **ORIGINS** Africa, occurring in stretches of the White Nile, the Chad basin, and in parts of Niger.
- **SIZE** 6 in (15 cm).
- **DIET** Prefers live foods, but will take tablets.
- **WATER** Temperature 72–79°F (22–26°C); soft to hard (50–200 mg/l) and acidic to alkaline (pH 6.5–7.5).
- **TEMPERAMENT** Placid.

Young Featherfin Synodontis have a pattern of dark-brown spots on a light-brown background. As the fish mature, the spots darken and the body usually takes on a bluish hue, but it can sometimes appear more gray. The name Featherfin derives from the featherlike extension on the dorsal fin. These active catfish need a fine substrate in which they can dig and clear areas in the tank where they can swim.

AN UPSIDE-DOWN WORLD

Synodontis catfish as a group are sometimes described as upside-down catfish, because of the way in which they occasionally swim. This behavior is more common in some species than others, including the Upside-Down Synodontis (*S. nigriventris*) seen below. By swimming upside down, the fish can reach algae growing in areas that would otherwise be out of reach. When inverted like this, their mottled underparts help to conceal their presence from any predators above. Young synodontids start to swim in this way when they are approximately two months old.

Synodontis decorus

Clown Synodontis

- **ORIGINS** Africa, reported from localities in Zaire and Cameroon.
- **SIZE** 10 in (25 cm).
- **DIET** Live foods and prepared diets.
- **WATER** Temperature 73–81°F (23–27°C); soft to hard (50–200 mg/l) and acidic to alkaline (pH 6.0–8.0).
- **TEMPERAMENT** Placid.

Bold, dark spots and blotches on a pinkish-white body, together with a much finer pattern of speckling on the head, characterize the Clown Synodontis. A spectacular long, narrow extension develops at the top of the dorsal fin in adult fish. These catfish should be caught with great care, because the spines on their pectoral fins can easily become enmeshed in a net. Their large size means that they often stir up sediment when digging in the aquarium. This tends to cover fine-leaved plants and impede their growth. It is therefore a good idea either to choose plastic plants, which will not be affected by the digging habits of these catfish, or to restrict the choice to tough, broad-leaved plants. Clown Synodontis do not swim upside down on a regular basis but may occasionally be observed feeding at the surface in this fashion, taking freeze-dried or fresh live foods, such as mosquito larvae.

Chaetostoma cf. *thomsoni*

Bulldog Catfish L146a

- **ORIGINS** South America, occurring in streams in the mountains of Colombia.
- **SIZE** 5 in (12.75 cm).
- **DIET** Omnivorous.
- **WATER** Temperature 73–82°F (23–28°C); soft (50–100 mg/l) and acidic to neutral (pH 6.5–7.0).
- **TEMPERAMENT** Usually placid.

Sail-like dorsal fin

Body plates

The Bulldog is called L146a under the L-numbering system (*see p.21*) for identifying loricariids that have yet to be confirmed as distinct species. Males have larger pelvic fins than females, possibly to stop sperm from washing away during spawning before fertilization occurs. The female lays about 80 eggs in the open, often on a vertical surface, and the male guards them until they hatch. Regular partial water changes are vital during the rearing period.

Pterygoplichthys gibbiceps

Leopard Pleco L083

- **ORIGINS** South America, found in the Amazon in parts of both Peru and Brazil.
- **SIZE** 18 in (45 cm).
- **DIET** Omnivorous.
- **WATER** Temperature 73–81°F (23–27°C); soft (50–100 mg/l) and acidic to neutral (pH 6.5–7.0).
- **TEMPERAMENT** Peaceful and social.

The large black spots that characterize this catfish extend over the fins and body. Sexing is possible in mature individuals, because males have a more prominent genital papilla. Breeding in the typical aquarium is unlikely, because these fish normally burrow into the riverbank, where they create a nesting chamber for their eggs. The species is bred commercially, however, and a pink-eyed albino form, showing slight traces of the spotted patterning, is sometimes available.

Pseudacanthicus pirarara

Scarlet Acanthicus L025

- **ORIGINS** South America, although its precise distribution is presently unclear.
- **SIZE** 18 in (45 cm).
- **DIET** Mainly meat-based foods.
- **WATER** Temperature 73–79°F (23–26°C); soft (50–100 mg/l) and acidic to neutral (pH 6.5–7.0).
- **TEMPERAMENT** Very territorial.

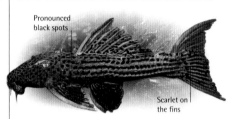

Pronounced black spots

Scarlet on the fins

These attractive, spotted plecos grow to a large size, with individuals displaying varying amounts of scarlet on their fins. Mature males are more slender-bodied than adult females. These fish can be aggressive and will defend their territory, becoming most active after dark. Although they prefer foods such as chopped shrimp and mussels, Scarlet Acanthicus are scavengers rather than active hunters, so they will take vegetable matter as well.

Baryancistrus xanthellus

Golden Nugget Pleco L018

- **ORIGINS** South America, although its precise distribution is presently unclear.
- **SIZE** 14 in (35 cm).
- **DIET** Thawed foods and vegetables.
- **WATER** Temperature 77–86°F (25–30°C); soft (50–100 mg/l) and neutral (pH 7.0).
- **TEMPERAMENT** Territorially aggressive.

Yellow spots on the body and yellow tips to the dorsal and caudal fins identify the Golden Nugget Pleco. Juveniles cannot be sexed, but adult females can be differentiated from males by their narrower, more rounded foreheads. When these fish are in breeding condition, they display by digging in the substrate. Spawning occurs in caves, and an airstone placed nearby will ensure that there is both gentle water movement and sufficient oxygen for the brood. The fry grow slowly, measuring less than 2 in (5 cm) after six months. Brine shrimp is a suitable first food for the young. Adults eat a variety of foods, including thawed bloodworm and shrimp, and will even gnaw on slices of cucumber or zucchini.

Hypostomus plecostomus

Pleco Hypostomus

- **ORIGINS** South America, widely distributed in northern parts; also occurs on Trinidad.
- **SIZE** 12 in (30 cm).
- **DIET** Mainly vegetarian.
- **WATER** Temperature 68–86°F (20–30°C); soft (50–100 mg/l) and neutral (pH 7.0).
- **TEMPERAMENT** Relatively peaceful.

The Pleco Hypostomus' mottled patterning varies throughout its range, making accurate identification difficult. Mature males have thicker pectoral fins, which turn reddish-pink in the spawning period. These fish tend not to occur in fast-flowing water, but they still need good filtration in their tank. In the wild, eggs are usually laid in a riverbank cavity below the water line, rarely in underwater caves.

Peckoltia species

Orange-Tipped Pleco L076

- **ORIGINS** South America, although its precise distribution is unclear.
- **SIZE** 6 in (15 cm).
- **DIET** Mainly meat-based foods.
- **WATER** Temperature 73–82°F (23–28°C); soft (50–100 mg/l) and acidic (pH 6.0).
- **TEMPERAMENT** Relatively peaceful.

The characteristic orange tips on the rear of the fins distinguish these Plecos from similar species. The other parts of the fins are a darker, olive-brown shade, and the body is slightly silvery-gray with dark markings. Orange-Tipped Plecos can be sexed by the rays on the pectoral fins; in males, they have a serrated edge with toothlike projections known as odontodes, while in females, the rays are smooth-edged. The female's body is also wider at this point than the male's. Pairs will spawn in caves and can be rather territorial during the breeding period. To reduce tensions, divide their aquarium into discrete sections, and include rockwork and bogwood in the tank. These plecos will not usually damage aquatic plants, and they generally ignore any algae growing in the aquarium. Orange-Tipped Plecos are most active toward dusk. They prefer to feed on items such as chopped shrimp. It is important not to allow the nitrate level in the tank to build up, so make regular partial water changes to keep this under control.

Male

Dekeyseria pulchra

Pretty Peckoltia L103

- **ORIGINS** South America, occurring in the Rio Negro, close to Moura, Brazil.
- **SIZE** 2¼ in (6 cm).
- **DIET** Mainly plant matter, including algae.
- **WATER** Temperature 75–82°F (24–28°C); hard (100–150 mg/l) and neutral (pH 7.0).
- **TEMPERAMENT** Peaceful.

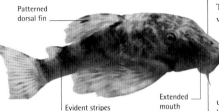

Patterned dorsal fin

Evident stripes

Extended mouth

Broad, wavy lines over the body and fins typify the Pretty Peckoltia, which is valued for its small size and docile nature. It is vital to establish favorable environmental conditions and profuse algal growth in the tank before acquiring these fish, because plant matter forms the bulk of their diet. An aquarium for Pretty Peckoltias should include a range of hiding places, such as partially buried flowerpots. More delicate plants may be eaten.

Dekeyseria species

Butterfly Peckoltia L052

- **ORIGINS** South America, occurring in parts of the Orinoco basin.
- **SIZE** 5 in (12.5 cm).
- **DIET** Omnivorous.
- **WATER** Temperature 73–82°F (23–28°C); soft (50–100 mg/l) and acidic (pH 6.0).
- **TEMPERAMENT** Territorial.

The coloration of these catfish, which can vary widely, is partly dependent on their background. Their patterning of alternating light and dark patches (seen in the example below) disappears rapidly if the fish are transferred to a tank with a dark substrate and may also be affected by the lighting. Pairs spawn in the relative safety of a cave, with the female laying up to 120 eggs. These may take 10 days to hatch, by which time the fry have used up virtually all of their yolk sacs. The young of this species can be reared on brine shrimp.

UNIQUE EYE-LOBE

The eye structure of loricariids is unique among vertebrates. On close examination, a protruding lobe can be seen on the iris. This lobe, which is often colorful, forms the outer ring of the eye. Loricariid catfish lack the pupillary reflex of other vertebrates, in which the size of the pupil adjusts in response to changes in the lighting conditions. Instead, they use their eye-lobe to regulate the amount of light entering the eye. In well-lit surroundings, the lobe enlarges to cover more of the pupil, while in dim light it recedes to allow in as much light as possible.

Hypancistrus species

Tiger Clown Pleco L066

- ORIGINS South America, where it is found in Brazil's Rio Xingu.
- SIZE 8 in (20 cm).
- DIET Omnivorous.
- WATER Temperature 73–82°F (23–28°C); soft (50–100 mg/l) and acidic (pH 6.0).
- TEMPERAMENT Territorial when mature.

A delicate patterning of white stripes on a black background distinguishes the Tiger Clown Pleco. These markings are highly variable, allowing individuals to be recognized easily. Females lack the spines on the pectoral and dorsal fins and have a broader body shape. Tiger Clown Plecos are most active after dark and eat a wide variety of foods. Shelled peas, either fresh or thawed, are a valuable source of vegetable matter, as are cucumber slices.

Ancistrus dolichopterus

Big-Fin Bristlenose L144a

- ORIGINS South America, recorded in Brazil's Rio Negro basin.
- SIZE 4 in (10 cm).
- DIET Omnivorous.
- WATER Temperature 73–77°F (23–25°C); hard (100–150 mg/l) and acidic (pH 6.5).
- TEMPERAMENT Actively territorial.

Sexing is straightforward in this species—only males develop the bristlelike projections on the head. In both sexes, the brown body carries a pattern of lighter spots. Small live foods help to trigger spawning behavior. The yellowish eggs are laid in a cave and watched over by the female; she often fans water over them with her fins. They hatch about five days later, and the fry are free-swimming within a week. They can then take powdered flake.

Male's "bristles"

Hypancistrus zebra

Zebra Pleco L106

- ORIGINS South America, where it is confined to the Rio Xingu in Brazil.
- SIZE 3¼ in (8 cm).
- DIET Relatively carnivorous.
- WATER Temperature 79–86°F (26–30°C); soft (50–100 mg/l) and neutral (pH 7.0).
- TEMPERAMENT Placid.

Circular white band behind the head

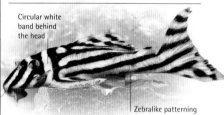

Zebralike patterning

The straight black bands across its body separate this catfish from the wavy-lined form, known as L098. Males have wider heads than females, when seen from above, and the first ray of the pectoral fin is broader. Include sand, rocks, and bogwood in the tank to mimic this pleco's natural habitat. The water must be well filtered and well oxygenated to encourage spawning, which occurs in a cave. The eggs are laid in batches and hatch in about a week.

Scobinancistrus aureatus

Gold-Spotted Pleco L014

- ORIGINS South America, where it appears to occur only in Brazil's Rio Xingu.
- SIZE 12 in (30 cm).
- DIET Catfish pellets and live foods.
- WATER Temperature 77–84°F (25–29°C); soft (50–100 mg/l) and acidic to neutral (pH 6.0–7.0).
- TEMPERAMENT Territorial when mature.

The coloration of burnished gold on the fins and contrasting yellower spots on the body is much brighter in young Gold-Spotted Plecos than in adults. Males have a broader first ray on the pectoral fin and a slightly bristly appearance on their heads. At present, no breeding records exist for these catfish, which were only officially described in 1994. They require well-filtered, moving water in their tank, with rocks and a sandy base to mimic their natural habitat. Gold-Spotted Plecos are not particularly difficult to keep, despite their large size, but they do become more aggressive as they grow larger. They feed near the substrate and can be persuaded to take catfish pellets and live foods. If there are retreats in the tank, it is important that uneaten food does not accumulate in them, since it will decrease the water quality and harm the fish.

RASPING SUCKERMOUTHS

Loricariids are sometimes called suckermouth catfish because of their powerful mouthparts, which allow them to anchor onto rockwork and submerged wood even when faced with a strong current. Many species also appear to need bogwood, which they eat, rasping off fragments with their teeth. If kept without bogwood, these catfish are unlikely to thrive. Aside from keeping their teeth in check, the bogwood may also aid the digestive process, perhaps by providing additional nutrients for the beneficial microbes in the fish's digestive tract, which help to break down plant matter.

Squaliforma emarginata

Black Hi-Fin Pleco L108

- **ORIGINS** South America, being common throughout much of the Amazon region.
- **SIZE** 7 in (18 cm).
- **DIET** Mainly vegetarian.
- **WATER** Temperature 73–82°F (23–28°C); soft (50–100 mg/l) and acidic (pH 6.0).
- **TEMPERAMENT** Territorial when mature.

The difficulty in naming plecos with certainty, even with the L-number system, is well illustrated by the Black Hi-Fin. Although it is often described as L108, this fish is assigned a variety of other L-numbers in different parts of its range. Exact coloration may vary, but all forms display a dense pattern of black spots on a darkish background. More than 600 loricariids are now known, making it a large and diverse family, and others still await discovery.

Squaliforma species

Longtail Pleco L131

- **ORIGINS** South America, being common throughout much of the Amazon region.
- **SIZE** 7 in (18 cm).
- **DIET** Mainly vegetarian.
- **WATER** Temperature 73–82°F (23–28°C); soft (50–100 mg/l) and acidic (pH 6.0).
- **TEMPERAMENT** Territorial when mature.

The elongated body of this fish, which narrows toward the caudal fin, displays a dense pattern of dark spots. Although the Longtail Pleco has been assigned the number L131, it is actually regarded as effectively being the same fish as L108 (*see top right*), but from a different area within their overall range. It has also been suggested that the loricariids assigned numbers L11, L035, L116, L153, L166, and L195 are, aside from minor variations in coloration and patterning, examples of this same species that have been described differently.

Panaque nigrolineatus

Royal Panaque L191

- **ORIGINS** South America, ranging from Colombia and Venezuela through central parts of the Amazon basin.
- **SIZE** 13½ in (34 cm).
- **DIET** Mainly vegetable matter.
- **WATER** Temperature 72–86°F (22–30°C); soft (50–100 mg/l) and neutral (pH 7.0).
- **TEMPERAMENT** Territorial.

The greenish background color of these catfish is marked with black lines, which appear straight on the face but are often more wavy elsewhere on the body. In common with other plecos, these fish need bogwood in their tank, on which they can rasp. Their diet can include shelled peas, but try experimenting with a range of plant foods, since these fish can be picky eaters.

Spectracanthicus punctatissimus

Peppermint Pleco L030

- **ORIGINS** South America, in the lower Rio Tocantins and the Rio Xingu in Para state, Brazil.
- **SIZE** 6 in (15 cm).
- **DIET** Mainly vegetarian.
- **WATER** Temperature 73–82°F (23–28°C); soft (50–100 mg/l) and acidic (pH 6.0).
- **TEMPERAMENT** Territorial when mature.

A black background decorated with white spots helps to identify this particular loricariid. It now seems that the Peppermint Pleco ascribed the number L030 is merely the juvenile form of the variety recognized as L031, which has a finer spotted pattern. L030 may appear more common simply because juveniles are caught more often than adults. Peppermint Plecos need to be kept in well-oxygenated water with a low nitrate reading.

Tall dorsal fin

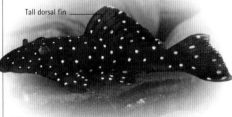

DIFFERING LIFESTYLES

Many pimelodid catfish are solitary by nature. They are predatory and grow to a large size, demanding an extensive area in which to hunt. However, a number of smaller species, including the Pictus Catfish (*Pimelodus pictus*) shown here, will associate in groups and can be kept together more easily in the home aquarium. Like various other pimelodids, Pictus Catfish are largely nocturnal. A special light that mimics the qualities of moonlight must be placed over the tank if the fascinating behavior of these fish is to be observed.

Microglanis iheringi

Bumblebee Catfish

- **ORIGINS** Northwest South America, where it is found in parts of Colombia and Venezuela.
- **SIZE** 4 in (10 cm).
- **DIET** Prepared foods and live foods.
- **WATER** Temperature 70–77°F (21–25°C); soft to hard (50–200 mg/l) and neutral (pH 7.0).
- **TEMPERAMENT** Will prey on smaller companions.

The orange and brownish-black bands on this fish resemble the markings of a bumblebee. It is often confused with two similar-looking species with the same common name, *Pseudomystus siamensis* from Asia and *Microglanis poecilus* from South America. Nocturnal by nature, this catfish can be kept safely with other nonaggressive species of a similar size, but it will prey on smaller companions, especially the fry of livebearing fish.

Hemisorubim platyrhynchos

Spotted Catfish

- **ORIGINS** South America, from Venezuela southward as far as Paraguay.
- **SIZE** 15 in (40 cm).
- **DIET** Prefers live foods of various types.
- **WATER** Temperature 72–77°F (22–25°C); soft to hard (50–250 mg/l) and neutral (pH 7.0).
- **TEMPERAMENT** Will prey on smaller companions.

The size of this catfish means that it requires suitably spacious accommodation from the outset. The Spotted Catfish has a large, rather bill-like mouth, with prominent barbels and an elongated body shape. Its blotched appearance includes a variable pattern of dark spots on its sides. Spotted Catfish naturally frequent cooler, deeper waters and will rest on a bed of pebbles or a similar raised area during the day, becoming active as darkness falls.

Platystomatichthys sturio

Sturgeon Catfish

- **ORIGINS** South America, where it is present throughout the entire Amazon region.
- **SIZE** 24 in (60 cm).
- **DIET** Worms and other live foods.
- **WATER** Temperature 72–81°F (22–27°C); soft to hard (50–200 mg/l) and neutral (pH 7.0).
- **TEMPERAMENT** Predatory.

The name of this fish comes from its snout, which curves slightly upward, rather like that of a sturgeon (*Acipenser* species). This adaptation enables it to dig in the substrate for its favorite food, which is worms of various types, although older individuals will also prey on other fish. A sandy floor covering in the aquarium and good filtration are essential, especially for bigger specimens. The maxillary barbels of the Sturgeon Catfish are remarkably long, extending in some cases not just along the length of the entire body but beyond the end of the caudal fin as well. The presence of a red spot roughly midway along the barbels is quite normal, and not a cause for concern. It is very important, however, that Sturgeon Catfish are able to extend their barbels fully, so the home aquarium must be more than twice as wide as the length of the individual barbels.

Pimelodus pictus

Pictus Catfish

- **ORIGINS** Northern South America, especially in the vicinity of Mitu, Colombia.
- **SIZE** 6 in (15 cm).
- **DIET** Prepared foods and live foods.
- **WATER** Temperature 72–77°F (22–25°C); soft to hard (50–100 mg/l) and neutral (pH 7.0).
- **TEMPERAMENT** Placid.

The Pictus Catfish is attractively patterned, with prominent black spots set against a silvery background. Its docile nature makes it suitable to be kept in a shoal, or mixed with other nonaggressive fish requiring similar water conditions. Feeding, too, presents no difficulties. Although these catfish are nocturnal, they may eat during the day under subdued lighting. Beware: their sharp pectoral fins may get stuck in netting.

Brachyrhamdia meesi

Pin Catfish

- **ORIGINS** South America, in the Amazon region; its precise distribution is presently unclear.
- **SIZE** 3¼ in (8 cm).
- **DIET** Live foods favored.
- **WATER** Temperature 73–81°F (23–27°C); soft to hard (50–200 mg/l) and neutral (pH 7.0).
- **TEMPERAMENT** Placid with unrelated fish of similar size.

A pale body, a dark vertical mark at the base of the dorsal fin, and a midline stripe along the flanks are the distinguishing features of this catfish. Its origins are mysterious, because it first appeared in a consignment of catfish exported from Belem, Brazil, during the mid-1980s, and the area where it naturally occurs has still not been clearly identified. These pimelodid catfish are likely to be territorial toward their own kind unless kept in groups of five or six individuals.

Leiarius pictus

Sailfin Marbled Catfish

- **ORIGINS** South America, found principally in the Amazon and its tributaries.
- **SIZE** 24 in (60 cm).
- **DIET** Live foods, especially worms.
- **WATER** Temperature 73–81°F (23–27°C); soft to hard (50–200 mg/l) and neutral (pH 7.0).
- **TEMPERAMENT** Predatory and intolerant.

The attractive patterning of young Sailfin Marbled Catfish fades as they grow older. This is a species for dedicated catfish enthusiasts, not just because of the size that these fish can attain, but also because of the length of their barbels. A very large aquarium or even an indoor pond is needed to provide them with enough swimming space and to ensure that their barbels can move freely in all directions. Sailfin Marbled Catfish are active during the day, retreating into underwater caves at night.

Pimelodus ornatus

Ornate Pimelodus

- **ORIGINS** South America, from Guyana and Surinam through the Amazon region, as far south as Paraguay.
- **SIZE** 11 in (28 cm).
- **DIET** Prepared foods and live foods.
- **WATER** Temperature 75–81°F (24–27°C); soft to hard (50–200 mg/l) and neutral (pH 7.0).
- **TEMPERAMENT** Can be disruptive.

The key features of this species are its two silvery stripes—one across the top of the body behind the gills, the other at the front of the dorsal fin (which also bears a grayish-black area). As with other pimelodids, the Ornate Pimelodus has three sets of barbels around its mouth. This very active catfish spends the day swimming, using its barbels to seek out food on the bottom of the aquarium. This fish cannot be trusted with significantly smaller companions, and breeding in aquarium surroundings is very unlikely.

Distinct lobes to the caudal fin

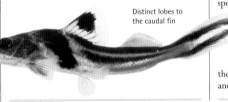

Phractocephalus hemioliopterus

Red-Tailed Catfish

- **ORIGINS** Northern parts of South America, including Guyana and Venezuela.
- **SIZE** Up to 42 in (106 cm).
- **DIET** Live foods preferred.
- **WATER** Temperature 73–79°F (23–26°C); soft (50–100 mg/l) and neutral (pH 7.0).
- **TEMPERAMENT** Predatory.

Characteristic red caudal fin

Mottled patterning on the head

It is easy to forget when seeing juveniles of this species that they grow rapidly into very large fish. Despite this caution, Red-Tailed Catfish enjoy a dedicated following, partly because these intelligent fish soon come to recognize their owners sufficiently to feed from the hand. Having fed, they will rest on the bottom to digest their meal. Efficient filtration and regular partial water changes are essential.

Auchenoglanis occidentalis

Giraffe Catfish

- ⊕ ORIGINS Africa, in the Nile, Congo, and Niger Rivers, plus Lake Tanganyika, Lake Chad, and others.
- ◷ SIZE 18 in (45 cm).
- ◑ DIET Prepared catfish foods and live foods.
- ◍ WATER Temperature 70–77°F (21–25°C); soft to hard (50–300 mg/l) and acidic to alkaline (pH 6.5–7.5).
- ⊕ TEMPERAMENT Not to be trusted with smaller fish.

The distinctive soft, mottled, brown-and-white patterning of this catfish resembles that of a giraffe. The actual markings may vary between individuals, with several dark dots apparent on both sides of the body. The Giraffe Catfish is active during the day, when it searches the substrate for edible items. It requires a fine floor covering, bogwood for retreats, and spacious accommodation in view of its likely adult size. Successful aquarium breeding is very unlikely.

Sail-like dorsal fin

Pseudomystus siamensis

Asian Bumblebee Catfish

- ⊕ ORIGINS Southeast Asia, occurring in both Thailand and Cambodia (Kampuchea).
- ◷ SIZE 8 in (20 cm).
- ◑ DIET Prepared catfish foods, algae, and live foods.
- ◍ WATER Temperature 68–77°F (20–25°C); soft to hard (50–300 mg/l) and acidic to alkaline (pH 6.5–7.5).
- ⊕ TEMPERAMENT Not social.

Black blotch on each caudal fin lobe

A coloration of light and dark bands characterizes these Asian catfish, which have relatively short barbels and a tubular body shape. They do not live well together, especially once mature, and are best housed individually. Asian Bumblebees are nocturnal by nature, so try to incorporate a variety of suitable retreats into their aquarium. They are quite adaptable in terms of their water chemistry needs—they even occur in brackish waters in parts of their range—but avoid making sudden changes, which will be stressful for the fish.

Mystus vittatus

Banded Mystus

- ⊕ ORIGINS Asia, occurring in parts of India, Myanmar (Burma), and Nepal, plus Thailand and Malaysia.
- ◷ SIZE 8 in (20 cm).
- ◑ DIET Prepared catfish foods and live foods.
- ◍ WATER Temperature 72–82°F (22–28°C); soft to hard (50–300 mg/l) and acidic to alkaline (pH 6.0–7.5).
- ⊕ TEMPERAMENT Reasonably peaceful.

Stripes running horizontally down the sides of the body help to identify this catfish, which also has long barbels and a blackish spot behind the gills. Banded Mystus are lively during the day and can be kept safely with other fish of similar size. When mating, the pair utter sounds resembling tweeting birds, and the large eggs are laid among aquatic vegetation. Unfortunately, breeding in aquarium surroundings is uncommon.

Mystus nigriceps

Two-Spot Catfish

- ⊕ ORIGINS Southeast Asia, extending from Thailand to the islands of Java, Sumatra, and Borneo.
- ◷ SIZE 6 in (15 cm).
- ◑ DIET Prepared catfish foods, algae, and live foods.
- ◍ WATER Temperature 68–77°F (20–25°C); soft to hard (50–200 mg/l) and acidic to alkaline (pH 6.5–7.8).
- ⊕ TEMPERAMENT Relatively social.

The Two-Spot can be differentiated from the Banded Mystus (*see top right*) by the presence of a dark spot on the caudal peduncle. The body is pinkish, and in healthy fish there is a golden hue around the black spot behind the gill cover. These active fish can be housed with their own kind, or with placid fish of a similar size, but smaller companions are likely to be eaten. Two-Spots prefer live foods, and will comb the floor of the aquarium seeking edible items, such as worms. They require relatively clear areas in their tank for swimming, as well as an efficient filtration system. Breeding is unusual, but placing several females (recognizable by their plumper appearance) in with a male and lowering the water temperature to 68°F (20°C) may trigger spawning behavior.

PROTECTIVE FIN RAYS

Many catfish of the Bagridae family, and some members of other families, too, are protected by sharp rays on their dorsal fins, which can be locked in an upright position. The small ray at the front of this fin is not conspicuous, but it plays a vital part in this process. When under threat, the catfish immediately raises its dorsal fin, causing this ray to hold the longer second ray upright. With this defensive mechanism in place, the catfish is harder for a predator to swallow. With luck, the catfish may be spat out largely unharmed, allowing it to swim away to safety.

Bathybagrus stappersii

Stappers' Catfish

- **ORIGINS** East Africa, found in Lake Tanganyika, within Zairean territory.
- **SIZE** 8 in (20 cm).
- **DIET** Prepared catfish foods, algae, and live foods.
- **WATER** Temperature 72–75°F (22–24°C); hard (150–300 mg/l) and alkaline (pH 8.0–9.0).
- **TEMPERAMENT** Not to be trusted with small fish.

This silver-bodied catfish has large eyes, which are especially prominent in juveniles of the species. Like many other African catfish, Stappers' Catfish is only occasionally available, but it makes an interesting addition to an aquarium housing larger Lake Tanganyika cichlids, with which it is often imported. The scientific name commemorates the discoverer of the fish, Dr. L. Stappers.

Chrysichthys ornatus

Mottled Catfish

- **ORIGINS** West Africa, with its distribution centered on the Congo and Zaire.
- **SIZE** 8 in (20 cm).
- **DIET** Prepared catfish foods and live foods.
- **WATER** Temperature 68–77°F (20–25°C); soft to hard (50–150 mg/l) and acidic to alkaline (pH 6.5–7.5).
- **TEMPERAMENT** Predatory, solitary as adults.

These catfish display a combination of blackish and silvery coloration, although these areas are not well defined, creating obvious mottling over much of the body. There are sharp spines on the dorsal and pectoral fins. Young Mottled Catfish are reasonably social and are also active during the day, but as they become adults, they develop into more solitary, nocturnal predators. Part of the area of the substrate should consist of sand, in which these catfish can dig. Little is known about the breeding habits of this bagrid.

Mottling extends to the fins

Notoglanidium macrostoma

Flatnose Catfish

- **ORIGINS** Found in West Africa, where it is restricted to Niger and Upper Volta.
- **SIZE** 10 in (25 cm).
- **DIET** Prepared catfish foods and live foods.
- **WATER** Temperature 73–81°F (23–27°C); soft to hard (50–150 mg/l) and acidic to alkaline (pH 6.5–7.5).
- **TEMPERAMENT** Predatory and solitary.

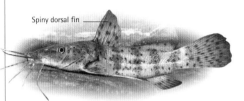

Spiny dorsal fin

The background of the Flatnose Catfish is pinkish in color, but brown spots, often overlapping, cover the entire body, including the fins. The sexes are similar in appearance. Shy by nature, the Flatnose Catfish requires a secluded environment, with the surface of the water covered by floating plants. Adequate retreats must be provided, and tank furniture should be rearranged to suit the requirements of the catfish as it grows.

Hyalobagrus ornatus

Ornate Bagrid

- **ORIGINS** Southeast Asia, occurring in Indonesia and Malaysia, where it frequents the Muar River.
- **SIZE** 1½ in (4 cm).
- **DIET** Prepared catfish foods and small live foods.
- **WATER** Temperature 68–77°F (20–25°C); hard (100–200 mg/l) and near neutral (pH 7.0).
- **TEMPERAMENT** Highly social by nature.

A dark streak running along each side of the body and a slightly golden area on the side of the head behind the eye help to distinguish this tiny catfish. The body is otherwise sufficiently transparent to allow the green eggs in the female's reproductive tract to be visible prior to spawning. The eggs are scattered among fine-leaved aquatic vegetation such as Java Moss (*Vesicularia dubyana*). These bagrids are very active swimmers, and they should always be kept in shoals. They can be housed safely with other nonaggressive species.

CICHLIDS

This large group of fish (the name of which is pronounced "sick-lids") originates mainly from Central and South America and Africa, although three species (including one recent rediscovery) are known to come from Asia. The characteristics and requirements of cichlids are as diverse as their origins, and most make excellent pets. Some individuals may learn to recognize their owner and even become tame enough to take food from the hand. The breeding behavior of many cichlids is fascinating, because these fish typically display strong parental instincts. Their care of the young can be observed closely in aquarium surroundings.

The **Ram or Butterfly Cichlid** (*Mikrogeophagus ramirezi*) is one of the smaller members of the group, growing to less than 3 in (7.5 cm) long.

Rocio octofasciata

Jack Dempsey Cichlid

- **ORIGINS** Central America, occurring on Mexico's Yucatán Peninsula and in Guatemala and Honduras.
- **SIZE** 8 in (20 cm).
- **DIET** Prepared cichlid foods, algae, and live foods.
- **WATER** Temperature 72–77°F (22–25°C); hard (100–150 mg/l) and neutral (pH 7.0).
- **TEMPERAMENT** Intolerant and aggressive.

Named after the late American boxer because of its pugnacious nature, the Jack Dempsey Cichlid displays a series of variable turquoise and yellowish markings over its body. Sexing can be done on the basis of the fins: in males, the anal and dorsal fins have pointed tips, with the latter displaying red edging, too. Pairs look after their young together, corralling them at first in special spawning pits excavated in the substrate. The eggs, numbering as many as 800, are laid on cleaned rockwork.

Amphilophus citrinellus

Midas Cichlid

- **ORIGINS** Central America, occurring in Costa Rica, Nicaragua, and Honduras.
- **SIZE** 12 in (30 cm).
- **DIET** Prepared cichlid foods and live foods.
- **WATER** Temperature 70–77°F (21–25°C); hard (100–150 mg/l) and neutral (pH 7.0).
- **TEMPERAMENT** Aggressive, destructive, and territorial.

The Midas gets its name from its gold coloration. Males develop a pronounced swelling, known as a nuchal hump, on the head. Pairs should be housed on their own in a large, sparsely decorated tank. Include rockwork, which will serve as a spawning site. Partial water changes will mimic the rains that trigger spawning activity in the wild. Females will lay up to 1,000 eggs. The fry feed on mucus on the flanks of the adults.

Young are duller in color than adults

Juvenile

Thorichthys meeki

Firemouth Cichlid

- **ORIGINS** Central America, occurring in Guatemala and on Mexico's Yucatán Peninsula.
- **SIZE** 6 in (15 cm).
- **DIET** Prepared cichlid foods and live foods.
- **WATER** Temperature 70–75°F (21–24°C); hard (150–200 mg/l) and neutral (pH 7.0).
- **TEMPERAMENT** Territorial and aggressive.

The fiery red on the throat and underside of the body distinguishes male Firemouths, pictured above, from females. A pair becomes aggressive when spawning, and will dig more frequently in the substrate. Firemouths are dedicated parents, watching over the eggs and then shepherding the young into a spawning pit. Brine shrimp make a valuable rearing food for the young at this stage.

Amphilophus labiatus

Red Devil

- ⊕ **ORIGINS** Central America, restricted to Lake Xiloa, Lake Nicaragua, and Lake Managua in Nicaragua.
- ⊜ **SIZE** 10 in (25 cm).
- ⊜ **DIET** Prepared cichlid foods and live foods.
- ⊜ **WATER** Temperature 75–79°F (24–26°C); hard (100–150 mg/l) and neutral (pH 7.0).
- ⊜ **TEMPERAMENT** Aggressive and territorial.

Like a number of other Central American cichlid species, the Red Devil is relatively adaptable in terms of its water chemistry needs. It is reddish, but the precise depth of coloration differs between individuals. The female of a pair, discernible by her blunt genital papilla and smaller size, may lay up to 700 eggs, guarding them until they hatch after about three days. It may take a further week for the fry to become free-swimming.

Cichlasoma severum

Severum

- ⊕ **ORIGINS** Northern parts of South America, extending throughout the Amazon basin.
- ⊜ **SIZE** 8 in (20 cm).
- ⊜ **DIET** Prepared cichlid foods and live foods.
- ⊜ **WATER** Temperature 73–77°F (23–25°C); soft (50 mg/l) and acidic (pH 6.0–6.5).
- ⊜ **TEMPERAMENT** Territorial.

Juvenile

The black band extending from the dorsal to the ventral fin is a key feature of the Severum. Young individuals show a series of such bands running down the sides of the body, but these fade as they grow older. Once mature, males can be identified by the elongated tips on the dorsal and ventral fins, and by the reddish-brown patterning on the head. Pairs may not always prove compatible.

CONFLICT RESOLUTION

Although many cichlids are aggressive, most disputes are resolved without actual physical conflict. The bright red of the Firemouth Cichlid (*Thorichthys meeki*), shown below, warns other fish to steer clear. If this does not work as a deterrent, a Firemouth will inflate its throat and flare out its gill covers. This makes the fish appear larger and more intimidating and may persuade a would-be rival to back down and swim away. In the aquarium, however, conflict is more likely because the fish cannot avoid one another.

Amatitlania nigrofasciata

Convict Cichlid

- ⊕ **ORIGINS** Central America, ranging from Guatemala southward to Panama.
- ⊜ **SIZE** 6 in (15 cm).
- ⊜ **DIET** Prepared cichlid foods, algae, and live foods.
- ⊜ **WATER** Temperature 68–77°F (20–25°C); hard (100–150 mg/l) and neutral (pH 7.0).
- ⊜ **TEMPERAMENT** Aggressive and territorial.

The black and bluish stripes on the body of this fish are not dissimilar to the pattern of old prison uniforms—hence the name Convict Cichlid. The female lacks the extensions to the dorsal and ventral fins seen in the male, but she is more colorful, with yellowish-orange underparts. A rare albino variant has also been bred. Provide a relatively bare aquarium for spawning purposes, but add a clay flowerpot and some slate to give a choice of egg-laying sites. As with related species, bloodworm and other live foods are important to keep these fish in good condition. Convict Cichlids also feed readily on vegetation, so they should only be housed with tough plants, which they are unlikely to destroy. Pairs will usually care for their young, but should they ignore them, it may be better to transfer the brood to a separate tank where they can be reared safely on their own.

Andinoacara pulcher

Blue Acara

- **ORIGINS** Central and northern South America, from Panama to Colombia and Venezuela. Also in Trinidad.
- **SIZE** 8 in (20 cm).
- **DIET** Prepared cichlid foods and live foods.
- **WATER** Temperature 64–77°F (18–25°C); soft (50 mg/l) to neutral (pH 7.0).
- **TEMPERAMENT** Territorial but not very aggressive.

Although the exact coloration of Blue Acaras varies between individuals, they all display obvious bluish markings set against a darker background. Mature males can usually be identified by the extensions at the rear of the dorsal and anal fins. Regular water changes to maintain water quality are very important for Blue Acaras, while raising the water temperature slightly, up to 82°F (28°C), should encourage spawning.

Cleithracara maronii

Keyhole Cichlid

- **ORIGINS** Northwestern South America, restricted to parts of Guyana.
- **SIZE** 6 in (15 cm).
- **DIET** Prepared cichlid foods and live foods.
- **WATER** Temperature 72–77°F (22–25°C); soft (50 mg/l) and acidic (pH 6.5).
- **TEMPERAMENT** Peaceful.

Captive-bred strains of the Keyhole Cichlid are often much smaller in size than wild stock. A black stripe passing through the eye and roughly circular black markings on the upper body adjacent to the dorsal fin are characteristic features of this cichlid. It is less destructive in aquariums than its relatives, rarely digging or damaging plants. The female will lay up to 300 eggs and then guard them until they hatch. She will also care for the resulting fry.

Andinoacara rivulatus

Green Terror

- **ORIGINS** Northwestern South America, occurring in western Ecuador and central Peru.
- **SIZE** 8 in (20 cm).
- **DIET** Prepared cichlid foods and live foods.
- **WATER** Temperature 68–77°F (20–25°C); soft (50 mg/l) to neutral (pH 7.0).
- **TEMPERAMENT** Territorial and aggressive.

Turquoise body markings separate these cichlids from related species. The male, pictured below, is more brightly colored than the female and usually larger. Most males acquire a nuchal hump on the forehead as they mature. It is usually better to keep pairs in a tank on their own, especially for breeding, offering plenty of retreats and spawning surfaces. Hatching can take four days; the fry become free-swimming one week later.

SUBSTRATE EXCAVATORS

Many of the New World cichlids are eager excavators of the substrate. This behavior is reflected in the name of one particular genus, Geophagus, which literally means "earth-eater." Such cichlids do not normally swallow the gravel that they pick up with their strong jaws but instead move it a short distance and then simply spit it out. As you can see from this picture of a Black Belt Cichlid (*Vieja maculicauda*), they can move relatively large amounts with each mouthful. Part of the reason for digging is undoubtedly to search for edible live foods, such as worms, that may be lurking in the substrate, but this behavior is also linked with the cichlids' breeding habits. The cichlid fry require a safe area when they first emerge from their eggs, while they are not yet free-swimming. In order to keep their brood together and safe from would-be predators, the adults dig a series of pits in the substrate where their offspring can shelter. Within an aquarium setup, this digging can lead to plants floating up to the surface if they are not set in pots. It may also compromise the workings of an undergravel filter. Rockwork, in particular, needs to be securely positioned to prevent it from being undermined by these excavations.

Astronotus ocellatus

Oscar

- **ORIGINS** South America, from the basins of the Amazon and Orinoco southward to Paraguay.
- **SIZE** 14 in (35 cm).
- **DIET** Prepared cichlid foods and live foods.
- **WATER** Temperature 72–77°F (22–25°C); soft (50 mg/l) and acidic (pH 6.0–6.5).
- **TEMPERAMENT** Not to be trusted with small companions.

The Oscar's dull, greenish-brown background color is offset with lighter, reddish-orange markings arranged in irregular patterns. There can be considerable variation between individuals. Soon becoming tame enough to feed from the hand, Oscars need an efficient filtration system to prevent any deterioration in water quality. Females develop a genital papilla prior to spawning, with pairs forming a strong pair bond.

Red Tiger Oscar Selective breeding has led to the development of Oscars in which large, bright orange areas predominate on the body.

Longfin Albino Oscar These Oscars display elongated fins, and this characteristic feature can be combined with any color. Note the red eye.

Geophagus brasiliensis

Pearl Cichlid

- **ORIGINS** Eastern South America, from the Atlantic coast of Brazil to Rio de la Plata in Argentina.
- **SIZE** 11 in (28 cm).
- **DIET** Prepared cichlid foods, algae, and live foods.
- **WATER** Temperature 68–77°F (20–25°C); soft (50 mg/l) and acidic (pH 6.5).
- **TEMPERAMENT** Territorial but not very aggressive.

Blue, pearl-like markings on many of the scales, set against a bluish-gray background, give this cichlid a very distinctive appearance. However, no two individuals have exactly the same patterning. Pearl Cichlids are very adaptable in terms of their water chemistry needs. They will busily excavate the substrate, especially as the time for spawning approaches. If a pair repeatedly eat their eggs after spawning, they are unlikely to be compatible.

Cichlasoma festae

Festivus

- **ORIGINS** Northern South America, occurring in western Guyana and parts of the Amazon basin.
- **SIZE** 8 in (20 cm).
- **DIET** Prepared cichlid foods and live foods.
- **WATER** Temperature 72–77°F (22–25°C); soft (50 mg/l) and acidic (pH 6.5).
- **TEMPERAMENT** Only territorial when breeding.

Festivus have a thick, uneven black area running from the eye toward the dorsal fin. Below the eye is a circular, orange-yellow blotch, along with smaller yellow spots. The dorsal fin is more elongated in males. These nervous fish need a well-planted aquarium, with a piece of slate for spawning. They make good companions for *Pterophyllum* angelfish (see pp. 140–141).

Pelvic fins have long, narrow extensions

Pterophyllum scalare

Angelfish

ORIGINS South America, occurring through much of the Amazon basin, eastward from Peru to Belem, Brazil.

SIZE 6 in (15 cm).

DIET Prepared foods, vegetable matter, small live foods.

WATER Temperature 75–82°F (24–28°C); soft (50–100 mg/l) and acidic (pH 6.5).

TEMPERAMENT Relatively peaceful, but territorial.

This graceful cichlid is among the most popular of all tropical fish. Young Angelfish are sometimes recommended for community aquariums, but their long fins make easy targets for fin-nippers such as Tiger Barbs *(see p.83)*. Furthermore, Angelfish will soon grow too large for the tank and may start to bully their companions. It is best to house these fish in a single-species setup, where a pair may be persuaded to breed. The only way of visually distinguishing the sexes is when the female swells with eggs prior to spawning.

In the early stages of pair-bonding, the fish lock jaws and engage in mouth-wrestling, which may be mistaken for aggression. The spawning site is usually a vertical surface, such as a piece of slate

Semitransparent fins

Coloration is brighter on upperparts

First band passes through the eye

Long, narrow pelvic fins

Silver Angelfish This variety most closely approximates to the wild form, displaying the characteristic four-banded patterning. The intensity of the black bands will fade somewhat if the fish are kept under bright light.

Golden Angelfish Originally known as the Butterball, the Golden Angelfish was first developed in the United States during the 1970s. The trend has since been to create individuals with a deeper, more orange appearance.

or a rigid leaf of one of the larger Amazon Swordplants (*Echinodorus* spp.) Surround the thermostatic heater with mesh to dissuade the fish from spawning nearby; heat will destroy the eggs. Angelfish eggs are susceptible to fungus, so you may need to add fungicide to the water. The female lays up to 1,000 eggs, which the pair guards until they hatch three days later. A young pair breeding for

the first time may produce a much smaller number of eggs and then eat them, but they will usually spawn again within a month.

The newly hatched fry are transferred to a pit excavated in the substrate, where they are watched over by both parents. Feed the young on fry food initially and then on brine shrimp. When they are free-swimming, they may nibble mucus off the flanks of the adults to supplement their diet.

Black Angelfish The black pigmentation on the fins and body of these fish keeps their striped patterning largely hidden. The body also shows green iridescence.

Marbled patterning is highly individual

Black evident on the fins

Touch-sensitive pelvic fins

Golden Marble Angelfish The black "marbling" is more extensive in some individuals than others. Gold coloration is displayed from the top of the head up to the dorsal fin.

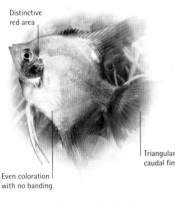

Distinctive red area

Even coloration with no banding

Triangular caudal fin

German Blue Blushing Angelfish First bred in Germany, this variety has a pale, silvery-blue body color and a contrasting bright-red area below the eyes. A darker Chocolate form also exists.

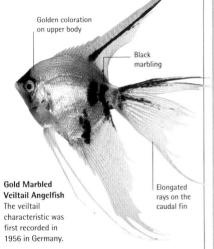

Golden coloration on upper body

Black marbling

Gold Marbled Veiltail Angelfish The veiltail characteristic was first recorded in 1956 in Germany.

Elongated rays on the caudal fin

Pterophyllum altum

Altum Angelfish

- **ORIGINS** South America, occurring in Colombia and Venezuela in the central part of the Rio Orinoco.
- **SIZE** 10 in (25 cm).
- **DIET** Prepared foods, vegetable matter, small live foods.
- **WATER** Temperature 82–86°F (28–30°C); soft (50–100 mg/l) and acidic (pH 6.0).
- **TEMPERAMENT** Relatively peaceful, but territorial.

The Altum Angelfish has a much taller body than the Angelfish itself. It also has a flattened area above the jaw and a steeper, less curved profile leading up to the dorsal fin. Altum Angelfish can be housed together in groups. As with the Angelfish, this species needs a relatively deep tank to accommodate its elongated shape. Altum Angelfish are difficult to breed, but maintaining good water chemistry may help to encourage spawning, as may raising the water temperature slightly and increasing the amount of live foods in the diet. Check fish for signs of white spot *(see p. 58)* before buying and also if water temperature falls significantly for any period.

THIN AND WEEDY

Angelfish live in relatively slow-flowing, reedy stretches of water, where their tall, narrow body shape allows them to weave in and out of the vegetation with ease. When danger threatens, they can dart in among the reeds to avoid detection. Even if they are spotted and pursued, they have an advantage over larger, bulkier predators, which cannot move as swiftly through the mass of plant stems. A wild Angelfish's body stripes confer a further advantage, helping to break up the outline of the fish so that it blends in with the reed stalks and shadows.

A tank for Angelfish should contain areas of thick plant growth into which the fish can retreat when nervous. Aquatic Amazonian plants with upright leaves, such as the Ruffled Amazon Swordplant (*Echinodorus major*), are ideal for this purpose. Taller Vallisnerias can also be used, because they thrive in the deeper aquariums that adult Angelfish require.

Symphysodon aequifasciatus

Blue Discus

- **ORIGINS** South America, from Rio Putumayo in Peru eastward through the Amazon basin in Brazil.
- **SIZE** 8 in (20 cm).
- **DIET** Discus foods and small live foods.
- **WATER** Temperature 75–84°F (24–29°C); soft (50–100 mg/l) and acidic (pH 6.0).
- **TEMPERAMENT** Quite placid and social.

The Blue Discus, named after its disk body shape, has grown in popularity over recent years thanks to the wide range of color varieties available. Four basic forms are known in the wild. The green form originates from the upper Amazon, while a brown variety is found around Belem and Manaus, closer to the river's mouth. Reddish fish occur near the Amazonian town of Alenquer, and a bluish strain was discovered in the Rio Purus, Rio Manacapura, and nearby lakes. Wild Heckel's Discus from the Rio Negro is a separate species (*Symphysodon discus*), but commercially available forms may be hybrids developed in breeding programs with the Blue Discus.

Turquoise Discus Much of the early development of this variety was undertaken in Germany. The facial markings on each of these fish are unique.

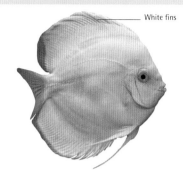

White fins

White Discus These fish will make a striking contrast, housed alongside more colorful individuals, although they are not very common.

These fish need a relatively tall tank, because of their body shape, and water conditions that mimic the blackwater environment that they naturally inhabit. If kept in less-than-ideal conditions, they are at risk from various diseases, including the parasitic illness known as hole-in-the-head *(see p.58)*. Blue Discus are best housed as a group in a single-species setup, although this demands a large tank with efficient filtration, supported by regular partial water changes.

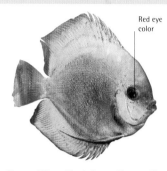

Red eye color

Blue Diamond Discus The darker markings on this fish are almost invisible; the blue coloration is intense over the entire body.

Once they are mature—by the time they are just over 4 in (10 cm) in length—the females can be identified (on close examination) by the rounded profile of the forehead, which has a more humped appearance in the males. The genital area behind the long, thin pelvic fin is a further aid to sexing this species, since the male's sperm duct is narrow and triangular in shape, while the female's egg-laying tube, or ovipositor, is broader and more rectangular in appearance.

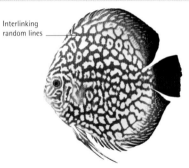

Interlinking random lines

Snakeskin Discus These attractive fish are individually patterned, with their markings said to resemble the scaled patterning of a snake.

Breeding pairs must be housed on their own in a tank that contains rockwork, such as slate, on which the fish can spawn. If the tank incorporates a thermostatic heater, this should be kept in a special heater guard; otherwise, any eggs that are laid near it will be destroyed. Prior to spawning, the fish clean their chosen site, where the female subsequently deposits 200–400 eggs. Pairs spawning for the first time may eat their eggs, particularly if they are disturbed during this period. Even so, the pair will probably spawn again before long, especially if live foods feature prominently in their diet. It can take up to ten attempts before they are successful. Blue Discus show great parental care and often help the fry to hatch from their eggs (see box, bottom right).

Red-Spotted Leopard Discus Vivid red markings set against a blue background typify this variety, but not all fish from the same spawning are necessarily well marked.

Blue and red fin coloring

Yellow Panda Discus The depth of coloration in this case is typically like that of a yellow canary, with some white markings in evidence.

Reddish-orange shade

Orange Discus The origins of this variety are thought to lie in Asia. The coloration of these discus can be intensified by color-feeding (see p.49).

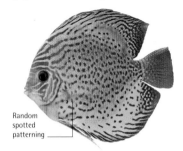

Even depth of color

Faint traces of banding on the flanks

Cobalt Blue Discus As its name suggests, this variety displays a rich blue color over its entire body and is largely free of darker markings.

Random spotted patterning

Blue Spotted Snakeskin Discus The lines of the snakeskin form are reduced to a pattern of separate spots here, on the flanks.

Slight hump indicates male

NOURISHING THE YOUNG

Discus fry are often seen swimming alongside their parents, nibbling at their flanks. They do this to obtain discus milk—a secretion that adults produce to nourish the fry. The "milk" may also contain immunoglobulins to protect the young fish from infections until their own immune systems are fully functioning. Discus fry that grow up with their parents develop at a faster rate than those reared in isolation, indicating the benefit of discus milk.

Dicrossus filamentosus

Chessboard Cichlid

- **ORIGINS** South America, in the Rio Orinoco basin in Colombia and the Rio Negro basin in Brazil.
- **SIZE** 3½ in (9 cm).
- **DIET** Live foods and prepared foods.
- **WATER** Temperature 73–77°F (23–25°C); soft (50–100 mg/l) and acidic (pH 6.0).
- **TEMPERAMENT** Relatively peaceful.

Alternating light and dark blotches along the side of the body help to identify this shy cichlid, as does the red line below the eye. The male, shown above, has more colorful fins and a more deeply forked caudal fin than the female. A single male should be housed with several females in a well-planted tank; rockwork will provide the fish with egg-laying sites. The female watches over her brood at first.

Nannacara anomala

Golden Dwarf Cichlid

- **ORIGINS** Widely distributed throughout northern parts of South America.
- **SIZE** 3 in (7.5 cm).
- **DIET** Prepared foods and small live foods.
- **WATER** Temperature 73–82°F (23–28°C); soft (50–100 mg/l) and acidic (pH 6.0).
- **TEMPERAMENT** Peaceful and social.

Sexing is straightforward with this species: females are plain yellow and smaller in size than the males, which are more colorful, with bluish markings on their flanks. Darker blotched markings may appear on the sides of the female's body just prior to spawning. A well-planted aquarium is required, with suitable retreats for breeding. This is a cave-breeding cichlid; the female guards the eggs and then watches over her offspring. Young fish are less colorful than adults.

Male

Apistogramma borellii

Umbrella Dwarf Cichlid

- **ORIGINS** South America, found in parts of the Mato Grosso, Brazil, and the Pantanal, Paraguay.
- **SIZE** 3 in (7.5 cm).
- **DIET** Prepared foods and small live foods
- **WATER** Temperature 72–77°F (22–25°C); soft (50–100 mg/l) and acidic (pH 6.0).
- **TEMPERAMENT** Males are territorial.

Male

Darker area along center of body

As with other *Apistogramma* dwarf cichlids, the Umbrella shows clear sexual dimorphism, with the male generally larger and more colorful than the female. This is a typical member of the group, displaying variable coloration. Umbrellas are rather nervous fish and settle better if housed with smaller, nonaggressive shoaling fish, such as tetras, which require similar water conditions.

Apistogramma macmasteri

Red-Tailed Dwarf Cichlid

- **ORIGINS** South America, occurring in the Rio Meta, close to Villavicencio, Colombia.
- **SIZE** 4 in (10 cm).
- **DIET** Prepared foods and small live foods.
- **WATER** Temperature 73–86°F (23–30°C); soft (50–100 mg/l) and acidic (pH 6.0).
- **TEMPERAMENT** Males are territorial.

The female Red-Tailed Dwarf, pictured here, lacks the red markings seen on the male's caudal fin. Keep these fish in a small group made up of a single male and several females. Partially buried clay flowerpots make ideal spawning sites. These should be spaced around the aquarium to provide a retreat for each female. A typical spawning results in up to 120 eggs, with the fry hatching in three days. The male sometimes helps to guard the fry.

Apistogramma agassizii

Agassiz's Dwarf Cichlid

- **ORIGINS** South America, occurring in many of the tributaries on the southern side of the Amazon.
- **SIZE** 4 in (10 cm).
- **DIET** Prepared foods and small live foods.
- **WATER** Temperature 72–77°F (22–25°C); soft (50–100 mg/l) and acidic (pH 6.0).
- **TEMPERAMENT** Males are territorial.

The appearance of these dwarf cichlids is variable, with a number of morphs recognized, all of them brightly colored. Yellow predominates in the individual pictured below, but others have reddish-orange fins. Weekly partial water changes of about 10 percent of total volume are important for all dwarf cichlids, to keep the nitrate level low. Water changes may trigger spawning and will lessen the risk of fungal attacks on the eggs.

Apistogramma nijsseni

Nijssen's Dwarf Cichlid

- **ORIGINS** South America, in the lower Rio Ucayali and the Rio Yavari in Peru.
- **SIZE** 3½ in (9 cm).
- **DIET** Prepared foods and small live foods.
- **WATER** Temperature 73–86°F (23–30°C); soft (50–100 mg/l) and acidic (pH 6.0).
- **TEMPERAMENT** Males are territorial.

Female

Rounded caudal fin

The coloration of the larger male is more variable than that of the female, which is predominantly black and yellow. Like the female, the male has a rounded caudal fin—unusual in male dwarf cichlids. As with other members of this group, Nijssen's is highly insectivorous and may prove reluctant to sample other foods. Mosquito larvae are particularly useful for encouraging spawning. The regular addition of aquarium peat to the filter (*see p.46*) is recommended. These cichlids need a tank well stocked with aquatic plants.

FACTORS INFLUENCING BREEDING

Suitable retreats in the aquarium are vital for successful breeding with *Apistogramma* dwarf cichlids, because these fish are cave-spawners. The female, such as the Cockatoo Dwarf pictured here, instinctively seeks out a site that affords her relative safety. In Nijssen's Dwarf (*see left*), environmental conditions have been shown to have a direct impact on breeding, and it may be that this is also the case with other group members. When the water temperature is above 84.4°F (29.1°C) only male fish result, while the offspring are all female when the water is 68–73°F (20–23°C). The influence of pH is relatively slight, but the percentage of eggs that hatch in naturally soft water is much higher than in hard water environments.

Apistogramma cacatuoides

Cockatoo Dwarf Cichlid

- **ORIGINS** South America, in parts of Peru and in adjacent areas of Brazil and Colombia.
- **SIZE** 3½ in (9 cm).
- **DIET** Prepared foods and small live foods.
- **WATER** Temperature 73–82°F (23–28°C); soft (50–100 mg/l) and acidic (pH 6.0).
- **TEMPERAMENT** Males are territorial.

When extended, the long rays at the front of this dwarf cichlid's dorsal fin resemble a crest. As in the other *Apistogramma* species, a single male should be housed with several females. The male frequents the middle layer of the tank, while the females establish small territories near the bottom. He visits their territories to breed but will remain outside the entrance to the spawning cave. When a number of females have broods at the same time, the young of different groups may join together.

Red Cockatoo Dwarf Cichlid These red morphs are among the most popular of the Cockatoo Dwarf variants. Fish available today have been extensively developed by selective breeding.

Dorsal fin "crest" is shown here folded back against the body

Cockatoo Dwarf Cichlid This species, like many dwarf cichlids, is highly variable in coloration. The fin rays are longer at both ends of the dorsal fin than at the middle.

Sunburst Cockatoo Dwarf Cichlid This variant gets its name from its brilliant yellow and red coloration. The female, shown above, can be identified by the more rounded shape of her caudal fin.

Etroplus maculatus

Orange Chromide

- **ORIGINS** Asia, found in western coastal areas of India and Sri Lanka.
- **SIZE** 3 in (7.5 cm).
- **DIET** Prepared cichlid foods and live foods.
- **WATER** Temperature 68–79°F (20–26°C); hard (150–200 mg/l) and neutral to alkaline (pH 7.0–7.5).
- **TEMPERAMENT** Not aggressive.

Dorsal fin almost reaches the caudal fin

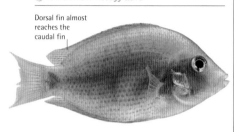

These pale-yellow cichlids have a regular pattern of orange spots running over the body, as well as faint blue markings. Selective breeding has led to the creation of a more orange variant. Unlike most cichlids, this species benefits from being kept in slightly brackish water, especially when breeding, since it protects the eggs from fungal attack. After spawning on rockwork, a pair will guard the site and watch over their young in special pits.

Etroplus suratensis

Green Chromide

- **ORIGINS** Asia, occurring in coastal parts of India and also Sri Lanka.
- **SIZE** 18 in (45 cm).
- **DIET** Prepared cichlid foods, algae, and live foods.
- **WATER** Temperature 72–79°F (22–26°C); hard (150–200 mg/l) and alkaline (pH 7.5–8.0).
- **TEMPERAMENT** Can be intolerant.

Green Chromides display variable coloration, with dark brown vertical stripes on the body, dominating the background green. These fish are less colorful and grow significantly larger than Orange Chromides *(see left)*. They also benefit from being kept in a brackish environment, although Greens can readily be maintained in fresh water. There is no way of distinguishing gender; the throats of both sexes turn black when breeding.

Pelvicachromis subocellatus

Eyespot Krib

- **ORIGINS** Western Africa, where it is widely distributed, from Gabon to the Congo River in Zaire.
- **SIZE** 4 in (10 cm).
- **DIET** Prepared cichlid foods, algae, and live foods.
- **WATER** Temperature 75–79°F (24–26°C); hard (100–150 mg/l) and acidic (pH 6.5).
- **TEMPERAMENT** Territorial when breeding.

These cichlids vary greatly in appearance, but they usually have blackish bands along the body, pinkish underparts, and often an eye-spot on the tail. Males are larger than females and have longer pelvic fins, and their dorsal and anal fins are more pointed. The female's color is at its finest just prior to spawning, which occurs in caves, with the male driving away fish that venture too close. Eyespot Kribs sometimes occur in brackish waters.

Pelvicachromis humilis

Yellow Krib

- **ORIGINS** West Africa, where it occurs in parts of Sierra Leone, Liberia, and southeastern Guinea.
- **SIZE** 5 in (13 cm).
- **DIET** Prepared cichlid foods, algae, and live foods.
- **WATER** Temperature 75–82°F (24–28°C); soft (50–100 mg/l) and acidic (pH 6.5).
- **TEMPERAMENT** Territorial when breeding.

The upperparts of the Yellow Krib's body are dark olive-green with darker vertical bars. Males are larger and have yellow underparts, while females are whitish beneath with violet or red on the belly. **Female** At least three different localized color variants are known. Kribs occur in areas of forest, so they need a relatively dark aquarium, with floating plants to diffuse the lighting. A number of caves should be included in the tank; clay flowerpots buried in the gravel are ideal for this purpose. Prior to spawning, these cichlids start digging in the substrate. When displaying, the female performs a series of shimmering movements to attract her mate. She stays with the brood until they are free-swimming and have emerged from the cave after about a week. They can then be left together as a family group; the young cichlids require foods such as brine shrimp.

Male

Female

Nanochromis parilus

Bar-Tailed Dwarf Cichlid

- **ORIGINS** Africa, occurring in the Zaire River basin, especially in the vicinity of Stanley Pool.
- **SIZE** 3 in (8 cm).
- **DIET** Prepared cichlid foods and live foods.
- **WATER** Temperature 75–82°F (24–28°C); soft (50–100 mg/l) and acidic (pH 6.5).
- **TEMPERAMENT** Territorial.

Male

Female

The red streaks on the caudal fin help to explain why these colorful and attractive dwarf cichlids are described as Bar-Tailed. These fish can be sexed quite easily, since the males are larger, and their dorsal and anal fins also taper to more evident points.
In addition, females tend to have more pinkish underparts. Their tank needs to be well planted, and the addition of aquarium peat to the filtration system is also recommended. Plenty of retreats are important because Bar-Tailed Dwarf Cichlids are cave-spawners. A female typically lays about 100 eggs in her chosen cave. She guards them while her mate patrols outside and aggressively chases off other males. Once the fry are free-swimming, both the male and female will watch over them. Small live foods are vital for successful rearing at this stage, with regular partial water changes becoming increasingly significant as the young grow larger.

Pseudocrenilabrus philander

South African Mouth-Brooder

- **ORIGINS** Southern Africa, from Angola, southern Zaire, and Mozambique down to South Africa.
- **SIZE** 4½ in (11 cm).
- **DIET** Prepared cichlid foods, flake, and live foods.
- **WATER** Temperature 68–79°F (20–26°C); hard (100–150 mg/l) and neutral (pH 7.0).
- **TEMPERAMENT** Aggressive and territorial.

Relatively long pectoral fins

The variations in the size and color of these fish reflect their wide distribution. In all cases, though, only the male, pictured above, shows a red spot near the rear of the anal fin and gold on the flanks; the female is much duller. Decorate the aquarium with plants set in pots and rockwork retreats. South African Mouth-Brooders burrow repeatedly into the base of the aquarium when spawning.

EGYPTIAN MOUTH-BROODER

Mouth-brooding is best documented in the cichlids of Africa's Rift Valley, but it is also a feature of other cichlids in East Africa, including the Egyptian Mouth-Brooder (*Pseudocrenilabrus multicolor*), shown below. In most mouth-brooders, the female collects the falling eggs in her mouth, where they are then fertilized by sperm from the male. In the Egyptian, however, the eggs are laid in a pit in the substrate and fertilized there, after which the female gathers them up. The Egyptian lays a relatively large number of eggs—typically about 100—while other species produce only as many as they can fit in their mouths.

Hemichromis bimaculatus

Jewel Fish

- **ORIGINS** West Africa, found in forested areas from southern Guinea to central Liberia.
- **SIZE** 6 in (15 cm).
- **DIET** Prepared cichlid foods, flake, and live foods.
- **WATER** Temperature 70–77°F (21–25°C); soft (50–100 mg/l) and neutral (pH 7.0).
- **TEMPERAMENT** Aggressive when spawning.

The reddish coloration of these cichlids becomes even more vibrant when they are spawning. Both sexes display the three black blotches along the body, but males may be identified by the presence of a pattern of fine spots on the sides of the head. Compatibility can be a problem, but established pairs will breed readily, with the female laying up to 500 eggs on a flat rock in a secluded part of the tank. These should hatch in two days.

Hemichromis lifalili

Lifalili Cichlid

- **ORIGINS** Western Central Africa, occurring in various lakes and rivers in Congo and Zaire.
- **SIZE** 6 in (15 cm).
- **DIET** Prepared cichlid foods, flake, and live foods.
- **WATER** Temperature 72–77°F (22–25°C); soft (50–100 mg/l) and neutral (pH 7.0).
- **TEMPERAMENT** Aggressive when spawning.

The reddish color of these cichlids is offset by pale blue spots. Colors become more vibrant in the male at spawning time, while the female turns darker. Breeding requirements are similar to those of the Lilalili Cichlid—well-oxygenated water and a diet of live foods will improve results. After hatching, the fry are closely guarded by their parents and are regularly moved from one spawning pit to another.

Labidochromis sp. Lupingo

Lupingo Labidochromis

- **ORIGINS** East Africa, in Lake Malawi, around Lupingo in the northeastern part of the lake.
- **SIZE** 6 in (15 cm).
- **DIET** Prepared cichlid foods and live foods.
- **WATER** Temperature 72–79°F (22–26°C); hard (150–200 mg/l) and alkaline (pH 7.5–8.0).
- **TEMPERAMENT** Quite peaceful.

A vivid, dark blue body coloration marked with black vertical bars, plus yellowish markings on the dorsal fin, typify this Lake Malawi cichlid. There is such confusion over the identity of many Lake Malawi cichlids that some of those that are common in aquariums do not have recognized scientific names. This is partly because even local populations of a species can differ dramatically in appearance. Consequently, these cichlids are often named after the area of the lake where a population was discovered. These descriptions are known as trade names. In *Labidochromis* species, the female carries the fertilized eggs in her mouth, releasing the free-swimming fry about three weeks later.

A HEAD START IN LIFE

Many Lake Malawi cichlids use mouth-brooding to increase the chances of survival of the next generation. While other species lay large numbers of eggs, many of which are eaten along with newly hatched fry, mouth-brooders produce fewer eggs but care for their offspring in their mouths and protect them during the critical early stages of life. Once the young cichlids are able to swim, the female opens her jaws and allows the fry to emerge. She still keeps a watchful eye on them at first, and whenever danger threatens, the youngsters will instinctively dart back into the safety of her mouth.

Labidochromis caeruleus

Labidochromis Electric

- **ORIGINS** East Africa, found only in Lake Malawi, typically on the northwest side of the lake.
- **SIZE** 6 in (15 cm).
- **DIET** Prepared cichlid foods, algae, and live foods.
- **WATER** Temperature 72–79°F (22–26°C); hard (150–200 mg/l) and alkaline (pH 7.5–8.0).
- **TEMPERAMENT** Quite peaceful.

Male

Dark fins

This species exists in a range of color morphs, such as the bright yellow form shown here. Both the male and the female display a stripe along the dorsal fin, but only the male has black ventral fins. *Labidochromis* cichlids are part of the mbuna group, whose members occur in rocky areas of Lake Malawi close to the shore, where they browse on algae. This species is one of the most placid, but a male should be kept with several females.

Protomelas fenestratus

Fenestratus

- **ORIGINS** East Africa, in Lake Malawi, where this species is found typically in the east-central region.
- **SIZE** 6 in (15 cm).
- **DIET** Prepared cichlid foods and live foods.
- **WATER** Temperature 72–79°F (22–26°C); hard (150–200 mg/l) and alkaline (pH 7.5–8.0).
- **TEMPERAMENT** Quite peaceful.

This cichlid can be identified by the thick, dark barring on its body. While the head is blue, the sides of the body have a yellowish-orange hue. Males are more brightly colored than females. Fenestratus is an active, mouth-brooding fish that occurs relatively close to the shore, in areas where there are both boulders and sandy stretches. Once free-swimming, the young Fenestratus need a diet that includes live foods, such as brine shrimp.

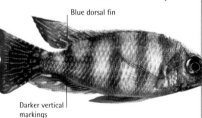

Blue dorsal fin

Darker vertical markings

Cyrtocara moorii

Malawi Blue Dolphin

- **ORIGINS** East Africa, all around Lake Malawi, especially in the south; also in Lake Malombe.
- **SIZE** 9 in (23 cm).
- **DIET** Prepared cichlid foods and live foods.
- **WATER** Temperature 72–79°F (22–26°C); hard (150–200 mg/l) and alkaline (pH 7.5–8.0).
- **TEMPERAMENT** Relatively peaceful.

The sides of this blue-bodied fish often show signs of darker vertical barring. Despite its extensive range in the lake, there are no recognized color morphs. The distinctive bulge on the head is more apparent in males. Blue Dolphins are found in sandy rather than rocky areas. They often shadow cichlids that dig in the substrate, in anticipation of finding edible items unearthed by the excavations.

Pseudotropheus socolofi

Eduardi

🌐 **ORIGINS** East Africa, occurring in the middle of the eastern side of Lake Malawi, in rocky coastal areas.

💧 **SIZE** 6 in (15 cm).

🍽 **DIET** Prepared cichlid foods and live foods.

🌊 **WATER** Temperature 72–79°F (22–26°C); hard (150–200 mg/l) and alkaline (pH 7.5–8.0).

😐 **TEMPERAMENT** Quite peaceful.

Both sexes of Eduardi are predominantly blue in color; yellow egg-spots at the rear of the anal fin serve to distinguish the male. The term "mbuna," which is applied to a number of cichlids from Lake Malawi, derives from a local Chichewa word for these fish, which feed on algae and associated invertebrates in rocky areas. The Eduardi is a mouth-brooding species, with the female caring for the eggs.

Maylandia barlowi

Golden Fuscoides

🌐 **ORIGINS** East Africa, occurring in the southern part of Lake Malawi, especially around the Maleri Islands.

💧 **SIZE** 4 in (10 cm).

🍽 **DIET** Prepared cichlid foods and live foods.

🌊 **WATER** Temperature 72–79°F (22–26°C); hard (150–200 mg/l) and alkaline (pH 7.5–8.0).

😐 **TEMPERAMENT** Males are quite aggressive.

Golden-yellow coloration predominates in these cichlids. This mbuna species has a highly fragmented distribution in Lake Malawi because fish will not stray from their rocky feeding grounds over adjacent sandy areas, so they never establish themselves in new habitats. The result is that they are restricted to isolated populations and thus show a considerable diversity in coloration. Males display such strong territorial instincts that they cannot be mixed safely in the aquarium.

Darker caudal fin

Labeotropheus trewavasae

Trewavas's Cichlid

🌐 **ORIGINS** East Africa, in Lake Malawi, where this species is widely distributed.

💧 **SIZE** 6 in (15 cm).

🍽 **DIET** Prepared cichlid foods and live foods.

🌊 **WATER** Temperature 72–79°F (22–26°C); hard (150–200 mg/l) and alkaline (pH 7.5–8.0).

😐 **TEMPERAMENT** Males are aggressive.

This blue cichlid is very similar to Fuelleborn's Cichlid *(see below)* in terms of its requirements and appearance, although numerous color morphs are recognized. Trewavas' Cichlid is slightly slimmer and smaller than its relative, with red markings sometimes apparent on its fins. Within Lake Malawi, it occurs in deeper water than Fuelleborn's, which prefers coastal shallows.

Labeotropheus fuelleborni

Fuelleborn's Cichlid

🌐 **ORIGINS** East Africa, in Lake Malawi, most common in the southwest and absent from the northeast.

💧 **SIZE** 7 in (18 cm).

🍽 **DIET** Prepared cichlid foods, algae, and live foods.

🌊 **WATER** Temperature 72–79°F (22–26°C); hard (150–200 mg/l) and alkaline (pH 7.5–8.0).

😐 **TEMPERAMENT** Males are not social.

This cichlid usually carries yellow markings on the sides of its body, but color otherwise varies considerably between individuals. Males are aggressive by nature, fighting with rivals and doggedly pursuing potential mates. For this reason, house just one male with a number of females. The males can, however, be kept safely in the company of various other mbuna cichlids, including *Melanochromis* species. When breeding, a mature male establishes a favored spawning ground, often inside a cave. The female lay her eggs and takes them into her mouth for protection. Attracted by the egg spots on the male's anal fin, she takes in sperm to fertilize the eggs in her mouth. The male takes no further part in caring for the brood, so it is best to transfer the female to a separate tank. She will release the young cichlids about three weeks later. Rearing foods can include powdered flake.

Nimbochromis livingstonii

Livingstoni

- **ORIGINS** East Africa, occurring only in Lake Malawi, where it is widely distributed throughout the lake.
- **SIZE** 10 in (25 cm).
- **DIET** Prepared cichlid foods and live foods.
- **WATER** Temperature 73–79°F (23–26°C); hard (150–200 mg/l) and alkaline (pH 7.5–8.0).
- **TEMPERAMENT** Highly predatory.

A mottled brown-and-silver pattern with a slight bluish cast identifies this mouth-brooding cichlid. Males can be distinguished by the red suffusion on their anal fin. The distinctive color scheme gives Livingstoni the appearance of a dead fish as it lies on the bottom of the lake. Any small fish or invertebrate that arrives to investigate the "corpse" is snapped up into the capacious mouth of this ambush hunter.

Sciaenochromis fryeri

Electric Blue

- **ORIGINS** East Africa, occurring only in Lake Malawi, widely distributed along the rocky shoreline.
- **SIZE** 7 in (18 cm).
- **DIET** Prepared cichlid foods and live foods.
- **WATER** Temperature 72–79°F (22–26°C); hard (150–200 mg/l) and alkaline (pH 7.5–8.0).
- **TEMPERAMENT** Predatory and territorial.

The rich blue of male Electric Blues takes up to a year to develop. Females are always paler. Like other mouth-brooding cichlids, eggs are fertilized in the female's mouth. Once they emerge, the young do not return to their mother's mouth, even if danger threatens. Males are aggressive toward other males and fish with a similar coloration. Electric Blues eat fry in the wild, so they will prey on smaller species in the tank.

Orange-brown iris

Faint traces of dark barring on the flanks

Male

Iodotropheus sprengerae

Rusty Cichlid

- **ORIGINS** East Africa, occurring only in Lake Malawi, where it is found in the southeastern part of the lake.
- **SIZE** 4 in (10 cm).
- **DIET** Prepared cichlid foods and live foods.
- **WATER** Temperature 72–79°F (22–26°C); hard (150–200 mg/l) and alkaline (pH 7.5–8.0).
- **TEMPERAMENT** Reasonably nonaggressive.

The background color of this cichlid is bluish, with whitish areas and darker stripes on the sides of the body, and a prominent black area on the dorsal fin. Some individuals display areas of rusty coloration. Caves and gravel are recommended for the tank to mimic the rocky areas where these fish naturally occur. The females are mouth-brooders, laying over 30 eggs; they may breed at an age of less than four months.

Haplochromis compressiceps

Malawi Eye-Biter

- **ORIGINS** East Africa, occurring throughout Lake Malawi, relatively close to the shore.
- **SIZE** 9 in (23 cm).
- **DIET** Live foods of various types.
- **WATER** Temperature 72–79°F (22–26°C); hard (150–200 mg/l) and alkaline (pH 7.5–8.0).
- **TEMPERAMENT** Predatory.

Large jaws and a narrow head characterize this aggressive hunter, which is bluish in color overall, with a dark stripe running along the midline. The long dorsal fin, which is black at the front, becomes taller and largely transparent toward the rear, where it is tipped with red. The common name derives from the way these cichlids strike at the eyes when attacking large prey. Their more usual prey—invertebrates and smaller fish—are simply seized whole. A large aquarium with open and rocky areas is recommended for this species, which, if necessary, can be accommodated with similar-sized lake cichlids. When breeding, the male Malawi Eye-Biter is territorial, creating hollow depressions in the sand on the floor of the lake, to which it attracts females for mating purposes. The female collects the eggs after spawning, and the young hatch in her mouth.

Copadichromis borleyi

Red Kadango

- **ORIGINS** East Africa, found only in Lake Malawi, where it is widely distributed.
- **SIZE** 6 in (15 cm).
- **DIET** Prepared cichlid foods and live foods.
- **WATER** Temperature 72–79°F (22–26°C); hard (150–200 mg/l) and alkaline (pH 7.5–8.0).
- **TEMPERAMENT** Quite peaceful.

The coloration of Red Kadangos is quite variable. While males tend to have blue heads and orange bodies, females are silvery with golden-yellow markings on their fins. In the wild, these fish are found in rocky areas with sand nearby. Include large rocks in the tank to provide sites where males can establish their territories. Females will produce 30–60 eggs, and the young emerge from their mother's mouth after about three weeks.

Pseudotropheus joanjohnsonae

Pearl of Likoma

- **ORIGINS** East Africa, in eastern Lake Malawi; also introduced into the southwest part of the lake.
- **SIZE** 4 in (10 cm).
- **DIET** Prepared cichlid foods, vegetable matter, live foods.
- **WATER** Temperature 72–79°F (22–26°C); hard (150–200 mg/l) and alkaline (pH 7.5–8.0).
- **TEMPERAMENT** Males are territorial but not destructive.

Sexing is simple in this largely bluish species, which is found off Likoma Island, because only the males display the black stripe that runs the length of the dorsal fin. Males also have yellow egg spots on the anal fin. These are absent in the mouth-brooding females, which have more pronounced reddish markings on their bodies. Males are aggressive toward one another and should be kept apart. House a single male with several females in a single-species setup.

Male

Melanochromis auratus

Auratus Cichlid

- **ORIGINS** East Africa, occurring in eastern Lake Malawi, where it is encountered around the rocky shoreline.
- **SIZE** 4 in (10 cm).
- **DIET** Prepared cichlid foods, algae, and live foods.
- **WATER** Temperature 72–79°F (22–26°C); hard (150–200 mg/l) and alkaline (pH 7.5–8.0).
- **TEMPERAMENT** Males are territorial and aggressive.

Blue morph from near Likoma Island

Male's blue stripe extends through the eye

This was one of the first Lake Malawi species to be kept by aquarists, and it remains popular today. Females of the species carry conspicuous golden bands on their flanks; in males, the band is blue. Tanks housing these fish should include rocky areas, since spawning naturally takes place in caves. Females are mouth-brooders, and several should be housed in the company of an individual male.

Maylandia zebra

Zebra Cichlid

- **ORIGINS** East Africa, in Lake Malawi, mainly in the north and northwest, but also on the eastern shoreline.
- **SIZE** 5½ in (12 cm).
- **DIET** Flake, live foods, and vegetable matter.
- **WATER** Temperature 72–79°F (22–26°C); hard (150–200 mg/l) and alkaline (pH 7.5–8.0).
- **TEMPERAMENT** Males are antisocial.

As with many of the lake cichlids, the coloration of the Zebra Cichlid varies throughout its range. The body is typically bluish, with males displaying yellow egg spots on the ventral fins *(see below)*. However, in the case of the

Tangerine morph, the dorsal fin has an orange shade. Individuals with heavily barred bodies usually originate from clearer waters. There is also a variation in size, with the largest fish found around Makulawe Point, off Likoma Island. These cichlids occur only in rocky areas, grazing on algae with their enlarged lips. In aquariums, they benefit from the addition of spirulina algae to their diet. This is another matriarchal, mouth-brooding species, with the young emerging about three weeks after mating has occurred.

MOTION DETECTORS

Some cichlids, notably the Aulonocaras (which means "pipe-heads"), have large sensory pores on their heads. These work in conjunction with the lateral line to detect movements in the water. Being inhabitants of relatively deep, dark water, the cichlids use the pores to locate invertebrate prey that may be near the head but not visible in the gloomy surroundings. This sensory system may also have other functions, such as helping the fish to home in on their eggs after spawning and helping females to keep track of their fry. The pores should not be confused with hole-in-the-head disease *(see p.58)*, an ailment sometimes encountered in cichlids.

Aulonocara baenschi

Sunshine Peacock

- **ORIGINS** East Africa, restricted to Lake Malawi, though widely distributed there.
- **SIZE** 4 in (10 cm).
- **DIET** Prepared cichlid foods, live foods, vegetable matter.
- **WATER** Temperature 72–79°F (22–26°C); hard (150–200 mg/l) and alkaline (pH 7.5–8.0).
- **TEMPERAMENT** Relatively peaceful.

Large eyes

Only the male of this species displays brilliant yellow coloration on its flanks and has a bluish head. The female is silvery overall and typically displays slight traces of dark barring on the sides of the body. A number of different color morphs of this cichlid have been described and are available to buy; the pictured example is the Yellow Regal morph, which is yellowish rather than blue on the top half of the head.

Aulonocara roberti

Orange-Shouldered Peacock

- **ORIGINS** East Africa, restricted to Lake Malawi, where it is widely distributed.
- **SIZE** 5½ in (12 cm).
- **DIET** Prepared cichlid foods, live foods, vegetable matter.
- **WATER** Temperature 72–79°F (22–26°C); hard (150–200 mg/l) and alkaline (pH 7.5–8.0).
- **TEMPERAMENT** Males are antisocial.

The striking, partially blue coloration associated with this group of cichlids is the reason they are known as Peacocks. In this particular species, there is orange coloration behind the head and along the underparts, and the caudal fin is vibrantly spotted with blue. Always pair Peacocks carefully, not just to prevent hybridization, but also to ensure that the different morphs from separate parts of the lake retain their individual characteristics.

Vertical blue bands

Aulonocara jacobfreibergi

Butterfly Peacock

- **ORIGINS** East Africa, widely distributed around the rocky shoreline of Lake Malawi.
- **SIZE** 6 in (15 cm).
- **DIET** Prepared cichlid foods, live foods, vegetable matter.
- **WATER** Temperature 72–79°F (22–26°C); hard (150–200 mg/l) and alkaline (pH 7.5–8.0).
- **TEMPERAMENT** Males are antisocial.

Male Butterfly Peacocks grow noticeably larger than females. They can also be identified by their more pointed dorsal and ventral fins, as well as their elongated pectoral fins. Although the male of this mouth-brooding species has no egg spots on its ventral fin to attract the female, fertilization of the eggs still occurs in the female's mouth. A male should be housed in the company of several females. He will display to them in a cave.

Aulonocara hansbaenschi

Blue Peacock Cichlid

- **ORIGINS** East Africa, on the eastern coast of Lake Malawi near Masinje; introduced around Thumbi Island.
- **SIZE** 4 in (10 cm).
- **DIET** Prepared cichlid foods, live foods, vegetable matter.
- **WATER** Temperature 72–79°F (22–26°C); hard (150–200 mg/l) and alkaline (pH 7.5–8.0).
- **TEMPERAMENT** Males are antisocial.

The head of the male Blue Peacock is a rich, royal blue, as is much of the body, which also bears dark patterning and often yellowish or red coloration near the head. The top of the dorsal fin has a whitish or pale blue border. Females are duller than males, displaying banded patterning on a whitish background. There is considerable difficulty in unraveling the relationships between the different Lake Malawi cichlids, and Blue Peacock Cichlids are often described mistakenly as *Aulonocara nyassae*—a related but distinct species. A tank for these mouth-brooding fish should incorporate a number of caves, since Blue Peacock Cichlids like to stay well concealed. A power filter will provide good surface movement to improve oxygenation, replicating the action of the waves on the surface of the lake.

COLOR AND SIGHT

The relatively large eyes of cichlids indicate that sight is an important sense for these fish. Cichlids generally have good color vision, and the differences in coloration between the various Lake Malawi cichlids help members of the same species to recognize potential mates in areas where several species occur together. Coloration also gives a clue to the depth at which these fish live. Blue cichlids tend to occupy shallower areas, while species from deeper regions are likely to be predominantly yellow—an adaptation that helps them spot their own kind in murky water.

Aulonocara stuartgranti

Regal Peacock

- **ORIGINS** East Africa, in northwest Lake Malawi, and ranging down the eastern side to Makanjila Point.
- **SIZE** 4³/₄ in (12 cm).
- **DIET** Flake, live foods, and vegetable matter.
- **WATER** Temperature 72–79°F (22–26°C); hard (150–200 mg/l) and alkaline (pH 7.5–8.0).
- **TEMPERAMENT** Males are antisocial.

These mouth-brooders can be sexed quite easily, thanks to the contrast between the bright blue of the males and the grayish appearance of the females (which are also slightly smaller). Males clearly display yellow egg spots on their ventral fin and yellow barring on the tail. In view of the extensive distribution of this cichlid, however, it is not surprising that a number of different color forms have been identified. The one shown here is known as the Blue Regal. This species has not proved to be aggressive, so it is suitable for inclusion in a community tank housing similar nonaggressive cichlids from the lake. The young can be reared easily on brine shrimp and powdered flake once they have emerged from their mother's mouth. Their scientific name acknowledges the biologist Stuart Grant, who pioneered the study of Lake Malawi cichlids.

Aulonocara korneliae

Blue-Gold Peacock

- **ORIGINS** East Africa, restricted to Lake Malawi, where it occurs in the vicinity of Chisumulu Island.
- **SIZE** 3³/₄ in (9 cm).
- **DIET** Prepared cichlid foods, live foods, vegetable matter.
- **WATER** Temperature 72–79°F (22–26°C); hard (150–200 mg/l) and alkaline (pH 7.5–8.0).
- **TEMPERAMENT** Males are antisocial.

Alternating vertical bars of blue and black, with a golden background color, characterize male Blue-Gold Peacocks; females are a dull shade of grayish-brown. Like many Lake Malawi cichlids, the Blue-Gold Peacock has become known to science only fairly recently, being identified in 1987. In the tank, use rocks and a sandy substrate to mimic the lake environment. Mosquito larvae are the favored live food for this species.

Cynotilapia afra

Afra Cichlid

- **ORIGINS** East Africa, restricted to Lake Malawi, where it occurs near the northern shoreline.
- **SIZE** 4¹/₂ in (12 cm).
- **DIET** Prepared cichlid foods and live foods.
- **WATER** Temperature 72–79°F (22–26°C); hard (150–200 mg/l) and alkaline (pH 7.5–8.0).
- **TEMPERAMENT** Males are antisocial.

There are many different color morphs of this mouth-brooding cichlid, which has a single tooth in each jaw. Afra Cichlids benefit from the provision of caves in the aquarium. They dig in the substrate in search of invertebrates, and the males also dig as part of courtship displays. As members of the mbuna group, they should be offered a diet that includes spirulina algae, which helps to maintain their coloration.

Long dorsal fin

Golden-orange morph

Chilotilapia rhoadesii

Rhoadesii Cichlid

- **ORIGINS** East Africa, in the south of Lake Malawi, where it usually occurs in muddy and often deep waters.
- **SIZE** 10 in (25 cm).
- **DIET** Prepared cichlid foods and live foods.
- **WATER** Temperature 72–79°F (22–26°C); hard (150–200 mg/l) and alkaline (pH 7.5–8.0).
- **TEMPERAMENT** Males are antisocial.

The male Rhoadesii Cichlid, shown above, is much brighter in color than the female, which is silvery with two dark stripes, one below the dorsal fin and the other along the midline behind the gills. This cichlid hunts for snails in the wild, and although it can be weaned onto alternative foods, any snails in the aquarium will be eaten. The large size of this mouth-brooder means that a breeding group requires a particularly spacious aquarium.

Tropheus moorii

Moorii

🌀 **ORIGINS** East Africa, restricted to Lake Tanganyika, where it occurs in the southern part of the lake.

🌀 **SIZE** 6 in (15 cm).

🌀 **DIET** Cichlid diets and vegetable matter, including algae.

🌀 **WATER** Temperature 72–81°F (22–27°C); hard (150–200 mg/l) and alkaline (pH 7.5–8.0).

🌀 **TEMPERAMENT** Somewhat territorial.

Yellow-Banded Moorii It is not possible to rely on the coloration of these Moorii to determine the sexes, because the banded patterning is very variable.

Striped Moorii This morph is from the southwestern coast of Lake Tanganyika. As with all Moorii, its tank must include rockwork.

These fish are similar in their habits to the mbuna cichlids of Lake Malawi, feeding on algae growing on rocks around the lake's perimeter. The many different morphs, which vary widely in appearance, are often named after the area of the lake in which they occur. Moorii are lively fish and thus may prove disruptive in a community tank. Difficult to sex, they are maternal mouth-brooders.

Xenotilapia flavipinnis

Yellow-Finned Xenotilapia

🌀 **ORIGINS** East Africa, restricted to the northern part of Lake Tanganyika.

🌀 **SIZE** 3 in (7.5 cm).

🌀 **DIET** Cichlid diets and live foods.

🌀 **WATER** Temperature 75–79°F (24–26°C); hard (150–200 mg/l) and alkaline (pH 7.5–8.0).

🌀 **TEMPERAMENT** Relatively peaceful.

The distinctive yellow markings on the fins of these large-eyed cichlids are more pronounced in males. Yellow-Finned Xenotilapias should be kept in groups, although disagreements may arise during the spawning period. The female collects and cares for the eggs in her mouth, with the male often sharing the mouth-brooding duties. These fish feed close to the substrate. Their aquarium should have little decor and a sandy base.

Julidochromis regani

Striped Julie

🌀 **ORIGINS** East Africa, around the rocky shoreline of Lake Tanganyika.

🌀 **SIZE** 12 in (30 cm).

🌀 **DIET** Cichlid diets and live foods.

🌀 **WATER** Temperature 72–77°F (22–25°C); hard (150–200 mg/l) and alkaline (pH 7.5–8.0).

🌀 **TEMPERAMENT** Territorial when breeding.

Long, low dorsal fin

Relatively slim body

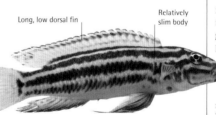

The chocolate-colored body of the Striped Julie has pale yellow stripes along its length, and there is often some blue on the lower fins. A tank for this relatively large cichlid needs a sandy base; there should be plenty of rocky areas and caves where the fish can hide. Sexing is difficult, but females swell noticeably with eggs prior to spawning, which occurs in caves. The eggs, which number up to 300, are guarded in the cave by both adults.

Cyphotilapia frontosa

Frontosa

🌀 **ORIGINS** East Africa, in Lake Tanganyika, where it occurs in slightly deeper water away from the shoreline.

🌀 **SIZE** 15 in (35 cm).

🌀 **DIET** Cichlid diets and live foods.

🌀 **WATER** Temperature 75–79°F (24–26°C); hard (150–200 mg/l) and alkaline (pH 7.5–8.0).

🌀 **TEMPERAMENT** Predatory.

Striped patterning and a distinctive hump on the forehead help to distinguish this cichlid, male fish generally have a larger hump. Its large size and predatory nature mean that the Frontosa should not be mixed with smaller companions. The female lays about 50 eggs in a cave and then incubates them in her mouth. At first, she also broods her young in a similar way at night. The male must be removed after spawning to prevent him from devouring the eggs and offspring.

WATCHFUL PARENTS

The care that adult cichlids lavish on their offspring frequently extends well beyond the hatching period. Like the Giant Tanganyika Cichlids (*Boulengerochromis microlepis*) shown below, the family swims together, and the watchful adults drive off potential predators. They warn their young of approaching danger by rippling their bodies in a distinctive way—a behavior known as jolting—or by flashing their brightly colored pelvic fins at their offspring. Should any of the fry lag behind, one of the parents may try to bring them back to the group, sometimes even retrieving them in the mouth. Aquarium studies suggest that behavior of this type is acquired rather than instinctive; indeed, where adult cichlids are faced with no risk of brood predation, their parental instincts decline over several generations. The addition of nonaggressive but active companions (often described as "dither fish") to the tank will cause the adults to become more protective toward their offspring again.

Brichardi

- **ORIGINS** East Africa, restricted to the rocky shoreline of Lake Tanganyika.
- **SIZE** 4 in (10 cm).
- **DIET** Cichlid diets, live foods, and vegetable matter.
- **WATER** Temperature 72–79°F (22–26°C); hard (150–200 mg/l) and alkaline (pH 7.5–8.0).
- **TEMPERAMENT** Social and peaceful.

Extended rays on the caudal fin give these elegant cichlids a lyre-tailed appearance; this feature is most pronounced in males. Unlike many cichlids, this species shoals readily. It is safe to keep a pair with their offspring. They spawn several times in succession, and the older fry may help the adults to guard their younger siblings.

Extended tip to dorsal fin

Bluish-white tips to fins

Lamprologus signatus

Signatus

- **ORIGINS** East Africa, restricted to deeper waters in the central part of Lake Tanganyika.
- **SIZE** 2¹⁄₂ in (5.5 cm).
- **DIET** Cichlid diets and live foods.
- **WATER** Temperature 72–79°F (22–26°C); hard (150–200 mg/l) and alkaline (pH 7.5–8.0).
- **TEMPERAMENT** Territorial.

The dark background color of this fish is interrupted by lighter vertical bands, which are more pronounced in the male. Adult females are about 1 in (2.5 cm) smaller than males. This enables them to occupy the empty shells of freshwater snails, in which they can lay their eggs in relative safety. Keep a single male in the company of several females and provide shells in the tank, allowing two or more per female. Spare shells may be used as retreats by newly hatched fry. Avoid mixing these bottom dwellers with other fish that frequent the lower levels of the tank; rainbowfish (*see pp.176–179*) are suitable companions. Small weekly water changes are advisable for all Lake Tanganyika cichlids.

Neolamprologus leleupi

Leleupi

- **ORIGINS** East Africa, found on both the western and eastern sides of Lake Tanganyika.
- **SIZE** 4¹⁄₄ in (11 cm).
- **DIET** Cichlid diets and live foods, such as shrimp.
- **WATER** Temperature 73–79°F (23–26°C); hard (150–200 mg/l) and alkaline (pH 7.5–8.0).
- **TEMPERAMENT** Intolerant of its own kind.

The color of this fish ranges from lemon through to rich yellow, with the northwestern race, *N. l. melas*, being a burnished brown. Carotene-rich foods will help to maintain the color. The female, which has shorter pelvic fins than the male, spawns on the roof of a cave. While she lays up to 150 eggs, the male fiercely defends the entrance to the nesting cave. If the young Lelupis are not removed before they are six weeks old, they are likely to be eaten by their parents.

Slim body

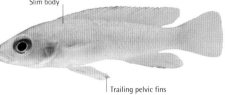

Trailing pelvic fins

LIVEBEARERS

These popular fish are characterized by their breeding habits. The eggs are fertilized internally; in most species, sperm are transferred to the female via the male's gonopodium—a tubelike projection of the anal fin—and the eggs subsequently develop in the relative safety of the female's body. Some livebearers, including limias, have a restricted distribution in the wild and are not widely kept. Others, including guppies, platies, and swordtails, are popular worldwide. The ease with which they can be bred has led to the development of these fish into a host of exhibition varieties.

Sexing livebearers is not difficult. The upper fish in this pair of **Red-Tailed Goodeids** (*Xenotoca eiseni*) is male, as evidenced by the hump on his head.

Anableps anableps

Four-Eyes

- **ORIGINS** Ranges from southern Mexico southward into northern parts of South America.
- **SIZE** 10½ in (27 cm).
- **DIET** Live foods preferred.
- **WATER** Temperature 72–86°F (22–30°C); hard (100–150 mg/l) and alkaline (pH 7.5).
- **TEMPERAMENT** Alert by nature.

Although dullish brown in color, these livebearers are fascinating to keep, particularly because they breed so readily. Males are much smaller than females and rarely exceed 6 in (15 cm) in length. The female can produce a batch of 6–13 offspring, each measuring up to 1½ in (4 cm) long, twice a year. An aquarium for Four-Eyes should not be filled to the top and must be covered to prevent them from leaping out. This species sometimes prefers brackish water.

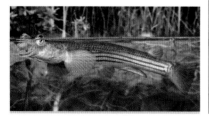

Gambusia sexradiata

Tropical Mosquitofish

- **ORIGINS** Central America, occurring in parts of Mexico, northern Guatemala, and northern Belize.
- **SIZE** 2½ in (6.5 cm).
- **DIET** Flake and small live foods.
- **WATER** Temperature 61–77°F (16–25°C); hard (100–150 mg/l) and alkaline (pH 7.5).
- **TEMPERAMENT** Social and nonaggressive.

Silver and blue feature strongly on the body of this fish, while the fins are marked with red and blue. Unlike other mosquitofish, this species does not tolerate brackish water. Include floating plants in its tank. Females are twice as large as males and have broader bodies. Some 10–35 young are born about a month after mating, and further broods follow at one-month intervals.

Gambusia affinis

Mosquitofish

- **ORIGINS** North America, occurring in the San Antonio River and the Rio Medina in Texas.
- **SIZE** 2½ in (6.5 cm).
- **DIET** Prepared foods and small live foods.
- **WATER** Temperature 61–77°F (16–25°C); hard (100–150 mg/l) and neutral (pH 7.0).
- **TEMPERAMENT** Peaceful and social.

Both sexes are brownish with silvery underparts, but females are significantly larger than males. In the wild, these fish feed on mosquito larvae, and they have been widely introduced in tropical areas to control mosquito numbers. Up to 60 young are born after a gestation of 24 days. The tank should be well planted to provide hiding places for the fry, or they will be eaten. The fry will take small live foods and are best reared in their own tank.

Swollen belly indicates a gravid female

Dorsal fin set well back

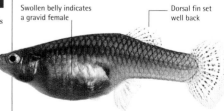

Heterandria formosa

Least Killifish

- 🌐 **ORIGINS** North America, where it is restricted to the state of South Carolina.
- ⭕ **SIZE** 1¾ in (4.5 cm).
- 🍽 **DIET** Prepared foods and small live foods.
- 💧 **WATER** Temperature 61–77°F (16–25°C); hard (100–150 mg/l) and alkaline (pH 7.5).
- 😐 **TEMPERAMENT** Social and nonaggressive.

Female

Variable black markings on the dorsal and anal fins

Silvery-white underparts

A broad but uneven black stripe running down the sides of the body, and an obvious black blotch on the dorsal fin, help to identify this tiny fish. The males, which are just ¼ in (2 cm) long, rank among the smallest of all vertebrates. Breeding is straightforward, but instead of giving birth to her brood of up to 20 fry all at once, the female produces offspring over an extended period of two weeks. A tank for Least Killifish needs to be densely planted.

Pseudoxiphophorus bimaculatus

Two-Spot Livebearer

- 🌐 **ORIGINS** Central America, from Mexico southward to parts of Guatemala, Belize, and Honduras.
- ⭕ **SIZE** 6 in (15 cm).
- 🍽 **DIET** Prepared foods and small live foods.
- 💧 **WATER** Temperature 72–77°F (22–25°C); hard (100–150 mg/l) and neutral (pH 7.0).
- 😐 **TEMPERAMENT** Aggressive.

These fish are often brownish-yellow, and many of the scales have darker borders. A dark spot is evident on the caudal fin. The male is smaller and can be identified by his gonopodium. Females may produce more than 100 fry every two months. After giving birth, the females must be removed, or they will eat their offspring.

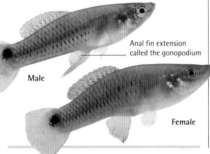

Anal fin extension called the gonopodium

Male

Female

Poecilia melanogaster

Blackbelly Limia

- 🌐 **ORIGINS** The Caribbean, where it is restricted to the island of Jamaica.
- ⭕ **SIZE** 2½ in (6 cm).
- 🍽 **DIET** Prepared foods, plant matter, and small live foods.
- 💧 **WATER** Temperature 72–82°F (22–28°C); hard (150–200 mg/l) and alkaline (pH 7.5).
- 😐 **TEMPERAMENT** Males often harass partners.

Bluish body color is characteristic of these livebearers. They can be kept in groups, ideally containing more females than males (to prevent individual females from being chased constantly by would-be partners). Algal growth in the tank is beneficial because it supplements the diet. Blackbelly Limias are not prolific breeders; the females produce no more than 25 fry every six weeks or so after reaching maturity.

Poecilia nigrofasciata

Black-Barred Limia

- 🌐 **ORIGINS** The Caribbean, where it is restricted to the Haitian part of Hispaniola.
- ⭕ **SIZE** 2½ in (6 cm).
- 🍽 **DIET** Prepared foods, plant matter, and small live foods.
- 💧 **WATER** Temperature 72–79°F (22–26°C); hard (100–150 mg/l) and alkaline (pH 7.1).
- 😐 **TEMPERAMENT** Peaceful by nature.

The distinctive black barring on the body of this livebearer shows best under bright light. The body coloration itself is variable, although it tends to be yellowish, especially around the head. Mature males, which are slightly smaller than females, develop a hump on their heads as they grow older. They also have a narrow keel, which extends back on the underside of the body between the anal and caudal fins. Like related species, Black-Barred Limias will take a wide variety of foods and benefit from vegetable matter in their diet. Females have relatively large broods, producing as many as 50 fry at a time; broods are born every six weeks or so. The percentage of male offspring apparently increases when females are kept at higher temperatures.

ALL-SEEING EYES

The surface-dwelling Four-Eyes *(see opposite)* is so called because each pupil is divided into two parts. This adaptation enables the Four-Eyes to see simultaneously both above and below the waterline, alerting it to feeding opportunities and predators in either environment. The eyes are positioned high on the head, so the fish can lie in the water with the rest of its body submerged, rather like a crocodilian. In the wild it occasionally clambers out of the water to rest on rocks, and it should be given the opportunity to do this in the aquarium, too.

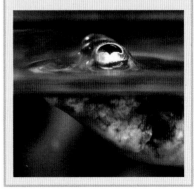

Girardinus

Girardinus metallicus

ORIGINS The Caribbean, where it is widely distributed in Cuba, except in the extreme east.

SIZE 3½ in (9 cm).

DIET Prepared foods, plant matter, and small live foods.

WATER Temperature 72–77°F (22–25°C); hard (100–150 mg/l) and alkaline (pH 7.1).

TEMPERAMENT Placid.

Both sexes have a metallic sheen, but the female (foreground of picture) is much larger than the male, which averages just 2 in (5 cm). A mature male can also be distinguished by his black gonopodium. Females can give birth every three weeks, with the average brood comprising around 50 fry. A densely planted tank will help to prevent the young from being cannibalized.

Malayan Halfbeak

Dermogenys pusilla

ORIGINS Southeast Asia, ranging from Thailand and the Malay Peninsula down to parts of Indonesia.

SIZE 3 in (7.5 cm).

DIET Mainly small live foods.

WATER Temperature 77–86°F (25–30°C); hard (100–150 mg/l) and neutral (pH 7.0).

TEMPERAMENT Males are quarrelsome.

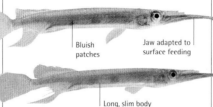

Bluish patches

Jaw adapted to surface feeding

Long, slim body

Known as Halfbeaks because the upper jaw is much shorter than the lower, these livebearers often occur in brackish water. Insects, such as wingless fruit flies (*Drosophila* sp.) and mosquito larvae, can be bred as food for them. These fish can jump well, so cover their tank. Raising the water temperature encourages breeding, but stillbirths are not uncommon. The broods, consisting of about 30 young each, are produced at eight-week intervals.

Pike-Top Minnow

Belonesox belizanus

ORIGINS Central America, ranging from southeastern Mexico down to Honduras.

SIZE 8 in (20 cm).

DIET Smaller live foods and fish.

WATER Temperature 79–90°F (26–32°C); hard (200–300 mg/l) and alkaline (pH 7.5).

TEMPERAMENT Aggressive.

This fish not only has a body shaped like a pike but also has similar predatory habits and hunts by ambushing its prey. Females are the larger sex and display yellowish or even orange coloration at the base of their anal fin. Pike-Top Minnows hide away in vegetation and may prefer brackish water. Their young are about 1 in (2.5 cm) long at birth and are rarely eaten by the female.

Merry Widow

Phallichthys amates

ORIGINS Central America, found in Guatemala and from Honduras to Panama.

SIZE 2 in (5 cm).

DIET Prepared foods, plant matter, and small live foods.

WATER Temperature 72–82°F (22–28°C); hard (150–200 mg/l) and neutral (pH 7.0).

TEMPERAMENT Placid.

A black stripe through the eye, a black spot near the vent, and black edging on the dorsal fin typify this fish; males are significantly smaller in size. The Orange Dorsal Livebearer (*P. a. pittieri*) is a related but slightly larger form, with different-colored edging on the dorsal fin. Females give birth every month, producing around 50 fry on each occasion—or sometimes as many as 150.

Spotted Skiffia

Skiffia multipunctata

ORIGINS Central America, restricted to the Mexican states of Jalisco and Michoacán.

SIZE 2½ in (6 cm).

DIET Prepared foods, plant matter, and small live foods.

WATER Temperature 68–77°F (20–25°C); hard (100–150 mg/l) and alkaline (pH 7.1).

TEMPERAMENT Placid.

A random pattern of black spots and blotches is evident on the body of the Spotted Skiffia. The difference in size between the sexes is much less apparent in this species than in some livebearers, but males can be distinguished easily by their notched anal fin and the yellow edging on the dorsal fin. This fin is also irregularly notched, which explains why these fish are also referred to as Sawfins. It is important not to allow the water in their aquarium to become too warm, because this may reduce the likelihood of successful breeding. The period between mating and birth can extend for up to 60 days, and there are often fewer than 20 fry in a brood. The young, which measure just under ½ in (about 1 cm) at birth, attain maturity at the age of about two months.

BEAK WRESTLING

Trials of strength are common between male Malayan Halfbeaks *(see opposite)*. The fish lock their mouthparts together and remain in this position for 20 minutes or more, until the weaker individual releases his grip and backs off. This can cause injury to the beak, so it is not to be encouraged, and male Halfbeaks should generally be kept apart from one another. The protruding lower jaw can also be injured if these fish become alarmed and swim wildly at the sides of the tank. This is most likely to occur when they are first transferred to new surroundings.

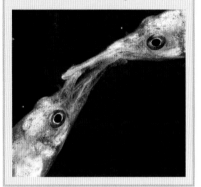

Ameca splendens

Butterfly Splitfin

- **ORIGINS** Central America, in the Rio Ameca and the Rio Teuchitlán in Jalisco State, Mexico.
- **SIZE** 4½ in (12 cm).
- **DIET** Prepared foods, plant matter, and small live foods.
- **WATER** Temperature 68–84°F (20–29°C); hard (100–150 mg/l) and neutral (pH 7.0).
- **TEMPERAMENT** Males are territorial and aggressive.

These livebearers can be sexed not only by their size but also by their coloration. The larger female has an irregular pattern of black spots over her body. In the more plainly colored male, shown below, the caudal fin is edged with black and then yellow, and the dorsal fin is more prominent. Females give birth to offspring that have an attachment resembling an umbilical cord; this disappears soon after birth. A typical brood consists of about 30 young, born about two months after mating. Young females giving birth for the first time have only a small number of offspring, which may vary noticeably in size. These goodeids require a well-lit tank with plenty of retreats, and they will browse readily on algae growing in their quarters.

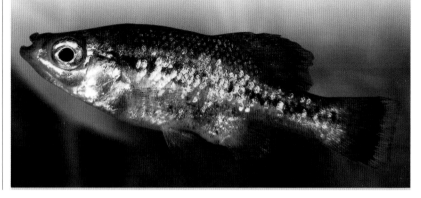

Allotoca dugesii

Golden Bumblebee Goodeid

- **ORIGINS** Central America, occurring in the Mexican states of Jalisco, Guanajuato, and Michoacán.
- **SIZE** 3 in (7.5 cm).
- **DIET** Prepared foods, plant matter, and small live foods.
- **WATER** Temperature 72–77°F (22–25°C); hard (100–150 mg/l) and alkaline (pH 7.1).
- **TEMPERAMENT** Can be a fin-nipper.

Yellowish-gold indicates a male

The body of the male Golden Bumblebee is black and yellowish-gold; the larger female has bluish lower parts. House a male with two females, because he will bully a solitary female. The female gives birth to up to 70 fry every two months or so. Although goodeids are easy to keep, they are not as widely available as most other livebearers, such as the more colorful guppies and swordtails.

Characodon lateralis

Rainbow Goodeid

- **ORIGINS** Central America, found mainly in the upper Rio Mezquital in Durango State, Mexico.
- **SIZE** 2¼ in (5.5 cm).
- **DIET** Prepared foods, plant matter, and small live foods.
- **WATER** Temperature 68–75°F (20–24°C); hard (100–150 mg/l) and alkaline (pH 7.1).
- **TEMPERAMENT** Relatively shy.

As the name suggests, this is a colorful goodeid; the adult male, shown below, displays areas of gold and red. The female Rainbow Goodeid produces fewer than 25 offspring per brood, with the brood interval being about eight weeks. Again, it is best to keep these goodeids in trios, but do not mix them with related species because of the risk of hybridization. Their aquarium should have plenty of vegetation.

Alfaro cultratus

Knife-Edge Livebearer

- **ORIGINS** Central America, found in parts of Nicaragua, Costa Rica, and western Panama.
- **SIZE** 4 in (10 cm).
- **DIET** Prepared foods, plant matter, and small live foods.
- **WATER** Temperature 75–82°F (24–28°C); soft (50–100 mg/l) and neutral (pH 7.0).
- **TEMPERAMENT** Shy, but may be aggressive.

Knifelike edge

Male's gonopodium

A row of scales projecting beyond the lower edge of the body gives these fish their distinctive knifelike appearance. Both sexes are pale yellowish-brown in color. Clear, clean water is very important for the well-being of Knife-Edge Livebearers; regular partial water changes are essential, since the fry in particular are prone to bacterial ailments. These fish can sometimes be kept in slightly brackish surroundings. Provide the fry with brine shrimp as a rearing food.

Xiphophorus hellerii

Swordtail

🌐 **ORIGINS** Central America, ranging down the Atlantic side from Mexico to northwestern Honduras.

〰️ **SIZE** 4 in (10 cm).

🍴 **DIET** Prepared foods and small live foods.

〰️ **WATER** Temperature 70–77°F (21–25°C); hard (100–150 mg/l) and alkaline (pH 7.5).

〰️ **TEMPERAMENT** Males are often quarrelsome.

In common with most domesticated livebearers, the Swordtail's color and pattern are very different from those of the wild type. Many of these fish have a hybrid ancestry, thanks to earlier crossings both with other types of swordtail and also with platies *(see p.162)*. Wild swordtails are usually quite plain-looking and greenish, blue, or reddish in coloration, while some display blotched patterning. Wild populations can vary significantly in appearance from place to place.

In aquarium surroundings, mature male swordtails are aggressive toward one another. It is relatively straightforward to identify the potential troublemakers, since only the males possess the impressive swordlike extension to the lower rays of the caudal fin. (The scientific name of these livebearers derives from *xiphos*, the Greek word for "sword.") Females, which are larger than males, sometimes develop the sword, too, as a result of hormonal changes. These individuals, however, are unable to reproduce as males.

Swordtails can prove to be quite prolific when breeding; larger females produce as many as 200 fry in a single brood and can repeat this feat every month or so. However, if the young are to survive, the female must be transferred to a breeding trap in a suitable nursery tank. Few livebearers are more notorious for hunting down and cannibalizing their offspring than swordtails. The young are relatively simple to rear on their own if provided with items such as fry foods and brine shrimp.

Female

Male

Golden Comet Swordtail The black tail streaks are called the comet characteristic. Golden Swordtails tend to be less brightly colored than red forms. The broader body of the female can be seen in this pair.

Reddish-spotted dorsal fin

Male's sword-like extension

Black Calico Swordtail This form was developed in the laboratory. The black gene is linked to tumor formation.

Dorsal fin has an extension

Lyre-tailed fin with symmetrical swords

Red Wag Lyre Swordtail The caudal fin has a sword on both its upper and lower rays, while its black coloration contrasts with the reddish color of the body. The lips are also black.

Gonopodium confirms that this is a male

Sword visible in this young male

Pineapple Swordtail The appearance of this very popular form varies from yellow to intense red. The depth of the Pineapple Swordtail's coloration can be improved by color-feeding.

Green Swordtail (left) These fish are the ancestors of today's domesticated varieties, but their subdued coloration means they are not as widely kept. The patterning varies throughout their natural range.

Xiphophorus birchmanni

Sheephead Swordtail

- **ORIGINS** Central America, occurring in various waterways around Hidalgo, Veracruz, Mexico.
- **SIZE** 3 in (7.5 cm).
- **DIET** Prepared foods and small live foods.
- **WATER** Temperature 70–77°F (21–25°C); hard (100–150 mg/l) and alkaline (pH 7.5).
- **TEMPERAMENT** Males are often quarrelsome.

This swordtail species is unusual, because the male has no trace of a sword on his caudal fin. He can still be easily distinguished from the female by his gonopodium and larger dorsal fin. Well-oxygenated water is important for this species, which has only been available to hobbyists since about 1990. The broods are small, typically consisting of 10–35 fry.

Mottled dorsal fin

Gonopodium of male

Slightly curved caudal fin

Xiphophorus montezumae

Montezuma Swordtail

- **ORIGINS** Central America, restricted to eastern Mexico, in the states of Tamaulipas and San Luis Potosi.
- **SIZE** 3 in (7.5 cm).
- **DIET** Prepared foods and small live foods.
- **WATER** Temperature 70–77°F (21–25°C); hard (100–150 mg/l) and alkaline (pH 7.5).
- **TEMPERAMENT** Males are often quarrelsome.

The key feature of the Montezuma Swordtail is the way in which the male's sword projects almost horizontally, rather than downward as in other swordtails. The length of the sword varies in fish from different populations, although wild-type stock is rarely available today. Males have a larger dorsal fin than females, although they are otherwise slightly smaller in size. Their broods average about 50 fry, with the interval between broods usually being about seven weeks, although seasonal factors may affect their breeding behavior. The Montezuma Swordtail is not especially brightly colored, although there can be individual exceptions. Typically, these fish are bluish-silver and display a pattern of dark spots on the flanks. The sword of the male is often greenish or yellow, with fairly prominent black edging. It is important not to keep these or other swordtails in the company of fin-nipping species, because their swords will inevitably be attacked. In addition, their aquarium needs to incorporate plenty of open swimming areas so that they do not damage their swords in confined spaces.

PREVENTING A SERIOUS DUEL

One way of reducing aggression between male swordtails is to house them in a mixed group. There is still likely to be intermittent squabbling, but the disputes will not be as fierce. The swordtails will also be less likely to bully other tank occupants. Avoid mixing swordtail species, because hybridization is likely and there will be conflict between the different males. Equally, they should not be kept alongside platies, with which swordtails will readily interbreed.

Xiphophorus maculatus

Platy

- ⚙ **ORIGINS** Central America, occurring in Veracruz state in Mexico and southward to Belize and Guatemala.
- ◈ **SIZE** 3 in (7.5 cm).
- ◈ **DIET** Prepared foods and small live foods.
- ◈ **WATER** Temperature 64–77°F (18–25°C); hard (100–150 mg/l) and alkaline (pH 7.5).
- ⚙ **TEMPERAMENT** Peaceful and social.

The Platy has a broad body and naturally rounded fins. Spotted patterning on the caudal peduncle and dorsal fin is common in the wild, and it has been maintained in a number of domestic strains. The caudal peduncle spot is more like a crescent moon in some cases, which explains why the Platy is occasionally called the Moonfish. The wild form is usually olive-brown, although some individuals display reddish coloration—a feature that is very evident in domestic strains. These strains have been created both by selective breeding and by hybridization with other *Xiphophorus* species, especially the Variegated Platy (*X. variatus*).

The different strains are described on the basis of their coloration, patterning, and fin type, and it is quite possible for two or even three of these characteristics to be evident in a single individual, as typified by a Blue Wagtail Hi-Fin Platy, for example. Such is the scope represented by the different characteristics identified in the Platy that in excess of 325,000 varieties could be created without resorting to hybridization with other Xiphophorus species. What sometimes occurs is that, in the quest to establish a particular feature, the breeding stock is too closely related, which reduces the fertility of the strain. Unless corrected by outcrossing to unrelated bloodlines, the strain may ultimately die out.

Pintail Red Wagtail Platy The Pintail first arose in Germany. Specimens in which the elongated part of the caudal fin has become wider are known as Plumetails.

The Platy is suitable for both a community aquarium and a single-species setup. A tank for these fish should incorporate a clear area at the front for swimming, with some specimen plants, and denser planting around the back and sides. It is important to include plants such as Java Moss (*Vesicularia dubyana*) if you hope to breed these livebearers successfully. Some of the young fish will survive long enough in this dense vegetation to grow to a size at which they will no longer be seen as a meal by other fish in the tank. The Platy benefits from being kept at a slightly higher temperature than its variegated relative.

Coffee and Ink Platy The spotted patterning on this variant is larger than on the so-called Salt and Pepper Platy, often resulting in a rather blotchy appearance. The caudal peduncle is largely black.

Rounded caudal fin

Even depth of coloration

Blue Platy The distinctive sky-blue body color of the Blue Platy contrasts with the dark spot on the caudal fin. A blue strain of platy with a red caudal fin now seems to have died out.

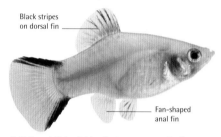

Black stripes on dorsal fin

Fan-shaped anal fin

Gold Comet Platy Gold variants were among the first to be created. This form combines peachy-gold coloration with black stripes on the top and bottom of the caudal fin. It is also known as the Twin-Bar Platy.

Rich, even, orange-red body coloration

Upturned mouth

Male's gonopodium

Blue Coral Platy Paler and more whitish than the Blue Platy, this form has a deep blue spot at the base of the caudal peduncle and displays black, crescent-shaped markings on the adjacent part of the caudal fin.

Yellow may shine through the black in places

Black Platy The amount of black on this strain is variable, sometimes covering the entire body except the face and throat.

High, flowing fin

Center of body is a lighter shade of orange

Pale head

Male's gonopodium

Sunset or Marigold Hi-Fin Platy The color of these fish becomes a more intense orange toward the rear of the body, with the caudal fin being darker in color than the others. The dorsal fin is long and trailing.

Underparts are also pale

BEARING LIVE YOUNG

Watching a female platy—such as this Salt and Pepper Platy—giving birth, one might get the impression that these fish have strong maternal instincts, but in reality they fail to acknowledge their offspring at all. As a result, the young fish are at risk of being eaten as soon as they are born. The safest option, therefore, is to transfer a gravid female to a breeding trap in a separate tank, where she can give birth in peace and cannot harm her offspring.

In platies and most other livebearing fish, there is nothing akin to the placental attachment between mother and young that is seen in mammals. The female platy's body simply serves as a shelter for her developing eggs, with the young uncurling from their egg sacs as they are born. By protecting their eggs in this way, livebearing fish ensure that their offspring will survive through to hatching at the very least, rather like mouth-brooders (see p.148). Females can produce a small brood of about 80 or so offspring every four to six weeks. She can do this without having to mate on each occasion, because she is able to store the male's sperm in her body.

Xiphophorus variatus

Platy Variatus

- **ORIGINS** Central America, occurring in eastern Mexico, from Tamaulipas state to Veracruz.
- **SIZE** 3 in (7.5 cm).
- **DIET** Prepared foods and small live foods.
- **WATER** Temperature 61–77°F (16–25°C); hard (100–150 mg/l) and alkaline (pH 7.5).
- **TEMPERAMENT** Placid and social.

As its name suggests, this fish exists in a variety of forms in the wild, where it occurs in ponds and slow-flowing stretches of water. Many wild populations are localized and so develop individual characteristics. Platy Variatus are adaptable, simple to maintain, and breed readily. Mature females can give birth to as many as 100 offspring each month. They can be sexed easily, since the female is larger and lacks the gonopodium, the male copulatory organ. It is difficult to obtain wild-type strains today because breeders have altered the coloration, patterning, and size of these fish so extensively. Platy Variatus will hybridize with other platies and also with swordtails.

Golden coloration most evident on the underparts

Greenish hue apparent on upper body

Blue Tuxedo Platy Variatus A Tuxedo Platy's body always combines black (at the rear) with another color, such as blue in the case of this variant.

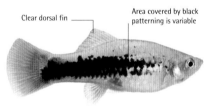

Clear dorsal fin

Area covered by black patterning is variable

Calico Platy Variatus This variant displays a mottled patterning. There is a Hawaiian strain, in which black covers almost the entire side of the body.

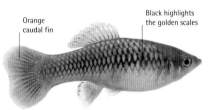

Orange caudal fin

Black highlights the golden scales

Green Platy Variatus The vibrant gold of this form is variable, with some individuals appearing more orange. This may reflect their diet as well as their breeding.

Variegated Platies will feed readily at the surface of the tank

Hi-Fin Golden Parrot Platy Variatus This attractive gold and green platy also displays the "hi-fin" characteristic, in which the dorsal fin is greatly enlarged.

Golden Platy Variatus This variety was probably developed from the Platy Variatus that occur in Mexico's Rio Axtla, which display similar markings and a red tail.

Poecilia reticulata

Guppy

🌐 **ORIGINS** South America, occurring in the Caribbean and in South America north of the Amazon.

🔵 **SIZE** 2 in (5 cm).

🔻 **DIET** Prepared foods and small live foods.

〰️ **WATER** Temperature 70–77°F (21–25°C); hard (100–150 mg/l) and alkaline (pH 7.5).

⚙️ **TEMPERAMENT** Placid and social.

One of the best known of all tropical fish, the Guppy is named after Rev. Thomas Guppy, who identified it on the Caribbean island of Trinidad. The several thousand varieties available to aquarists today are far removed from their wild ancestors and can be found in a wide range of colors and with many different body patterns and fin types. The different color variations are displayed most impressively in male fish, which are naturally more colorful as well as smaller than females. (All the fish illustrated here are male.) Guppies show well as a group in a single-species tank, but they can also be kept with other nonaggressive fish as part of a community aquarium. The female Guppy gives birth to live offspring, but unfortunately these are likely to be cannibalized soon after birth, even in a breeding tank setup, unless the young can escape out of reach. Various breeding traps are available for this purpose *(see p.66).* When buying these fish, it is worth remembering that the largest females give birth to correspondingly bigger broods.

Wild Guppy These fish may be found in brackish water, and the addition of salt to their aquarium is recommended. Wild Guppies are not readily available today, and domestic strains will prove far more adaptable.

Characteristic caudal fin marking

Contrasting bluish body color

German Yellow Guppy Named after its country of origin, this is a particularly striking variety in which the enlarged caudal fin and the dorsal fin are both yellow. Fin shape, as well as color, is important in fancy guppies.

Gonopodium

Caudal fin has unbroken red coloration, with paler edges

Blonde Guppy The red caudal fin and rear part of the body contrast with the lighter blonde coloration seen on the underparts near the head. The use of color food is often recommended for red strains of guppies.

Enlarged red caudal fin contrasts with black coloring

Red Tail Half-Black Guppy As with other strains created by selective breeding, fertility may be impaired if these fish are heavily inbred. Not all such strains are commercial products; some are bred by enthusiasts only.

Silver-Backed Tuxedo Guppy The distinctive black area on the flank varies in size and density between individuals.

Sloping head

Rainbow Cobweb Delta Guppy This is another "composite variety," so called because of its varied coloration, cobweblike pattern, and the delta shape of its caudal fin.

Poecilia reticulata (continued)

One of the most significant factors to consider when breeding guppies is that a female needs to mate only once in order to continue producing young throughout her life—potentially giving birth to seven or more broods using sperm stored in her body. This is why even if you choose a well-marked male and female from the same tank in a pet store, the likelihood will be that at least some of the young will not be the offspring of that particular male (although the majority of offspring are likely to be the result of the most recent mating). This also explains how females kept on their own can give birth to young.

The only way to be sure of the parentage of guppies is to separate the sexes as early as possible. As soon as the young males can be recognized by their gonopodium—usually when they are about three weeks old—they should be transferred to a separate tank. The females will be noticeably larger than the males from three months onward, by which time some of the males will already be sexually mature. A female guppy will have her first brood approximately a month after mating. The number of offspring produced is likely to be small at first, sometimes no more than 10, but it increases to between 50 and 100 per brood as the female grows bigger. Some strains are more prolific breeders than others.

Golden Snakeskin Delta Guppy
This is one of a number of guppy varieties that have become very popular in Russia. It is thought that the famous Moscow Blue strain may have evolved from the Snakeskin line.

The snakeskin patterning can be combined with other colors, but it is usually associated with varieties displaying a broad caudal fin

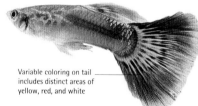

Variable coloring on tail includes distinct areas of yellow, red, and white

The reptilian markings extend over the entire body

Cornflower Blue Delta Guppy (below)
The exact patterning on the broad tail varies between individuals, allowing them to be distinguished quite easily, but the body coloration should be a consistent feature.

Red Varitail Guppy (above) Broad-tailed guppies like the Red Varitail tend to be more popular than those with narrow tails, because the wide caudal fin allows for some striking tail patterns to be developed.

Gold Cobra Delta Guppy Of American origin, Cobra Guppies are now popular internationally. The male (seen above) is always more colorful than the female, although she may display a patterned caudal fin in some strains.

Even projections on the caudal fin

Golden coloration on the body

Golden Lyretail Guppy The upper and lower rays of this guppy's caudal fin are greatly extended, creating a sleek appearance. It is important in the lyre-tailed forms that these two areas are symmetrical and even in size.

Tuxedo Rainbow Delta Guppy When selecting these and other delta-tailed guppies, be sure to check that there is no significant damage to the broad caudal fin, which could lead to a life-threatening infection.

Selective breeding of guppies began during the 1950s, but unfortunately, a number of strains—even some that are carefully maintained—are not stable. This means that many of the resulting offspring may not display the most desirable characteristics of their parents. Cobra patterning is one of the most stable characteristics in terms of markings. This is a dominant genetic characteristic, so well-marked individuals are always likely to pass their cobra patterning on to the next generation. However, recessive characteristics, such as tuxedo patterning, may disappear for several generations of a particular bloodline and then reemerge unexpectedly at a later stage.

Tuxedo Multicolored Delta Guppy The tuxedo characteristic (the black area toward the rear of the body) can be combined with different colors to striking effect. This variety also has the broad delta-tail.

Relatively even, blackish tip on the broad caudal fin

Blue Tuxedo Guppy This particular variety is also described as the Blue Delta-Tail Half Black, based on its body coloration. The broad caudal fin is predominantly blue, with variable black markings.

Irregular patterning running across the caudal fin

Green Variegated Delta Guppy Like other guppies with a delta-shaped caudal fin, the Green Variegated Delta Guppy is a slower swimmer than sleeker-tailed forms. This variety should be housed in a single-species tank.

Flamingo Guppy The breeding trend with this strain has been to improve the depth of the red coloration. Early examples were considerably paler than those being bred today. Flamingo Guppies are also called Golden Redtails.

Poecilia reticulata

Endler's Livebearer

ORIGINS South America, where this attractive wild form of the Guppy is found in parts of Venezuela.

SIZE 2 in (5 cm).

DIET Prepared foods and small live foods.

WATER Temperature 70–77°F (21–25°C); hard (100–150 mg/l) and alkaline (pH 7.5).

TEMPERAMENT Peaceful and social.

These spectacularly colored fish are a wild form of the guppy. Despite this, Endler's Livebearers must not be housed with guppies, because they hybridize readily with them and soon lose their distinctive characteristics. The body colors of Endler's Livebearers range from blues and greens through to oranges and reds. These fish are prolific breeders, but they have a relatively short life span of no more than a year.

THE GRAVID SPOT

Breeders of popular livebearers have relied on the appearance of the dark gravid spot to indicate that a female is about to give birth. This spot is formed by the dark lining of the abdomen, known as the peritoneum, which bulges against the sides of the female's body just before she gives birth, pushed out by the increasing size of her brood. The gravid spot is less apparent in swordtails, simply because the female's abdominal wall is more muscular. Once the gravid spot has appeared, the female should not be moved; otherwise, she is likely to abort her brood.

Poecilia butleri

Pacific Mexican Molly

- 🌐 **ORIGINS** Central America, ranging from northern Mexico into Guatemala and Honduras.
- 🌀 **SIZE** 3½ in (8.5 cm).
- 🍽 **DIET** Prepared foods, vegetables, and small live foods.
- 💧 **WATER** Temperature 75–81°F (24–27°C); hard (100–150 mg/l) and alkaline (pH 7.5).
- 😊 **TEMPERAMENT** Placid and social.

The coloration of Pacific Mexican Mollies in the wild is variable, ranging from shades of blue to yellowish and silvery tones. Darker coloration is seen in some individuals, notably in the race known as Limantour's Molly (*P. m. limantouri*). The wild Pacific Mexican Molly is the ancestor of the Black Molly, the most popular of the domesticated varieties available today. The Black Molly was created in the 1930s by a breeder in New Orleans, although the exact route of its development is now unclear. Its distinctive matt-black coloration almost certainly

derives from melanistic examples of the wild form. Like the Pacific Mexican, the Black Molly has a low dorsal fin, a relatively elongated body, and a compact, rounded caudal fin. The black coloration has also been combined with marbled and balloon characteristics. Black varieties look good in an aquarium alongside bright orange swordtails *(see p.160)*. These fish often fare better when kept in slightly brackish water. This seems to make them less vulnerable to the parasitic illness known as "ich," or white spot *(see p.58)*, which shows up clearly against the color of their bodies. Females typically produce 40–300 offspring per brood.

Marbled Lyretail Molly In this form, the upper and lower rays of the caudal fin are elongated. However, changes to the structure of the caudal fin are not common in mollies.

Black marbled patterning extends onto the fins

Upturned mouth indicates that these fish are naturally surface feeders

Marbled Molly A relative of the Black Molly and another descendant of the Pacific Mexican Molly, this fish has a varied pattern of large, irregular black blotches, making it easy to distinguish from spotted or dotted forms.

CLONED FISH

The Amazon Molly (*P. formosa*) is believed to be a wild hybrid of the Yucatán and Pacific Mexican species. Its common name refers not to its distribution (which extends from parts of the southern U.S. into Central America) but to the fabled all-female tribe of South America. Only one in every 10,000 of these fish is a biological male, and the arrangement of the males' chromosomes suggests that they are sterile. The offspring produced by female mollies are therefore clones of their mothers, making the Amazon Molly the first unisexual vertebrate to be discovered.

Female Amazons still need to mate in order to give birth, so they attract males of other molly species, by interrupting their courtship rituals, and also simply by living alongside them so that the males eventually come to identify with them. While sperm produced by these males triggers the development of an Amazon Molly's eggs, it does not fertilize them. The eggs contain the complete genetic blueprint of the young, rather than just the half normally contributed by female vertebrates that reproduce sexually.

Poecilia velifera

Sailfin Molly

- **ORIGINS** Central America, restricted to the Yucatán Peninsula of Mexico.
- **SIZE** 7 in (18 cm).
- **DIET** Prepared foods and small live foods.
- **WATER** Temperature 73–82°F (23–28°C); hard (100–150 mg/l) and alkaline (pH 7.5).
- **TEMPERAMENT** Generally social.

The Sailfin Molly has a long, tall dorsal fin that runs almost the entire length of the fish's back. Sailfin Mollies have been used extensively in the development of today's domesticated variants, which exist in a wide range of colors. These relatively large mollies must be kept under good water conditions; if conditions are less than ideal, they display their unhappiness by rocking gently back and forth in the same spot—a behavior known as shimmying. Wild Sailfin Mollies usually require a brackish environment, and this may also be of benefit to their domesticated relatives. When acquiring these fish, always check whether they have been kept in brackish water previously so that you can adjust the conditions in their tank accordingly to ensure a trouble-free period of acclimatization. Female Sailfin Mollies may produce a brood of more than 100 fry as regularly as every six weeks. The height of the dorsal fin in

their offspring can vary greatly, because of crossings involving the Pacific Mexican Molly *(see opposite)* and the other sailfin form, *P. latipinna*, which has a shorter, less impressive dorsal fin. Like other mollies, Sailfins will live in groups, but it is important to house several males together; if there are just two, the weaker individual will inevitably be bullied.

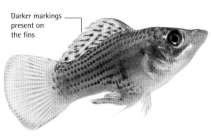

Spotted Silver Molly This variant displays black spots on its silvery body. The distribution of the spots is entirely random, allowing individuals to be distinguished by their appearance.

Darker markings present on the fins

Green Molly (above) This variety is easily distinguished by its pale, greenish body color and the lines of darker dots running along its sides.

Height and shape of dorsal fin varies

Lyre-Tailed Black Balloon Molly These fish have a compact, rounded body shape and sometimes encounter difficulty in swimming as a consequence.

Marmalade Molly (below) These mollies are a vibrant orange, with some individuals resembling platies *(see p.162)* in their depth of coloration.

Rounded caudal fin

Upturned mouth

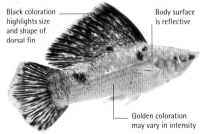

Black coloration highlights size and shape of dorsal fin

Body surface is reflective

Golden coloration may vary in intensity

Goldfin Metallic Marble Sailfin Sometimes called the Starburst Molly, the black areas on this variety are mainly restricted to the dorsal and caudal fins.

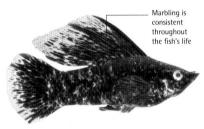

Marbling is consistent throughout the fish's life

Brown Marble Sailfin Molly This individual is heavily marbled, although there is little pigmentation on the outer parts of the dorsal and caudal fins.

KILLIFISH

These narrow-bodied, often highly colorful fish deserve to be far more popular. Their small size means that they do not need to be accommodated in a large, expensive tank. In addition, many of the more widely available killifish are easy to maintain, and they can often be persuaded to spawn in home aquariums. Their main drawbacks are that some are highly aggressive toward their own kind, and, since they feed naturally on small invertebrates in the wild, it can be difficult to wean them onto substitute diets. Sexing is simple, since the fins of the males are more pointed than those of the females.

Some killifish, such as the **Red-Striped Killie** (*Aphyosemion striatum*), can be kept at near room temperature. Keeping them too warm dulls their color and shortens their life span.

Common Lyretail

Aphyosemion australe

- ORIGINS West Africa, found in parts of Gabon, Cameroon, and Zaire.
- SIZE 2½ in (6 cm).
- DIET Live foods and prepared diets.
- WATER Temperature 64–75°F (18–24°C); soft (50–100 mg/l) and acidic (pH 6.0–6.5).
- TEMPERAMENT Peaceful and social.

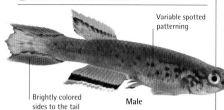

Variable spotted patterning

Brightly colored sides to the tail

Male

Male Common Lyretails are much more brightly colored than females and have pronounced caudal-fin rays that give the tail a lyrelike shape. They need a slightly brackish environment and a densely planted tank. They are usually kept in a single-species setup, partly because their small size limits the choice of tankmates, but also so they do not hybridize with other killifish. Lyretails tend to live for about three years—longer than most types of killifish.

Red-Striped Killie

Aphyosemion striatum

- ORIGINS Occurs in West Africa, restricted to pools and swamps in northern Gabon.
- SIZE 2 in (5 cm).
- DIET Live foods and prepared diets.
- WATER Temperature 64–72°F (18–22°C); soft (50–100 mg/l) and acidic (pH 6.0–6.5).
- TEMPERAMENT Nonaggressive.

Bright red horizontal stripes running along the body and a blue background help to distinguish male Red-Striped Killies. Females are a dull golden-brown shade, with rounded fins. Like other killifish, they eat small live foods, even freeze-dried items. The water should be slightly brackish, and their tank must be densely planted. It must also be kept covered, since these fish can jump. Java Moss (*Vesicularia dubyana*) is a good spawning medium.

Tall dorsal fin

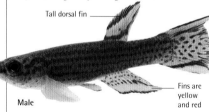

Male

Fins are yellow and red

Two-Striped Killie

Aphyosemion bivittatum

- ORIGINS West Africa, found in parts of Cameroon, Nigeria, and Togo.
- SIZE 5 cm (2 in).
- DIET Live foods and prepared diets.
- WATER Temperature 22–24° C (72–75° F); soft (50–100 mg/l) and acidic (pH 6.0–6.5).
- TEMPERAMENT Peaceful.

Dorsal fin folded down

Male

Lower stripe is less well defined

The color of these killifish is highly variable, but in all cases males have more elaborate fins than females. The upper stripe on the fish's body is more prominent than the lower, which runs along the underparts. As in other related aphyosemions, spawning is a lengthy process that takes place over the course of several days amid aquatic plants. For breeding purposes, several males should be housed with a single female. The eggs start to hatch after a period of three weeks.

Aphyosemion bitaeniatum

Twin-Banded Killifish

- **ORIGINS** West Africa, occurring in coastal parts of Nigeria, Benin, and Togo.
- **SIZE** 2 in (5 cm).
- **DIET** Live foods and prepared diets.
- **WATER** Temperature 72–75°F (22–24°C); soft (50–100 mg/l) and acidic (pH 6.0–6.5).
- **TEMPERAMENT** Males can be quarrelsome.

With their tall dorsal fin, the males of this species are exceptionally eye-catching. Their coloration is variable, although red may dominate, and there are usually two darker stripes down the sides of the body. Females are a dull brown, with less elaborate fins. Male Twin-Banded Killifish will fight, so they should be kept apart. House a single male with two females for spawning, which usually occurs among plants.

Male

Fundulopanchax gardneri

Clausen's Steel-Blue Killie

- **ORIGINS** West Africa, recorded in parts of Nigeria and western Cameroon.
- **SIZE** 3 in (7.5 cm).
- **DIET** Live foods and prepared diets.
- **WATER** Temperature 72–79°F (22–26°C); soft (50–100 mg/l) and acidic (pH 6.0–6.5).
- **TEMPERAMENT** Males are quarrelsome.

Brilliant blue coloration and a variable red-spotted patterning along the body identify the male of this species, seen below. The female Clausen's Steel-Blue Killie has brown dots on its flanks and a duller coloration on the fins. This species does not require brackish water. Spawning in the aquarium occurs on fine-leaved plants or on a spawning mop. If the tank is left full of water, the fry will hatch after two weeks.

Fundulopanchax sjostedti

Blue Gularis

- **ORIGINS** West Africa, found in shallow waters in Ghana, Cameroon, and Nigeria.
- **SIZE** 5 in (13 cm).
- **DIET** Live foods and prepared diets.
- **WATER** Temperature 73–79°F (23–26°C); soft (50–100 mg/l) and acidic (pH 6.0–6.5).
- **TEMPERAMENT** Relatively aggressive.

The color on the flanks of the Blue Gularis is variable—yellowish in some individuals, as seen below, and greenish-blue in others. Keep just one male in the company of several females, and make sure the tank has a peat base, since these killifish are substrate-spawners. Remove the adults after spawning, and carefully drain the tank before removing the egg-laden peat and partially drying it. Refill the tank after laying a new peat base.

Dorsal fin is set well back

Irregular spotting

Trident-shaped caudal fin

Fundulopanchax walkeri

Walker's Aphyosemion

- **ORIGINS** West Africa, found in southern Ghana and southeastern Ivory Coast.
- **SIZE** 2 in (5 cm).
- **DIET** Live foods and prepared diets.
- **WATER** Temperature 68–73°F (20–23°C); soft (50–100 mg/l) and acidic (pH 6.0–6.5).
- **TEMPERAMENT** Aggressive and active.

Stunning bluish coloration on the flanks, red spots on the sides of the body, and yellowish fins are the key features of these fish. They spawn either among plants or on a peat base. Peat with eggs buried in it should be removed and stored in a warm place in a plastic bag so that it partially dries but does not become powdery. After about six weeks, put the peat back into the tank. The eggs will hatch quite rapidly, and the fry can be reared on brine shrimp.

ISOLATION AND DIVERSITY

Some killifish, notably members of the *Nothobranchius* and *Cynolebias* genera, and a few *Aphyosemion* species, are known as "annual killifish," because they die in the dry season when the temporary shallow pools in which they live evaporate. Others, including Lyretails, inhabit permanent bodies of water. Killifish populations tend to be isolated from one another, so these fish often display a wide range of color morphs. The diversity that exists within a single species can be seen by comparing the pair of Lyretails below with the example illustrated opposite *(see bottom left)*.

Nothobranchius palmqvisti

Palmqvist's Notho

- ⊕ **ORIGINS** East Africa, found in coastal regions of Kenya and Tanzania.
- ◎ **SIZE** 2 in (5 cm).
- ◎ **DIET** Small live foods and prepared diets.
- ◎ **WATER** Temperature 64–68°F (18–20°C); soft (50–100 mg/l) and acidic (pH 6.0).
- ◎ **TEMPERAMENT** Aggressive.

Nothobranch killifish can be distinguished from Aphyosemions by their broader bodies. Palmqvist's Notho is an annual killifish *(see p.171)* with a red caudal fin and a bluish body covered with a meshlike pattern of red lines. Females are smaller and duller in color than the males, which are quarrelsome by nature. The care of these fish is straightforward, although they often do better in slightly brackish water. When breeding Palmqvist's and other nothos, place the fish in a relatively small aquarium; it should be only about half full of water and have a peat base. Offer the fish plenty of live foods, and allow the water to evaporate naturally. The falling water level should trigger spawning behavior, and the female will lay her eggs in the peat substrate.

LONG-LASTING EGGS

Life is a brief affair for annual killifish *(see p.171)*, because the adult fish die each year when their pools dry up. However, the population itself survives, thanks to the way in which the killifish spawn on the substrate as the water level falls. Drying mud encases the eggs (shown magnified below) and protects them until the rains return, perhaps as much as two years later, filling the pools and allowing the next generation of killifish to hatch. The young fish grow rapidly, feeding largely on insects, and may reach maturity in just six weeks.

Nothobranchius korthausae

Korthaus' Notho

- ⊕ **ORIGINS** East Africa, restricted to swamp areas of Mafia Island, Tanzania.
- ◎ **SIZE** 2½ in (6 cm).
- ◎ **DIET** Small live foods and prepared diets.
- ◎ **WATER** Temperature 73–79°F (23–26°C); soft (50–100 mg/l) and acidic (pH 6.0).
- ◎ **TEMPERAMENT** Aggressive.

There are several morphs of this killifish. The example above is one of two red-tailed forms; the other has a blue body. In brown morphs, the body lines are brown rather than red, and the caudal fin is brown and yellow. Males are always brighter than females. The dependence of notho killifish on seasonal pools of water in the wild means that the water chemistry in the tank is crucial for these fish. They can live for a couple of years in aquariums.

Nothobranchius guentheri

Gunther's Nothobranch

- ⊕ **ORIGINS** East Africa, from Mombasa, Kenya, to the Pangani River in Tanzania, and on the island of Zanzibar.
- ◎ **SIZE** 2 in (5 cm).
- ◎ **DIET** Small live foods.
- ◎ **WATER** Temperature 73–79°F (23–26°C); soft (50–100 mg/l) and acidic (pH 6.0).
- ◎ **TEMPERAMENT** Aggressive.

Highly colorful, yet variable in appearance, these killifish should be kept in groups of several females and one male. They will spawn in peat but often die soon afterward. To hatch the eggs, dry the peat, rewet, and dry again before placing in a tank. Some eggs only hatch on a second contact with water; this strategy helps to protect wild populations from being wiped out if their pool dries up prematurely.

Epiplatys annulatus

Banded Panchax

- ⊕ **ORIGINS** West Africa, in streams in parts of Sierra Leone, Nigeria, Liberia, and Guinea.
- ◎ **SIZE** 1¾ in (4.5 cm).
- ◎ **DIET** Small live foods and prepared diets.
- ◎ **WATER** Temperature 73–79°F (23–26°C); soft (50–100 mg/l) and acidic (pH 6.0).
- ◎ **TEMPERAMENT** Relatively peaceful.

Male

Scales evident in darker areas | Red horizontal lines on caudal fin

The distinctive patterning of Banded Panchax—four broad, dark bands separated by paler areas—allows them to be identified with ease. Males have more colorful fins than females and also possess blue irises. Banded Panchax should be kept in a well-planted tank, under subdued lighting. They tend to swim close to the surface and will spawn among the vegetation, or on a spawning mop. These killifish will not eat their eggs, which should hatch after eight days.

Epiplatys sexfasciatus

Six-Barred Panchax

- **ORIGINS** West Africa, where its distribution is restricted to Cameroon.
- **SIZE** 4¼ in (11 cm).
- **DIET** Small live foods and prepared diets.
- **WATER** Temperature 73–82°F (23–28°C); soft (50–100 mg/l) and acidic (pH 6.0).
- **TEMPERAMENT** Relatively peaceful.

Male

This killifish is named after the six vertical bands that encircle its body, although these bands tend to be fairly inconspicuous. There are several localized forms, but in all cases the males are larger and more colorful than the females, which have more rounded fins. A breeding group of a male and two or three females will spawn among vegetation such as Java Moss (*Vesicularia dubyana*), or on a spawning mop. The eggs hatch within two weeks, and the fry can be reared largely on brine shrimp.

Aplocheilus blockii

Dwarf Panchax

- **ORIGINS** Asia, occurring in southern India in the vicinity of Madras; may also occur on Sri Lanka.
- **SIZE** 2 in (5 cm).
- **DIET** Live foods and prepared diets.
- **WATER** Temperature 73–79°F (23–26°C); soft (50–100 mg/l) and acidic (pH 6.0).
- **TEMPERAMENT** Peaceful.

The yellow markings on the bodies of these killifish, which are the smallest members of their genus, are more vibrant in the males. Dwarf Panchax are reasonably tolerant, both of their own kind and of other nonaggressive fish of a similar size. As with other plant-spawning killifish, the eggs should be removed from the plants in the breeding tank each day and hatched separately. The young Dwarf Panchax will require infusoria at first.

Aplocheilus dayi

Day's Green Panchax

- **ORIGINS** Asia, occurring in southern India in the vicinity of Madras; may also occur in Sri Lanka.
- **SIZE** 2 in (5 cm).
- **DIET** Small live foods and prepared diets.
- **WATER** Temperature 73–82°F (23–28°C); soft (50–100 mg/l) and acidic (pH 6.0).
- **TEMPERAMENT** Peaceful.

The flanks of Day's Green Panchax are bluish and bear dark markings. Females have smaller, more rounded fins than males. These lively killifish, which can be kept with other fish of a similar size, tend to occupy the upper levels of the aquarium. A cover for the tank is vital to prevent them from leaping out. They spawn quite readily, and the eggs, which are clearly visible, take 12–14 days to hatch.

Aplocheilus panchax

Blue Panchax

- **ORIGINS** Asia, from southern India eastward through Thailand and the Malay Peninsula to parts of Indonesia.
- **SIZE** 3 in (7.5 cm).
- **DIET** Small live foods and prepared diets.
- **WATER** Temperature 68–77°F (20–25°C); soft (50–100 mg/l) and acidic (pH 6.0).
- **TEMPERAMENT** Rather aggressive by nature.

The Blue Panchax is one of the most widely distributed killifish species, with a number of variants found throughout its extensive range. The greenish body has a blue color along each side that often highlights the outline of the scales. The fins of the male (the lower of the two fish shown here) are more colorful than those of the female (the upper fish). Blue Panchax will spawn quite readily among fine-leaved vegetation in a breeding tank. The eggs are laid in batches, which can be hatched in small containers filled with mature water. The young typically emerge 10–14 days later; the length of this period is partly influenced by the water temperature. Partial water changes are very important for the subsequent well-being of the fry as the young Blue Panchax grow in size.

UNUSUAL KILLIFISH

Killifish are seen less often in aquatic outlets than other groups of fish, simply because their reproductive cycle makes it unprofitable to farm them commercially in large numbers. Instead, killifish are most often available as eggs, which enthusiasts trade among themselves. For this reason, some odd varieties, with anomalous requirements, are sometimes seen. One example

is the Blue Tanganyika Killifish (*Lamprichthys tanganicanus*). While most killifish live in temporary pools with relatively soft, acidic water, this species inhabits Lake Tanganyika, where the water is very hard and alkaline. The Blue Tanganyika is the largest of all the African killifish, with males reaching 5½ in (13 cm) long. It lives in schools around the lake's rocky shores.

Epiplatys fasciolatus

Orange Panchax

● **ORIGINS** West Africa, occurring in Liberia, Nigeria, and Sierra Leone.
● **SIZE** 3¼ in (8 cm).
● **DIET** Live foods and prepared foods.
● **WATER** Temperature 73–82°F (23–28°C); soft (50–100 mg/l) and acidic (pH 6.0).
● **TEMPERAMENT** Relatively peaceful.

These killifish do not display any well-defined pattern of banding. Males have red and greenish speckling on their bodies, with some populations being more colorful than others. Females are a much duller shade of yellowish-brown, with a rounded rather than elongated caudal fin. These killifish require a well-planted aquarium and will spawn among vegetation, such as moss. Orange Panchax may be persuaded to take flake sprinkled on the surface.

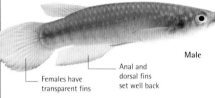

Male

Females have transparent fins

Anal and dorsal fins set well back

Austrolebias adloffi

Banded Pearlfish

● **ORIGINS** South America, in the lowlands between southern Brazil and eastern Uruguay.
● **SIZE** 2 in (5 cm).
● **DIET** Small live foods and prepared foods.
● **WATER** Temperature 68–75°F (20–24°C); hard (100–150 mg/l) and neutral (pH 7.0).
● **TEMPERAMENT** Males aggressive to unresponsive females.

About 10 vertical dark bands run down the greenish-gray body of the male Banded Pearlfish. The male's dorsal and anal fins are more pointed than those of the female, whose coloration is made up of dark blotches on a brown background. Less aggressive than some related pearlfish, this species may spawn almost continually, so a removable peat tray is useful. The eggs will hatch after being out of the water for eight weeks.

Austrolebias alexandri

Entre Rios Pearlfish

● **ORIGINS** South America, in southern Brazil, western Uruguay, and northeastern Argentina.
● **SIZE** 2 in (5 cm).
● **DIET** Small live foods and prepared foods.
● **WATER** Temperature 68–75°F (20–24°C); hard (100–150 mg/l) and neutral (pH 7.0).
● **TEMPERAMENT** Can be aggressive.

Males are more brightly colored than females, with a yellowish-green body color and vertical bars along the flanks. In females, the body is blotched with brown and black. A red-eyed, pink-bodied strain also exists, in which the males have darker pink barring on the body and pale spotted fins. Although these pearlfish prefer small live foods, they may also be persuaded to sample flaked foods quite readily.

Austrolebias nigripinnis

Black-Finned Pearlfish

- **ORIGINS** South America, in Argentina and also in parts of southern Brazil and Uruguay.
- **SIZE** 2 in (5 cm).
- **DIET** Small live foods and prepared foods.
- **WATER** Temperature 68–72°F (20–22°C); soft (50–100 mg/l) and acidic (pH 6.0).
- **TEMPERAMENT** Can be aggressive.

Mature males of this species are blackish with a variable pattern of iridescent greenish spots and a green band across the top of the dorsal fin. Females can be recognized by their light gray coloration. The base of the spawning tank needs a thick layer of peat for spawning purposes. Store the eggs in moist peat in a plastic bag. They can be returned to the water after two to three months and may remain viable for up to three years.

Cynopoecilus melanotaenia

Fighting Gaucho

- **ORIGINS** South America, occurring in both southeastern Brazil and northern Uruguay.
- **SIZE** 2 in (5 cm).
- **DIET** Small live foods and prepared foods.
- **WATER** Temperature 64–75°F (18–24°C); soft (50–100 mg/l) and acidic (pH 6.5).
- **TEMPERAMENT** Aggressive toward its own kind.

Fins have serrated edges

A white throat plus two yellow bands along each side of the body, separated by a darker stripe, are defining features of this pearlfish. The top of the body is brownish, with reddish-brown color on the fins. Females are less brightly colored. The quarrelsome nature of Fighting Gauchos means that their tank must be well planted and have many retreats; however, they can be housed safely with unrelated fish. Pairs spawn on a peat substrate. The resulting fry can be reared on brine shrimp.

Simpsonichthys bokermanni

Bokerman's Cynolebias

- **ORIGINS** South America, restricted to pools close to Ilheus, Bahia, near the Brazilian coast.
- **SIZE** 2 in (5 cm).
- **DIET** Small live foods and prepared foods.
- **WATER** Temperature 68–75°F (20–24°C); soft (50–100 mg/l) and acidic (pH 6.0).
- **TEMPERAMENT** Can be aggressive.

The identifying characteristics of the Bokerman's Cynolebias are its greenish-gold body and blue gill covers, with vertical blue lines behind. Pale blue speckling is also apparent on the fins of males. Females are duller, with a black spot on the flank. These fish spawn in the substrate. As with other species, the peat carrying the eggs must be stored in a plastic bag at room temperature for about two months, before being returned to the aquarium so that the eggs may hatch.

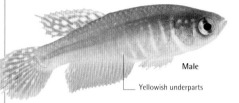

Male

Yellowish underparts

Jordanella floridae

American Flagfish

- **ORIGINS** Extends from Florida around the Gulf Coast and down to Mexico's Yucatán Peninsula.
- **SIZE** 2¹⁄₂ in (6.5 cm).
- **DIET** Small live foods and prepared diets.
- **WATER** Temperature 66–77°F (19–25°C); hard (100–150 mg/l) and neutral (pH 7.0).
- **TEMPERAMENT** Males are aggressive.

The deep yet narrow body of the American Flagfish is decorated with a combination of bluish-green and reddish-orange speckles; vibrant red is displayed on the dorsal fin. The female of the species can be distinguished from the male by a blackish spot at the rear of the dorsal fin and her generally duller appearance. Inhabiting densely vegetated stretches of water in the wild, American Flagfish require similar surroundings in their aquarium; otherwise, they will prove to be nervous tank occupants. Vegetable matter is an important component of the diet of these fish, and they will browse readily on algae growing in the aquarium. Males will become quite aggressive as the time for spawning approaches. The female can produce up to 70 eggs, either scattering them among the vegetation or burying them in the substrate. The male will usually guard the eggs for a week or so until they hatch.

RAINBOWFISH

The brilliantly colored rainbowfish have grown rapidly in popularity in recent years as they have become more readily available. A significant number of new species have been discovered and introduced to the aquarium hobby since the 1980s, and there are almost certainly others awaiting discovery, especially in New Guinea. Local populations within a species may look very different, and it is important to maintain their individual features. Unfortunately, some rainbowfish breeders are now hybridizing their stock in an attempt to create even more colorful offspring. There is some evidence that these fish may themselves occasionally hybridize in the wild.

Threadfin Rainbowfish (*Iriatherina werneri*) vary slightly across their range, with those from New Guinea being darker than their Australian relatives.

Iriatherina werneri

Threadfin Rainbowfish

- **ORIGINS** New Guinea, between the Fly and Merauke rivers; also occurs in northern Australia.
- **SIZE** 2 in (5 cm).
- **DIET** Eats both prepared diets and small live foods.
- **WATER** Temperature 72–84°F (22–29°C); soft (50–100 mg/l) and acidic to alkaline (pH 6.5–7.5).
- **TEMPERAMENT** Peaceful.

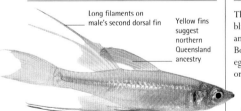

Long filaments on male's second dorsal fin

Yellow fins suggest northern Queensland ancestry

Mature males display with their elaborate fins during the courtship ritual, holding them erect and then flicking them up and down. An aquarium planted with fine-leaved vegetation is essential for spawning; females lay over the course of several days. The eggs can take a week or longer to hatch, and the fry can be reared on tiny rotifers. Adult fish prefer to eat small live foods, but will take flake powdered onto the water's surface.

Bedotia madagascariensis

Bowfish

- **ORIGINS** East Africa, where it is found only in upland areas on the island of Madagascar.
- **SIZE** 6 in (15 cm).
- **DIET** Eats both prepared diets and small live foods.
- **WATER** Temperature 68–75°F (20–24°C); hard (150–200 mg/l) and neutral (pH 7.0).
- **TEMPERAMENT** Peaceful.

The male bowfish is multicolored, with shades of blue apparent on the flanks; the female is yellow and has a rounded, rather than jagged, dorsal fin. Bowfish thrive in groups and will not eat their eggs (which become tangled among vegetation) or their fry. Young bowfish are unusual in that they swim immediately after hatching, initially in a somewhat vertical stance, although they soon adopt a more horizontal posture.

Glossolepis incisus

Salmon-Red Rainbowfish

- **ORIGINS** New Guinea, where it is restricted to Lake Sentani in Irian Jaya.
- **SIZE** 6 in (15 cm).
- **DIET** Small live foods preferred.
- **WATER** Temperature 77–86°F (25–30°C); hard (150–200 mg/l) and neutral (pH 7.0).
- **TEMPERAMENT** Peaceful.

Silvery overlay on some scales

Sexing is straightforward because the male is salmon-pink while the female is yellowish and olive-green. Juveniles tend to resemble adult females in appearance, but the young males change color once they are about 2 in (5 cm) in length. In the wild, Salmon-Red Rainbowfish feed on small invertebrates and associate in areas of aquatic vegetation. Females lay batches of between 100 and 150 eggs, which usually hatch within 10 days.

Chilatherina bleheri

Bleher's Rainbowfish

🌐 **ORIGINS** New Guinea, occurring in Lake Holmes and associated streams, and in Mamberamo, Irian Jaya.

💧 **SIZE** 4³/₄ in (12 cm).

🍤 **DIET** Prefers small live foods.

🌊 **WATER** Temperature 77–86°F (25–30°C); hard (150–200 mg/l) and neutral (pH 7.0).

😊 **TEMPERAMENT** Peaceful.

A deep-bodied appearance is a characteristic feature of Bleher's Rainbowfish, with males attaining a larger size than females and being more vibrantly colored. The front of the body is an iridescent silvery-green, becoming yellowish and then purplish-red toward the tail. It is possible to tell when males are in spawning condition because they display a yellow-orange stripe on the forehead. The species is named after Heiko Bleher, a well-known German aquarist.

Pseudomugil signifer

Australian Blue-Eye

🌐 **ORIGINS** Australia, from northern Queensland southward to Narooma in New South Wales.

💧 **SIZE** 2¹/₂ in (6 cm).

🍤 **DIET** Eats both prepared diets and small live foods.

🌊 **WATER** Temperature 59–82°F (15–28°C); soft (50–100 mg/l) and acidic to alkaline (pH 5.5–7.8).

😊 **TEMPERAMENT** Territorial when spawning.

Greenish upperparts typify southern populations

Characteristic blue iris

These rainbowfish vary significantly in appearance throughout their range. Northern specimens have very long filaments on their dorsal and anal fins and are larger than their southerly counterparts. The males' coloration changes with age and becomes more vibrant when they are in spawning condition. Females are less colorful than males. Australian Blue-Eyes are found in fresh and brackish waters, and even in the sea in some areas.

Pseudomugil furcatus

Forktailed Blue Eye

🌐 **ORIGINS** Eastern New Guinea, occurring between Collingwood and Dyke Ackland Bays.

💧 **SIZE** 2¹/₂ in (6 cm).

🍤 **DIET** Small live foods preferred.

🌊 **WATER** Temperature 73–79°F (23–26°C); soft to hard (50–150 mg/l) and neutral to alkaline (pH 7.0–8.0).

😊 **TEMPERAMENT** Peaceful.

The male Forktail, shown below, is more brightly colored than the female, has a larger dorsal fin, and often shows red on its pelvic and pectoral fins. Forktails occur in clear, fast-flowing streams where there is a good covering of aquatic plants, so their tank should be well planted but include clear areas for swimming. Spawning may occur on the substrate or among plants. The eggs develop slowly and can take up to two weeks to hatch.

Raised pectoral fins

Marosatherina ladigesi

Celebes Rainbowfish

🌐 **ORIGINS** Indonesia, where it occurs on the island of Sulawesi (formerly known as Celebes).

💧 **SIZE** 3 in (7.5 cm).

🍤 **DIET** Eats both prepared diets and small live foods.

🌊 **WATER** Temperature 68–82°F (20–28°C); hard (150–200 mg/l) and neutral to alkaline (pH 7.0–7.5).

😊 **TEMPERAMENT** Peaceful.

Male Celebes Rainbowfish are more colorful than females and have longer fin rays. These fish are sensitive to water quality, so it is vital to check the parameters to which they are acclimatized. Change conditions slowly, by making a partial water change of 25 percent of the tank's volume each week. Under favorable conditions, adults may spawn almost nonstop for months, although they often eat their eggs. Hatching takes up to 11 days.

SPLIT-FIN DESIGN

One of the most distinctive characteristics of this group of fish is the structure of the dorsal fin, which is divided into two parts, with an obvious gap between. This arrangement enables each part of the dorsal fin to be raised or lowered largely independently of the other. It is usually the front part of the fin that is kept lowered, probably so that it does not interfere with the swimming ability of the rainbowfish. This suggests that the function of the split fin is primarily for display. The shape of the fin can also be useful in distinguishing between the sexes. The rear part of the second dorsal generally tapers to a point in the males, although this characteristic is less evident in the rainbowfish of Lake Tebera, New Guinea.

Melanotaenia boesemani

Boeseman's Rainbowfish

- **ORIGINS** New Guinea, occurring in the Ajamaru Lakes area of the Vogelkop Peninsula.
- **SIZE** 4³/₄ in (12 cm).
- **DIET** Eats prepared diets and small live foods.
- **WATER** Temperature 77–86°F (25–30°C); soft (50–100 mg/l) and acidic to alkaline (pH 6.5–8.0).
- **TEMPERAMENT** Peaceful.

In terms of its coloration, Boeseman's Rainbowfish is a fish of two halves. The head and front of the body are bluish, while the rear half is yellowish-orange, sometimes bordering on red. Males are brighter in color, display more elongated rays on their dorsal fins, and have a deeper body. Their aquarium needs to incorporate open areas for swimming as well as vegetation such as Java Moss (*Vesicularia dubyana*), which can serve as a spawning site. Females typically lay between 100 and 200 eggs, with hatching occurring approximately a week later. Originally collected by Dr. Marinus Boeseman on a Dutch expedition in 1954, this species was then rediscovered in 1982. Some live specimens were sent to Europe by Heiko Bleher, where they bred successfully, thus providing the foundation for today's aquarium strains.

Melanotaenia maccullochi

Dwarf Rainbowfish

- **ORIGINS** Australia, in parts of Queensland and also in the Northern Territory.
- **SIZE** 2³/₄ in (7 cm).
- **DIET** Eats both prepared diets and small live foods.
- **WATER** Temperature 72–86°F (22–30°C); soft (50–100 mg/l) and acidic to alkaline (pH 5.5–7.5).
- **TEMPERAMENT** Peaceful.

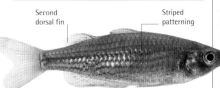

Second dorsal fin

Striped patterning

Dark horizontal stripes along the sides of the body characterize this rainbowfish; the stripes become more wavy and broken on the lower parts. Females are less colorful but grow larger than males. There are color differences between isolated populations; the orange fin markings that typify fish around Cairns are absent in those from the Jardine River region. Dwarf Rainbowfish occurring near the Litchfield National Park in the Northern Territory have more yellowish bodies and are decidedly smaller in size.

Melanotaenia parkinsoni

Parkinsoni Rainbowfish

- **ORIGINS** Eastern New Guinea, between Milne Bay and the Kemp Welsh River.
- **SIZE** 5¹/₂ in (14 cm).
- **DIET** Eats both prepared diets and small live foods.
- **WATER** Temperature 77–84°F (25–29°C); hard (100–150 mg/l) and alkaline (pH 7.6).
- **TEMPERAMENT** Peaceful.

The attractive appearance of this rainbowfish is best appreciated under subdued lighting, which emphasizes the yellow, golden, and blue areas on the body and fins. Individuals display some variation in coloration—some are more orange than yellow. Females resemble males but are generally duller in color. Some males have enlarged anal and dorsal fins, and the fins in some individuals have a slightly ragged appearance.

Melanotaenia lacustris

Lake Kutubu Rainbowfish

- **ORIGINS** Central New Guinea, in Lake Kutubu and the Soro River, which flows out from the lake.
- **SIZE** 4³/₄ in (12 cm).
- **DIET** Eats both prepared diets and small live foods.
- **WATER** Temperature 72–77°F (22–25°C); hard (100–150 mg/l) and alkaline (pH 8.0–9.0).
- **TEMPERAMENT** Peaceful.

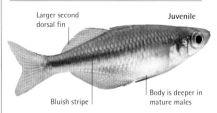

Larger second dorsal fin

Juvenile

Bluish stripe

Body is deeper in mature males

The blue coloration of these rainbowfish varies according to water conditions and diet. An orange area on the nape of the neck indicates that they are in spawning condition. Spawning itself can be a protracted affair, with the females laying eggs in batches over the course of several days. In the wild, these fish are found in parts of Lake Kutubu where there is floating vegetation near the surface. In aquariums, they should be kept in single-species groups to avoid hybridization with other fish.

Melanotaenia herbertaxelrodi

Lake Tebera Rainbowfish

- **ORIGINS** Central highlands of New Guinea near Lake Tebera, part of the Purari River system.
- **SIZE** 5 in (13 cm).
- **DIET** Eats both prepared diets and small live foods.
- **WATER** Temperature 75–81°F (24–27°C); hard (100–150 mg/l) and alkaline (pH 7.5).
- **TEMPERAMENT** Peaceful.

The yellowish body of these fish has a dark central band along the midline. The middle section of the first part of the dorsal fin is longer in male Lake Teberas. The males' appearance alters at the start of the spawning period, when they develop a blue or white stripe that runs from the dorsal fin and down over the head. This stripe gradually becomes much darker in color. Lake Teberas live in shallow, vegetated areas, so they will benefit from having a well-planted area in their tank. Females lay up to 150 eggs, which hatch about 10 days later.

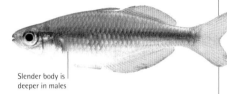

Slender body is deeper in males

Melanotaenia affinis

New Guinea Rainbowfish

- **ORIGINS** Northern New Guinea, including the Sepik River (Pagwi), and Madang (Blue Water Creek morph).
- **SIZE** 5½ in (14 cm).
- **DIET** Eats both prepared diets and small live foods.
- **WATER** Temperature 70–82°F (21–28°C); hard (100–150 mg/l) and neutral (pH 7.0).
- **TEMPERAMENT** Peaceful.

Three separate varieties of this species are known, with the so-called Standard having the widest range. In all cases, males are more colorful, with elaborate dorsal and anal fins, and grow slightly larger than females. New Guinea Rainbowfish have proved to be very adaptable, reflecting the fact that in the wild their environment can change markedly though the year. Like related species, they should be kept in shoals. Females lay up to 200 eggs when spawning.

Melanotaenia splendida

Splendid Rainbowfish

- **ORIGINS** Northern Australia, Aru Island, and between Etus Bay and the Aramia River in southern New Guinea.
- **SIZE** 6 in (15 cm).
- **DIET** Eats both prepared diets and small live foods.
- **WATER** Temperature 72–82°F (22–28°C); soft (50–100 mg/l) and acidic to alkaline (pH 5.6–7.4).
- **TEMPERAMENT** Peaceful.

There are between four and six subspecies of this rainbowfish, all of which show reddish-brown speckling on the body. Males are generally more brightly colored and have deeper bodies than the females. The young of both sexes are duller than the adults, taking up to a year to acquire their full coloration. It is vital to keep the nitrate level in the tank low, so make sure the filtration system is efficient and carry out partial water changes every week or so.

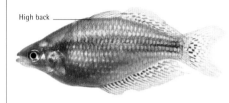

High back

Melanotaenia praecox

Peacock Rainbowfish

- **ORIGINS** New Guinea, occurring in small areas of the Mamberamo River system, near Iritoi, Dabra, and Siewa.
- **SIZE** 2½ in (6 cm).
- **DIET** Eats both prepared diets and small live foods.
- **WATER** Temperature 75–81°F (24–27°C); soft (50–100 mg/l) and acidic (pH 6.5).
- **TEMPERAMENT** Peaceful.

The brilliant neon blue of these rainbowfish is displayed to best effect when they are kept in aquariums with floating plants to diffuse the light.

Well-oxygenated water is also important, since Peacock Rainbowfish naturally inhabit fast-flowing waters. It has been suggested that only males display red edges to their fins, although this characteristic can in fact be seen in both sexes. Yellow-fin edges are more common in aquarium strains. One recently discovered wild population has red stripes on the body. Females spawn over several days, producing up to 50 eggs per day; the eggs stick to vegetation. Hatching takes about a week, and the fry can be reared on brine shrimp. Feeding the adults well makes them less likely to eat their eggs or offspring.

DIVERSIFICATION

Rainbowfish form an extremely adaptable group. This is illustrated not only by differences in appearance between individuals of the same species but also by the way in which these fish have colonized localized habitats, such as Lake Tebera in New Guinea, which is now home to unique species of rainbowfish. Their adaptability is further confirmed by the Mountain Rainbowfish (*Melanotaenia monticola*) shown here. Occurring at high altitude in New Guinea, it has adapted to life in water that is typically just 61–64°F (16–18°C)—far colder than the lowest temperatures tolerated by other New Guinea rainbowfish.

LOACHES

Often confused with catfish, loaches belong to the family Cobitidae and occur in both tropical and temperate regions. All loaches have barbels around the mouth and, even though it is not always obvious, a spine below each eye. When a loach feels threatened, it will raise its spines, making itself more difficult for a predator to swallow. The spines can easily become entangled in a net, so extra care is called for when catching these fish. Loaches are mostly shy, nocturnal fish, and so may have to be coaxed into view with live foods.

The Clown Loach (*Chromobotia macracanthus*) is more conspicuous in the aquarium than many of its relatives, not only because of its attractive patterning, but also because it is more active in the daytime.

Pangio kuhlii

Coolie Loach

- **ORIGINS** Southeast Asia, on the Malay Peninsula, as well as Java, Sumatra, and other nearby islands.
- **SIZE** 4 in (10 cm).
- **DIET** Prefers worms, but will take tablets.
- **WATER** Temperature 75–86°F (24–30°C); soft (50 mg/l) and acidic (pH 6.0–6.5).
- **TEMPERAMENT** Relatively social.

The beautiful banding of Coolie Loaches is highly variable, ranging from predominantly yellow to reddish between individuals. These fish burrow readily into the substrate and may even disappear completely out of sight beneath the undergravel filter plate if they can gain access to this part of the tank. Spawning is possible in aquarium surroundings, with the green eggs simply being scattered among the tank vegetation.

Banding may not encircle the underparts

Eel-like body

Acantopsis dialuzona

Horse-Face Loach

- **ORIGINS** Southeast Asia, from Vietnam and Myanmar (Burma), via the Malay Peninsula, to Borneo and Sumatra.
- **SIZE** 9 in (23 cm).
- **DIET** Prepared foods and live foods.
- **WATER** Temperature 75–86°F (24–30°C); soft (50 mg/l) and acidic (pH 6.0–6.5).
- **TEMPERAMENT** Nonaggressive.

Long, thin body

Horselike head shape

The pattern of dark spots on a brownish-yellow background is variable, probably reflecting the different local habitats in which this fish naturally occurs. In the aquarium, very little will be seen of the Horse-Faced Loach, since it spends most of its time either on the bottom of the tank or buried in the substrate with just its face protruding. A fine-grained covering over at least part of the tank's base is recommended for burrowing purposes. Aquarium plants should be set in pots, to prevent them from being uprooted by the loach's activities.

Yasuhikotakia morleti

Skunk Loach

- **ORIGINS** Southeast Asia, ranging eastward from northern India to Thailand.
- **SIZE** 4 in (10 cm).
- **DIET** Eats a wide variety of foods.
- **WATER** Temperature 79–86°F (26–30°C); soft (50 mg/l) and acidic (pH 6.0–6.5).
- **TEMPERAMENT** Peaceful.

The golden-yellow coloration of the Skunk Loach is broken by a black stripe running along the entire length of the body and a black band that encircles the caudal peduncle. Like other Botia loaches, it has two pairs of sensory barbels on its snout, which help it to locate food as it digs in the substrate. The lighting for this fish should be subdued, and the tank decor must include retreats where the fish can shelter.

Longitudinal stripe

Chromobotia macracanthus

Clown Loach

- **ORIGINS** Southeast Asia, where it occurs in various parts of Indonesia, including Sumatra, and on Borneo.
- **SIZE** 12 in (30 cm).
- **DIET** Prepared foods and live foods.
- **WATER** Temperature 75–86°F (24–30°C); soft (50 mg/l) and acidic (pH 6.0–6.5).
- **TEMPERAMENT** Usually active and lively.

Dark dorsal fin

Males may have larger tails

Broad bands of orange and black on the body of the Clown Loach allow this species to be identified without difficulty. Clown Loaches thrive in groups, although large specimens will occasionally prey on much smaller companions. Clown Loaches have a disconcerting habit of floating on their sides. This is not usually a sign of their imminent demise— it is simply the position in which these fish sometimes choose to rest.

Botia striata

Zebra Loach

- **ORIGINS** Asia, restricted to muddy waters in parts of southern India.
- **SIZE** 4 in (10 cm).
- **DIET** Prepared foods of all types.
- **WATER** Temperature 75–86°F (24–30°C); soft (50 mg/l) and acidic (pH 6.0–6.5).
- **TEMPERAMENT** Nonaggressive.

The patterning of the Zebra Loach consists of alternating pale yellow and brownish vertical stripes, with the paler bands especially being variable in width. These loaches should be kept in groups. They will often choose to hide away for long periods under bogwood and in other retreats. In common with other *Botia* loaches, breeding is unknown in aquarium surroundings. Feeding is very straightforward, with small worms being a favorite food of these loaches.

Yasuhikotakia modesta

Redtail Loach

- **ORIGINS** Asia, including northeast India, Vietnam, Thailand, and the Malay Peninsula.
- **SIZE** 9½ in (24 cm).
- **DIET** Prepared foods and live foods.
- **WATER** Temperature 75–86°F (24–30°C); soft (50 mg/l) and acidic (pH 6.0–6.5).
- **TEMPERAMENT** Shy.

All the fins have an orange hue

Sensory barbels around the mouth

Although the body coloration of this species sometimes appears to be bluish-gray, in optimum water conditions it will change to a much brighter shade of blue, making a striking contrast with the orange on the fins. These loaches are adapted to group living, but they may be less well disposed toward similar species sharing their tank. Nocturnal by nature, they will hide away during the day, often under the substrate, and emerge to feed at night.

Ambastaia sidthimunki

Dwarf Loach

- **ORIGINS** Asia, occurring in muddy lakes from northern India to northern Thailand.
- **SIZE** 2¼ in (5.5 cm).
- **DIET** Flake and small live foods.
- **WATER** Temperature 75–86°F (24–30°C); soft (50 mg/l) and acidic (pH 6.0–6.5).
- **TEMPERAMENT** Peaceful.

This small species has dark horizontal and vertical bands on its upper body that interconnect to create a distinctive chainlike pattern. Dwarf Loaches are lively and active fish that look best when kept together in a group, and make an ideal choice for a community setup. Cryptocoryne plants are a good addition to their aquarium, since the fish like to rest on their broad leaves. Dwarf Loaches will comb the tank floor in search of edible items.

DESIGNED FOR BURROWING

The wormlike shape that characterizes many loaches makes them efficient burrowers. They are able to wriggle their slender bodies down into the substrate with surprising speed. Numerous species rely on this technique to escape from danger, rather than trying to swim away to safety. The mottled patterning of burrowing loaches helps to conceal their presence as they lie partially buried in sediment on the bottoms of rivers and streams. The eyes of loaches are typically located high on the head. This is another adaptation to a burrowing lifestyle, since it enables them to survey the water above while remaining hidden from view themselves.

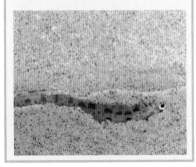

OTHER AMERICAN SPECIES

This varied selection of popular American species demonstrates the amazing adaptability of fish. For example, the Amazon Puffer, whose relatives are confined to marine and brackish waters, has changed its body chemistry to enable it to exploit salt-free environments. In other species, such as Knifefish, Ocellated Stingrays, and Violet Gobies, the body shape has altered to help the fish merge into the background or enable them to burrow in the substrate. The streamlined Arawana, which lurks much closer to the surface, has developed the ability to breathe atmospheric air directly, an adaptation that enables it to catch prey by leaping dramatically out of the water.

The Ocellated Stingray (*Potamotrygon motoro*) swims with a rippling motion of its greatly enlarged pectoral fins.

Colomesus asellus

Amazon Puffer

- **ORIGINS** Ranges widely through tropical parts of South America, including Brazil.
- **SIZE** 10 in (25 cm).
- **DIET** Various live foods.
- **WATER** Temperature 73–79°F (23–26°C); hard (100–150 mg/l) and neutral (pH 7.0).
- **TEMPERAMENT** Active.

This puffer has a bright-yellow iris, a random pattern of gold and brownish-black on its upper body, and white undersides. Young individuals have a more burnished appearance, with less intense dark markings. Amazon Puffers are predatory, and any snails in their aquarium are likely to be eaten, since mollusks form part of their natural prey. These fish may also dig in the substrate for worms, so the tank should have a sandy base.

Gobioides broussonnetii

Violet Goby

- **ORIGINS** Ranges from the southern United States down through Central America to Santa Catarina, Brazil.
- **SIZE** 24 in (60 cm).
- **DIET** Small live foods.
- **WATER** Temperature 73–77°F (23–25°C); hard (100–150 mg/l) and neutral (pH 7.0).
- **TEMPERAMENT** Territorial and aggressive.

The long, grayish body of this fish is tinged with violet and has a reflective sheen. The tank should have a thick layer of sand in which the fish can burrow, with rockwork and caves to provide extra hiding places. The Violet Goby must be kept alone. It may benefit from the addition of some marine salt to the water.

Osteoglossum bicirrhosum

Arawana

- **ORIGINS** South America, occurring in the Amazon and western Orinoco regions, extending to Guyana.
- **SIZE** 40 in (100 cm).
- **DIET** Meat-based foods.
- **WATER** Temperature 75–86°F (24–30°C); soft (50–100 mg/l) and acidic (pH 6.0).
- **TEMPERAMENT** Predatory.

These striking fish have a distinctive outline when seen in profile. Their fins appear to merge around the rear end of their body; adult males have longer anal fins than females. The problem with these fish is that they grow very large, and unless they can be moved to an indoor pond (in temperate areas), it may be impossible to keep them. Arawanas are incredibly agile, able to jump out of the water to seize invertebrates on overhanging branches. They will also prey on fish.

Apteronotus albifrons

Black Ghost Knifefish

- **ORIGINS** South America, occurring in parts of Venezuela, Ecuador, Peru, Brazil, and Guyana.
- **SIZE** 20 in (50 cm).
- **DIET** Live foods.
- **WATER** Temperature 73–82°F (23–28°C); soft (50–100 mg/l) and acidic (pH 6.5).
- **TEMPERAMENT** Shy, sometimes aggressive.

The dense black body of this fish is punctuated by two contrasting white areas—one on the head, the other at the rear, often on the caudal peduncle. The Black Ghost Knifefish generates a weak electrical current that helps it to find its prey in murky water, acting rather like a form of radar. An aquarium for these fish should be shaded, with a dark substrate and a well-planted interior that offers the fish a range of hiding places.

Eyes blend in with body color

Knifelike shape

Eigenmannia virescens

Glass Knifefish

- **ORIGINS** South America, from Colombia's Rio Magdalena south to the Plate River in Argentina.
- **SIZE** 13½ in (35 cm).
- **DIET** Live foods.
- **WATER** Temperature 73–82°F (23–28°C); soft (50–100 mg/l) and acidic (pH 6.0).
- **TEMPERAMENT** Nervous yet social.

There is a significant difference in size between the sexes in this species, with females rarely growing larger than 8 in (20 cm). Both sexes have a long, narrow body shape with neither a caudal fin nor a dorsal fin. Glass Knifefish live in groups with a well-organized social structure, which helps to reduce conflict. The dominant male is usually the largest in the group. These Knifefish are most active after dark, although spawning usually occurs early in the morning, with up to 200 eggs being laid.

Semitransparent body

STING IN THE TAIL

The arrangement of fins in rays, such as this Venezuelan stingray (*Potamotrygon hystrix*), is very different from that seen in other freshwater fish. The unusual pectoral fins, which extend all around the sides of the body, are responsible for the distinctive swimming motion. Stingrays lack a dorsal fin and derive no propulsive power from the caudal fin. However, the caudal fin does have a defensive function: it is equipped with a stinger in the form of a raised, venomous spine. The spine pierces an attacker's flesh, triggering a localized infection around the wound, while venom is pumped directly into the bloodstream. The effects of the venom on humans vary between the different stingray species, but they are always painful and in some people they can even be life-threatening.

These ancient relatives of sharks rely on their sting for protection against potential predators, including crocodilians and even jaguars, which sometimes hunt them in the shallows. Stingrays are also well camouflaged to help them avoid detection and can modify their coloration to a certain degree to blend in with the river bottom.

Potamotrygon motoro

Ocellated Stingray

- **ORIGINS** South America, where it is restricted to the rivers of Paraguay.
- **SIZE** 12 in (30 cm).
- **DIET** Meat-based diets.
- **WATER** Temperature 73–82°F (23–28°C); soft (50–100 mg/l) and acidic (pH 6.0).
- **TEMPERAMENT** Fairly placid, but has a dangerous sting.

The spots on the body of this ray consist of a light center surrounded by a dark outer circle. Patterning is highly variable, perhaps to match the different habitats in which these fish occur. It is also affected by the level of lighting in the aquarium. There is a very real danger associated with keeping these fish—take great care to avoid contact with the stinger when servicing the tank. Always wear sturdy, unperforated gloves, and keep disturbance to an absolute minimum. Whenever possible, use tongs rather than your hands to carry out tank maintenance. In spite of their venomous sting, these fish are not inherently aggressive, and groups can be kept together in a large aquarium. The floor covering of the tank should be deep and sandy. This will enable the fish to burrow into the substrate, where they will lie with just their eyes exposed. Mating is unknown in aquariums, but in the wild it occurs in September and October, with the young rays being born about five months later.

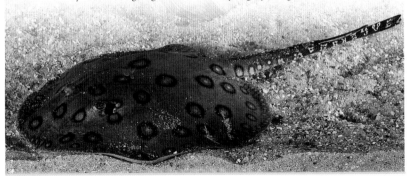

OTHER ASIAN SPECIES

Some Asian species are far more popular in their home region than elsewhere in the world, often because of the myths and folklore that surround them. The Dragon Fish of Southeast Asia, for example, is believed to bestow good fortune on its keeper. Such unusual species are sporadically available from specialist dealers and may be worth seeking out for their novelty and beauty. These fish vary greatly in size and care requirements. Some, such as the Knight Goby, have close marine relatives and may benefit from slightly brackish water conditions.

The predatory **Dragon Fish** (*Scleropages formosus*) is endangered in the wild. Stocks available to aquarists today are captive-bred, in order to safeguard the future of the species.

Mogurnda mogurnda

Purple-Striped Gudgeon

- **ORIGINS** From Southeast Asia to New Guinea and northern and central parts of Australia.
- **SIZE** 7 in (17.5 cm).
- **DIET** Live foods and prepared foods.
- **WATER** Temperature 75–86°F (24–30°C); hard (50–100 mg/l) and neutral to alkaline (pH 7.0–7.2).
- **TEMPERAMENT** Territorial.

The striped patterning that characterizes the Purple-Striped Gudgeon is most apparent on the head and gill covers, while dark spots are scattered over the attractive sky-blue flanks. Visual distinctions between the sexes are slight, but female Purple-Striped Gudgeons grow noticeably larger than males.

A tank for these fish should incorporate a number of retreats in the form of rock caves or upturned flowerpots. Pairs should ideally be kept on their own for spawning purposes, since males become more territorial and aggressive at this time. A pair will spawn in a cave chosen by the male, with up to 200 eggs being laid. After driving away the female, the male will keep watch over the eggs. Hatching occurs within about two days, depending on the water temperature. The male loses interest in the fry once they are free-swimming and may eat the young if he is not removed. The fry can be reared on small live foods.

Oryzias latipes

Medaka

- **ORIGINS** Asia, from China to Japan and South Korea; may also occur on some Indonesian islands.
- **SIZE** 1½ in (4 cm).
- **DIET** Prepared diets and small live foods.
- **WATER** Temperature 68–79°F (20–26°C); hard (100–150 mg/l) and neutral (pH 7.0).
- **TEMPERAMENT** Placid and quite social.

The Medaka, which frequents the upper levels of the tank, exists in a number of color forms; the golden morph (below) is one of the most common. Females have shorter, more rounded anal fins and fewer reflective scales on the rear of the body. Immediately after spawning, a string of eggs resembling a bunch of grapes can be seen trailing from the female's vent. The eggs are soon deposited among fine-leaved vegetation.

Stigmatogobius sadanundio

Knight Goby

- ⊛ **ORIGINS** Mainland Southeast Asia and Indonesian islands, including Java, Sumatra, and Borneo.
- ◉ **SIZE** 3½ in (8.5 cm).
- ◉ **DIET** Live foods and algae.
- ◉ **WATER** Temperature 68–79°F (20–26°C); hard (100–150 mg/l) and neutral (pH 7.0).
- ◉ **TEMPERAMENT** Quite territorial.

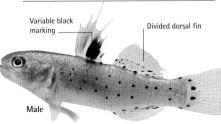

Variable black marking

Divided dorsal fin

Male

The pale-silver body of the male Knight Goby is dotted with black spots that become more numerous toward the rear. The female has similar markings but is more yellow and has smaller fins. Knight Gobies may benefit from being kept in slightly brackish water conditions. They require an aquarium with plenty of cavelike retreats in which the fish can spawn. The male will guard the eggs, which are laid on the underside of the cave roof.

Tateurndina ocellicauda

Eye-Spot Sleeper

- ⊛ **ORIGINS** Southeast Asia, occurring in New Guinea, where it is restricted to eastern Papua.
- ◉ **SIZE** 3 in (7.5 cm).
- ◉ **DIET** Small live foods preferred.
- ◉ **WATER** Temperature 68–79°F (20–26°C); soft (50–100 mg/l) and acidic (pH 6.0–6.5).
- ◉ **TEMPERAMENT** Nonaggressive.

These striking gobies display red markings on a blue background and yellowish underparts. Females are more rotund than males and have more pronounced yellow banding along the edges of the dorsal and anal fins. A pair seeks out a spawning site, and the male guards the eggs until they hatch. Fry can be reared on brine shrimp once they are free-swimming.

Xenentodon cancila

Silver Needlefish

- ⊛ **ORIGINS** Asia, ranging from India and Sri Lanka east to Myanmar (Burma), Thailand, and Malaysia.
- ◉ **SIZE** 12 in (30 cm).
- ◉ **DIET** Live foods, typically fish.
- ◉ **WATER** Temperature 72–82°F (22–28°C); hard (150–200 mg/l) and neutral (pH 7.0).
- ◉ **TEMPERAMENT** Predatory.

Jaws lined with teeth

These slender fish have a dark stripe along each side of the body. Although they are difficult to sex, males generally have a dark edge on their dorsal and anal fins. Silver Needlefish need to be kept in a group under relatively subdued lighting. The tank should include plants and plentiful open spaces for swimming. It must also be covered, to prevent the fish from jumping out. Silver Needlefish naturally prey on small fish, but in aquariums it may be possible to wean them onto larger invertebrates.

Toxotes jaculatrix

Archer Fish

- ⊛ **ORIGINS** A wide distribution, from the Middle East through southern parts of Asia to Australia.
- ◉ **SIZE** 10 in (25 cm).
- ◉ **DIET** Live foods.
- ◉ **WATER** Temperature 77–86°F (25–30°C); hard (100–150 mg/l) and neutral (pH 7.0).
- ◉ **TEMPERAMENT** Smaller individuals may be bullied.

Very short upper jaw

The predatory Archer Fish has a silvery body and a yellowish tinge to the caudal fin. The sides of the body display black blotches, as do the fins. This reasonably social fish needs a deep, covered tank, and it may benefit from the addition of some marine salt to the water. Archer Fish have yet to be bred successfully in aquariums. This may be linked to the fact that the adults are believed to head out to sea to spawn on reefs.

SHARP-SHOOTING FISH

The hunting ability of the Archer Fish is remarkable. Rather than seeking food in the water, it patrols just below the surface, searching with its sharp eyes for invertebrates on overhanging vegetation. Using its powerful jaws, the Archer Fish sucks water into its mouth and carefully targets its intended quarry from below the water line. Then it raises its mouth up slightly, just breaking through the surface, and fires a powerful jet of water at the unsuspecting victim. If its aim is true, the creature is caught off guard and knocked into the water below, where the Archer Fish snaps it up. The water jet is a formidable hunting weapon, being effective over distances up to 5 ft (1.5 m).

Young Archer Fish begin to fire water jets at quite an early age, but it takes practice to perfect their shooting skills. The biggest problem to overcome is refraction, which causes an object to appear to be in a different position when seen from below the water line. The Archer Fish deals with this by careful positioning. It shoots from almost directly under its quarry and adopts an upright posture in the water, thereby reducing refraction to a minimum and significantly increasing its chances of hitting the target.

OTHER AFRICAN SPECIES

Africa is home to a wide range of unique and fascinating fish. The fossil record shows that some of these, notably the African lungfish *(see opposite)*, have altered relatively little over millions of years. Lungfish and similar species, such as Cuvier's Bichir, are most likely to be seen in specialist aquatic stores, partly because they can be difficult to keep. The long-snouted mormyrids, or "elephant fish," rank among the most popular of the other African groups, thanks to their bizarre appearance. They also have the ability to generate weak electrical currents.

Like the lungfish, Africa's **Reed Fish** (*Erpetoichthys calabaricus*) can breathe out of water thanks to its swim bladder, which takes over the job of oxygenating the blood when the gills cannot function as normal.

Gnathonemus petersii

Peter's Elephant-Nose

- **ORIGINS** Western and central parts of Africa, ranging from Nigeria and Cameroon to Zaire.
- **SIZE** 9 in (23 cm).
- **DIET** Mainly live foods.
- **WATER** Temperature 72–82°F (22–28°C); soft (50–100 mg/l) and neutral (pH 7.0).
- **TEMPERAMENT** Territorial.

The elongated lower jaw of this mormyrid fish resembles an elephant's trunk. In the wild, it is used to dig for food in the muddy substrate. Mainly nocturnal in habit, these fish can be identified by two vertical white stripes extending down each side of the body from the dorsal fin. Elephant-Noses can generate electrical impulses, which help them to navigate in murky water. They do not get along well with their own kind, and their breeding behavior in home aquariums has yet to be documented.

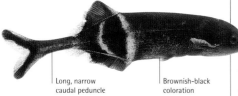

Long, narrow caudal peduncle

Brownish-black coloration

Campylomormyrus rhynchophorus

Down Poker

- **ORIGINS** West Africa, in Angola and the Congo basin, especially around Kinshasa.
- **SIZE** 8¹⁄₂ in (22 cm).
- **DIET** Mainly live foods.
- **WATER** Temperature 72–75°F (22–24°C); soft (50–100 mg/l) and acidic (pH 5.0).
- **TEMPERAMENT** Territorial.

The Down Poker has a speckled body and a long, thick proboscis formed by its elongated jaws. This mormyrid requires subdued lighting, along with a fine substrate in which it can dig for invertebrates, such as small worms. It is difficult to wean off live foods, but in time this fish may eat flaked food as well. Down Pokers should not be kept together, but they can be housed singly with other nonaggressive species.

Pantodon buchholzi

Butterflyfish

- **ORIGINS** West Africa, where it occurs in parts of Nigeria, Cameroon, and Zaire.
- **SIZE** 4 in (10 cm).
- **DIET** Mainly live foods.
- **WATER** Temperature 73–86°F (23–30°C); soft (50–100 mg/l) and acidic (pH 6.5).
- **TEMPERAMENT** Will not mix with other surface dwellers.

With its elegant pectoral fins, this fish resembles a butterfly in flight. The straight back and upturned mouth indicate that it spends much of its time at the surface. A tank for Butterflyfish needs to be shallow and covered, with floating plants over part of the surface. Butterflyfish congregate beneath vegetation, darting out to obtain food. Their eggs, numbering up to 200, also float on the surface.

Polypterus senegalus

Cuvier's Bichir

- **ORIGINS** Africa, in parts of Senegal, Gambia, Chad, and Niger, and in the White Nile and Lakes Rudolf and Albert.
- **SIZE** 12 in (30 cm).
- **DIET** Live foods.
- **WATER** Temperature 77–82°F (25–28°C); soft (50–100 mg/l) and neutral (pH 7.0).
- **TEMPERAMENT** Aggressive and quarrelsome.

Light greenish coloration and pale underparts typify these eel-like fish. They also have a number of small dorsal fins running down their back and supplementary gills to help them survive in the wild in poorly oxygenated water. Cuvier's Bichirs require a large yet relatively shallow tank, with planting kept to a minimum but with plenty of retreats, such as bogwood. As members of the pike family, they are highly predatory by nature.

Xenomystus nigri

African Knifefish

- **ORIGINS** West Africa, occurring in the the Niger River in Liberia, Niger, Zaire, and Gabon; also in the upper Nile.
- **SIZE** 12 in (30 cm).
- **DIET** Live foods.
- **WATER** Temperature 73–82°F (23–28°C); soft (50–100 mg/l) and acidic (pH 6.5).
- **TEMPERAMENT** Becomes territorial with age.

Shaped like a knife blade, this fish is brown to brownish-gray, occasionally with vertical stripes down the body. Knifefish require a well-planted aquarium and must not be kept with smaller companions. They tend to be nocturnal and may make bell-like sounds by compressing their swim bladders. The female is more colorful when she is in spawning condition, often turning reddish-brown. She may produce up to 200 eggs.

COCOONED IN MUD

Lungfish get their name from their ability to breathe air out of water. Air is taken up via folds in the swim bladder, which increase the organ's surface area and allow more blood to flow through it, facilitating gas exchange. Thanks to this auxiliary method of respiration, lungfish can survive through the dry season, when the pools that they inhabit evaporate. As water levels fall, a lungfish burrows down into the substrate and becomes sealed in the drying mud. With its body covered in mucus to prevent water loss, and using its swim bladder to breathe, it remains inert in its muddy cocoon until the rains return, typically four to six months later. The lungfish then frees itself from its cocoon and reverts to its aquatic existence.

Protopterus annectens

African Lungfish

- **ORIGINS** Africa, from Senegal eastward to Nigeria, and from Zaire southward to Zambia and Mozambique.
- **SIZE** 26 in (65 cm).
- **DIET** Carnivorous.
- **WATER** Temperature 79–86°F (26–30°C); soft (50–100 mg/l) and neutral (pH 7.0).
- **TEMPERAMENT** Aggressive and predatory.

An elongated gray body, narrow, trailing pectoral fins, and a crestlike dorsal fin running down the back make the African Lungfish unmistakable. This fish needs to be housed on its own in a large, well-planted aquarium with relatively shallow water. The tank must have a soft, fine substrate and include bogwood to provide a suitable retreat. Feeding this unpicky predator is straightforward—even aquarium snails sharing its quarters are likely to be eaten. It has not yet proved possible to breed these lungfish in aquarium surroundings. This is partly because of their intolerant nature, which makes pairings difficult to achieve. Lungfish are not fussy about the pH and relative hardness of the water in their tank, but good filtration is essential to keep ammonia levels low.

BRACKISH WATER SPECIES

A brackish-water tank lets you keep a number of interesting, but often overlooked, species that are native to estuarine waters. The home aquarium can be a mixed community setup, or a recreation of a specialized habitat such as a swamp. It could incorporate mangrove plants and mudskippers *(see p.43)*, which venture out of water more regularly than any other fish. Brackish-water tanks are not difficult to maintain; correct water conditions are initially achieved by adding a small amount of sea salt. Some of the more adaptable aquarium plants will grow in these surroundings, although plastic substitutes can also be used.

The Spotted Puffer (*Dichotomyctere fluviatilis*) will thrive in a brackish environment, but other members of this family occur in freshwater and marine habitats.

Monodactylus argenteus

Mono

- **ORIGINS** Range extends from East Africa eastward as far as parts of Indonesia and Australia.
- **SIZE** 10 in (25 cm).
- **DIET** Live foods and vegetable matter.
- **WATER** Temperature 72–77°F (22–25°C); alkaline (pH 7.6–8.0) with SG 1.002–1.007.
- **TEMPERAMENT** Peaceful, but avoid small companions.

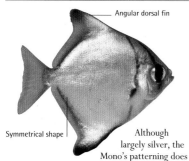

Angular dorsal fin

Symmetrical shape

Although largely silver, the Mono's patterning does differ to some extent between individuals. The prominence of the black banding through the eye and behind the gills is variable, as is the amount of yellow coloration on the fins. Monos look most effective if kept in a group. They are elegant but rather nervous fish by nature.

Datnioides microlepis

Siamese Tigerfish

- **ORIGINS** Southeast Asia, in parts of Cambodia (Kampuchea), Thailand, and on Sumatra and Borneo.
- **SIZE** 16 in (40 cm).
- **DIET** Live foods of various types.
- **WATER** Temperature 72–79°F (22–26°C); alkaline (pH 7.6–8.0) with SG 1.002–1.007.
- **TEMPERAMENT** Predatory.

The vertical yellow and black stripes of these fish help to explain their name. Their dorsal fin has a decidedly spiky appearance. Although Siamese Tigerfish are unlikely to grow as big in aquariums as they do in the wild, juveniles will still ultimately need a large tank. Siamese Tigers cannot be trusted with smaller companions, since other fish naturally form part of their diet.

Paramsassis ranga

Glassfish

- **ORIGINS** Asia, found throughout India, Myanmar (Burma), and Thailand.
- **SIZE** 3 in (8 cm).
- **DIET** Live foods of various types.
- **WATER** Temperature 77–79°F (25–26°C); alkaline (pH 7.6–8.0) with SG 1.002–1.007.
- **TEMPERAMENT** Instinctively rather nervous.

Lateral line

Skeleton

Swim bladder

The internal organs of Glassfish are visible through their skin. Males have a more pointed swim bladder and bluer edges on their fins. Spawning can be induced by raising the water temperature. Up to 150 eggs are scattered among plants. They hatch in a day, and the fry are free-swimming soon after. These fish are sometimes injected with bright dyes and sold as "painted glassfish"; even if they survive, their garish colors will fade after a few months.

Macrognathus aculeatus

Lesser Spiny Eel

- **ORIGINS** Southeast Asia, in brackish environments close to the shoreline.
- **SIZE** 14 in (35 cm).
- **DIET** Live foods of different types.
- **WATER** Temperature 73–82°F (23–28°C); alkaline (pH 7.6–8.0) with SG 1.002–1.007.
- **TEMPERAMENT** Predatory and aggressive.

The distinguishing features of this species are the alternate light and dark horizontal bands along the side of the body and the false eye-spots on the dorsal fin. As with other spiny eels, there are a number of spines in front of the dorsal fin itself. Nocturnal by nature, and eager burrowers into the aquarium substrate, Lesser Spiny Eels do not rank among the most conspicuous of aquarium occupants. They will hunt for food on the floor of the tank, so provide them with foods that sink to the bottom. Spiny eels tend not to agree with one another when kept in small groups.

Periophthalmus barbarus

Mudskipper

- **ORIGINS** East Africa, south to Madagascar, and east to parts of southeast Asia and Australia.
- **SIZE** 6 in (15 cm).
- **DIET** Live foods of all types.
- **WATER** Temperature 77–86°F (25–30°C); alkaline (pH 7.6–8.0) with SG 1.002–1.007.
- **TEMPERAMENT** Territorial.

An aquarium for these fish needs to be relatively large and have a raised area like a beach at one end to allow the fish to hop onto land, just as they would at low tide in the wild. Tree roots and plants, too, will be needed, and an external power filter is essential to maintain the water quality. Keeping the aquarium covered will help to ensure that the air inside is warm and humid when the mudskippers emerge from the water.

SANCTUARY IN THE SWAMP

Mangrove swamps occur in tidal areas, often close to estuaries, where salt-tolerant plants grow in the mud. The tangled mass of the plants' submerged roots provides sanctuary for the young of fish, such as the Tiger Scat *(see bottom left)*, which slip easily between the roots and out of reach of larger predators. Mudskippers in particular have adapted to the mangroves, because they can survive temporarily out of water on the exposed mudflats at low tide, as visible below. When the tide comes in, the mudskippers' brown, mottled coloration helps to camouflage them as they lie on the muddy bottom.

Scatophagus argus

Tiger Scat

- **ORIGINS** Indo-Pacific region, from the coasts of India extending eastward to the Pacific islands.
- **SIZE** 12 in (30 cm).
- **DIET** Largely vegetarian.
- **WATER** Temperature 77–79°F (25–26°C); alkaline (pH 7.6–8.0) with SG 1.002–1.007.
- **TEMPERAMENT** Social and nonaggressive.

The spotted patterning that characterizes young Scats alters as they mature, with the background color becoming silvery rather than golden. Being vegetarian, these fish will damage or destroy living plants in their aquarium. Java Fern (*Microsorum pteropus*) must not be incorporated, since it may be toxic to Scats if they eat it. Scats are active fish by nature, and a group will require a large, spacious aquarium.

Brachygobius xanthozonus

Bumblebee Goby

- **ORIGINS** Southeast Asia, in brackish waters close to the shoreline.
- **SIZE** 2 in (5 cm).
- **DIET** Small live foods preferred.
- **WATER** Temperature 77–79°F (25–26°C); alkaline (pH 7.6–8.0) with SG 1.002–1.007.
- **TEMPERAMENT** Territorial by nature.

The yellowish-orange and black banding on these small gobies resembles that of a bumblebee. Males tend to be more brightly colored and thinner than females. Bumblebee Gobies spend their time close to the bottom of the tank. Their eggs, which are susceptible to fungus even in brackish water, are hidden under a rock and guarded by the male until they hatch about four days later. Provide retreats to lessen displays of territorial aggression.

DIRECTORY OF
FRESHWATER
PLANTS

FLOATING PLANTS

Floating plants are chosen less for their appearance than those growing in the main body of the tank and more for their function. They provide spawning sites and food for many fish species and also give cover and help to diffuse harsh aquarium lighting. Floating plants vary widely in both size and leaf shape, and some grow on land as well as in water. Currents in the tank have a marked effect on the distribution of floating plants, so you may have to adjust the filter outlet to achieve an even spread.

The choice of aquatic plants may be affected by legislation depending on where you live, because of fears that some could become invasive if introduced to native waterways. **Fairy Moss** (*Azolla filiculoides*) is shown here.

Pistia stratiotes

Water Lettuce

- **ORIGINS** Abundant in waterways in tropical and subtropical parts of the world.
- **SIZE** Leaves can be up to 4 in (10 cm) in length.
- **WATER** Temperature 72–86°F (22–30°C); soft (50–100 mg/l) and around neutral (pH 6.5–7.5).
- **PROPAGATION** Break off the plantlets that develop on the stemlike stolons.

The leaf clusters of this floating plant are lettucelike in appearance. The tiny flowers emerge in the axils, between the leaf and stalk, while the trailing roots—which extend down to a depth of 12 in (30 cm)—may be used as spawning sites by some egg-laying fish. Good lighting is vital for the successful spread of the plant, which grows fast under favorable conditions. Prevent condensation from dripping on to the leaves because the plant will rot.

Azolla cristata

Carolina Fairy Moss

- **ORIGINS** From the US to South America; introduced to Europe in the 1870s, and now found wild in some areas.
- **SIZE** Leaves each measure about ½ in (1.5 cm).
- **WATER** Temperature 68–86°F (20–30°C); hard (100–150 mg/l) and around neutral (pH 6.5–7.5).
- **PROPAGATION** Reproduces asexually, so simply divide an existing clump.

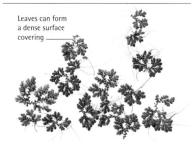

Leaves can form a dense surface covering

Fairy Moss can appear bright red under intense lighting, but more usually it will be bright green, as seen above. This plant will soon spread over the surface of an aquarium and is a useful supplement to the diet of vegetarian fish. It also provides a support for the nests of bubble-nesting species, such as gouramis, and serves as a retreat for fry, which may find food among its trailing roots.

Limnobium laevigatum

Amazon Frogbit

- **ORIGINS** From Mexico through Central America to Paraguay in South America.
- **SIZE** Leaf diameter is 1–2 in (2.5–5 cm).
- **WATER** Temperature 68–86°F (20–30°C); soft (50–100 mg/l) and around neutral (pH 6.5–7.5).
- **PROPAGATION** Split up existing plants; the divisions will grow rapidly.

The rosette-shaped leaves of Amazon Frogbit are paler underneath and sometimes display dark markings on top. They are able to float thanks to their spongy, air-filled structure, which gives them a slightly swollen, convex appearance. Aquarium strains are female, and although they flower quite readily, there is no likelihood that they will set seed in the absence of the male flower's pollen.

Salvinia auriculata

Butterfly Fern

- **ORIGINS** Found widely in waterways from Mexico south as far as Paraguay in South America.
- **SIZE** Leaves are 1 in (2.5 cm) long, ½ in (1.25 cm) wide.
- **WATER** Temperature 64–77°F (18–25°C); soft (50–100 mg/l) and around neutral (pH 6.8–7.2).
- **PROPAGATION** Simply break up the branches formed by this fern.

This is another species for which bright lighting is very important. If the tank has a glass cover, it must be tilted slightly so that condensation droplets do not fall on to the ferns, because this will cause the plants to rot. It may occasionally be necessary to thin out the growth, because Butterfly Fern can spread rapidly into a dense mat that will prevent light from reaching other plants beneath.

Lemna minor

Duckweed

- **ORIGINS** Found throughout the world in both temperate and tropical regions.
- **SIZE** Leaves are small, measuring about ⅕ in (5 mm).
- **WATER** Temperature 41–86°F (5–30°C); soft (50–100 mg/l) and around neutral (pH 6.5–7.2).
- **PROPAGATION** Split off a few pieces from a mat; they will soon start to replicate.

Duckweed grows under a wide range of conditions, even in an unlit aquarium, provided that it receives some natural light. This plant is a useful addition to a rearing tank and can serve as a food source for vegetarian fish. If there are no plant eaters to keep its growth in check, remove some of the duckweed with a net to prevent it from choking the surface.

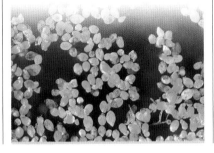

Wolffia arrhiza

Watermeal

- **ORIGINS** Found throughout the world, in both temperate and tropical regions outside polar areas.
- **SIZE** Tiny leaves measure about ⅟₂₅ in (1 mm).
- **WATER** Temperature 59–82°F (15–28°C); soft (50–100 mg/l) and around neutral (pH 6.5–7.5).
- **PROPAGATION** Split off a few pieces from a mat; they will soon start to replicate.

Watermeal is the smallest flowering plant known. Although the genus *Wolffia* comprises eight recognized species, they are all very similar in appearance. Watermeal is easy to grow, with its green coloration turning reddish under bright light. A ready supply of trace elements in the water will encourage rapid growth.

Riccia fluitans

Liverwort

- **ORIGINS** Another widely distributed species, occurring in parts of the Americas, Asia, and Europe.
- **SIZE** Leaves typically no more than ⅟₁₂ in (2 mm).
- **WATER** Temperature 68–82°F (20–28°C); hard (100–150 mg/l) and around neutral (pH 6.5–7.5).
- **PROPAGATION** Break off a piece from an existing clump to add to a new tank.

Liverwort differs from other floating plants in that it grows just below the surface, where it ultimately forms large balls. It can also be grown out of water in damp mud. It will thrive under relatively hard water conditions in the aquarium. Liverwort is especially valuable in tanks housing livebearers, providing fry with a safe refuge from the predatory attentions of other tank occupants.

Ceratopteris thalictroides

Watersprite

- **ORIGINS** Grows widely throughout the world's tropical regions.
- **SIZE** Up to 24 in (60 cm) tall when rooted.
- **WATER** Temperature 72–86°F (22–30°C); soft (50–100 mg/l) and slightly acidic (pH 6.5).
- **PROPAGATION** Young plants develop on the edges of existing leaves.

This is one of the most versatile of all aquarium plants, since it can either be rooted in the substrate or float on the surface. When allowed to float, it has a more flattened appearance, with the leaves appearing slightly broader, while the white roots simply trail down in the water. Watersprite's coloration is variable, depending not only on the lighting conditions but also on the levels of dissolved nutrients in the water. Bright lighting is essential if it is to thrive, but if the illumination is too intense, the leaves may become scorched and die back. Watersprite is not a long-lived plant, being effectively an annual. Mature leaves produce buds that ultimately give rise to new plants. These may be separated from the parent plant once they are about 1½ in (4 cm) across, but they can also be left to detach themselves. They will then float up to the surface and develop there naturally, sometimes protruding above the waterline.

SUBSTRATE PLANTS

Hundreds of plants are available to the freshwater aquarist, and selecting the right species means matching the optimum conditions for the plants to the needs of your fish. Plants and fish should share water chemistry needs and thrive under the same lighting conditions. Themed tanks, in which the geographical origins of fish and plants are matched, can work well. If your tank fish browse on vegetation, avoid slow-growing plants, which will not recover quickly enough; similarly, avoid prolific species that will soon outgrow a small tank.

The **cryptocoryne group** includes some of the most popular aquarium plants. It contains 60 different species, although only about half are cultivated for aquarium use. This is one of the smaller species, known as the Dwarf Crypt (*C. nevillii*).

Vesicularia dubyana

Java Moss

- **ORIGINS** Parts of southern Asia, ranging from India through the Malay Peninsula to Java.
- **SIZE** Forms strands up to 4 in (10 cm) long.
- **WATER** Temperature 72–77°F (22–25°C); soft (50–100 mg/l) and acidic to neutral (pH 6.5–7.0).
- **PROPAGATION** Break off pieces of the moss and fix them in position as required.

This moss will help to create a natural aquascape, growing readily over rockwork and on bogwood. In the first instance, attach it with an elastic band and allow it to put down its rootlike hapterons, which will bind it in place. Take care to ensure that the lighting is not too bright; otherwise, the Java moss will be overgrown by algae in the water.

Aponogeton madagascariensis

Madagascar Laceplant

- **ORIGINS** The island of Madagascar, off the Southeast coast of Africa, and also on nearby Mauritius.
- **SIZE** Leaves can be 20 in (50 cm) in length.
- **WATER** Temperature 72–77°F (22–25°C); soft (50–100 mg/l) and around neutral (pH 6.5–7.2).
- **PROPAGATION** Divide a rhizome. A root above the substrate may also produce a plant. Rarely seeds.

Madagascar Laceplants will grow readily from rhizomes, but they can be tricky to maintain. Shade them from bright light and keep them cool. Well-filtered water will stop debris from clogging their open leaf structure and prevent contamination by algae.

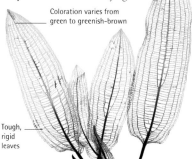

Coloration varies from green to greenish-brown

Tough, rigid leaves

Microsorum pteropus

Java Fern

- **ORIGINS** Occurs widely throughout Southeast Asia, ranging into southern parts of China.
- **SIZE** Leaves may grow as large as 12 in (30 cm).
- **WATER** Temperature 72–86°F (22–30°C); hard (100–150 mg/l) and around neutral (pH 6.5–7.8).
- **PROPAGATION** Divide mature specimens, or remove the tiny plantlets that form on older leaves.

The tough nature of Java Fern enables it to survive in tanks where most plants would be destroyed by the fish, although there have been suggestions that it is toxic to a few species. The rhizome should be attached to the decor with an elastic band, rather than set in the substrate. The fronds will develop transparent areas if the lighting is too bright.

Rotala macrandra

Giant Red Rotala

- **ORIGINS** Asia, occurring in India and on the nearby island of Sri Lanka.
- **SIZE** Typically about 8 in (20 cm) tall in aquariums.
- **WATER** Temperature 72–79°F (22–26°C); soft (50–100 mg/l) and acidic (pH 6.5–6.9).
- **PROPAGATION** Split off shoots and plant these in the substrate, where they will root easily.

Bright lighting accentuates the red coloration of this attractive species. Unfortunately, Giant Red Rotala is difficult to establish in aquariums, and it can be damaged easily by rough handling. However, it is worth the effort, because it makes a striking contrast with green plants. To create the best effect, plant shoots in groups.

Side shoots can be broken off and rooted

Leaves are greenish above and reddish below

Crinum thaianum

Onion Plant

- **ORIGINS** Southeast Asia; particularly abundant in southern Thailand.
- **SIZE** Leaves can be up to 60 in (1.5 m) long.
- **WATER** Temperature 64–81°F (18–27°C); hard (100–150 mg/l) and around neutral (pH 6.5–7.5).
- **PROPAGATION** May occasionally develop offsets on the bulb; these can be taken off and replanted.

Numerous types of Onion Plant are available, all of which reach a relatively large size. The straplike leaves can be up to 3 in (7.5 cm) wide and are variable shades of green. In the wild, the Onion Plant will often grow above the surface, and it is then that it produces its characteristic white flower. The flower reveals that this is not a member of the onion family but a relative of the popular Amaryllis houseplant. The Onion Plant looks best in a large, deep aquarium, especially when planted toward the back of the tank. If the bulb is set deep in the substrate, with just its shoulder visible, it should soon establish itself and start sprouting leaves. Onion Plants are quite tough, so they can be incorporated successfully in aquariums housing large vegetarian fish, where more delicate plants would be consumed. They are also unfussy about their water conditions and do not require brightly lit surroundings.

Nymphaea maculata

African Tiger Lotus

- **ORIGINS** Found naturally in parts of western Africa, notably in Gabon and Congo.
- **SIZE** Spread may be up to 18 in (45 cm) across.
- **WATER** Temperature 68–86°F (20–30°C); soft (50–100 mg/l) and around neutral (pH 6.5–7.5).
- **PROPAGATION** It may be possible to take shoots off the tuber. Can also be grown from seed.

The Tiger Lotus, a broad-leaved relative of the water lily, has two distinct forms. The African Tiger Lotus, shown here, has greenish leaves with purple blotches and pale green undersides. The Red Tiger Lotus has reddish leaves, again marked with purple. Once planted in the substrate, tubers should grow rapidly under bright light. The leaves, up to 6 in (15 cm) in diameter, provide retreats for small fish. Tiger Lotuses may flower in the tank, producing white blooms above the surface that open at night. If the resulting seeds are left to fall into the tank, they may germinate on the substrate.

Barclaya longifolia

Orchid Lily

- **ORIGINS** Southeast Asia, occurring from Myanmar (Burma) to parts of Thailand and Malaysia.
- **SIZE** Up to 12 in (30 cm) across.
- **WATER** Temperature 77–86°F (25–30°C); soft (50–100 mg/l) and around neutral (pH 6.8–7.2).
- **PROPAGATION** Small plantlets on the rhizome can be split off and planted. May also be propagated from seed.

The Orchid Lily produces a series of attractive upright leaves and may even bloom on occasion. The red flower produced is able to self-fertilize. Under brightly lit conditions, the leaves will be green, but relatively subdued lighting will bring out a more brownish tone. The substrate must be nonalkaline for this plant to thrive.

Bacopa caroliniana

Giant Red Bacopa

- **ORIGINS** Found naturally from southern parts of the United States into northern Mexico.
- **SIZE** Stems can grow up to 12 in (30 cm) long.
- **WATER** Temperature 68–75°F (20–24°C); hard (100–150 mg/l) and around neutral (pH 6.5–7.5).
- **PROPAGATION** Strip off the lower two pairs of leaves from the stem and plant in the substrate as a clump.

The leaves of this plant have no stalk but attach directly to the stems. They display a rich coppery color in bright light but appear more green under subdued illumination. A variegated form is also sometimes available. In the wild, the Giant Red Bacopa often grows as a bog plant above the water, but it thrives equally well submerged in a tank.

Glossostigma elatinoides

Glosso

- **ORIGINS** Found in Australia, in New South Wales and Tasmania, and also in New Zealand.
- **SIZE** About 1 in (2.5 cm) in height.
- **WATER** Temperature 68–82°F (20–28°C); soft (50–100 mg/l) and around neutral (pH 6.5–7.5).
- **PROPAGATION** Clumps can be split up and used as cuttings. Alternatively, runners can be used.

This small plant is ideal for the foreground of the tank, creating a pleasing carpet of growth that provides a refuge for fish fry. It spreads through the substrate and benefits from small amounts of aquarium plant fertilizer. If its surroundings are not well lit, Glosso will become taller and rather straggly in appearance.

Alternanthera lilacina

Red Telanthera

- **ORIGINS** Grows widely throughout tropical regions of South America.
- **SIZE** Can grow up to 12 in (30 cm) high.
- **WATER** Temperature 72–86°F (22–30°C); soft (50–100 mg/l) and acidic (pH 6.0–6.5).
- **PROPAGATION** Easily propagated by means of cuttings, which will root readily in the substrate.

This plant may be found growing above the water's surface in its native habitat, but it adapts well to cultivation underwater in an aquarium. Red Telanthera will be seen in its full depth of color only if the tank is brightly lit. The upper surface of the leaves tends to be greenish with red hues, while purplish-red coloration is concentrated on the undersides. This plant is a good choice for a themed Amazon tank, although it can also be used in a community aquarium. Red Telanthera is best placed at the sides of the tank, or toward the rear. Set cuttings into the substrate so that they grow to form a dense clump, and place green plants of a similar height nearby in order to emphasize the contrasting leaf colors. Make sure you strip off the lower leaves from Red Telanthera before planting, because they will rot if they are buried. Use small rocks to weigh down the bases of the cuttings until they root and become established.

Ludwigia mullertii

Red Ludwigia

- **ORIGINS** This plant does not grow naturally in the wild, but its ancestors occur in North America.
- **SIZE** Can grow to a height of 15 in (38 cm).
- **WATER** Temperature 68–86°F (20–30°C); soft (50–100 mg/l) and around neutral (pH 6.5–7.5).
- **PROPAGATION** Very easily propagated from cuttings, although it may also be grown from seed.

This vigorously growing plant is probably a natural hybrid between Marsh Ludwigia (*L. palustris*) and a different Red Ludwigia (*L. repens*). The key requirement of Red Ludwigia is bright light, which will maintain its distinctive red coloration. Because of its rather elongated shape, Red Ludwigia looks at its best when it is planted in clumps. Regular trimming back of the stems should help to ensure a denser, less straggly appearance and as a result provide more cover for the fish in the tank. This hardy plant is especially useful in tanks housing fish that require relatively low water temperatures, because it will grow well in such surroundings. If it is included in an uncovered tank, it may grow above the water's surface and subsequently flower, although the white blooms it produces are tiny and inconspicuous.

Nymphoides aquatica

Banana Plant

- **ORIGINS** Occurs naturally in the eastern states of the United States, extending along the Atlantic coastline.
- **SIZE** Can grow up to 12 in (30 cm).
- **WATER** Temperature 68–77°F (20–25°C); soft (50–100 mg/l) and around neutral (pH 6.5–7.5).
- **PROPAGATION** Divide the rootstock, remove small plantlets, or split off runners.

Although this adaptable plant will tolerate being permanently submerged in a tank, it prefers shallower waters similar to those of its natural habitat. The leaves of the Banana Plant spread out over the surface in the aquarium, and it may produce white flowers on stalks above the water level. In the wild, the bananalike roots act as a water reservoir to sustain the plant in times of drought.

Bright light encourages good leaf growth

Leaf shape indicates a close relationship with water lilies

Cabomba species

Cabomba

- **ORIGINS** From the southeastern United States along the eastern side of Central America down to Argentina.
- **SIZE** Branches can grow to well over 20 in (50 cm).
- **WATER** Temperature 68–79°F (20–26°C); soft (50–100 mg/l) and around neutral (pH 6.5–7.2).
- **PROPAGATION** Roots easily from cuttings placed into the substrate.

There are five different species of Cabomba: some have larger whorls of fine leaves than that pictured, and others have mauve-tipped leaves. Good lighting is important for these plants, which are best planted in groups near the back of the tank, using pieces of slate to weigh them down until they take root. Cut back leggy plants to encourage vigorous and compact new growth.

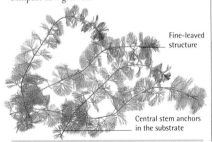

Fine-leaved structure

Central stem anchors in the substrate

Ceratophyllum submersum

Hornwort

- **ORIGINS** Grows widely throughout the world's tropical and subtropical regions.
- **SIZE** May reach a length of 18 in (45 cm).
- **WATER** Temperature 68–82°F (20–28°C); hard (100–150 mg/l) and around neutral (pH 6.5–7.5).
- **PROPAGATION** Very easy to propagate; pieces break off naturally, giving rise to new plants.

This attractive plant has fine green foliage. It needs to be held down in the substrate with rocks, because it has lost the ability to anchor itself with roots. Hornwort is also fragile, and pieces break off easily, usually from the crown. These may then grow at the water's surface as floating plants. Hornwort fares well under bright light.

Cardamine lyrata

Bitter Cress

- **ORIGINS** Asia, found naturally in parts of eastern China, as well as in Korea and Japan.
- **SIZE** Can grow to about 12 in (30 cm) in height.
- **WATER** Temperature 59–77°F (15–25°C); soft (50–100 mg/l) and around neutral (pH 6.5–7.5).
- **PROPAGATION** Take cuttings, which will rapidly establish themselves in the substrate.

This is an ideal choice for tropical aquariums. House it with Paradise Fish *(see p.108)* and other species that share the same waters in the wild. This plant prefers cool surroundings and may not thrive at temperatures above 68°F (20°C). The leaf form ranges from circular to kidney-shaped. Above the surface, the leaves are pointed, and small, white flowers are produced.

Planted in clumps, the stems will entwine

Anubias barteri

Barter's Anubias

- 🌐 **ORIGINS** West Africa, occurring in Nigeria, Gabon, Ivory Coast, and Cameroon.
- 🌿 **SIZE** Leaves may be up to 12 in (30 cm) long.
- 💧 **WATER** Temperature 72–77°F (22–25°C); hard (100–150 mg/l) and around neutral (pH 6.0–7.5).
- 💧 **PROPAGATION** Divide the rhizome and transplant the pieces to different areas of the tank.

A number of local strains of this plant occur throughout West Africa. The smallest is the Dwarf Anubias (*A. b.* var. *nana*), pictured right, which is widely cultivated for aquarium use because of its compact shape. The relatively thick leaves are about 2 in (5 cm) across, and the stalks are of a similar length. The largest variety, the Lance-Leaf Anubias (*A. b.* var. *lanceolata*), has long, narrow leaves. Barter's Anubias is slow-growing and benefits from a substrate fertilizer. It grows from a rhizome, which should not be buried but simply left on the substrate, where its roots will spread out. It is thus possible to anchor this plant to tank decor such as bogwood. The plant's low height makes it ideal for the front of a tank, and it will thrive under subdued lighting. Its spathe flower, which is produced above the water's surface, is unlikely to yield fertile seed.

Hygrophila polysperma

Dwarf Hygrophila

- 🌐 **ORIGINS** Occurs in southern Asia; particularly common in many parts of India.
- 🌿 **SIZE** May reach 10 in (25 cm) or so in height.
- 💧 **WATER** Temperature 59–86°F (15–30°C); hard (100–150 mg/l) and around neutral (pH 6.5–7.5).
- 💧 **PROPAGATION** Take cuttings using the lower leaves that are trimmed off before the plants are set in place.

Dwarf Hygrophila is one of the most adaptable and easily cultivated of all aquarium plants. The long, green leaves sometimes develop red tips when the plant is kept in brightly lit surroundings. This fast-spreading plant will provide valuable cover in the tank, although its growth may be curbed if snails attack the leaves before it becomes established.

Shinnersia rivularis

Mexican Oak-Leaf Plant

- 🌐 **ORIGINS** Central America, where its distribution is restricted to Mexico.
- 🌿 **SIZE** May reach 12 in (30 cm) in height.
- 💧 **WATER** Temperature 68–79°F (20–26°C); hard (100–150 mg/l) and around neutral (pH 6.5–7.5).
- 💧 **PROPAGATION** Can be grown either from cuttings or by transplanting runners.

The stems of this plant, which grows upright when submerged, are thick and robust. The tooth-edged, oaklike leaves vary in color from light to dark green. Groups planted near the back of the tank look very attractive. Easy to establish, the Mexican Oak-Leaf Plant thrives not only in brightly lit conditions but also under more subdued lighting.

Vallisneria tortifolia

Twisted Vallisneria

- 🌐 **ORIGINS** Probably southern Europe. Now occurs widely in tropical and subtropical localities.
- 🌿 **SIZE** Leaves typically measure up to 8 in (20 cm) long.
- 💧 **WATER** Temperature 72–82°F (22–28°C); hard (100–150 mg/l) and around neutral (pH 6.5–7.5).
- 💧 **PROPAGATION** Split off the runners produced by established plants.

It is unclear whether this plant is a hybrid or a natural variant of the Straight Vallisneria (*V. spiralis*). Allow space between the plants so that light can penetrate: this is vital for their growth. A larger form is the Asiatic Vallisneria (*V. asiatica*), which has serrated leaf edges.

Long, twisted, ribbonlike leaves

Vallisneria gigantea

Giant Vallisneria

- **ORIGINS** Mainland Southeast Asia and various islands, including New Guinea and the Philippines.
- **SIZE** Leaves may be up to 40 in (1 m) long.
- **WATER** Temperature 72–82°F (22–28°C); soft (50–100 mg/l) and around neutral (pH 6.5–7.2).
- **PROPAGATION** Split off and transplant the runners produced by established plants.

This large, straight-leaved *Vallisneria* species makes a striking centerpiece for a large aquarium. Cultivated strains that develop a reddish hue under bright light are particularly attractive. Changes in water quality may cause this sensitive plant to die back.

Tall, straight leaves

Leaves may be over 1 in (2.5 cm) wide

Echinodorus tenellus

Pygmy Chain Swordplant

- **ORIGINS** Occurs widely through the Americas, from the state of Michigan to southern Brazil.
- **SIZE** Grows to a height of about 6 in (15 cm).
- **WATER** Temperature 59–79°F (15–26°C); hard (100–150 mg/l) and around neutral (pH 6.2–7.0).
- **PROPAGATION** Separate and transplant runners. Can also be grown from seed.

Open-weave plastic pots constrain root growth

Groups of these small swordplants form attractive foreground cover in the tank. Cultivated strains vary in height, so adjustments may be needed once the plants are established. If allowed to grow above the surface in shallow water, the leaves will be broader, and flowers will be produced and may give rise to fertile seed.

Sagittaria subulata

Dwarf Sagittaria

- **ORIGINS** North America, where it occurs on the eastern side of the continent, down the Atlantic coast.
- **SIZE** About 6 in (15 cm) in height.
- **WATER** Temperature 55–79°F (13–26°C); hard (100–150 mg/l) and around neutral (pH 6.5–7.5).
- **PROPAGATION** Split off runners produced by mature plants and transplant them elsewhere.

A large expanse of this hardy, adaptable plant resembles a grass lawn and provides a safe retreat for fry. To achieve this effect, place several plants in the mid-ground area, with gaps between them. They will soon spread out and fill in the gaps, especially on a coarse gravel substrate.

Lilaeopsis novae-zelandiae

False Tenellus

- **ORIGINS** This plant is native to Australia and New Zealand.
- **SIZE** Can grow to 3 in (7.5 cm) in height.
- **WATER** Temperature 64–82°F (18–28°C); soft (50–100 mg/l) and acidic to neutral (pH 6.4–7.0).
- **PROPAGATION** Readily produces runners that can be split off and transplanted elsewhere.

Similar in appearance to the Pygmy Chain Swordplant *(see above)*, False Tenellus is a popular choice for the foreground of the tank, partly because it grows well in a wide range of water temperatures. This plant establishes itself readily and spreads well, especially when planted in a substrate that contains added nutrients. In fact, False Tenellus can be so prolific that its growth may sometimes need to be curbed to prevent it from dominating the tank, since this could impair the efficiency of the undergravel filter and lead to a deterioration in water quality. False Tenellus is often sold in bunches, with each plant consisting of one to three narrow leaves that taper to a point and lack a petiole (the part that usually connects a leaf to the stem). Other *Lilaeopsis* species may also become available to aquarists from time to time, but they all look very much alike and have similar growth characteristics and requirements.

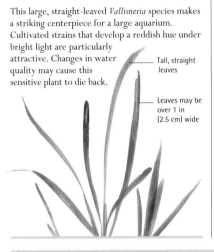

Saururus cernuus

Lizard's Tail

- **ORIGINS** Range extends down the eastern side of North America, from Canada to Florida.
- **SIZE** Leaves may grow to 8 in (20 cm) in length.
- **WATER** Temperature 64–75°F (18–24°C); soft (50–100 mg/l) and around neutral (pH 6.5–7.2).
- **PROPAGATION** Divide the rhizome, take cuttings, or split off and transplant runners.

The name of this swamp-dwelling plant derives from the spiked arrangement of its yellowish-white flowers. The leaves, which are a variable shade of green, appear slightly hairy at first, but they become smoother as they age. Lizard's Tail grows from a rhizome but thrives only when it can spread both above and below the surface. It is a temperate plant so must not be kept too warm. In a pond, it should be containerized so that it does not damage the lining.

Heart-shaped leaves

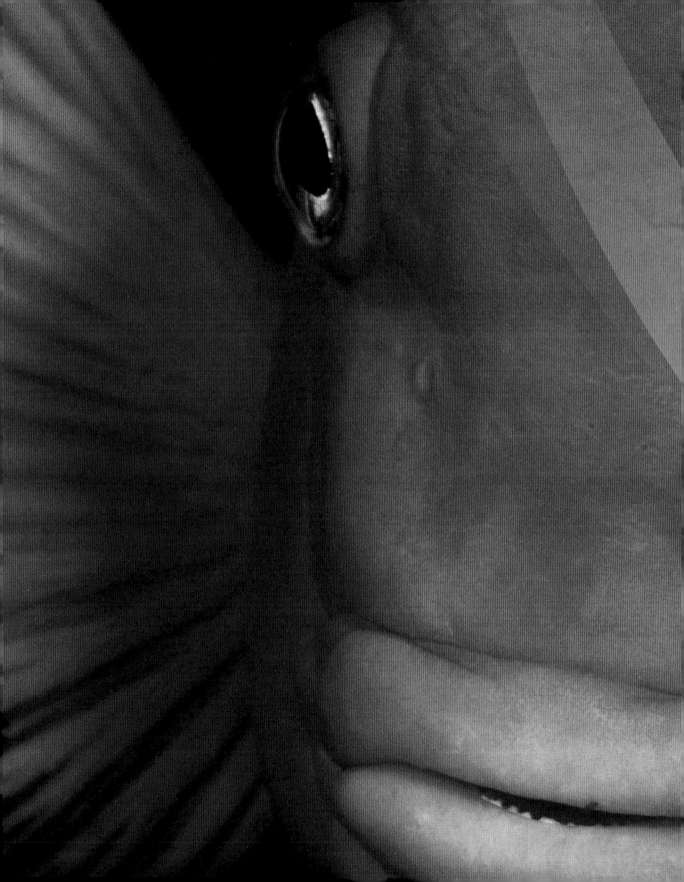

INTRODUCTION TO
MARINE FISH

What to consider

A marine aquarium, populated with stunningly patterned reef fish and invertebrates in crystal-clear water, makes a stunning centerpiece for any room. Historically, marine tanks have been considered more difficult to establish and maintain than freshwater setups, but today's aquarium technology and breeding methods put them within reach even of novice aquarists.

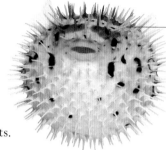

Body inflated in a defensive posture

Porcupinefish *(see p.251)* cannot be housed with invertebrates because they are likely to prey on them. Tankmates for these fish need to be chosen with care.

There are plenty of good reasons to keep marine fish in the home aquarium. They are diverse, often beautifully colored, and their biology and behavior are endlessly fascinating; you will never tire of watching a cleaner wrasse at work, for example, or a clownfish weaving between the tentacles of an anemone. Marine fishkeeping can also present real challenges, even for the experienced freshwater aquarist, so keeping a successful saltwater aquarium is particularly rewarding.

Marine fish are considered challenging because seawater is very stable in both composition and temperature. Unlike freshwater fish, most marine species have little tolerance for fluctuations in water quality, and so keeping them in a home aquarium demands more monitoring, more attention to detail, and more patience, especially in the early stages, because the tank may take up to three months to become fully established. Tanks are available in all shapes and sizes *(see pp.30–31)*.

A marine tank is often more costly to maintain than its freshwater equivalent, not least because larger tanks (of at least 48 gallons, or 180 liters) are preferred. This is because

The amount of free swimming space required by a marine fish depends partly on the species. Triggerfish, for example, often show aggression to each other when several are kept in a smaller tank, but they can sometimes live together harmoniously, especially when young, if kept in very spacious surroundings.

Stocking densities in marine tanks are generally lower than in freshwater setups *(see p.30)* and depend on type of fish kept, maintenance, filtration, and feeding regimens.

A clownfish (right) swims between the protective tentacles of a sea anemone. Clownfish are the most widely bred marine fish.

MARINE CHOICES

● Marine tanks need careful planning for long-term success. Take time before buying to learn about the fish, invertebrates, and equipment.

● Consider whether you can afford the extra time commitment involved in keeping marine species.

● If you wish to breed fish, choose marine species carefully—many will not reproduce in the aquarium.

● Bigger is better where marine tanks are concerned.

it is much easier to ensure the stability of water composition for the tank occupants in a greater volume of water. The tank will also need more power, because brighter lighting is needed to support marine algae and invertebrates, and power heads should be used for effective biological filtration *(see p.211)*.

Types of marine aquariums

There are two basic categories of marine aquariums: those in which the fish are kept on their own and those that include various types of invertebrate as well; the latter are often described as reef tanks. A fish-only tank is usually recommended for novice fishkeepers because it avoids

Reef tanks (above), containing a mixture of fish and invertebrates, provide an ongoing challenge for more experienced fishkeepers.

REEF CONSERVATION

Nearly all marine fish seen in home aquariums originate from coral reef areas (below), rather than from the open seas. Up to 30 million fish are caught for the trade every year in countries including Indonesia, the Philippines, Vietnam, Brazil, and Sri Lanka. The aquarium trade, together with most national governments, is working hard to promote sustainable reef management and outlaw practices such as the use of cyanide to catch fish and the capture of species with low survival rates. When carefully regulated, the trade in fish can provide a revenue for local people that far exceeds that from fishing and so can encourage care for these fascinating but highly fragile environments.

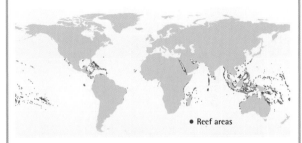

• Reef areas

the pitfalls of compatibility that can arise when fish are mixed with starfish, anemones, shrimp, or other invertebrates that may form their prey. In many other respects, the factors relating to the choice and siting of a marine tank are similar to those that apply to freshwater setups *(see pp. 30–31)*. The tank itself may be glass or acrylic (the latter is lighter in weight but scratches more easily). It must be sited on a suitable stand or piece of furniture on a strong, level floor, preferably adjacent to a load-bearing wall and electrical outlets. The tank should be located well away from windows, radiators, or any other environmental hazards and surrounded by an uncluttered area to allow access for maintenance and water changes.

Stocking the tank

It is easy to make mistakes when stocking your marine tank, and a little research before buying will pay great dividends later. As a general rule, start with a few inexpensive fish in a fish-only tank—you can always graduate to a reef tank later. A great variety of species are available through the aquarium trade, the majority of which are collected from the oceans rather than raised in captivity. This can make some species hard to feed, because they are accustomed to live foods and reluctant to take substitute diets, even when these contain all the ingredients needed to keep them in good health. At first, avoid species with very specialized diets (seahorses, for example, need a diet based on brine shrimp *nauplii*); instead, choose fish that can be kept on marine flake or similar prepared foods, and, if possible, select tank-raised marine fish, such as clownfish, which are particularly easy to keep.

Telling sexes apart is difficult in most marine species; when buying a shoal, you may well end up with a preponderance of males, which may display aggressive behavior.

Buying policy

If possible, try to view the fish yourself before buying; this will allow you to make sure they are healthy and feeding well. Look for lively, alert specimens, with good, clean colors and no obvious blemishes. The fish should not appear abnormally thin (for the species), swim at a strange angle, or display abnormal gill movements. Try to purchase young fish—not just because they are less expensive than larger, mature specimens, but because they are likely to adapt better to aquarium life than adults. It is, however, hard to tell the age

DANGEROUS FISH

Some marine fish have powerful teeth that can inflict a painful bite or sharp spines that can injure the unwary fishkeeper. Other species can inflict venomous stings that are not only dangerous to humans but can kill other tank inhabitants. In some cases, chemicals released into the water cause toxic poisoning of many or all of the tank inhabitants. Although they may be highly decorative, these fish, which include scorpion-, rock-, and lionfish, as well as boxfish, pufferfish, and squirrelfish, are probably best avoided by novice aquarists.

Spines inflict a burning, painful sting

Lionfish produce a venom that is a nerve poison, or neurotoxin. When injected into a fish, it paralyzes its muscles, including the heart.

of marine fish, unless they are of a species that has a distinctive juvenile pattern or are clearly much smaller in size than adults in the same tank.

Planning for the future

If you are setting up a community aquarium, try to select fish that occupy different areas or levels within the tank, because this adds visual interest, and also lessens the potential for territorial disputes. The stocking levels in a marine aquarium need to be built up gradually *(see p. 216)*, so it is a good idea to plan the evolution of your tank from the outset, taking into account the compatibility of fish to be introduced. Adding the fish in groups rather than haphazardly as individuals will help reduce the likelihood of bullying.

The Blue Ribbon Eel is one of the largest species that can be kept in a home aquarium. It can grow to 3 ft (1 m) in length but is sedentary, occupying the lower part of the tank.

MARINE FISH

SETTING UP
THE TANK

Lighting considerations

Lighting the marine aquarium is both an art and a science. In an ideal setup, the light should appear natural and should enhance the colors and forms of the tank occupants, but it must also be of the correct intensity and quality to sustain life. Achieving this balance requires some planning, especially in reef tanks.

The marine fishkeeper is presented with an apparently bewildering variety of lighting alternatives; choosing the right one depends largely on the types of marine organism housed in the aquarium. Most fish are tolerant of a wide range of lighting conditions, so for a fish-only setup, it is usually enough to provide lighting that displays the fish most effectively. Light levels should not, however, be set too low, or the growth of undesirable red/brown algae will be encouraged.

Lighting a reef tank is a very different matter. Many invertebrates in reef aquariums, such as corals and anemones, survive only because they form partnerships with tiny photosynthetic algae that live inside their bodies *(see box, below)*. If the algae do not receive sufficient light, they die,

Lighting arrays or hoods often include two fluorescent tubes—one creating good viewing conditions and the other providing blue light.

along with their hosts. In their natural setting—shallow reefs in tropical seas—these organisms are exposed to bright light from the sun for 8 to 10 hours per day, and these conditions must be replicated with artificial lighting if they are to survive in a tank. Using sunlight to illuminate the aquarium is not a viable option. Instead, special tubes and bulbs, usually mounted in a specially made hood, are used to simulate both the intensity and the quality of light falling on a reef.

A natural coral reef has many different zones of light. Colored corals predominate in the sunlight zone. Deeper down, leather corals, anemones, tubeworms, and others are more prevalent. Darker areas are occupied by soft corals, sponges, and invertebrates that lack zooxanthellae.

LIGHT AND INVERTEBRATES

Certain invertebrates, such as various sea anemones, corals, and some mollusks, contain photosynthetic algae called zooxanthellae in their fleshy bodies. This is a symbiotic relationship in which the algae supply their host invertebrate with food and oxygen and in return receive shelter and take up some by-products of the animals. When a sea anemone (bottom left) opens its tentacles, the maximum amount of light reaches its algal partners; the tiny greenish bodies of the zooxanthellae are visible in the close-up of a coral polyp (below right). However, not all invertebrates in a marine tank thrive under high light levels, and there should be suitable retreats in an aquarium if it is to house crabs and sponges.

A reef tank changes in appearance between day (left) and night (right). When lit, corals and anemones open; in the dark, they close up and fish may appear duller in color.

The algae within corals and anemones need light at the blue end of the visible spectrum *(see box, right)* to photosynthesize. For this reason, marine aquarists tend to light their tanks with fluorescent actinic tubes that strongly emit blue wavelengths. Often, a more neutral daylight-simulating tube is used alongside the actinic tube to replicate the viewing conditions under sunlight and eliminate any bluish cast. Regular domestic (tungsten or halogen) bulbs are not suitable, because the light quality is inappropriate, and because they generate excessive heat, which tends to increase water temperature and cause evaporation.

Fluorescent tubes are available in a range of lengths to suit almost any size of tank. They have a long life span (up to two years), and specialized tubes are designed to deliver a consistent high output throughout their life. In the case of marine invertebrate setups, however, powerful metal halide bulbs may be the best option but must incorporate an ultraviolet filter for safety. Mercury vapor lights are another possibility but are costly and run very hot so need to be carefully mounted in order to disperse the heat produced.

SPECTRAL OUTPUT OF LIGHTS

Natural daylight is made up of a mixture of wavelengths (colors of light)—literally all the colors of the rainbow. However, most fluorescent tubes and light bulbs emit light at some wavelengths in preference to others. In the marine aquarium, it is vital to select lighting that supplies the wavelengths of light that are needed by plants and by symbiotic algae. If you are in any doubt, consult your aquarium dealer.

Natural daylight

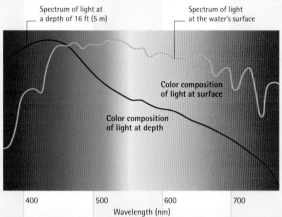

Spectrum of light at a depth of 16 ft (5 m)

Spectrum of light at the water's surface

Color composition of light at surface

Color composition of light at depth

Wavelength (nm)

Sunlight contains more or less equal proportions of all wavelengths of light. As it passes through water, red and yellow components are filtered out, which is why reefs appear to be bathed in blue light. To set up a reef aquarium, it is essential to duplicate these lighting conditions using bulbs or fluorescent tubes (below).

Artificial light

Balanced daylight tubes give out a powerful penetrating light that matches the spectrum of natural light for optimal viewing.

Actinic tubes give out a predominantly violet-blue light, which is required by zooxanthellae—the algae that live symbiotically with corals.

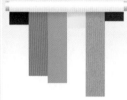

Specialist aquarium tubes are designed for power and consistency of light output. They promote the growth of invertebrates and algae.

Grow lights are used principally in freshwater tanks; the red-rich light they emit enhances the appearance of many animals.

Siting and substrate

There are no firm rules about where to position a marine aquarium in the home, but following a few simple guidelines will help maximize the health of the fish and ensure human safety. The choice of substrate (such as gravel or sand) greatly influences the overall appearance of the tank and is more than just cosmetic. Substrate composition directly affects water chemistry and so influences the long-term welfare of the fish.

Positioning the tank

Aquariums should never be moved if they contain water, sand, or gravel because their great weight makes them prone to shattering. The larger the tank, the longer it takes to empty and strip down for moving; so for marine aquariums, which tend to be larger than their freshwater counterparts, getting the location right the first time is particularly important.

As a general rule, set up the tank in the room where you spend most time, and position it at eye level for the best views of the fish. Taller tanks provide eye-level interest whether seated or standing and are a good choice for compact rooms where there is space only for a tank with a small base. Support the aquarium on a specially built stand or cabinet; if you use an existing piece of furniture, make sure it is strong enough to take the weight of the filled tank and will not be damaged by spillages. Allow enough space around the tank for routine maintenance—you should be able to reach all inner and outer surfaces of the glass without stretching.

Marine aquariums may be used architecturally, built into walls, or set up as room dividers. They should not, however, be sited in rooms where cigarette smoke can build up, because this can diffuse into the water and harm fish and invertebrates.

FILTER AND SUBSTRATE

Most marine aquaria are fitted with an undergravel filter, in addition to a power filter *(see p.211)*. The filter medium is the substrate itself—typically, crushed coral or shell, covered with finer coral sand—which becomes colonized by beneficial aerobic bacteria. A mesh net called a gravel tidy separates the two layers, thus maintaining the flow of water through the filter bed. Always buy prepared substrate from a reputable aquarium dealer, and check it thoroughly for foreign bodies, such as fragments of plastic, metal, and glass, before placing it in the tank.

❶ Place the tank on sponge matting
Wash out the tank to remove dust or glass spicules. Glass aquariums need to be rested on special sponge matting to absorb any unevenness in the surface beneath.

❷ Fit the undergravel filter
Lay the corrugated plastic of the filter plate, with uplift tube attached, on the base of the tank. The plate can be cut to size and should cover the whole base area.

Avoid placing the aquarium where it will be exposed to direct sunlight, because you will then lose control over the light intensity and temperature in the tank.

Water and electricity don't mix, so it is essential to keep cabling short and neat; avoid using messy extensions and always consult a professional electrician if you have any doubts about your system. Never plug pumps or filters into switched outlets—it is all too easy inadvertently to flip the wrong wall switch and shut off the tank's life support systems.

Substrate matters

The substrate in a marine aquarium is not just for decoration. Some fish, such as jawfish (see p.282), like to burrow, so the sand or gravel used must be of a suitable texture. The substrate is also important in maintaining water chemistry; thanks to its calcium carbonate content, it acts as a buffer, helping to counter the progressive acidification of the water (see p.221). And, when an undergravel filter is used, the substrate also serves as a filter bed (see p.211). In this case, the size and depth of the substrate particles are key; the substrate needs to be deep enough to be effective as a filter, and water must be able to pass between the particles. Usually, the filter plate is covered with a layer of coarse material (see below), such as crushed coral, shell, or dolomite chips (all of which are high in calcium carbonate). Coral sand or aragonite sand is then laid on top of this to create a more natural appearance.

Acrylic admits about 15 percent more light than glass of comparable thickness, and it can be shaped into more unusual forms with rounded corners.

Acrylic tanks are preferred by some marine aquarists. They are lighter and easier to handle than glass, and holes may be drilled through them to conceal inlet and outlet pipes. However, they do scratch more readily than conventional glass tanks and are more expensive.

DECORATIVE BACKDROPS

A tank's inlet and outlet tubes are rather unsightly but are easily hidden behind a backdrop, stuck to the outside rear of the tank. Commercially available backdrops made from fade-resistant, waterproof plastics feature all sorts of images, from reef scenes, which create a good illusion of depth, to tropical beaches and even lunar landscapes.

❸ **Add coarse substrate**
Place a layer of calcareous substrate— washed in aquarium disinfectant and well rinsed—onto the filter plate to a depth of about 2 in (5 cm), and spread it out evenly.

❹ **Fit the gravel tidy**
Lay the mesh net over the coarse substrate layer, turning the edges down. This will prevent the sand from sinking and filling in the spaces between the coarser grains.

❺ **Cover with coral sand**
Pour fine coral sand onto the mesh to a depth of about 1 in (2.5 cm). Shape the sand layer to the desired form, typically sloping it forward toward the front of the tank.

Heating and filtration

Tropical marine fish thrive at water temperatures of between 77°F (25°C) and 81°F (27°C). Their natural reef environment changes little from day to day, so they are poorly adapted to fluctuations in temperature and water quality. As a result, reef fish are far less tolerant of change than their freshwater cousins: creating and maintaining a constant environment is the key challenge when keeping such species in a home aquarium.

Unless you live in a particularly warm climate, you will need a heater to maintain the water temperature in your tank. Standard heaters contain a glass-encased heating element and have an integral thermostat, which switches the unit on and off to maintain a preset temperature. These heating units are available in a range of sizes and power ratings (wattages). You should allow a rating of about 1 watt per ¼ gal (1 liter) of water, or up to 2 watts per ¼ gallon (1 liter) if the ambient temperature is particularly low. Many aquarists prefer to use two slightly underpowered heaters to keep the water at the desired temperature, rather than a single, more powerful unit. The principle is that if one heater fails, the other will be able to keep the temperature at a reasonable level; and conversely, if one unit fails to switch off, it will be insufficiently powerful to overheat the tank.

Currents created on the surface of the marine aquarium (above) by a power filter outlet replicate the motion of waves over a reef (left), which helps to keep the water well oxygenated.

HEAT DISTRIBUTION

The heater should be fixed where water can flow readily around it and so distribute warmed water around the tank; avoid parts of the tank cluttered by rockwork and other fixtures. The sensible aquarist is always skeptical about thermostat settings on heater units—even the best units can become unreliable, or fail completely, with disastrous consequences. To guard against this, fit the tank with a separate thermometer *(see p.33)*, and check the temperature regularly.

The heater unit is held in place with suction caps. A small light on the unit shows when it is operating.

Heater safety

The heater should be installed after all the rockwork and tank decor are in place—this will minimize the risk of accidental damage to its outer glass casing. It is also a good idea to fit a heater guard—a ventilated shield around the heater—to prevent any direct contact between fish and the body of the heater, which can cause serious burns. Never switch on the power until the heater is completely submerged, and always turn off the power before placing your hand in the water.

A few species, notably boxfish *(see pp.250–251)*, have extremely sharp teeth and will bite through the plastic casing of electrical cables that carry power to the heater. For these fish, it is safer to use undergravel heating units *(see p.33)*.

Types of filtration

Filtration is needed to rid the tank of toxic wastes, undesirable particles, and other dissolved chemicals. There are many different designs of filters, and their mode of action may be biological, mechanical, or chemical, or a combination of these.

Biological filters remove nitrite and ammonia from the water *(see p.34 for a full description of the nitrogen cycle)*. They work by providing a home

FILTRATION CHOICES

Two or more filters are often used in the same tank to maximize water quality and eliminate ammonia and nitrite, which reef fish cannot tolerate. Biological filters, such as undergravel and trickle designs (right), are often teamed with external power filters, which pump water through an external canister containing filter media such as sponge, filter wool, or activated carbon. The filtered water is then sprayed back into the tank through the fine holes of a spray bar—a process that helps oxygenate the water. Some aquariums feature ozonizers—units that produce bubbles of ozone gas to oxidize waste matter—or protein skimmers (below), which use yet another method to remove potentially harmful organic waste.

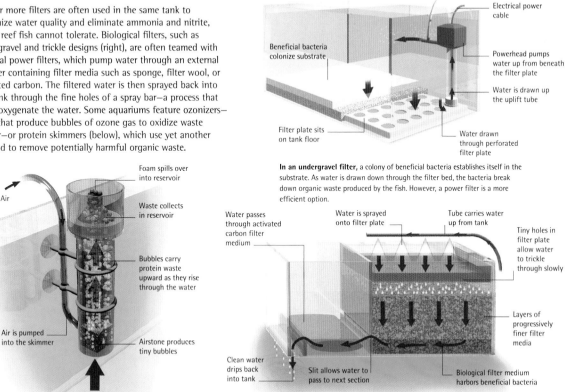

Air enters to aerate water

Electrical power cable

Beneficial bacteria colonize substrate

Powerhead pumps water up from beneath the filter plate

Water is drawn up the uplift tube

Filter plate sits on tank floor

Water drawn through perforated filter plate

In an undergravel filter, a colony of beneficial bacteria establishes itself in the substrate. As water is drawn down through the filter bed, the bacteria break down organic waste produced by the fish. However, a power filter is a more efficient option.

Foam spills over into reservoir

Air

Waste collects in reservoir

Bubbles carry protein waste upward as they rise through the water

Air is pumped into the skimmer

Airstone produces tiny bubbles

The protein skimmer works in a different way from conventional filters. A stream of electrically charged air bubbles rises through a plastic tube; proteins and other organic wastes stick to the bubbles and rise to the surface, where they form a thick foam. This must be regularly collected for disposal, preferably twice a week.

Water passes through activated carbon filter medium

Water is sprayed onto filter plate

Tube carries water up from tank

Tiny holes in filter plate allow water to trickle through slowly

Layers of progressively finer filter media

Clean water drips back into tank

Slit allows water to pass to next section

Biological filter medium harbors beneficial bacteria

The trickle filter provides sophisticated biological and mechanical filtration. Water is drawn up from the tank and sprayed over a stack of different filter media, through which it trickles before flowing back into the aquarium. Spraying also oxygenates the water, improving bacterial action within the filter.

for bacteria that convert these natural waste products into harmless compounds. Mechanical filters remove particles by forcing water through some kind of filter cartridge. Some of these cartridges contain filter media that trap particles as small as 3 microns across and can be used periodically to scrub the water of bacteria and algal blooms. Chemical filters remove dissolved substances from the water, such as ozone, chlorine, heavy metals, and medications. Most work by forcing the water through a filter medium of activated carbon (a manufactured form of carbon that is highly porous). Chemical filters are useful for eliminating the yellow coloring that often develops in aquarium water.

WATER STERILIZATION

Ultraviolet (UV) light is a powerful sterilizing agent, capable of killing bacteria, parasites, and even tough algal spores. Some aquarists use sterilizing units that pass water from the filter over a UV lamp before returning it to the tank. There is some evidence that use of these lamps reduces the incidence of disease.

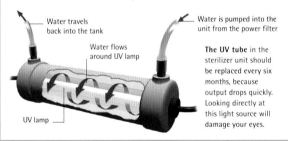

Water travels back into the tank

Water flows around UV lamp

Water is pumped into the unit from the power filter

The UV tube in the sterilizer unit should be replaced every six months, because output drops quickly. Looking directly at this light source will damage your eyes.

UV lamp

Selecting the decor

It is possible to keep marine fish in a bare tank. However, even the simplest fish-only setup will benefit from judicious decoration, which will accentuate the colors of the fish, provide them with places to shelter and spawn, and set out territorial spaces in the tank. In more advanced reef aquariums, the choice of decor is vital, because it transforms the tank into a dynamic ecosystem capable of supporting a diversity of life.

There is a huge range of tank decor available to the home aquarist. Real and artificial rocks, corals and sea fans, shells, amphoras, and even plastic novelty items such as shipwrecks and sharks can be used to provide three-dimensional interest in the tank. Decoration is a matter of taste, but it is vital to consider the welfare of the fish when making a selection. Marine organisms are sensitive to pollutants; even traces of metal, especially copper, may be toxic. Always buy tank decor from a reputable marine dealer and never be tempted to use household objects or items intended for a freshwater setup (such as bogwood). Do not overwhelm the aquarium; every item you place in the tank displaces valuable water. This reduces available swimming space and concentrates dissolved waste.

SAFETY CONCERNS

● Keep rock structures simple and stable. Collapsing rocks can crack the glass of the tank. Use large, flat-faced rocks for the foundations.

● Wear gloves when handling dead corals and artificial rocks, such as lava rock. Some have razor-sharp edges that can inflict cuts.

Porous rocks, such as tufa, are preferred for the marine aquarium because they displace less water than solid rocks. Smooth rocks, such as slate, are used as spawning surfaces by some species, while artificial corals and sea fans add visual interest.

BUYING LIVE ROCK

Live rock is available from aquarium suppliers. It is usually shipped in plastic bags so that it stays moist, which keeps its complement of attached living organisms alive. Most suppliers hold the live rock in tanks for a period to "cure" it before putting it on sale. The curing process involves repeated, careful cleaning of the rock. This type of live rock, which is described as "seeded," can be introduced into the aquarium without further treatment.

Live rock from the Red Sea and the Caribbean is generally preferred to that from the Indo-Pacific origins, because it supports a wider diversity of life.

Rocks and corals

The basic component of a naturalistic marine tank is rock. Calcium-rich rocks, such as tufa, are ideal because, like crushed coral *(see p.209)*, they have a buffering effect and help to control water acidity. Tufa has the additional advantage of being soft enough to carve into almost any shape. It also provides an excellent surface for colonization by marine invertebrates and algae. In fact, it can be difficult to distinguish well-colonized tufa from live rock *(see opposite)*.

Rock arrangements should ideally include some overhangs and bridges where the fish can shelter in semi-natural safety; the rocks should not be piled in solid walls, or against the sides of the tank, because this will impede water circulation, which is essential for effective oxygenation. Some aquarists like to add dead coral skeletons and shells to the tank; these can provide useful hiding places for smaller fish and

Artificial sea fan — Artificial table coral — Slate — Petrified wood — Tufa rock — Calcium-rich ocean rock

invertebrates but are soon colonized and discolored by algae, which are hard to remove. Dead coral skeletons and shells carry toxins or even undesirable dormant organisms, so they should be sterilized by boiling for at least 30 minutes and then cooled before introduction to the tank. Many hobbyists prefer synthetic corals, which look far more convincing and natural than dead skeletons, and are free from these problems.

Live rock

The most beneficial of all tank decorations is live rock, which is made up of the compacted calcium-rich skeletons of long-dead corals and other hard-shelled marine animals. Live rock naturally accumulates in areas adjacent to coral reefs, where its porous structure provides a home for diverse bacteria, invertebrates, and algae. When placed in an aquarium, live rock brings with it its population of beneficial organisms, which can significantly improve water quality in the tank. With its many pores and pits colonized by microbes, the rock serves as a highly efficient biological filter, removing waste products, such as ammonia and nitrite, from the solution; algae and photosynthetic corals on the rock also take up the resulting nitrates. Indeed, live rock is so good at cleaning the water that, when used in conjunction with a protein skimming device *(see p.211)*, it can eliminate the need for more complex filtration systems. Using live rock as the basis of filtration in a marine tank is known as the Berlin method.

Live rock has other benefits, too: it provides an ideal substrate for larger algae and invertebrates, such as anemones and sponges, and it is a living pantry for fish, which can browse on the teeming life it houses. Live rock benefits the simplest fish-only marine setup but is almost essential for the larger reef tank, where it provides the basis for any re-creation of the reef habitat.

A reef aquarium is never entirely stable because it depends on dynamic interactions between many life forms. Watching its evolution over the years is a large part of its appeal.

DECORATIVE ALGAE

Microscopic algae will naturally colonize a marine aquarium. Their excessive growth is undesirable *(see p.220)*, but in limited numbers they help to maintain diversity in the tank and provide additional sources of food for reef fish. Larger algae, or macroalgae, by contrast, are usually bought specially to decorate the aquarium. There is a great diversity of colors and growth habits from which to choose, a small selection of which is shown below. When buying, select specimens that are firm, well colored, and, if possible, attached to substrate (such as live rock).

Bladed Sand Moss (*Caulerpa prolifera*) is a popular and fast-growing green alga.

Fan Weed (*Avrainvillea* sp.) has a fan-shaped head held above the sand by a central stalk.

Grape Caulerpa (*Caulerpa racemosa*) resembles small bunches of grapes.

Flagweed (*Halymenia floresia*) is a prized, slow-growing red alga with ragged margins.

Turtleweed (*Chlorodesmis fastigiata*) grows in emerald-green tufts. It is browsed by tangs.

Shaving Brush (*Penicillus capitatus*) has a smooth stalk topped with a brushlike head.

Preparing the tank

Assembling the components of the marine aquarium is straightforward, but there are a few important factors, such as the quality of your water supply, that must be considered if you are to achieve a successful result. Patience is vital; even under ideal conditions, it can take several months for conditions in the tank to become stable enough for the most delicate marine species.

When setting up a marine tank, care must be taken not only to create the desired visual effect but also to consider the varied needs of the marine species you wish to keep. Once the substrate has been prepared *(see pp.208–209)*, the next step is to arrange tank decor *(see also pp.212–213)*, taking great care to ensure that it is clean and firmly supported in the aquarium. If necessary, pieces can be held together with silicone sealant, to make them more secure. Arrange the rockwork to contain niches and crevices for the fish to use as retreats. Position rockwork toward the back of the aquarium, leaving a clear

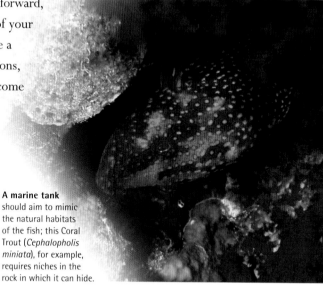

A marine tank should aim to mimic the natural habitats of the fish; this Coral Trout (*Cephalopholis miniata*), for example, requires niches in the rock in which it can hide.

swimming area at the front—this will make the fish and other marine life more visible. Live rock *(see p.213)* should be added only once the tank is full and the water conditions have stabilized *(see p.216)*.

Heating and filtration components *(see pp.210–211)* can then be fitted around the rocks. When making your arrangement, be careful not to create "dead spots"—areas where current from the filter does not reach—because uneaten food and debris can accumulate here, which will lead to a gradual deterioration in water quality.

CALCULATING THE VOLUME OF YOUR TANK

To calculate the volume of a rectangular tank, multiply the dimensions of the tank in inches (height x width x depth), then multiply the result by 0.0043 to get the volume of the tank in gallons. If measuring in centimeters, again, multiply the dimensions together, but divide the result by 1,000, to give the volume in liters. Whichever method you use, the final figure must then be reduced by 10 percent, to take account of the rockwork and other decor.

SETTING UP THE TANK

Special commercial salt mixes make it easy to create the necessary water conditions for a marine tank. It is best to mix the water with the salt before pouring it into the tank so that you can be sure that it is thoroughly dissolved. A second option is to add the salt to a prescribed volume of water in the tank, allowing the filtration system to mix the solution, but it can be harder to be sure that it has dissolved fully.

❶ Measure the marine salt
Read the instructions on the packaging of the marine salt carefully before you start. Using a measuring cup, pour out the appropriate amount of salt.

❷ Prepare the saltwater solution
Measure out the required volume of water in a watering can or bucket, and treat it with a dechlorinator before slowly stirring in the salt with a wooden spoon.

ADDING SESSILE INVERTEBRATES

Before placing sessile invertebrates, such as anemones and corals, in the tank, make sure the rockwork is securely positioned; the delicate bodies of these animals are easily damaged. Position them in a well-lit part of the aquarium and relatively close to a powerhead, where there is a good flow of water that will waft food to them and carry away their waste. Sponges prefer a more shady area in the aquarium. Always transfer these creatures in water to minimize the risk of structural damage, which can lead to bacterial infection.

Gloves guard against stinging tentacles

Allow a gap equivalent to at least their own width between anemones and corals when placing them in the aquarium. If they are too close together, they are likely to sting each other with their extended tentacles.

Water quality

Marine tanks must be filled with a specially prepared salt-and-water mixture (see below). Never use water from the hot-water system in your home; it may be contaminated with copper from the pipework, which can be deadly to invertebrates. Even if using water from the cold-water system, it is always a valuable precaution to test for copper, using a suitable test kit. A further potential problem in some areas is the high level of nitrates present in the domestic water supply. This typically occurs in agricultural areas, where nitrate fertilizers leach through the soil and contaminate the water supply. Test your supply for nitrates (see p.221), or contact your water company for information. There are various options available for removing nitrates; the simplest is to run the tap water through a special nitrate filter. A reverse osmosis (RO) unit is a more expensive option that removes not only nitrate but also other pollutants, including phosphates. Finally, the water must be treated with a dechlorinator or water conditioner, which neutralizes chlorine and chloramine.

Marine salt is available from aquarium stores and comes with detailed instructions on how to make up a saltwater solution of the correct salinity. It also contains all the key ingredients, notably calcium and magnesium, to ensure that the solution is sufficiently hard, and to enhance its buffering capacity. Salinity is measured on the specific gravity (SG) scale. This compares the density of the salt solution to pure water; the more concentrated the solution, the more dense it is relative to water and the higher the SG reading. Specific gravity can be measured using a hydrometer; the SG reading is taken from the floating hydrometer's position at the surface of the water.

Once the tank has been filled (see below), the system is ready to be switched on. The thermostatic heater will raise the water temperature gradually to the preset figure.

❸ Build up the decor
Add rockwork and other basic decor to the dry tank. First create a secure base with larger rocks, and then build up the rockwork, leaving plenty of holes and crevices.

❹ Fill the tank
Place a clean saucer on the substrate and then carefully pour the salt and water solution onto it, to avoid disrupting the base layers of substrate in the tank.

❺ Measure salinity
Check the salinity of the water using an instrument that measures specific gravity, such as this hydrometer.

The maturation process

Coral reefs are very stable environments, and the fish living there are not adapted to significant shifts in water parameters such as temperature, salinity, or water chemistry. In the aquarium, therefore, conditions need to be stabilized before the tank can be stocked with more delicate fish and invertebrates. Before the biological filter is fully functional, levels of ammonia and nitrites can rise to dangerous levels *(see p.222)*. One way to speed up the maturation process is to introduce hardier species, such as damselfish *(see pp.240–243)*, into the new tank; these fish can endure the fluctuating water quality, and the waste they produce encourages populations of beneficial bacteria to develop within the biological filter. In addition, cultures of beneficial bacteria are available that can be added to the tank. Regular testing *(see pp.220–222)* will reveal when the water conditions have stabilized; at this point, ammonia and nitrite levels should be virtually undetectable.

Stocking the tank

A vast range of colorful and interesting species are available today from aquarist suppliers, but it is important to take time to plan and research the numbers and species of fish that are appropriate for your particular setup before you make any purchases. A typical rectangular aquarium can support 1 in (2.5 cm) of fish per 4 gallons (15 liters) of water in the first six months, increasing to 1 in (2.5 cm) of fish per 2 gallons (7.5 liters) thereafter. If you introduce very fast-growing species into the aquarium, however, their eventual size must be taken into account when calculating the stocking density, in order to avoid overstocking problems at a later date.

Make sure the fish you choose are compatible with one another; if you are planning to create a reef tank, check that they will not harm invertebrates. Fish to be introduced should ideally first be quarantined in a separate tank for up to two weeks, to allow any signs of illness to become apparent.

INTRODUCING THE FISH

Wait for a few days after setting up the tank before obtaining and introducing the first few hardy fish, to be sure that the system is functioning properly. When choosing a fish for your tank, always ask the supplier to let you examine it closely; carefully inspect both sides of its body for any signs of illness or injury. Also ask to see the fish feeding, because this is a good guide to its general state of health. The supplier will catch your chosen fish and transfer it to a plastic bag; it should be kept here for the minimum possible time before introducing it to the tank (right).

❶ **Equalize water temperatures**
Float the bag in the aquarium for about 15 minutes. This allows the temperature in the bag to slowly rise to match that in the tank, thus minimizing the stress on the fish.

❷ **Catch the fish**
Net the fish inside the plastic bag, being careful that the water in the bag, which may contain medication or harmful microbes, does not spill into the tank.

❸ **Release the fish**
Allow the fish to swim out of the net and into the tank. Newly introduced fish will often hide away at first, retreating into crevices in the rockwork.

Finished tank with damselfish

Damselfish make ideal first occupants of a new tank

MAINTENANCE

Food and feeding

Providing your fish with a balanced, healthy diet demands a little research. It is vital to establish not only the dietary preferences of each species but also the way in which it feeds. Some, for example, feed only on floating foods, while others will feed exclusively from the substrate. Today, there are commercially prepared marine foods that suit all feeding styles and meet the requirements of almost every species.

Marine fish naturally seek food at different levels in the water. In the surroundings of the coral reef, this prevents direct competition for food from other fish and allows each fish to occupy its own space without conflict. Some species are confident and will feed in open water or on the seabed. Others eat algae from the rocks or hide within crevasses waiting for an opportunity to emerge. Herbivorous fish spend much of their time browsing to meet their nutritional needs, while predatory species, such as eels, may not feed every

FEEDING INVERTEBRATES

SESSILE FEEDERS
Some invertebrates, such as the Flowerpot Coral shown here, spend their lives attached to rock and cannot move to reach food. Consequently, they must be offered food in their immediate vicinity.

MOBILE FEEDERS

Invertebrates that can move to seek food will eat a varied diet, and even scavenge on food provided for aquarium fish. The starfish pictured here has engulfed a pile of mussels.

day. These diverse feeding strategies are also evident in the aquarium and must be addressed if the fish are to thrive. Suppliers of aquarium fish are able to offer advice on the optimum feeding strategies for each species.

Sources of food
Some reef fish are opportunistic and eat a variety of foods in the wild. This is useful to the aquarist because these species can also be persuaded to eat a varied diet in the tank. However, their diet should be underpinned by a commercially prepared staple food, which will ensure that they get the correct mix of vitamins, minerals, proteins, and carbohydrates that is

Sea horses are specialist feeders and need to be provided with a regular supply of live brine shrimp larvae (shown here) in aquarium surroundings.

essential for good health. There are varieties of commercially prepared foods that float, sink, or even stick to the sides of the tank, to suit the feeding styles of different species. Store the food in sealed containers to keep it dry and fresh. It should not be used after the manufacturer's expiration date, after which the vitamin content declines rapidly. A range of freeze-dried, frozen, and vegetarian marine fish food is also available. Frozen foods, such as shellfish, are subjected to gamma irradiation, which makes them free from the pathogens that may be carried in fresh foods.

Feeding time provides an opportunity to monitor the well-being of the fish; if any loss of appetite is detected, it may be an early indicator of illness. It is important not to overfeed, because uneaten food may pollute the water and can poison the fish. Excess food should be removed from the tank as soon as the inhabitants are finished eating.

The new tank

Not all marine fish will feed readily after being moved to a new aquarium. If you are able to, observe your chosen fish feeding prior to purchase; if they feed readily, you can be reasonably confident that they will settle down and regain their appetite in a day or two, especially if provided with familiar food to encourage their appetites. If they ignore commercially prepared foods, try to disguise them in fresh foods. For predatory species, place small pellets in the mouths of frozen fish, such as lancefish, or insert them into the bodies of krill. Once the fish have gained an appetite for commercially prepared foods, they will take them independently. When choosing fish for a new tank, avoid vegetarian species, because at this early stage, there is little algal growth in the tank for them to browse. Acclimatization of such species is likely to be easier in a mature tank, once algal growth is established.

Herbivorous fish, like this Yellow Sailfin Tang *(see p.238),* will eat some leaf vegetables. Blanch the leaf to aid digestion then secure it in the tank using a plastic clip or weight. This makes it easier for the fish to eat and for leftover food to be removed.

TYPES OF FOOD

There are four main types of marine food available: dry food with added nutrients; freeze-dried and frozen live foods, such as fish and crustaceans; and vegetable or vegetable-based products.

DRY FOOD

Tablets are suitable for most marine fish. They can be stuck to the side of the tank, to allow fish to nibble them, or dropped to the bottom.

Pellets come in two varieties: those that float, for surface-feeding fish, and those that sink, for fish that dwell at lower levels.

Flakes, which float and then sink, are ideal for midtank fish. Varieties are available for carnivorous, vegetarian, and omnivorous fish.

FREEZE-DRIED LIVE FOODS

Krill feature in the natural diet of many marine creatures. Suitable for larger fish, they are high in protein and beta-carotene.

Bloodworms are high in protein. Their small size means they will be eaten not only by fish but also by some invertebrates.

Brine shrimp are suitable for smaller fish, including young fry. They provide high levels of nutrients and essential fatty acids.

FROZEN FOOD

Frozen krill has a higher moisture content than the freeze-dried form. In comparison, this makes it more palatable for the fish.

Cockles feature in the diet of many species in the wild. As with other frozen foods, only thaw the required quantity for each feed.

Lancefish are ideal for larger predatory marine fish. All frozen food must be completely defrosted before it is placed in the tank.

VEGETARIAN FOOD

Dried green algae supplement the natural diet of herbivorous fish, especially in new tanks where algal growth is not well established.

Vegetable wafers are a commercially prepared food supplemented with vitamins. They are ideal for bottom-feeding species that eat algae.

Peas are a good source of vitamins and fiber. They should be shelled and (if frozen) defrosted before they are given to the fish.

Monitoring and adjusting

Although the upkeep of a marine aquarium—especially a reef tank—is more demanding than a freshwater system, a few routine maintenance tasks and the judicious use of test kits to check water quality will ensure a healthy environment for the tank occupants. Watch for signs of algal overgrowth, and carefully observe the fish, since abnormal behavior may be an early indicator of deteriorating conditions.

Algae are key ingredients of reef aquariums, but if they become rampant, they look unsightly and may smother corals, causing them to die. While you can physically remove the algal overgrowth, you should also address the underlying cause of the problem, which could be excessive or inadequate lighting, or high levels of nitrate and phosphate.

Water chemistry

Monitor pH closely, and regularly check the water's buffering capacity—its ability to resist a change in its pH. Buffering depends largely on the concentration of carbonate in the tank water, which neutralizes any acidifying substances present.

Herbivorous fish, such the Red Sea Clown Surgeon (above, foreground), help to keep algal growth in check. A healthy reef aquarium contains a range of algae (left); if one type predominates, it indicates that conditions in the tank are not ideal.

REGULAR MAINTENANCE TASKS

DAILY

● Feed the aquarium occupants in the morning and evening, as required, taking care not to overfeed them.

● Watch the fish feed, because a loss of appetite may be a sign of illness or declining water conditions.

● Check the water temperature, as shown by the thermometer. Any fluctuation suggests a heater malfunction.

● Be sure that you actually see the fish every day. A sudden, undetected death will have a serious impact on water quality.

WEEKLY

● Carry out water tests, recording the results to create an ongoing record of the conditions in the tank.

● Add buffering solutions and calcium or trace element supplements as required, based on the results of the water tests.

● Top off the aquarium with dechlorinated tap water, to replace evaporative loss.

● Clean the sides of the tank to remove any algal growth, using a magnetic or long-handled cleaner.

FORTNIGHTLY

● Carry out a partial water change of about 20 percent of the aquarium volume, using a gravel cleaner to remove mulm from the substrate at the same time.

● Keep an eye on the specific gravity reading and other test parameters. Review figures recorded previously.

● Service filters, rinsing sponge components in water siphoned from the tank. Clean the protein skimmer, and make sure that airstones are not blocked.

● Adjust the level of lighting if algal growth is starting to get out of control.

CHANGING THE WATER

Partial water changes not only reduce harmful accumulations of nitrate, phosphate, and other chemicals by dilution but also replenish levels of carbonate (reinforcing the buffering capacity) and trace elements, which are vital to the well-being of the tank occupants. When setting up the aquarium, make an inconspicuous mark on the side of the tank with a felt-tip pen to show the water level when the tank is full. This makes it easier to fill up the tank with the correct amount of water, both when making partial water changes and when replacing evaporated water.

❶ Check the salinity and temperature
A conductivity meter gives readings in millisiemens per centimeter (mS/cm). At 77°F (25°C), 50.1 mS/cm corresponds to an SG reading on a hydrometer of 1.023.

❷ Drain the water and clean the gravel
Fix a gravel cleaner to the siphon and suck up mulm from the substrate while draining the water. This will prevent the undergravel filter from becoming clogged with waste.

❸ Add more water
Replace the drained water with a fresh, dechlorinated salt solution of the correct temperature and salinity. Test the water for toxic copper before adding it to the tank.

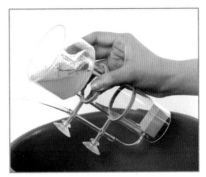

❹ Clean out the protein skimmer
Carefully remove the accumulated debris from the cup. Then rinse the cup with warm, dechlorinated water to remove fat deposits, which make the skimmer less efficient.

CHECKING SALINITY

The correct salinity, in terms of specific gravity (SG), will be in the range of SG 1.020–1.025, depending on the species in your tank. Salinity can be tested with a hydrometer *(see p.215)* or a conductivity meter, which determines the water's salt content from its ability to conduct electricity. With a hydrometer, you may need to adjust the reading to take account of the water temperature: cold water is denser than warm water, so it gives a slightly lower SG reading. The instructions provided with the hydrometer should enable you to make the right adjustments.

TESTING THE WATER

Tank samples can be tested with reagents to monitor a range of water parameters, including pH and levels of chemicals such as iron, nitrate, phosphate, carbonate, calcium, strontium, iodine, and copper. Read the instructions on the kits carefully, store them appropriately, and use them before they are out of date; otherwise, they will give inaccurate readings that may endanger the health of both fish and invertebrates. Electronic meters give more accurate results for many of these parameters, but they are far more expensive.

Test kits use reagents that cause the water sample to change color. The sample is then compared to a color chart that gives the numerical figure.

The calcareous substrate bolsters buffering because it contains calcium carbonate, which gradually dissolves and replenishes the water's carbonate content. The typical pH range is 8.0–8.3, but pH inevitably falls as carbonate is used up. A carbonate hardness test (also called an alkalinity test) measures the level of carbonate in milliequivalents per liter (meq/l). If the result is outside of the range of 3.0–3.5 meq/l, take remedial action; partial changes of gravel and water help to restore the buffering capacity, as does the addition of a commercial buffering solution.

In an established tank with good filtration, there should be no ammonia or nitrite. The nitrite level increases slightly if you add new fish or invertebrates, reflecting extra waste output, but it soon falls as the nitrogen cycle *(see p.34)* converts nitrite into nitrate. Use partial water changes to reduce nitrate levels—which should be close to zero and never above 20 milligrams per liter (mg/l)—preferably in conjunction with a protein skimmer *(see p.211)* to remove waste before it decomposes.

Heat from the lighting causes evaporation from the water's surface

Before you add replacement water, make sure it is at the same temperature as the water in the tank

Evaporation of tank water is the main cause of changes in salinity. As water evaporates, the concentration of salt in the tank (and thus the SG figure) rises. To replace lost water and restore the correct salinity, top off the tank with fresh, dechlorinated water; do not use salt solution, which will make the problem worse.

You should also test the concentrations of trace elements, especially in a reef tank. Calcium, strontium, iron, and iodine are vital for a healthy reef community. The ideal levels are 400–475 mg/l for calcium, 8 mg/l for strontium, 0.05 mg/l for iron, and 0.5 mg/l for iodine. You can correct these levels by adding commercial aquarium preparations. Phosphate, too, is essential, but if it exceeds 0.01 mg/l, it can lead to a proliferation of unwanted hair algae (*Derbesia* sp.).

Redox meters

An excellent way of monitoring the general health of the tank is to install a redox meter, which measures the water's oxidation-reduction potential—that is, the ease with which chemical reactions occur in the water. The reading, in millivolts (mV), should ideally lie within the range of 320–380 mV, indicating

A collection of marine fish, such as the damsels, angelfish, and tangs in this tank, can only be built up safely over the course of many weeks.

that the nitrogen cycle is working well, the water is relatively pure, and oxygenation is good. The redox potential naturally fluctuates over any 24-hour period, due to the biological processes taking place in the tank, so take readings at the same time each day. A sharp decline may signify that the airstone is blocked. Live rock and some types of algae help to raise the redox potential; however, if the redox figure exceeds 400 mV, which can happen if your ozonizer *(see p.211)* is too large for the aquarium, there may be fatalities among the tank occupants.

NEW TANK SYNDROME

Ammonia and nitrite can rise to dangerous levels in a new tank, before the colonies of beneficial bacteria that break down these toxic waste products have developed in the biological filter. Use test kits to take weekly readings of ammonia and nitrite in a new tank to monitor the progress of this maturation process. Some hardy species, notably damselfish *(see pp.240–243)*, can be introduced at this time, but most marine species should be added to the aquarium only when the system has stabilized.

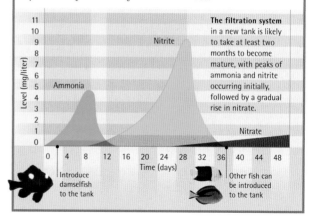

The filtration system in a new tank is likely to take at least two months to become mature, with peaks of ammonia and nitrite occurring initially, followed by a gradual rise in nitrate.

Introduce damselfish to the tank

Other fish can be introduced to the tank

VACATION ISSUES

● Try not to be away when a tank is in the early stages of maturing or immediately after adding new occupants to an existing setup.

● Carry out a partial water change a few days before you leave, to ensure all is well.

● Leave very clear feeding instructions, in writing.

● Be sure to leave sufficient food and a replacement lighting tube or bulb.

MARINE FISH

ILLNESS AND TREATMENT

Health concerns

Most marine aquarium fish are natural inhabitants of coral reefs, which are among the most stable and unchanging of all ecosystems on the planet. Many of the health problems suffered by marine aquarium fish therefore stem from fluctuations in water quality in the home tank. Maintaining suitable water conditions will keep the fish in the best condition to repel pathogens and parasites.

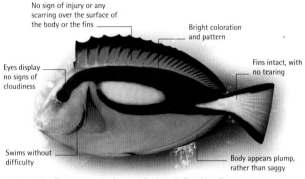

No sign of injury or any scarring over the surface of the body or the fins

Bright coloration and pattern

Eyes display no signs of cloudiness

Fins intact, with no tearing

Swims without difficulty

Body appears plump, rather than saggy

A healthy fish appears active and feeds well. This Blue Tang (*Paracanthurus hepatus*) demonstrates what to look for in a well-kept specimen.

Overgrown teeth make eating difficult, leading to a loss of condition

Appropriate food is essential for good health. Pufferfish deprived of their natural, hard-shelled invertebrate foods soon develop long, overgrown teeth.

Isolation and establishment

A long-established, well-maintained marine tank rarely succumbs to disease. Most health problems experienced by marine fish are seen when a tank is immature, or when new animals or plants are introduced, along with unwanted pathogens. Marine species are vulnerable to a range of bacterial, viral, fungal, and especially parasitic diseases, which spread quickly if fish are kept in suboptimal conditions—for example, during transportation. For this reason, newly acquired fish should always be transferred to a separate tank for two weeks before entering the main tank. This isolation tank should be set up and run from the outset in parallel with the main tank. It need not be large, and tank decor should be kept to a minimum for ease of cleaning. The fish in the isolation tank should be fed well and encouraged to take foods such as marine flake, which contain all the key vitamins and minerals. When you are satisfied that the fish are well nourished and free from disease, they can be transferred carefully to the main tank.

Health problems can also occur when changes take place in the environment (*see opposite*). In a reef tank, for example, corals rapidly take up elements such as calcium, which they incorporate into their hard skeletons. The depletion of calcium affects the health not only of corals but also of other tank inhabitants, so regular monitoring and partial water changes help to maintain a healthy environment.

Choosing fish

Most problems can be avoided by careful selection at the outset. Avoid individuals with any signs of ill health; juveniles are preferable to mature individuals, which are hard to age and therefore may be very old and more susceptible to illness. Young fish acclimatize better to aquarium life and are more likely to be compatible with their tankmates.

UNWANTED GUESTS

Many organisms "hitch a ride" into your tank when you acquire new fish, rocks, invertebrates, or algae. Some of these unwanted guests can be seen with the naked eye; most will die or be eaten by your fish, but some can be a nuisance and should be removed.

MANTIS SHRIMP
These predatory crustaceans will attack desirable invertebrates and fish in the tank. They have claws capable of cracking open the body casing of crabs.

BRISTLEWORMS
Similar in appearance to caterpillars, these worms will attack mollusks in the tank. Take care when removing—these worms have sharp spines.

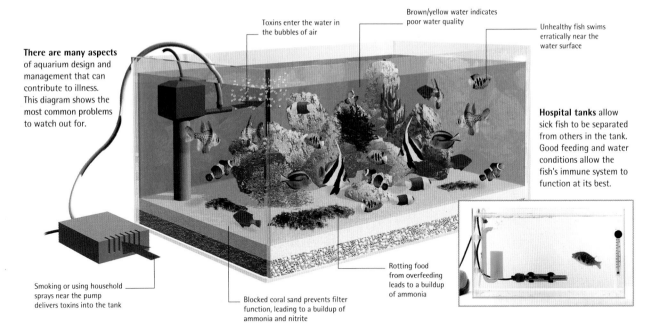

There are many aspects of aquarium design and management that can contribute to illness. This diagram shows the most common problems to watch out for.

Toxins enter the water in the bubbles of air

Brown/yellow water indicates poor water quality

Unhealthy fish swims erratically near the water surface

Hospital tanks allow sick fish to be separated from others in the tank. Good feeding and water conditions allow the fish's immune system to function at its best.

Smoking or using household sprays near the pump delivers toxins into the tank

Blocked coral sand prevents filter function, leading to a buildup of ammonia and nitrite

Rotting food from overfeeding leads to a buildup of ammonia

INVERTEBRATE HEALTH

Bacterial, viral, and parasitic diseases can affect tank invertebrates, but home diagnosis and treatment of these animals is not highly developed. More often, health problems arise from environmental shortcomings, such as a lack of suitable lighting above the aquarium or lack of compatibility between tankmates. Even apparently sessile invertebrates, such as corals and anemones, must be well spaced so there is no risk of neighbors coming into contact and stinging each other to death.

A number of marine invertebrates can suffer from light starvation. Powerful illumination is needed by those which contain beneficial algae, and a strong water current is necessary to waft food within reach and remove waste.

Treatment and medication

Drugs are available to treat many conditions of marine fish; a good retailer will guide you to the most suitable. Treatments should be carried out in an isolation tank or bath, rather than in the main aquarium itself; this is because the remedies may contain copper or other elements that are toxic to invertebrates. Before moving the fish back into the main tank, remove any copper by using carbon in the filtration system of the isolation tank; check the water with a copper test kit.

Water treatment

Some conditions of marine fish, especially parasitic infestations, can be treated rapidly and effectively by exposure to fresh water, with no medication necessary. These treatments work by upsetting the internal balance of water and body salts within parasites—a process called osmotic shock. Bear in mind that they will affect the patient in a similar way so must be carried out very carefully to minimize stresses.

Treatment should be administered in a basic acrylic tank filled with dechlorinated tap water and heated to the same temperature as the water in the main aquarium. The pH of the tank should be raised to the appropriate level by adding one teaspoonful of sodium bicarbonate to 1 gallon (4 liters) of water. All water parameters should be checked before introducing the sick fish.

Do not leave the fish during treatment, and monitor and watch its behavior closely. It is normal for marine fish to lie on their sides and breathe heavily when being treated, but if you think the fish is in distress, return it to the saltwater tank without delay. As a general rule, fish should be immersed for no longer than five minutes for successful treatment.

DANGEROUS FISH

● Take care with lionfish (see pp.264–265), which are armed with poisonous spines.

● Fish such as eels may bite your fingers if fed by hand.

● Sharp projections on the bodies of surgeonfish can inflict a painful cut if they brush against your hand.

● Some marine fish can bite through electrical cables. Avoid danger by using reinforced cables.

DIAGNOSIS OF COMMON PROBLEMS

Watching fish in your home aquarium every day gives you an instinctive awareness of developing health problems. Assessing the fitness of an unfamiliar fish—when browsing in a store, for example—is more difficult. In general, active fish that show good appetites are likely to be reasonably healthy. The primary causes of illness, especially in marine species, are just as likely to be environmental as pathogenic, so always check the water parameters carefully.

The tables on the following pages will help you diagnose the most common conditions affecting marine species and point you toward appropriate treatments. First, in the table below, identify the part of the fish's body that appears to be affected. Next, find the signs of illness that most closely match those displayed by the fish. Note the number(s) associated with the relevant signs, and refer to pages 227 and 228 for a fuller description of the possible conditions and their treatment.

SIGNS OF ILLNESS

EYES

- Eye missing from the socket, with no other symptoms or abnormalities evident ⑮
- Bulging eye or eyes, together with swollen belly and raised scales ⑭
- Eyes become cloudy in appearance ① ② ⑥ ⑨

- Fish persistently rubs its head on rocks or on the walls of the aquarium ⑤
- Small white spots evident on the eyes and possibly also elsewhere on the body ①
- One eye appears white in color, while the other is normal ⑭

SKIN

- Ulceration develops on the surface of the body ⑤ ⑨
- Scales disappear from the body ① ⑤ ⑧ ⑮
- Velvetlike patches on the skin ①
- Gas bubbles visible under the skin ⑭

- Lumps or more distinct cauliflower-like swellings develop on the body ⑫
- Skin starts to slough off the body ⑧
- Skin has a more slimy appearance than normal ① ② ③ ④ ⑤ ⑧ ⑯

FINS

- Fins appear frayed at their edges ⑨
- Fins display prominent tears ⑮
- Fins have evident reddish bases; most apparent in pale-colored fish ⑨ ⑩
- Fin rays exposed; fish loses appetite ⑩

- Pieces missing from the fins; especially evident in larger-finned species ⑮
- Golden-brown spots are evident on the fins, as well as on the body ①
- Fin posture changes, with caudal fin clamped shut. Ability to swim and activity are reduced ⑰ ⑲

COLOR

- The fish darkens in color, accompanied by a range of other symptoms, such as loss of activity ⑤ ⑰
- Blood visible; signs of hemorrhaging into the water from the body ② ③
- Duller than usual, often with tiny golden spots ①

- White spots appear over the surface of the body ① ②
- Dark, blackish spots are evident over the body ④
- Irregular patches indicating loss of color ⑨
- Abnormal patches of color, paler than the surrounding areas. These may enlarge and coalesce ① ⑪ ⑬

SHAPE

- Fish becomes swollen, notably in the vicinity of the belly ⑦ ⑨
- Relatively large, flattened object(s) evident on the flanks of the fish ⑤
- Profile of the fish starts to alter, with its spine becoming abnormally curved ⑪

- Fish develops an emaciated appearance ⑦ ⑩ ⑪ ⑬ ⑳
- Holes start to develop in the vicinity of the head and on the sides of the body ⑤
- Lips of the fish appear swollen, and there may be some loss of appetite ⑲

BEHAVIOR

- One individual is persecuted by one or more other individuals in the tank ⑮
- Fish displays abnormal swimming pattern ⑬ ⑰ ⑱ ⑲
- Gasping at the surface, with the fish sometimes trying to jump out of the tank ⑱ ⑲

- Fish scrapes its body against rockwork ② ④ ⑥ ⑯
- Fish appears to be disoriented and has difficulty maintaining its balance ⑬ ⑰ ⑱
- Respiratory rate alters markedly; the fish often has obvious difficulty breathing ① ② ③ ⑥ ⑨

PARASITES

CONDITION	AT RISK	SYMPTOMS	TREATMENT
① Marine velvet disease	All fish	Rapid gill motion in early stages, because the gills are typically the first site of infestation. About four days later, velvety patches become apparent on the body, and the eyes may become cloudy. The velvety look will spread over the entire body. Advanced cases display small, white spots, as if the fish has been rolled in powdered sugar. Fish may die within two days of initial exposure.	Isolate and treat rapidly with commercial remedy, usually one containing copper sulfate. The free-swimming parasites can be destroyed by using an ultraviolet sterilizer; dropping specific gravity down to around 1.010 will also kill parasites in the water.
② Marine white spot	All fish	White spots, around 1/32 in (1 mm) across, spread over the fish's body and fins. Infected fish typically try to relieve irritation by rubbing themselves on the tank decor. Other symptoms can include clamping of the fins, cloudy eyes, and even hemorrhaging. Caused by the protozoan *Cryptocaryon irritans*.	Isolate and treat with a commercial remedy. Continue treatment for at least a week after the fish appears to have recovered to reduce the likelihood of return outbreaks. Ultraviolet sterilization can help to curb spread of the parasite. Watch out for secondary fungal and bacterial infections.
③ Brooklynellosis	Anemonefish	Typically excessive mucus production, resulting from the protozoan's irritating effect. The excess mucus may make the fish appear duller in color. Other signs often include labored breathing as the gills become damaged and increased loss of color. The gills may hemorrhage and small red spots may be observed. Heavy infections will usually cause the death of the fish.	Use a commercial remedy. The condition responds well to treatments based on formalin or malachite green. Freshwater bath often beneficial. Maintain good water quality and low stress levels to prevent brooklynellosis in aquariums.
④ Black spot (tang turbellarian disease)	All fish, including tangs	Affected fish develop a series of blackish cysts no more than 1/16 in (2 mm) in diameter on the body and fins. The spots are clearly visible on light-colored fish but inconspicuous on darker species. The spots cause irritation, and the fish may rub repeatedly against the tank decor. Other symptoms include reddened skin and listlessness. The cause of this is a flatworm, which remains on the host for about six days and then drops off on to the substrate. Five days later the body wall of the adult worm ruptures, releasing hundreds of young.	A freshwater bath combined with a commercial remedy (usually based on formalin) will destroy these parasites, which are the larval stages of flatworms. Watch for signs of secondary infection.
⑤ Head and lateral line erosion disease	All fish, with tangs most vulnerable	Obvious pitted appearance on the head of affected fish, with erosion of the area over the lateral line. Fish become lethargic, and open wounds provide access for other bacterial and fungal pathogens, which may eventually kill the fish. Also known as hole-in-the-head disease.	Causes of the condition are unclear. Possible factors include poor water quality, high nitrate levels, poor nutrition, or infestation by the protozoan parasite *Octomita necatrix*. If affected, treat rapidly with medicated food. Addition of vitamin C to the diet may help in long-term prevention.
⑥ Flukes	All fish	Symptoms include rapid breathing, in the case of gill flukes, and irritation. Fish suffering from skin flukes will often rub themselves on tank decor. Cloudy eyes and color changes may also indicate infestation. There are many species of flukes, and these parasites are common on imported fish. Microscopic examinations of skin scrapings and gill clippings are required for positive identification of parasite species.	Use a commercial formalin-based treatment. A freshwater bath can help to overcome these parasites rapidly.
⑦ Intestinal worms	All fish	Not easy to identify with certainty, but worm infestation is possible if a recently acquired fish starts to lose weight rapidly or develops a swollen abdomen. The internal gut parasites deprive the fish of nutrition, so affected individuals appear quite lively but tend to display a ravenous appetite.	Specially medicated foodstuffs containing an anthelminthic will overcome these parasites, which are then voided from the fish's body.
⑧ Uronemosis	All fish	Early sign of infection is skin discoloration, leading to ulceration; may be confused with bacterial disease at this stage. Skin sloughs and may also become pitted, distinguishing this from brooklynellosis. Some fish, such as seahorses, may, however, show no external signs, simply developing respiratory complications.	Rapidly fatal if internal organs, such as the kidneys, are affected. Treat the fish in a freshwater bath, or in a medicated bath containing a combination of formalin and malachite green treatments.

BACTERIAL, VIRAL, AND FUNGAL DISEASES

CONDITION	AT RISK	SYMPTOMS	TREATMENT
⑨ Vibriosis	All fish	Variable symptoms, ranging from loss of body color, frayed fins, and cloudy eyes to sudden death. Caused by *Vibrio* bacteria that are naturally present in the gut.	Outbreaks often occur in fish that are in poor condition or that are new introductions. Treatment with an antibiotic bath can be effective.
⑩ Fin rot	All fish	Reddening and erosion at the edges of the fins, which may become ragged and allow entry of secondary infections. Caused by any one of a range of bacteria.	Usually linked with poor water conditions, so carry out a partial water change. Make sure fish are not being overfed. Treat infections with antibiotics.
⑪ Piscine tuberculosis	All fish	Weight loss, pale body color, declining appetite, and, ultimately, scale loss. This condition results from infection by *Mycobacterium*.	Hard to detect at first. No effective treatment. Review diet, because the bacteria may be introduced in fresh shellfish and fish foods.
⑫ Cauliflower disease	All fish	Whitish, raised growths on the body, which can develop a cauliflower-like branched appearance. New arrivals to the tank are most vulnerable. The disease—also known as lymphocystis—is caused by a viral infection.	A short freshwater dip may help. No treatment is available, but the disease is neither highly infectious nor (usually) grave; left alone, the nodules will fall off within about a month. Do not try to remove the nodules, because the rate of reinfection is high.
⑬ Marine fungus (whirling fungus)	All fish	Disorientation (hence the name "whirling fungus"). The fungus attacks body organs and commonly results in loss of color and weight and ulceration.	Treatment is virtually impossible. Separate suspected cases immediately and treat with antifungal agents.
⑭ Pop-eye	All fish	One or both eyes protrude abnormally from their sockets and sometimes turn cloudy. Can result from infection or poor water quality. Also known as exophthalmia.	Try antibacterial treatments and improve water conditions.

ENVIRONMENTAL CONDITIONS AND INJURIES

CONDITION	AT RISK	SYMPTOMS	TREATMENT
⑮ Bullying	All fish	Physical injury, including damaged fins and eyes, and scales missing from the body. Fish may be seen being chased and attacked, although this may occur after dark.	Separate and treat affected individual as required. Watch for opportunistic infections on any damaged parts of the body.
⑯ Excessive mucus production	All fish	Slimy appearance to the body. Fish frequently rubs itself against tank decor. Excessive mucus production indicates skin inflammation.	Try to establish underlying cause and treat accordingly. If no symptoms of parasites are evident, this can be the result of trauma rather than infection.
⑰ Loss of balance	All fish	Affected individual swims at an abnormal angle in the water and is often incapable of swimming in a straight line.	Typically caused by chilling. Add warm water to tank and closely monitor temperature. Can also result from swim bladder disorders, for which little can be done.
⑱ Poisoning	All fish	Depends on nature of poison. Fish may cluster at the surface, where they appear to struggle for breath. May also float on their sides, hang at abnormal angles, or try to jump out of the water. Often, fish die very rapidly.	Ascertain cause as quickly as possible. Immediately carry out a water change of up to 40 percent of the tank volume. Add activated carbon to the filtration system to remove harmful substances.
⑲ Salt imbalance	All aquarium occupants	Numerous behavioral changes, such as swimming abnormally, gasping, and loss of appetite, resulting from an increase in tank salinity.	Check SG readings regularly and be sure not to increase the salinity in the aquarium by incorrect dilution of water.
⑳ Starvation	Many	Fish not seen feeding, but may appear healthy. Affected individuals develop an indented lower body line, commonly described as "pinched up." Fish with highly specific diets are at greatest risk.	Offer a good variety of suitable foodstuffs. Try to house with other fish that feed readily, which should encourage the more nervous fish to sample unfamiliar foodstuffs.

MARINE FISH
BREEDING

Reproduction and breeding

The breeding habits of tropical marine fish are not well documented because only a small number of species have reproduced in captivity. A few groups, however, such as clownfish, gobies, and sea horses, breed with relative ease. Others, such as angelfish, may spawn in aquariums, but their young are so poorly developed when they hatch that it is extremely difficult to raise them successfully.

Ensuring that you have at least one mixed-sex pair from which you can breed is problematic with marine fish, since in most species the males and females are visually alike. However, if the fish can be kept in a group, their behavior should give you some clues to the sex of the fish. Two similar-looking fish that fight are probably males involved in a territorial dispute. Conversely, two fish that get on amicably may well be a compatible male and female. Unusual swimming motions or postures could be signs of courtship, which will indicate that you have a pair in the tank. Slight physical differences sometimes become apparent toward spawning time, when females develop a noticeably plumper belly as they swell with eggs. In addition, the males' color often intensifies and their patterning becomes better defined at this time.

More than 15 different species of clownfish have now been bred in aquariums. If you do not want to wait for young clownfish to grow to sexual maturity, you can buy a pair of adults from an aquarium store.

Some species can change sex, which actually helps rather than hinders the process of finding breeding pairs. All young clownfish (see pp.244–245), for example, are male, but if you have a small group—even just two—the dominant fish will change into a female and pair up with the next dominant male. The sex change goes the other way in hawkfish (see p.278) and some angelfish (see pp.252–255), with the dominant individual in an all-female group turning into a male.

Regardless of the species, you will need to be patient if you want to breed from fish you acquire as juveniles. The onset of sexual maturity is generally determined by size, rather than by age. It can take three months for young gobies (see pp.260–261) to reach breeding size, and with clownfish, you may be waiting for up to a year before they are sexually mature.

Spawning at sea

Marine fish have spawning habits broadly similar to those of freshwater species (see pp.64–66). Some marine fish form long-term or temporary pairs; others spawn in small groups or even in mass gatherings. Fertilization is usually external. Most tropical reef fish, such as angelfish and butterflyfish (see pp.256–259), are egg-scatterers, releasing floating eggs into the open ocean to rise to the surface and be swept along with the current. The fry are not fully developed when they hatch, which is why they are often called larvae. Lacking fins, they cannot swim and simply drift in the surface waters, feeding on microscopic plant and animal life called plankton. Only when they are larger and fully formed do they swim back to the reef.

CONDITIONING IN AQUARIUMS

The factors that trigger spawning in the wild are largely unknown, so there is little you can do to encourage marine fish to breed except to ensure that conditions in their tank are as close as possible to those in their natural habitat. This means that the water must be at the correct temperature and salinity, and also of good quality, so efficient filtration and regular partial water changes are essential. Make sure that the fish get enough light—typically about 14 hours per day. The breeding stock should be mature, healthy, and well fed on a protein-rich diet. If they need rocks, shells, or caves for spawning sites, be sure that these are included in their tank.

Keen observation is needed to tell whether fish like these gobies have spawned, because they hide their eggs in shells or small caves.

This is a risky breeding strategy, since a large proportion of eggs and fry perish. Females lay almost daily during the spawning period, producing hundreds of thousands of eggs to increase the chances that at least some of their offspring will survive.

Parental care in marine species

Some species, such as clownfish, gobies, and damselfish *(see pp. 240–243)* are egg-depositors. These fish lay their eggs close to the substrate or on the reef, attaching them to rocks or laying them in caves or shells. They produce far fewer eggs than egg-scatterers but ensure a higher survival rate by guarding the eggs until they hatch. The well-formed fry then swim to the surface, where they feed and develop before returning to the reef. Other species, including jawfish *(see p.282)*, are mouth-brooders, collecting the eggs in their mouths and incubating them there. Mouth-brooding species often lay fewer than 50 eggs, simply because they cannot fit any more in their mouths. Sea horses and pipefish *(see p.279)* carry their eggs.

Powder Blue Tangs naturally live in large shoals. One benefit of shoaling is that it makes it easier to find a partner of the opposite sex.

INVERTEBRATE BREEDING

Crustaceans and mollusks are rarely bred successfully in aquariums: the factors that trigger breeding are difficult to duplicate in tanks, their offspring are devoured by other tank occupants; and the ultraspecific food needs of the young make rearing nearly impossible. Invertebrates such as sponges, anemones, starfish, and corals fare better, since they can reproduce asexually, with pieces dividing off from an existing colony or individual and developing into new organisms. In the wild, they also reproduce sexually, releasing eggs and sperm into the open water, as seen in the corals below.

REPRODUCTION IN CLOWNFISH

If two clownfish spawn successfully, they are likely to continue breeding throughout the year. In fact, well-fed specimens may spawn as regularly as every month or six weeks. Not surprisingly, tank-bred clownfish tend to reproduce more readily in aquariums than wild-caught specimens.

The clownfish choose a spawning site near the base of their host anemone so that they can retreat among its tentacles if danger threatens. After they have cleaned the site, the female lays her eggs. The male releases his sperm over the eggs to fertilize them.

An adhesive coating glues the eggs to the spawning site. The male is largely responsible for guarding the eggs and will attack any would-be egg-stealers. He fans the eggs with his fins to improve water flow so that they receive a good supply of oxygen.

The color of the eggs changes from orange to dark brown during the incubation period. With a diameter of $^1/_{32}$ in (1 mm), the eggs are among the largest laid by any marine fish, and far larger than those of egg-scatterers. The eggs hatch after 7 to 10 days, usually at night.

The adults show no interest in the fry, which should be moved to a rearing tank until they are large enough to survive in the main aquarium. It may take three weeks for the fry to obtain their full coloration. The juveniles shown here are about three months old.

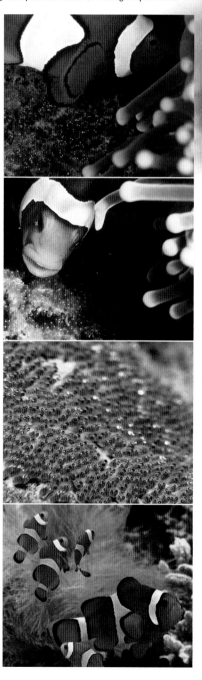

The male carries the eggs until they hatch, which is usually about a week after spawning

In Yellow-headed Jawfish, the male undertakes the task of mouth-brooding the eggs. The fry initially live close to the surface but swim down to the bottom when they are about three weeks old. Jawfish inhabit burrows excavated in the substrate, where they spawn during the warmer months of the year.

The female lays her eggs in the male's abdominal pouch, where they are fertilized. The fry hatch in the pouch and emerge as free-swimming young. A few marine species practice internal fertilization (*see p.61*) and give birth to live young.

Removal of eggs and fry

Eggs and fry make tasty meals for many tank occupants, fish and invertebrates alike, and generally stand a better chance of survival when they are moved to less hostile surroundings. This is not as simple as it sounds. Egg-scatterers spawn at night, so if you are not prepared for the event, most of the eggs may be devoured by the time you check the tank in the morning. If you manage to reach the eggs in time, scoop them out with a cup and transfer them to a hatching/rearing tank. Clownfish and other egg-depositors guard the spawning site, so there is no need to move the eggs, and the protective behavior of the adults is fascinating to watch. If you insist on moving clownfish eggs, avoid exposing them to the air, and do not dislodge them from the rock to which they are attached. The fry hatch tail-first, and if an egg is not firmly glued in place, the fry will be unable to wriggle free of its shell. With its gills still covered by the egg casing, the fish will suffocate.

The eggs of clownfish and other egg-depositors typically hatch after dark. If you shine a light on the water's surface, the fry will flock to this spot, and you can collect them in a cup. (Do not use a net, which will damage their delicate fins and gills.) Transfer the fry to a pre-prepared rearing tank, with water conditions identical to those in the main tank. The tank floor should be bare, so that waste food can be removed easily. Place a guard on the heater to protect the fry. Good water flow is vital to oxygenate the water sufficiently and to circulate particles of food. A power filter is unsuitable, since it may trap fry; a sponge filter, supported by an airstone, will be sufficient.

Feeding regimens for young fish

Your breeding efforts will fail if you do not have enough food of the right size for the fry during their first few days. The young fish will be able to consume only tiny food items, which must be evenly distributed throughout the rearing tank at a relatively high density, since the fry are not able to swim far at this early stage in life. In the wild, newly hatched fish eat plankton. Catching plankton is feasible if you live near the sea, but you will need a special net with a mesh of approximately 50 microns. Although this may be appealing because it is a natural feeding option, harvesting plankton from the sea increases the risk of introducing disease into the aquarium.

More commonly, fry are fed on the larvae of marine organisms called rotifers, which you can culture at home. If this is not convenient, use frozen rotifers as an alternative, but always make sure that they are thawed before you add them to the tank. You should be able to see the fry feeding—the young fish will curl their bodies into an S-shape and then lunge at their quarry. As their gut starts to fill up with food, the underside of the body takes on a whitish hue.

A culture of brine shrimp will also need to be set up as a follow-up food. Introduce small amounts of brine shrimp to the diet of the young fish after the first three days or so. Brine shrimp eggs can be bought complete with a hatching kit *(see p.68)*. When the white stomachs of the fry turn reddish, you will know that they are eating brine shrimp rather than rotifers.

Other foods, including powdered marine flake, can be introduced as the fish grow larger, typically once they start to gain adult coloration. When the young are about ¼ in (6 mm) long, they should be transferred to a tank with more effective filtration.

ROTIFERS FOR MARINE FRY

The most popular rotifer for rearing marine fry is *Brachionus plicatilis*. Culture rotifers in a small tank on unicellular marine algae, which requires good lighting, or on yeast. Regularly check water samples from the culture tank with a hand lens. When there are at

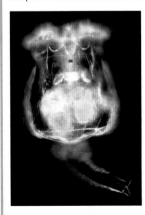

least 25 rotifers per teaspoon, you will have a dense enough concentration to sustain the fish. Feed the fry two or three times each day. Gentle currents in the rearing tank will keep the rotifers afloat so that they are accessible to the fry. Siphon the tank floor daily to prevent any buildup of uneaten dead rotifers, which will impair the water quality.

A rotifer, shown here magnified 50 times, is half the size of a brine shrimp. This female is carrying eggs, which are the gray smudges at the base of her tail. Some female rotifers are mature within a day of hatching.

Expelled by muscular contractions, up to 100 free-swimming sea horse fry start to emerge from an adult male's brood pouch, six weeks after the eggs were deposited there by the female. The male takes no further interest in the fry. Sea horse fry will take brine shrimp as a first food.

DIRECTORY OF
MARINE FISH

SURGEONFISH AND TANGS

These fish commonly include significant amounts of algae in their diet—in fact, the word "tang," which is applied to smaller members of the group, is an abbreviation of the German *Seetang*, meaning "seaweed." Just like land herbivores, these fish have beneficial populations of bacteria in their alimentary tracts to assist the breakdown of plant matter. Medications, especially those containing copper, need to be used very carefully so as not to wipe out these organisms and compromise the fish's ability to digest its food.

Color can indicate mood in this group. For example, if **Yellow Sailfin Tangs** (*Zebrasoma flavescens*) develop a vertical white band on their bodies during the day, it indicates that they have been frightened.

Acanthurus japonicus

White-Faced Surgeonfish

- **ORIGINS** The tropical western Pacific, ranging from northern Australia, via Indonesia, and up to Japan.
- **SIZE** 8 in (20 cm).
- **DIET** Predominantly vegetarian.
- **WATER** Temperature 79–82°F (26–28°C); alkaline (pH 8.1–8.3) with SG 1.021–1.024.
- **TEMPERAMENT** Intolerant of its own kind.

This brightly colored surgeonfish has yellow, blue, and green areas on its tail, as well as a prominent reddish-orange stripe that arcs around the rear of the dorsal fin. The white on the face allows this tang to be distinguished from the related Powder Brown (*A. nigricans*). The White-Faced is easier to keep than the Powder Brown because it adapts more readily to an artificial diet.

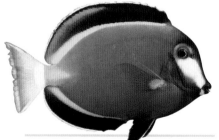

Acanthurus achilles

Achilles Tang

- **ORIGINS** Occurs in more temperate, as well as tropical, parts of the western Pacific.
- **SIZE** 11 in (28 cm).
- **DIET** Predominantly vegetarian.
- **WATER** Temperature 79–82°F (26–28°C); alkaline (pH 8.1–8.3) with SG 1.021–1.024.
- **TEMPERAMENT** Intolerant of its own kind.

Curved dorsal fin

The vibrant red markings of the Achilles Tang make an attractive contrast with its overall matt-black coloration. The red blotch at the rear of the body extends to the caudal peduncle, while a red bar runs across the tail, and red stripes bordered by white extend along the bases of the dorsal and anal fins. White is also evident on the gill covers. The head has a very rounded profile, with the lips being prominent to help rasp algae from rocks.

Acanthurus pyroferus

Chocolate Surgeonfish

- **ORIGINS** From northern Australia to the southern and eastern Asiatic coasts, and out into the Pacific.
- **SIZE** 8 in (20 cm).
- **DIET** Predominantly vegetarian.
- **WATER** Temperature 79–82°F (26–28°C); alkaline (pH 8.1–8.3) with SG 1.021–1.024.
- **TEMPERAMENT** Intolerant of its own kind.

In spite of its name, the Chocolate Surgeonfish can be decidedly yellow, bearing some resemblance to the Yellow Sailfin Tang (*see p.238*). These species can be distinguished quite easily, however, since the mouthparts of the Chocolate Surgeonfish are shorter and more compact, while the dorsal fin has a more rounded and less angular shape.

Acanthurus leucosternon

Powder Blue Tang

- **ORIGINS** East Africa, via the Indo-Pacific, to Southeast Asia, including Indonesia and the Philippines.
- **SIZE** 9 in (23 cm).
- **DIET** Predominantly vegetarian.
- **WATER** Temperature 79–82°F (26–28°C); alkaline (pH 8.1–8.3) with SG 1.021–1.024.
- **TEMPERAMENT** Intolerant of its own kind.

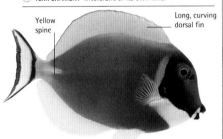

Yellow spine

Long, curving dorsal fin

This is one of the most stunningly colored tangs, but it does not always acclimatize to aquarium conditions, being prone to parasitic infections. The body is primarily blue, but the face is darker, and there is a white area under its mouth that extends to the base of the jaws. Try culturing marine algae, especially at the outset, to guarantee a constant supply for this fish.

Acanthurus olivaceus

Orange-Shoulder Surgeonfish

- **ORIGINS** Ranges from the coast of Southeast Asia eastward out into the Pacific.
- **SIZE** 10 in (25 cm).
- **DIET** Predominantly vegetarian.
- **WATER** Temperature 79–82°F (26–28°C); alkaline (pH 8.1–8.3) with SG 1.021–1.024.
- **TEMPERAMENT** Intolerant of its own kind.

The juvenile of this species *(see below)* is yellow, but its body develops a greener hue with age. All individuals display the characteristic orange marking that extends back from the top of the gills and are blue near the jaws. The "shoulder patch" becomes more prominent in older fish. Good water quality and effective filtration, supported by regular partial water changes, are essential to ensure good health in this species.

Acanthurus triostegus

Convict Surgeonfish

- **ORIGINS** Extends from East Africa through the Indo-Pacific and eastward out into the Pacific.
- **SIZE** 11 in (27 cm).
- **DIET** Predominantly vegetarian.
- **WATER** Temperature 79–82°F (26–28°C); alkaline (pH 8.1–8.3) with SG 1.021–1.024.
- **TEMPERAMENT** Intolerant of its own kind.

Sixth stripe runs down forehead to mouth

The Convict Surgeonfish gets its name from the pattern of bold, black stripes on a bluish background, which is reminiscent of an old-fashioned jail uniform. There are six black stripes in total, not including the black marking on the caudal peduncle. Although Convict Surgeonfish are seen in shoals on the reef, they will not thrive if housed as a group in the aquarium, often disagreeing violently.

Acanthurus lineatus

Clown Surgeonfish

- **ORIGINS** Extends from East Africa, via the Indo-Pacific and Indonesia, into the Pacific.
- **SIZE** 15 in (38 cm).
- **DIET** Predominantly vegetarian.
- **WATER** Temperature 79–82°F (26–28°C); alkaline (pH 8.1–8.3) with SG 1.021–1.024.
- **TEMPERAMENT** Intolerant of its own kind.

The colorful, striped pattern of the Clown Surgeonfish allows this species to be identified easily. The stripes run in roughly horizontal lines along the sides of the body. Individual variations in the patterning occur, especially near the caudal fin, where vertical markings are likely to be evident. The stripes themselves are blue and yellow, separated by black lines. They do not extend over the entire body, however, with the keel of the body being a contrasting shade of light blue. Bright, vibrant coloration suggests a healthy individual. The Clown Surgeonfish requires a large aquarium, in view of its potential size when full grown. This fish should be offered a range of supplementary plant matter on a regular basis, since it cannot sustain itself on just the algae in the aquarium.

SURGEON'S SCALPEL

The name of this family—Acanthuridae—is derived from the Greek words for "spine" and "tail," referring to the distinctive spines on these fish. The spines are located on each side of the body near the caudal peduncle, although not all are as colorful as the example shown here on a Sohal Surgeonfish *(see p.238).* It is because of the sharpness of these spines, which resemble the blade of a scalpel, that the popular name of surgeonfish was coined for this group. In the wild, the spines have a defensive role, protecting the fish from attack, and their bright, contrasting colors serve as a warning to would-be predators.

Acanthurus sohal

Sohal Surgeonfish

- **ORIGINS** Also known as the Red Sea Clown Surgeon because it is confined to the waters of the Red Sea.
- **SIZE** 15³/₄ in (40 cm).
- **DIET** Primarily plant-based foods.
- **WATER** Temperature 79–82°F (26–28°C); alkaline (pH 8.1–8.3) with SG 1.021–1.024.
- **TEMPERAMENT** Intolerant of its own kind.

Closely set, horizontal, yet slightly wavy lines along the body help to identify this fish, along with a vibrant orange spine that stands out clearly near the base of the caudal peduncle. In spite of its large size, the Sohal Surgeon is essentially vegetarian. It feeds on algae in the wild, and persuading it to take artificial substitutes may be problematic. Do not mix this fish with other related species, since it is potentially aggressive.

Paracanthurus hepatus

Blue Tang

- **ORIGINS** Ranges widely from East Africa through the Indian Ocean and across much of the Pacific.
- **SIZE** 10 in (25 cm).
- **DIET** Mainly vegetarian.
- **WATER** Temperature 79–82°F (26–28°C); alkaline (pH 8.1–8.3) with SG 1.021–1.024.
- **TEMPERAMENT** Do not mix with any related fish.

The striking coloration and patterning of the Blue Tang, as well as the relative ease with which it can be kept, have made it a firm favorite with aquarists. The markings are distinctive, with a black stripe running through the eye to the tail on each side of the body, and looping around beneath to create a blue oval bordered by black. The tail is yellow, with black edging at top and bottom. On the reef, small shoals of Blue Tangs can be seen feeding in areas of profuse algal growth. However, they will be less social in the confines of an aquarium, unless the tank is very large. It is usually possible to wean these tangs onto a variety of foods, but initially they will instinctively seek out vegetable matter. Their aquarium should therefore be established well in advance of their introduction. It also needs to be very well lit, in order to ensure continued growth of marine algae on the rockwork and elsewhere.

Zebrasoma xanthurum

Sailfin Tang

- **ORIGINS** Another member of this group with distribution restricted to the Red Sea.
- **SIZE** 8¹/₂ in (22 cm).
- **DIET** Mainly vegetarian.
- **WATER** Temperature 79–82°F (26–28°C); alkaline (pH 8.1–8.3) with SG 1.021–1.024.
- **TEMPERAMENT** Intolerant of its own kind.

The Sailfin Tang has a relatively oval, disklike shape. The upper part of the body and the anal fin have a distinctive blue coloration, while the caudal fin is bright yellow. The remainder of the body is darker, with speckling on the face, and the pectoral fin behind the gills is edged with yellow. Like other tangs, the Sailfin Tang will spend the day busily seeking food, before finding a suitable cave where it can shelter during the night.

Pale lips

Zebrasoma flavescens

Yellow Sailfin Tang

- **ORIGINS** From the Great Barrier Reef, off Australia's eastern coast, across much of the Pacific.
- **SIZE** 6 in (15 cm).
- **DIET** Mainly vegetable matter.
- **WATER** Temperature 79–82°F (26–28°C); alkaline (pH 8.1–8.3) with SG 1.021–1.024.
- **TEMPERAMENT** Usually compatible with invertebrates.

High dorsal fin

Beaklike mouth

The brilliantly colored Yellow Sailfin Tang has a narrow, flattened body shape. It naturally feeds on algae and will require vegetable substitutes to supplement its diet in aquarium surroundings. When servicing the aquarium, be careful not to get caught by the white spine, which is located near the base of the caudal peduncle.

Zebrasoma veliferum

Pacific Sailfin Tang

- **ORIGINS** Much of the tropical Pacific region, eastward from southeast Asia and Australia's Great Barrier Reef.
- **SIZE** 16 in (40 cm).
- **DIET** Vegetable matter, plus some meat-based foods.
- **WATER** Temperature 79–82°F (26–28°C); alkaline (pH 8.1–8.3) with SG 1.021–1.024.
- **TEMPERAMENT** Not well disposed toward its own kind.

The striped patterning of the Pacific Sailfin Tang varies markedly through its range, with some individuals being more brightly colored than others. Age is also significant: young fish are predominantly yellow with dark banding, while in adults, the color scheme is reversed. The tall, backward-sloping dorsal fin above and the curved anal fin below give the Pacific Sailfin a disklike appearance.

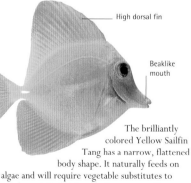

Adult coloration

LOST IN THE CROWD

Aquarists may be puzzled that tangs such as these Yellow Sailfins *(see opposite)* can live in groups on the reef, even spawning communally in some cases, without serious quarrels developing, yet they cannot be housed together safely in the aquarium. Associating in groups helps to decrease the tangs' natural levels of aggression. The large number of fish means that particular individuals come into contact with each other less frequently, so the risk of conflict is reduced. In the tank, where usually only two or three tangs are housed together, the scope for bullying is greater.

Zebrasoma scopas

Brown Tang

- **ORIGINS** From the Red Sea eastward through the Indo-Pacific through to the east Pacific.
- **SIZE** 8 in (20 cm).
- **DIET** Predominantly vegetarian.
- **WATER** Temperature 79–82°F (26–28°C); alkaline (pH 8.1–8.3) with SG 1.021–1.024.
- **TEMPERAMENT** Young fish rarely get along well together.

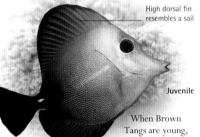

High dorsal fin resembles a sail

Juvenile

When Brown Tangs are young, they are yellow and may be confused with the Yellow Sailfin *(see opposite)*. The coloration of these fish darkens as they mature, except for the white spine on each side of the caudal peduncle. The teeth in the small mouth are for grazing on algae and plankton, which form the basis of their diet in the wild. Healthy specimens have hearty appetites.

Naso unicornis

Unicorn Tang

- **ORIGINS** Extends from the Red Sea through the Indo-Pacific region and into the eastern Pacific.
- **SIZE** 28 in (70 cm).
- **DIET** Primarily requires vegetable matter.
- **WATER** Temperature 79–82°F (26–28°C); alkaline (pH 8.1–8.3) with SG 1.021–1.024.
- **TEMPERAMENT** Intolerant of its own kind.

The distinctive horn extending from between the eyes explains the common name of this fish. The horn is not evident in young individuals, which are also not so brightly colored. Well-oxygenated water is essential for unicorn fish, just as it is for other tangs. A lack of oxygen will cause them to sink to the bottom, where they will lie on their side and breathe heavily. Unicorn Tangs are very active and need plenty of space in order to thrive.

Horn length increases with age

Spine

Naso lituratus

Naso Tang

- **ORIGINS** Extends from the Red Sea through the Indian Ocean to the eastern Pacific.
- **SIZE** 18 in (45 cm).
- **DIET** Primarily vegetable matter.
- **WATER** Temperature 79–82°F (26–28°C); alkaline (pH 8.1–8.3) with SG 1.021–1.024.
- **TEMPERAMENT** Do not mix with similar fish.

This large tang has striking yellow patterning on the forehead, with stripes running down the sides of the face and on the outer trailing edge of the caudal fin. The spine and the adjacent area around the tail are orange. Juveniles are duller in color, being primarily grayish with a plain-colored caudal fin. Naso Tangs occur not only on the reef but also in the open ocean. They need a large tank with a good circulation of water. They can be accommodated in a reef aquarium, but bear in mind that despite what seems like almost constant foraging, there will not be enough plant matter growing there to sustain these fish. They must eat large quantities of food in order to maintain good bodily condition, since their natural food has a relatively low nutritional value. Dietary supplements can include some meat-based foods as well as vegetable matter.

DAMSELFISH

These hardy fish are often recommended as the initial occupants of a newly established marine aquarium. Many species of damselfish adapt well to the fluctuations in water chemistry that will inevitably occur until the filtration system is fully established. In addition, damselfish are easy to feed, and they can also be kept in small groups when young. However, once settled in their quarters, they are likely to become more territorial by nature, which can lead to the bullying of weaker individuals. There is a large number of species, some of which are very similar, and distinguishing between them is not always straightforward.

These **Staghorn Damselfish** (*Amblyglyphidodon curacao*) originate from Pacific waters. Other damselfish are found in the Caribbean.

Dascyllus trimaculatus

Domino Damselfish

- **ORIGINS** The Pacific, from southeast Asia to southern Japan and eastern Australia, extending to Oceania.
- **SIZE** 2½ in (6.25 cm).
- **DIET** Eats most marine foods, including flake.
- **WATER** Temperature 77–79°F (25–26°C); alkaline (pH 8.1–8.3) with SG 1.020–1.024.
- **TEMPERAMENT** Territorial by nature.

The white markings on a black background are reminiscent of the spots on a domino, but this characteristic is seen only in juveniles. As the fish mature, they become grayer and the spots fade. Young Dominos often seek the protection of sea anemones, forming a relationship similar to but less permanent than that seen in their relatives, the clownfish (*see pp.244–245*).

White spot on side of dorsal fin

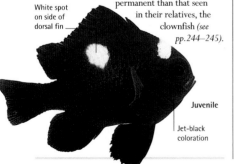

Juvenile

Jet-black coloration

Dascyllus carneus

Cloudy Damsel

- **ORIGINS** The Pacific region adjacent to southeast Asia, north to southern Japan and south to eastern Australia.
- **SIZE** 3¼ in (8 cm).
- **DIET** Takes fresh and dried marine foods.
- **WATER** Temperature 77–79°F (25–26°C); alkaline (pH 8.1–8.3) with SG 1.020–1.024.
- **TEMPERAMENT** May quarrel with its own kind.

The body patterning of this fish, which is predominantly grayish with blue spots, is less distinct than in most other damselfish, giving rise to its common name. This species can sometimes be confused with the Reticulated Damselfish (*D. reticulatus*), although the latter can be identified by a dark band that extends vertically down the rear edge of the body.

Dascyllus aruanus

Three-Stripe Damsel

- **ORIGINS** Extends from the Red Sea eastward through the Indo-Pacific region to the east coast of Australia.
- **SIZE** 3¼ in (8 cm).
- **DIET** Marine foods, including flake.
- **WATER** Temperature 77–79°F (25–26°C); alkaline (pH 8.1–8.3) with SG 1.020–1.024.
- **TEMPERAMENT** Intolerant of other damselfish.

This damselfish resembles the Blacktail Damsel (*see opposite, top left*), but it can be distinguished by the black edging on the dorsal fin and its white caudal fin. Three-Striped Damselfish can be quarrelsome among themselves, especially as they grow older. If an individual is harassed, it should be removed from the aquarium.

Clear white banding

Dascyllus melanurus

Blacktail Damsel

- ⊕ **ORIGINS** The Pacific region, extending from eastern Asia and Australia eastward to the reefs of Oceania.
- ⟳ **SIZE** 4 in (10 cm).
- 🍽 **DIET** All types of marine food.
- ≋ **WATER** Temperature 77–79°F (25–26°C); alkaline (pH 8.1–8.3) with SG 1.020–1.024.
- ⊕ **TEMPERAMENT** Bullying can be a problem.

Black blotch on caudal fin

The Blacktail is one of several damselfish with a predominantly black-and-white striped pattern. It will sample virtually all types of food, but take extra care not to overfeed this and other damselfish in a newly established tank, since rotting food scraps are likely to seriously decrease the water quality. Regular monitoring of the water is essential to safeguard the health of the fish.

Neoglyphidodon oxyodon

Black Neon Damsel

- ⊕ **ORIGINS** The Pacific region off the shores of eastern Asia down to Australia, and north to southern Japan.
- ⟳ **SIZE** 3 in (7.5 cm).
- 🍽 **DIET** A varied range of marine foods.
- ≋ **WATER** Temperature 77–79°F (25–26°C); alkaline (pH 8.1–8.3) with SG 1.020–1.024.
- ⊕ **TEMPERAMENT** Territorial by nature.

The Black Neon has horizontal wavy blue lines on its head and also farther down the body on its flanks. A contrasting yellow vertical stipe extends over the back. Although the blue stripes may become less vivid with age, overall loss of color in *Abudefduf* damsels is not necessarily a sign of poor health, since it can also be associated with male fish coming into spawning condition.

SAFETY IN SHOALING

Shoaling in a particular area of the reef, typically close to a "head" of coral, is an important survival technique used by damselfish. Rather than relying on just its own senses for survival, each individual benefits from the combined alertness of the group. When danger is detected, the shoal darts back into the shelter of the coral, almost as if it were a synchronized movement. The sudden flash of stripes or contrasting colors gives added protection, making it difficult for predators to select a victim from the mass of fleeing bodies.

Abudefduf sexfasciatus

Scissor-Tail

- ⊕ **ORIGINS** Northeast coast of Africa, in the northern and western parts of the Indian Ocean and in the Red Sea.
- ⟳ **SIZE** 6 in (15 cm).
- 🍽 **DIET** Flake and other prepared foods.
- ≋ **WATER** Temperature 77–79°F (25–26°C); alkaline (pH 8.1–8.3) with SG 1.020–1.024.
- ⊕ **TEMPERAMENT** Not to be trusted with its own kind.

These damsels display a series of bold, vertical black stripes down the body. There is also a small, broad stripe at the base of the caudal peduncle, with black stripes running from here in a more horizontal fashion along the tail forks. These features help to distinguish the Scissor-Tail from similarly patterned species. In addition, the dark head stripe tends to extend only to the eye, rather than down to the throat.

Abudefduf saxatilis

Sergeant Major

- ⊕ **ORIGINS** Circumtropical range through the Indian and Pacific Oceans, as well as the Atlantic.
- ⟳ **SIZE** 6 in (15 cm).
- 🍽 **DIET** Thawed, fresh, and dried foods.
- ≋ **WATER** Temperature 77–79°F (25–26°C); alkaline (pH 8.1–8.3) with SG 1.020–1.024.
- ⊕ **TEMPERAMENT** Forms a dominance hierarchy.

The huge range of these striped damselfish means that they have been split into three distinct species, based essentially on their distribution, although the appearance of the adults is virtually identical around the globe. Differences are most marked in juveniles, with those from the Atlantic being silvery and having yellow upperparts to their body. As the name suggests, Sergeant Majors may bully other fish, so keeping two together in a relatively small aquarium is likely to result in persistent territorial disputes. Sergeant Majors are highly valued by marine aquarists, because their readiness to feed on a wide variety of foods can stimulate other fish in the aquarium to sample unfamiliar foods. Including fresh food in the diet of Sergeant Majors may help to encourage breeding behavior.

Chromis cyanea

Blue Chromis

- **ORIGINS** Caribbean region, ranging from Florida down to the coast of northern South America.
- **SIZE** 2 in (5 cm).
- **DIET** Eats all types of marine fish food.
- **WATER** Temperature 77–79°F (25–26°C); alkaline (pH 8.1–8.3) with SG 1.020–1.024.
- **TEMPERAMENT** Has a territorial disposition.

A long, highly forked caudal fin gives the Blue Chromis a very elegant appearance and also sets it apart from other damselfish. The body, too, is narrower, emphasizing the streamlined shape. The top of the body is blackish, with dark coloration also evident on some of the fins, while on the flanks there are black markings on the individual blue scales, creating a speckled appearance. Blue Chromis can be sexed visually, although this is not easy. Just prior to spawning, females have an orange egg-laying tube, or ovipositor, that protrudes out of the vent. When a pair of these damselfish are ready to spawn, they prepare a site, usually on rocks, where the female lays up to 50 eggs, which the male then fertilizes. The eggs are guarded by the male until the fry hatch about two days later. The young, free-swimming Blue Chromis should be fed on rotifers as a first food.

Chromis viridis

Green Chromis

- **ORIGINS** Occurs in the Pacific Ocean, east of the Philippines, New Guinea, and eastern Australia.
- **SIZE** 2½ in (6.5 cm).
- **DIET** Will take a varied diet, including flake.
- **WATER** Temperature 77–79°F (25–26°C); alkaline (pH 8.1–8.3) with SG 1.020–1.024.
- **TEMPERAMENT** Relatively social.

Chromis species, such as the Green Chromis, generally rank among the most placid of all damselfish and are unlikely to be disruptive in a reef aquarium. Even so, they should not be mixed with other similar species, since they may be harried by more belligerent companions. At night, they retreat out of sight into a favorite nook or cranny, emerging again the following morning.

Pomacentrus alleni

Allen's Damselfish

- **ORIGINS** Ranges from the Red Sea through the Pacific; common around the Similian Islands adjoining Thailand.
- **SIZE** 4 in (10 cm).
- **DIET** Eats both fresh and dried marine fish foods.
- **WATER** Temperature 77–79°F (25–26°C); alkaline (pH 8.1–8.3) with SG 1.020–1.024.
- **TEMPERAMENT** Cannot be kept in groups.

Well-defined scales

Named after a famous ichthyologist, Dr. Gerald Allen, this damselfish can be identified by its fin markings. The lower part of the caudal fin bears a distinctive black area, with yellow extending from the anal fin along the ventral side of the body. The dorsal fin is dark, becoming whitish at the rear. Allen's Damselfish has care needs similar to other damsel species, and, like many damselfish, it will be aggressive toward its own kind in the aquarium.

Pomacentrus coelestis

Neon Damselfish

- **ORIGINS** Widely distributed off the eastern coast of Asia and Australia, including the Great Barrier Reef.
- **SIZE** 5 in (12.5 cm).
- **DIET** Fresh, thawed, and freeze-dried foods; also flake.
- **WATER** Temperature 77–79°F (25–26°C); alkaline (pH 8.1–8.3) with SG 1.020–1.024.
- **TEMPERAMENT** Individuals are territorial.

The behavior of these damselfish changes as they mature. The young associate in groups but become more aggressive and territorial with maturity. Weaker individuals may have to be removed. To reduce the risk of disputes, design the aquarium with plenty of retreats, and use rocks to help divide the tank into different territories. Neon Damselfish are not likely to be aggressive toward unrelated fish sharing their tank.

Yellow coloring present on the flanks

Pomacentrus caeruleus

Yellow-Bellied Blue Damsel

- **ORIGINS** From the east coast of Africa through the Indo-Pacific region to eastern Asia and south to Australia.
- **SIZE** 4 in (10 cm).
- **DIET** A varied range of marine fish foods.
- **WATER** Temperature 77–79°F (25–26°C); alkaline (pH 8.1–8.3) with SG 1.020–1.024.
- **TEMPERAMENT** Intolerant toward its own kind.

Yellow underparts

Identifying these fish can be difficult, because their wide distribution means that there are regional differences between individual populations. In addition, their coloration can change with age; young fish are essentially blue but develop more pronounced yellow markings as they mature. Although territorial, these fish are not aggressive toward invertebrates, so they are suitable for inclusion in a reef aquarium.

Pomacentrus moluccensis

Lemon Damsel

- **ORIGINS** From Indonesia, including the Moluccas, east into the Oceania region of the Pacific.
- **SIZE** 4½ in (11 cm).
- **DIET** Eats all types of marine fish foods.
- **WATER** Temperature 77–79°F (25–26°C); alkaline (pH 8.1–8.3) with SG 1.020–1.024.
- **TEMPERAMENT** Adults are territorial.

This attractive yellow Asiatic species has fine blue markings on its body and fins. Lemon Damsels can be quarrelsome with each other. To reduce the likelihood of aggression when introducing them to the tank, obtain an odd number of fish and choose specimens of a similar size. Also, choose juveniles rather than adults, because they are likely to settle better in aquarium surroundings.

Microspathodon chrysurus

Yellowtail Damselfish

- **ORIGINS** Found in the Caribbean region, from Florida down to the northern coast of South America.
- **SIZE** 8 in (20 cm).
- **DIET** Will take prepared and fresh marine foods.
- **WATER** Temperature 77–79°F (25–26°C); alkaline (pH 8.1–8.3) with SG 1.020–1.024.
- **TEMPERAMENT** Tends to quarrel with its own kind.

This large damselfish changes color with age, its blue spots fading as it matures. The caudal fin, which is transparent in juveniles, turns bright yellow at this stage. Yellowtail Damselfish often associate with Fire Corals (*Millepora* spp.) in the wild, being immune to the invertebrate's stinging cells. When threatened, the fish dart back among the coral, where predators are unlikely to follow.

Stegastes planifrons

Three-Spot Damsel

- **ORIGINS** In the Caribbean region, where it is commonly encountered from Florida to northern South America.
- **SIZE** 6 in (15 cm).
- **DIET** Eats almost any marine fish food.
- **WATER** Temperature 77–79°F (25–26°C); alkaline (pH 8.1–8.3) with SG 1.020–1.024.
- **TEMPERAMENT** Watch for territorial disputes.

Three-Spots are among the largest of damselfish. The juvenile, pictured below, is easily identified by the orange-yellow coloration, offset with black spots—one on the caudal peduncle and one on each side of the dorsal fin. Unfortunately, this attractive coloration does not last, and adults become dark gray. Mature Three-Spot Damsels are relatively aggressive and less compatible together than smaller species. They are easy to feed, however, taking flake and freeze-dried foods readily. Although damselfish have a reputation for being hardy and adaptable, there is no excuse for letting conditions in the tank deteriorate, even if they are the only fish in residence. Keep the nitrate reading low, carrying out regular water tests and making partial water changes as necessary.

DISPUTES AND DOMINANCE

In aquariums, most damsels prove to be aggressive as they mature, but on the reef they live in larger groups, which lessens the level of aggression between individuals. An order of dominance is established within each group, resulting in fewer challenges. The fish are also less confined in the wild, and there are many more retreats, which further reduces the likelihood of conflict. Some damsels lay claim to specific feeding grounds, where algae are plentiful. They will defend these sites to ensure that they have access to an ongoing food supply.

CLOWNFISH

Clownfish are also known as anemonefish because of their close relationship with this group of marine invertebrates. This association may have developed because the anemones' stinging cells kill the *Oodinium* skin parasites to which these fish are susceptible. Clownfish are one of the easiest marine species to breed in the home aquarium. Sexing, too, is no problem, because if you buy two juveniles, measuring no more than ½ in (13 mm) long, then you can be sure of acquiring a true pair. Although all juveniles are male, the dominant individual of the two will change into a female as the fish mature.

Clownfish, such as this **False Percula** (*Amphiprion ocellaris*), are so called because their markings resemble the face paint of a clown.

Amphiprion ephippium
Red Saddleback Clown

- **ORIGINS** Indo-Pacific region, occurring on reefs from the Andaman and Nicobar Islands eastward to Java.
- **SIZE** 5 in (12 cm); 3 in (7.5 cm) in aquariums.
- **DIET** Prepared foods and small live foods.
- **WATER** Temperature 77–79°F (25–26°C); alkaline (pH 8.1–8.3) with SG 1.020–1.024.
- **TEMPERAMENT** Intolerant of its own kind.

Red Saddleback Clowns have a rich, tomato-red coloration with dark patches on the flanks. Young fish often display a white stripe behind the eyes that disappears as they mature. The female typically lays up to 200 eggs, close to her host anemone. As with every clownfish species, it helps to match the fish with a sea anemone found in its natural range.

Dark, "saddle-back" marking

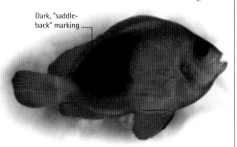

Amphiprion bicinctus
Two-Band Clownfish

- **ORIGINS** Distributed in reefs from the Red Sea through to the Indo-Pacific region.
- **SIZE** 5 in (12 cm); 3 in (7.5 cm) in aquariums.
- **DIET** Chopped, thawed live foods and flake.
- **WATER** Temperature 77–79°F (25–26°C); alkaline (pH 8.1–8.3) with SG 1.020–1.024.
- **TEMPERAMENT** Not usually aggressive.

The appearance of the Two-Band Clownfish changes with age. Instead of the adult patterning of two white stripes running vertically down the sides of their bodies, juveniles have three, with the third stripe being close to the tail. Anemonefish are potentially long-lived, with a life expectancy of up to 18 years in aquarium surroundings.

Amphiprion akallopisos
Skunk Clownfish

- **ORIGINS** Occurs on reefs in the western part of the Indo-Pacific region.
- **SIZE** 3 in (7.5 cm).
- **DIET** Prepared foods and small live foods.
- **WATER** Temperature 77–79°F (25–26°C); alkaline (pH 8.1–8.3) with SG 1.020–1.024.
- **TEMPERAMENT** Peaceful but fairly shy.

This clownfish forms very close associations with anemones, so it should always be kept alongside them. Being a relatively sluggish swimmer, it usually remains close to its host anemone, retreating into the tentacles at any hint of danger. Such behavior means that this species, with its plain, yellowish-orange body, is less conspicuous in the aquarium than other, bolder clownfish.

Amphiprion clarkii

Clarkii Clown

- **ORIGINS** Widely distributed throughout the Indo-Pacific region.
- **SIZE** 4 in (10 cm).
- **DIET** Mixed, varied diet, including vegetable matter.
- **WATER** Temperature 77–79°F (25–26°C); alkaline (pH 8.1–8.3) with SG 1.020–1.024.
- **TEMPERAMENT** Bold.

The yellow caudal fin allows the Clarkii Clown to be distinguished easily from other species with similar body patterning. This robust species serves as a good introduction to the group. An unusual feature of the behavior of these and other clownfish is the way they grunt, typically if they are threatened or spawning. These grunts may be audible if the room is very quiet. Young fry must be reared initially on rotifers then brine shrimp.

MUTUAL BENEFITS

Clownfish are generally immune to the stinging tentacles of their sea anemone hosts, thanks to the protective covering of sugar-based mucus on the surface of their bodies, which acts like a shield. Both host anemone and clownfish benefit from the association, which biologists describe as commensalism. The fish finds protection from predators, while in return, the sea anemone receives pieces of food dropped by the fish when it returns to eat a meal within the relative safety of the anemone's tentacles.

Amphiprion perideraion

Pink Skunk Clownfish

- **ORIGINS** The Pacific region, from the coast of Thailand east to Samoa. Also occurs on the Great Barrier Reef.
- **SIZE** 3 in (8 cm); 3 in (7.5 cm) in aquariums.
- **DIET** Prepared foods and small live foods.
- **WATER** Temperature 77–79°F (25–26°C); alkaline (pH 8.1–8.3) with SG 1.020–1.024.
- **TEMPERAMENT** One of the more shy clownfish.

The white vertical bar behind the eyes is characteristic of this species. Males can be sexed visually by the presence of orange areas at the top and bottom of the caudal fin. As in other species, the overall depth of coloration may differ between individuals. When introducing clownfish to a new aquarium, do not worry if they will not adopt an unfamiliar anemone immediately, as this can take some time. Like many marine species, they tend not to grow as big in aquariums as they do in the wild.

Amphiprion ocellaris

False Percula

- **ORIGINS** On reefs throughout the Indo-Pacific region, including Papua New Guinea and the Great Barrier Reef.
- **SIZE** 3 in (8 cm); 2 in (5 cm) in aquariums.
- **DIET** Will eat a mixed diet based on prepared foods.
- **WATER** Temperature 77–79°F (25–26°C); alkaline (pH 8.1–8.3) with SG 1.020–1.024.
- **TEMPERAMENT** Occasionally territorial.

False Perculas can be kept and even bred in an aquarium without the presence of a sea anemone, but a better insight into their behavior will be gained if they are housed with a suitable anemone. The differences in the coloration and patterning of False Perculas may help them to blend in with the different species of sea anemone found across their range. Tank-raised specimens, such as the one shown below, tend to more yellow.

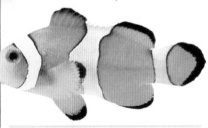

Premnas biaculeatus

Maroon Clown

- **ORIGINS** Occurs on reefs in the Pacific Ocean, between Indonesia, Taiwan, and the northern Great Barrier Reef.
- **SIZE** 6 in (15 cm); 4 in (10 cm) in aquariums.
- **DIET** Small livefoods and prepared foods, such as flake.
- **WATER** Temperature 77–79°F (25–26°C); alkaline (pH 8.1–8.3) with SG 1.020–1.024.
- **TEMPERAMENT** Can be aggressive on occasion.

This is one of the largest and darkest of the clownfish. Maroon Clowns are unusual in that females can grow up to three times as large as males. Care must be taken when catching this species because of the spines at the back edge of its gill covers. These can become stuck in the material of a net and may then be damaged.

TRIGGERFISH

A relatively stocky, oval body shape characterizes triggerfish. Their caudal fins are quite small, so these fish rely mainly on the rippling movement of the rear part of their dorsal fin, combined with their anal fin, to propel themselves through the water. Their front dorsal spine interlocks with a smaller second spine and can be used to anchor the fish in a rocky crevice. Once the locking mechanism has been triggered, it is virtually impossible to dislodge them. It may be necessary to buy a triggerfish in its chosen retreat, should it seek sanctuary there while being caught. Triggerfish are ideal for those seeking a fish with "personality." They can be tamed sufficiently to feed on hand-offered items.

Triggerfish, such as the **Red-tooth Triggerfish** (*Odonus niger*) shown here, are found in tropical seas worldwide, usually in association with reefs.

Odonus niger

Niger Triggerfish

- **ORIGINS** From the Red Sea eastward, through the Pacific to Japan and Australia's Great Barrier Reef.
- **SIZE** 16 in (40 cm).
- **DIET** Animal-based foods, including shrimp.
- **WATER** Temperature 77–79°F (25–26°C); alkaline (pH 8.2–8.4) with SG 1.023–1.027
- **TEMPERAMENT** Placid.

These triggerfish are equipped with a formidable array of red teeth at the front of their mouths. The body color varies throughout their range, from a bluish hue to green. It can also be affected by their mood. Despite their scientific name (*niger* means "black"), these fish are never black. When danger threatens, they usually swim headfirst into a suitable reef crevice. Adequate retreats must be therefore included in the tank.

Lyre-shaped caudal fin

Pseudobalistes fuscus

Blueline Triggerfish

- **ORIGINS** From East Africa and the Red Sea through the Pacific to Samoa and Micronesia.
- **SIZE** 22 in (56 cm).
- **DIET** Fish and invertebrates.
- **WATER** Temperature 77–79°F (25–26°C); alkaline (pH 8.2–8.4) with SG 1.023–1.027.
- **TEMPERAMENT** Unpredictable and always territorial.

Tough and hardy, these fish typify the requirements of triggerfish. They need to be kept singly, because they will prey on smaller fish, and generally do not like the company of their own kind. Blueline Triggerfish can also be very destructive within the aquarium, since they dig in the substrate, and must not be housed with invertebrates.

Rhinecanthus aculeatus

Picasso Triggerfish

- **ORIGINS** From East Africa throughout the Indo-Pacific region to Japan and Hawaii.
- **SIZE** 10 in (25 cm).
- **DIET** Fish and animal-based foods.
- **WATER** Temperature 77–79°F (25–26°C); alkaline (pH 8.2–8.4) with SG 1.023–1.027.
- **TEMPERAMENT** Territorial.

Patterned like an abstract painting, the Picasso Triggerfish has yellow markings along the sides of its face, suggesting that the jaws can open wide. In reality, however, the mouth is small, as in other triggerfish. The Picasso's pelvic fins are reduced to a small projection under the body, and it is a slow, slightly clumsy swimmer. This triggerfish often grunts, especially when being caught.

Erect dorsal fin

TROUBLESOME TEETH

Although triggerfish have small mouths, their powerful jaws have rows of teeth that can inflict a very painful bite. Take particular care when offering food directly, and always feed the fish before you put your hands into the tank to service it. A triggerfish is even capable of biting through the heater cable in the aquarium—electrocuting itself as a consequence—so protect cabling accordingly. In the wild, triggerfish rely on their powerful teeth to crush the shells of crustaceans, mollusks, and even coral, which they eat. They also communicate by grinding their teeth.

Balistoides conspicillum

Clown Triggerfish

- **ORIGINS** From East Africa through the Pacific to Japan, Samoa, and the eastern side of Australia.
- **SIZE** 20 in (50 cm).
- **DIET** Animal-based foods.
- **WATER** Temperature 77–79°F (25–26°C); alkaline (pH 8.2–8.4) with SG 1.023–1.027.
- **TEMPERAMENT** Not to be trusted with other fish.

Individual spotted pattern

The spotted patterning of the Clown Triggerfish is unmistakable. The pattern effectively conceals the eyes of this fish, so disorienting potential predators. The dorsal fin is split into two separate parts, with the darker front portion often being folded down into a groove running along the back. An aquarium for this and other triggerfish should include caves to which the fish can retire at night.

Sufflamen bursa

Triggerfish

- **ORIGINS** Widely distributed throughout the Indo-Pacific region.
- **SIZE** 10 in (25 cm).
- **DIET** Omnivorous; very easy to feed.
- **WATER** Temperature 77–79°F (25–26°C); alkaline (pH 8.2–8.4) with SG 1.023–1.027.
- **TEMPERAMENT** Antisocial.

A pale body, largely transparent fins, and two stripes on each side of the head identify this triggerfish. Males are more brightly colored than females. Like related species, the Triggerfish is relatively tolerant in terms of water quality, but it needs an aquarium to itself. Start out with a lively juvenile of about 4 in (10 cm), which will be easier to tame. Young Triggerfish grow at a surprisingly fast rate.

Large eyes

Balistapus undulatus

Orange-Lined Triggerfish

- **ORIGINS** The Indo-Pacific region, although it does not extend as far as Hawaii.
- **SIZE** 12 in (30 cm)
- **DIET** A range of animal foods, including river shrimp.
- **WATER** Temperature 77–79°F (25–26°C); alkaline (pH 8.2–8.4) with SG 1.023–1.027.
- **TEMPERAMENT** One of the most aggressive triggerfish.

In the Orange-Lined Triggerfish of the Pacific, shown below, the caudal fin is yellow-green, but in Indian Ocean specimens it is orange. One of the appealing features of triggerfish is their eyes, which move independently. When anchoring themselves in a crevice, triggerfish may inflate their bodies slightly, although not as much as pufferfish.

Balistes vetula

Queen Triggerfish

- **ORIGINS** Found through tropical parts of the western Atlantic, typically in Caribbean waters.
- **SIZE** Up to 20 in (50 cm); 10 in (25 cm) in aquariums.
- **DIET** Fish and invertebrates.
- **WATER** Temperature 77–79°F (25–26°C); alkaline (pH 8.2–8.4) with SG 1.023–1.027.
- **TEMPERAMENT** Antisocial.

These large, attractive triggerfish are relatively easy to cater for, since they feed on a variety of meat-based foods and do not need company. Unfortunately, their reproductive habits rule out breeding in the home aquarium. In the wild, the larger and more colorful male mates with several females in turn. The females then guard their eggs in individual spawning pits until they hatch, even to the extent of biting divers who come too close.

FILEFISH

Closely related to the triggerfish, filefish are not so boisterous in aquarium surroundings as their cousins, nor as destructive. Most species are smaller than triggers, but they share with them the split configuration of the dorsal fin (the first part forms a spine used to anchor the fish in a cave). The ventral fin on the underside of the body is similarly modified, providing additional support. Filefish have small mouths, with teeth designed for nibbling at foods such as algae. They must not be housed in a reef tank, because they naturally feed on coral polyps. Although usually tolerant of their own kind, some filefish may occasionally harry other fish.

A large dorsal spine, clearly visible on this **Plainhead Filefish** (*Stephanolepis hispidus*), is a characteristic feature of this group.

Monacanthus chinensis

Fan-Bellied Leatherjacket

- 🌐 **ORIGINS** The Pacific region, north to southern Japan, and south via Malaysia to Australia's Great Barrier Reef.
- 🔄 **SIZE** 10 in (25 cm).
- 🍴 **DIET** Omnivorous, but prefers algae and meat-based foods.
- 🌡 **WATER** Temperature 77–79°F (25–26°C); alkaline (pH 8.1–8.4) with SG 1.020–1.025.
- 😐 **TEMPERAMENT** Not normally aggressive.

The blotched coloration of this species may vary between individuals. It is also affected by the fish's surroundings—the Fan-Bellied Leatherjacket can alter its appearance to some extent to blend in with the background. Since it is not a strong swimmer, its survival relies more on avoiding detection rather than fleeing from predators.

Monacanthus tuckeri

Slender Filefish

- 🌐 **ORIGINS** The Caribbean region, extending from Florida down to the coast of Venezuela.
- 🔄 **SIZE** 3½ in (9 cm).
- 🍴 **DIET** Fresh and thawed meat-based foods, plus algae.
- 🌡 **WATER** Temperature 77–79°F (25–26°C); alkaline (pH 8.1–8.4) with SG 1.020–1.025.
- 😐 **TEMPERAMENT** Placid.

Rounded
caudal fin

A narrow, rather elongated body sets the Slender Filefish apart from related species. It displays variable coloration, depending partly on its environment, although the body patterning is often blotchy. These filefish will not only browse on marine algae in the aquarium but also feed quite readily on prepared diets featuring invertebrates, such as cockles and crustaceans, especially once they are established.

Cantherhines macrocerus

White-Spotted Filefish

- 🌐 **ORIGINS** The Atlantic, from the eastern U.S. and down through the Caribbean to northern South America.
- 🔄 **SIZE** 18 in (46 cm).
- 🍴 **DIET** Algae and meat-based foods.
- 🌡 **WATER** Temperature 77–79°F (25–26°C); alkaline (pH 8.1–8.4) with SG 1.020–1.025.
- 😐 **TEMPERAMENT** Not normally aggressive.

It is the juveniles of this species that most strongly display the characteristic patterning of white spots on a dark background. Orange coloration is apparent, too, typically toward the rear of the body. Breeding has yet to be accomplished in aquariums. In the wild, filefish lay green eggs, which sink to the bottom. The fry develop in the upper layer of the water, feeding on plankton.

Juvenile

Pervagor melanocephalus

Orange-Head Filefish

- **ORIGINS** Throughout the Indo-Pacific region, including the coastal area of Australia and eastward to Hawaii.
- **SIZE** 4 in (10 cm).
- **DIET** Will take meat-based foods and algae.
- **WATER** Temperature 77–79°F (25–26°C); alkaline (pH 8.1–8.4) with SG 1.020–1.025.
- **TEMPERAMENT** Placid.

Dorsal spine

Fan-shaped caudal fin

This filefish tends to have a light-colored body and a bluish head, although its appearance can differ throughout its range. Hawaiian specimens are the most colorful, with red bodies. House this filefish in an aquarium with decor that is as bright as the fish itself, because in the wild it frequents colorful areas on the reef, where its slow movements help to disguise its presence.

Chaetodermis penicilligerus

Tasseled Filefish

- **ORIGINS** The Pacific region, from Malaysia to Japan in the north, and Australia's Great Barrier Reef in the south.
- **SIZE** 10 in (25 cm).
- **DIET** Feeds on algae and meat-based foods.
- **WATER** Temperature 77–79°F (25–26°C); alkaline (pH 8.1–8.4) with SG 1.020–1.025.
- **TEMPERAMENT** Not normally aggressive.

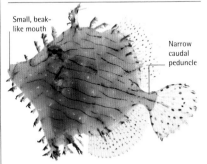

Small, beak-like mouth

Narrow caudal peduncle

The horizontal body stripes and the unusual tassel-like appendages of this slow-swimming filefish help to conceal it among the seaweed fronds on which it often feeds. The black blotches on each side of the body behind the eyes become less prominent as the Tasseled Filefish grows older.

ROUGH DEFENSE

The small scales on a filefish's body are arranged randomly, producing raised areas that give a rough, ridged texture like the surface of a file. The scales form a tough, protective covering—hence the alternative common name of Leatherjacket, which is given to various filefish species. Because of their rough skin and their dorsal and ventral spines, it is better to use a container rather than a net when trying to catch them.

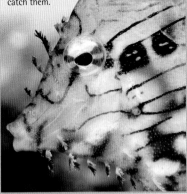

Pervagor spilosoma

Fantail Filefish

- **ORIGINS** The Pacific region, north and east of Indonesia, and north of New Zealand to Hawaii.
- **SIZE** 5 in (13 cm).
- **DIET** Eats meat-based foods and algae.
- **WATER** Temperature 77–79°F (25–26°C); alkaline (pH 8.1–8.4) with SG 1.020–1.025.
- **TEMPERAMENT** Not normally aggressive.

This species should not be confused with the Hawaiian Redtail Filefish (*P. aspricaudus*), in which the blotches on the body are replaced by fine dots. In order to feel secure, the Fantail Filefish—like all filefish—needs plenty of small caves in the aquarium for use as retreats. When threatened, it anchors itself in place with its spines and inflates its body slightly, becoming difficult to dislodge.

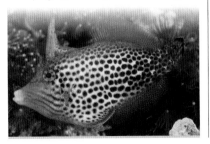

Oxymonacanthus longirostris

Long-Nosed Filefish

- **ORIGINS** The Red Sea and eastward through the Indo-Pacific region to the vicinity of Hawaii.
- **SIZE** 4 in (10 cm).
- **DIET** Algae and meat-based foods.
- **WATER** Temperature 77–79°F (25–26°C); alkaline (pH 8.1–8.4) with SG 1.020–1.025.
- **TEMPERAMENT** Shy and docile.

Long-Nosed Filefish can be identified by their striking patterning of orange spots set against a sky-blue background. They are regarded as the most colorful of all the filefish, with attractive alternating spokes of blue and yellow surrounding the pupil of each eye. These filefish often prove difficult to wean off of coral polyps—especially the *Acropora* species, which they consume in the wild—and onto substitute diets.

BOXFISH AND PORCUPINEFISH

These fish have a distinctive appearance and can become surprisingly tame in aquarium surroundings. Characterized by a lack of streamlining, they cannot swim quickly to escape danger. They do, however, have other means of protection, such as the sharp spines of the porcupinefish. Take special care when catching and handling fish of this group, since many will release a poisonous mucus that, in the confines of an aquarium, will be deadly both to themselves and to the other tank inhabitants. The powerful jaws and teeth reflect their natural diet of marine crustaceans and mollusks. Aquarium specimens need similar foods in order to prevent their teeth from becoming overgrown.

The square body shape of the **Scribbled Boxfish** (*Ostracion solorensis*), combined with weak fins, makes this fish a sluggish swimmer.

Diodon holocanthus

Spiny Puffer

- **ORIGINS** Circumtropical, occurring in the Pacific and Eastern Atlantic; also Florida and the Bahamas to Brazil.
- **SIZE** 9 in (22.5 cm).
- **DIET** Plant matter and larger live foods.
- **WATER** Temperature 77–79°F (25–26°C); alkaline (pH 8.3–8.4) with SG 1.023–1.027.
- **TEMPERAMENT** Do not mix with invertebrates.

The sharp teeth of these porcupinefish mean that, as with other larger members of this group, all electrical aquarium cabling should be protected, in case they attempt to bite through it. Cloudy eyes are a sign that porcupinefish have been kept in suboptimal water conditions. The total nitrate level must be kept below 20 ppm, with a water change of about 20 percent being advisable every two weeks.

Ostracion cubicus

Spotted Cube

- **ORIGINS** East Africa and the Red Sea through the Pacific to the Ryukyu Islands, east to Hawaii and Tuamotu.
- **SIZE** 18 in (45 cm).
- **DIET** Algae and various marine foods.
- **WATER** Temperature 77–79°F (25–26°C); alkaline (pH 8.3–8.4) with SG 1.023–1.027.
- **TEMPERAMENT** Individuals can live with docile tankmates.

Juvenile coloration

Only young fish of this species (pictured) display the stunning yellow coloration. Adults are dramatically different, with bluish bodies and pink-humped noses. In mature specimens, the yellow areas are limited to mere lines, mainly on the sides of the face and at the base of the caudal fin. Do not mix this boxfish or similar species with cleaner fish, such as gobies or wrasse, since the cleaners will damage their thin skin.

Ostracion meleagris

Blue Boxfish

- **ORIGINS** From East Africa across the Pacific, via Australia's Great Barrier Reef, and Hawaii to Mexico.
- **SIZE** 6¼ in (16 cm).
- **DIET** Omnivorous; fresh and thawed food, plus algae.
- **WATER** Temperature 77–79°F (25–26°C); alkaline (pH 8.3–8.4) with SG 1.023–1.027.
- **TEMPERAMENT** Can live with nonaggressive companions.

Male and female Blue Boxfish look significantly different, with only males displaying the blue coloration. In contrast, females have white spots on a brownish-black background. Little is known about their reproductive behavior, and breeding in the tank is unlikely, as with other boxfish. They are easy to maintain but must not be housed with invertebrates, which form part of their diet in the wild.

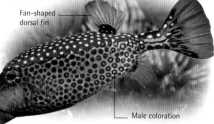

Fan-shaped dorsal fin

Male coloration

Lactoria cornuta

Long-Horned Cowfish

- **ORIGINS** Red Sea through the Pacific to Japan, the eastern coast of Australia, and the Marquesas Islands.
- **SIZE** 18 in (45 cm).
- **DIET** Omnivorous; prefers algae and livefoods.
- **WATER** Temperature 77–79°F (25–26°C); alkaline (pH 8.3–8.4) with SG 1.023–1.027.
- **TEMPERAMENT** Aggressive to their own kind.

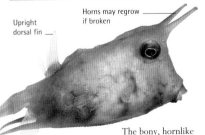

Upright dorsal fin

Horns may regrow if broken

The bony, hornlike projections on the head of this fish explain its common name. There is also a similar, less prominent projection at the rear of the body. The slow motion of the Long-Horned Cowfish means that it may lose out when competing for food with more agile fish, especially since it likes to feed near the tank bottom. If the flanks are drawn in, this indicates a serious loss of condition.

INFLATABLE ARMOR

When threatened, porcupinefish can defend themselves by inflating their bodies. This makes them harder to swallow, reinforcing the effectiveness of their armor of sharp spines. The teeth, normally used by the fish to obtain their food, can also deliver a painful bite to an attacker's flesh. If lifted out into air after being caught, porcupinefish may inflate themselves with air instead of water, which can prove fatal. For this reason, they should be steered into a suitable submerged container of water, in which they can be taken safely out of the aquarium.

Canthigaster valentini

Saddled Toby

- **ORIGINS** Red Sea through the Indo-Pacific region to Japan and eastern Australia, as far as Tuamotu Islands.
- **SIZE** 4 in (10 cm).
- **DIET** Omnivorous; prefers algae and meaty foods.
- **WATER** Temperature 77–79°F (25–26°C); alkaline (pH 8.3–8.4) with SG 1.023–1.027.
- **TEMPERAMENT** Nonaggressive.

These small fish make good home aquarium specimens because they can be housed in a modest tank. However, they should not be kept with invertebrates, which form their natural prey. The Saddled Toby can be distinguished from the Four-Barred Toby (*C. coronata*) by the presence of just three white horizontal bands across its back. Note, though, that a filefish called the Valentini Mimic (*Paraluteres prionurus*) has similar markings.

Chilomycterus schoepfii

Striped Burrfish

- **ORIGINS** Inhabits coastal waters, ranging from the north of Florida southward as far as Brazil.
- **SIZE** 20 in (50 cm).
- **DIET** Omnivorous; prefers algae and animal-based foods.
- **WATER** Temperature 77–79°F (25–26°C); alkaline (pH 8.3–8.4) with SG 1.023–1.027.
- **TEMPERAMENT** Will attack invertebrates.

Tetrosomus gibbosus

Humpback Turretfish

- **ORIGINS** Indo-Pacific region, with specimens often originating from Sri Lanka.
- **SIZE** 16 in (40 cm).
- **DIET** Algae, plus prepared and fresh foods.
- **WATER** Temperature 77–79°F (25–26°C); alkaline (pH 8.3–8.4) with SG 1.023–1.027.
- **TEMPERAMENT** Incompatible with invertebrates.

In common with related species, the Humpback Turretfish usually grows much smaller in aquariums than it does in the wild. Its body shape, combined with the way it swims, gives the impression that it is hovering in the water. Like other boxfish, it is protected by a rigid bony casing, augmented in the Humpback by various sharp projections along its sides, which help to deter would-be predators.

The spines of these fish offer good protection against predators, and the so-called "ocelli," or false eyes, on their bodies also act as a deterrent against attack. Striped Burrfish must not be caught in a net, because their spines can easily become entangled in the mesh. As with similar fish, they can be greedy when feeding, which may place an extra burden on the tank's filtration system, so water quality must be carefully monitored.

ANGELFISH

The angelfish family includes some of the most attractively patterned of all reef fish. The size of angelfish can vary greatly between species. While dwarf angelfish (*Centropyge* species) attain a maximum length of 5 in (12.5 cm), their larger relatives may grow as large as 24 in (60 cm) in the wild. The young of these bigger species often differ significantly in appearance from the adults. Bear this in mind when choosing young angelfish for the aquarium, and never base your choice simply on size, because you could easily end up buying a juvenile that rapidly outgrows its accommodation. Larger species are also more quarrelsome than their smaller relatives and not generally suited to being housed with invertebrates.

Dazzling coloration, as typified by the **Regal Angelfish** (*Pygoplites diacanthus*) shown here, is a feature displayed by many members of the group.

Centropyge bispinosa

Coral Beauty

- **ORIGINS** The Indo-Pacific region, including areas of Southeast Asia, Australia, and the Philippines.
- **SIZE** 3½ in (9 cm).
- **DIET** Plant matter and some animal matter.
- **WATER** Temperature 77–79°F (25–26°C); alkaline (pH 8.1–8.3) with SG 1.020–1.024.
- **TEMPERAMENT** Typically docile.

The Coral Beauty varies throughout its range, with stunning shades of blue and green as well as rich yellow hues all being evident. Juveniles have red flanks, broken by fine purple lines, rather than the yellow background of the adults (this is also seen in older individuals from the Philippines). As with all dwarf angelfish, algae and other plant matter must feature significantly in their diet.

Serrated fin

Centropyge vrolikii

Half-Black Angelfish

- **ORIGINS** Found in the Indo-Pacific region, as with the vast majority of Centropyge species.
- **SIZE** 4¾ in (12 cm).
- **DIET** Plant matter and prepared diets.
- **WATER** Temperature 77–79°F (25–26°C); alkaline (pH 8.1–8.3) with SG 1.020–1.024.
- **TEMPERAMENT** Not harmful to invertebrates.

The markings on the scales of this dwarf angelfish resemble tiny pearls—hence its common name. The lack of stripes across the body distinguishes it from Eibl's Angelfish (*see opposite*). The Half-Black Angelfish's subdued coloration means that it is less popular than other, more colorful members of the group, but it is easy to care for, thriving in an established tank where it can graze on algae.

Centropyge loriculus

Flame Angelfish

- **ORIGINS** East of the Philippines to Samoa, occasionally Hawaii, and south to the Great Barrier Reef.
- **SIZE** 4 in (10 cm).
- **DIET** Plant matter plus other foods.
- **WATER** Temperature 77–79°F (25–26°C); alkaline (pH 8.1–8.3) with SG 1.020–1.024.
- **TEMPERAMENT** Rather shy.

This is one of the hardiest of the dwarf angelfish and its care, in terms of its diet, is straightforward. However, Flame Angelfish are not very tolerant of their own kind, so if you want to keep more than one, it is important to include plenty of retreats. To pair them up, obtain two juveniles. These will initially both be female, but one will develop into a male, becoming bigger and more colorful.

Variable blotches

Centropyge bicolor

Bicolor Angelfish

- **ORIGINS** The Indo-Pacific region, extending from Malaysia to northwestern Australia, Japan, and Samoa.
- **SIZE** 5 in (12.5 cm).
- **DIET** Eats animal protein and plant matter.
- **WATER** Temperature 77–79°F (25–26°C); alkaline (pH 8.1–8.3) with SG 1.020–1.024.
- **TEMPERAMENT** Usually placid.

A yellow and purple-blue coloration gives the Bicolor Angelfish a striking appearance. It may be possible to keep these dwarf angelfish in pairs or small groups. Females are able to change into males if the male of a group dies or is removed. Spawning normally occurs at dusk, with the eggs floating and being dispersed on the current, although successful breeding in aquariums is still a rare event. The young initially feed on plankton.

Centropyge flavissima

Lemonpeel Angelfish

- **ORIGINS** North of the Philippines to the Marianas, south to Rapa Island; also Cocos-Keeling and Christmas Island.
- **SIZE** 5½ in (14 cm).
- **DIET** Mainly plant matter, plus prepared foods.
- **WATER** Temperature 77–79°F (25–26°C); alkaline (pH 8.1–8.3) with SG 1.020–1.024.
- **TEMPERAMENT** Tolerant.

Blue fin edging distinguishes this species from juvenile Chocolate Surgeonfish *(see p.236)* and the closely related Herald's or Golden Dwarf Angelfish (*C. heraldi*). Rather confusingly, however, Lemonpeel Angelfish from Christmas Island also lack the blue markings. The Lemonpeel, like other dwarf angelfish, frequents the middle and lower levels of the tank and thrives in a reef-type setup.

Centropyge eibli

Eibl's Angelfish

- **ORIGINS** The Indo-Pacific region. (Only two of about 33 *Centropyge* species occur in the Caribbean.)
- **SIZE** 4 in (10 cm).
- **DIET** Plant matter, plus other marine foods.
- **WATER** Temperature 77–79°F (25–26°C); alkaline (pH 8.1–8.3) with SG 1.020–1.024.
- **TEMPERAMENT** Nonaggressive.

One of least brightly colored members of the group, Eibl's Angelfish is certainly not drab, with an attractive wavy body pattern and distinctive circles of gold and blue around the eyes. Maintaining good water quality in the aquarium is vital, just as it is for all angelfish. Replace up to a quarter of the volume every two weeks, especially in a new tank. Otherwise, the fish are likely to succumb to parasitic and bacterial diseases.

Centropyge potteri

Potter's Angelfish

- **ORIGINS** Occurs in the Pacific Ocean, especially in the area around Hawaii.
- **SIZE** 4 in (10 cm).
- **DIET** Prefers plant matter, and grazes on algae.
- **WATER** Temperature 77–79°F (25–26°C); alkaline (pH 8.1–8.3) with SG 1.020–1.024.
- **TEMPERAMENT** Not usually aggressive.

These attractive dwarf angelfish have a reputation for being one of the harder species to acclimatize successfully to aquarium surroundings. They are unlikely to thrive in a new tank setup, which will lack the algae they need in their diet. They should, however, settle well into an established reef aquarium with stable water chemistry. Most invertebrates are safe with these angelfish, although they occasionally prey on tubeworms.

CORAL RETREATS

Angelfish are naturally encountered among the lower levels of coral reefs, rather than close to the surface. They instinctively seek out snug hiding places where they can retreat and hide from any possible danger, and rest at night, as well. The tall, narrow body shape of angelfish means that they are able to slip easily into narrow crevices, beyond the reach of most predators, where the bright colors on their flanks are hidden from view. It is important to replicate these retreats within the aquarium in order to give the fish a sense of security. In the tank, an angelfish will tend to favor one particular crevice as its retreat.

Pomacanthus maculosus

Yellow Bar Angelfish

- **ORIGINS** Throughout the Persian Gulf and Red Sea, extending into northwestern parts of the Indian Ocean.
- **SIZE** 12 in (30 cm).
- **DIET** Animal-based foods and plant matter.
- **WATER** Temperature 77–79°F (25–26°C); alkaline (pH 8.1–8.3) with SG 1.020–1.025.
- **TEMPERAMENT** Bold and fearless.

Although it ranks among the most expensive of all angelfish, the Yellow Bar is one of the most easily kept of the larger species, thanks partly to its curiosity. Its inquisitive nature means that it can be weaned quite easily onto a range of suitable prepared diets, rather than the sponges and corals that form a substantial part of the natural diet of large angelfish. The yellow banding on the body of the adult Yellow Bar is variable both in size and shape. The young look very different from their elders, displaying a series of blue, white, and black bands across the body. It typically takes about 18 months for a small juvenile to acquire its adult coloration. Delayed development of adult color suggests that the water conditions in the aquarium may not be ideal.

Pomacanthus semicirculatus

Koran Angelfish

- **ORIGINS** Widely distributed, ranging from the Red Sea throughout the Indo-Pacific region.
- **SIZE** 16 in (40 cm).
- **DIET** Eats both plant and animal matter.
- **WATER** Temperature 77–79°F (25–26°C); alkaline (pH 8.1–8.3) with SG 1.020–1.024.
- **TEMPERAMENT** Territorial.

The juvenile form of the Koran Angelfish has a pattern of curving lines running along its body. As this angelfish matures, its blue background coloration is mainly replaced by brown, with blue edging to the fins and stripes across the face becoming evident. The pale markings that develop on the caudal fin are regarded as being similar to the Arabic script used in the holy Koran, explaining the unusual name of this fish.

Juvenile in transitional coloring

Pomacanthus navarchus

Majestic Angelfish

- **ORIGINS** Widely distributed throughout the Pacific Ocean.
- **SIZE** 10 in (25 cm).
- **DIET** Eats both animal foods and plant matter.
- **WATER** Temperature 77–79°F (25–26°C); alkaline (pH 8.1–8.3) with SG 1.020–1.024.
- **TEMPERAMENT** Not very social.

These angelfish need plenty of swimming space, as well as retreats where they can hide. The adults are very different from the young, which display white stripes on a dark blue background. Unfortunately, it is not always easy to persuade Majestic Angelfish to take substitute diets, so you should only buy specimens that you have seen feeding.

Girdle of blue stripes

Arusetta asfur

Asfur Angelfish

- **ORIGINS** Occurs throughout the Persian Gulf and the Red Sea.
- **SIZE** 20 in (50 cm).
- **DIET** Will eat both animal foods and plant matter.
- **WATER** Temperature 77–79°F (25–26°C); alkaline (pH 8.1–8.3) with SG 1.020–1.025.
- **TEMPERAMENT** Lively; can be aggressive.

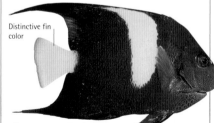

Distinctive fin color

This species is closely related to the Yellow Bar Angelfish *(see top left)*. While the young of both species are similar, an adult Asfur can be identified by its bright yellow caudal fin. It is not advisable to mix larger angelfish, especially those with similar patterning, because weaker individuals are likely to be bullied. Finding other suitable tankmates for these fish is difficult.

Pomacanthus annularis

Annularis Angelfish

- **ORIGINS** Widely distributed throughout the Indo-Pacific region.
- **SIZE** 16 in (40 cm); 10 in (25 cm) in aquariums.
- **DIET** Omnivorous; likes to graze on algae.
- **WATER** Temperature 77–79°F (25–26°C); alkaline (pH 8.1–8.3) with SG 1.020–1.024.
- **TEMPERAMENT** Adults are antisocial in aquariums.

In addition to the distinctive blue circle located above and behind the gills, adult Annularis have a series of blue lines running across the body and fins. Slight differences in patterning are evident between individuals. Young fish display a series of near-horizontal white lines against a blue background. Annularis sometimes become sufficiently tame to feed from the hand.

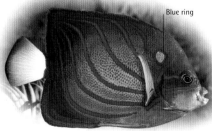

Blue ring

Pomacanthus paru

French Angelfish

- **ORIGINS** Northern area of the Gulf of Mexico south to Brazil. Also recorded in the vicinity of Ascension Island.
- **SIZE** 15 in (38 cm).
- **DIET** Animal-based foods plus plant matter.
- **WATER** Temperature 77–79°F (25–26°C); alkaline (pH 8.1–8.3) with SG 1.020–1.024.
- **TEMPERAMENT** Nonaggressive.

Young French Angelfish are black with yellow stripes. The adults are a duller grayish-black flecked with yellow. The pictured fish is starting to develop this coloration as its stripes fade. In the wild, juveniles act as "cleaners," removing parasites from the adults. Their different appearance means their advances are less likely to be seen as a threat.

Pomacanthus xanthometopon

Blue-Faced Angelfish

- **ORIGINS** Occurs on coral reefs throughout the Indo-Pacific region.
- **SIZE** 15 in (38 cm); 12 in (30 cm) in aquariums.
- **DIET** Both plant matter and animal foods.
- **WATER** Temperature 77–79°F (25–26°C); alkaline (pH 8.1–8.3) with SG 1.020–1.024.
- **TEMPERAMENT** Territorial.

The Blue-Faced Angelfish is difficult to wean onto an artificial diet, so be prepared to offer a wide range of foods to assist this process. As with other species, breeding in the home aquarium is unlikely to be successful. In the wild, external fertilization occurs, with the eggs floating and the fry then developing in plankton near the surface. After a month, they make their way back down to the reef.

Blue on the face

DISRUPTIVE CAMOUFLAGE

To human eyes, the colorful striped patterning associated with many angelfish is both striking and highly attractive. But in the wild, such colorful patterning is crucial to the survival of these fish because it breaks up the outline of their body shapes. As a result, especially when viewed from a distance, an angelfish's distinctive markings help it to blend in with the background of its reef habitat, concealing its presence from the eyes of predators. This phenomenon, which is the same as that used by animals such as tigers, is described by biologists as disruptive camouflage.

Pomacanthus imperator

Emperor Angelfish

- **ORIGINS** From the Red Sea eastward through the Indo-Pacific to Japan, Hawaii, and Tahiti.
- **SIZE** 16 in (40 cm).
- **DIET** Consumes both animal and plant foods.
- **WATER** Temperature 77–79°F (25–26°C); alkaline (pH 8.1–8.3) with SG 1.020–1.024.
- **TEMPERAMENT** Domineering and territorial.

There is no obvious similarity in appearance between the adult Emperor Angelfish, shown here to the left of the picture, and the juvenile pictured on the bottom right, which has dark blue coloring and a pattern made up of concentric, semicircular lines. Indeed, they appear so different that up until the 1930s, young Emperor Angelfish were believed to be a separate species. Their darker coloration may help them to merge into the background of coral, while mature individuals are much more conspicuous, swimming readily over the reef. It is best to start with young individuals, which should adjust well to aquarium surroundings. As with other large angelfish, Emperors should not be kept in a reef aquarium, because they naturally prey on a wide variety of invertebrates, from corals and tubeworms to sea anemones.

BUTTERFLYFISH

The elegant swimming motion of these fish has been likened to the flight of butterflies fluttering around flowers in the garden. As with their insect counterparts, butterflyfish also tend to be brightly colored, making them an extremely popular choice for the home marine aquarium. However, some species that subsist entirely on coral polyps in the wild are difficult to switch to substitute diets. Good water quality is absolutely vital when keeping any butterflyfish, since they will readily show signs of illness if the pH starts to fall, long before other fish sharing the same tank become affected.

The **Philippine Butterflyfish** (*Chaetodon adiergastos*) in this shoal display the rather flattened, oval body shape that is characteristic of the group as a whole.

Chaetodon rafflesii

Latticed Butterflyfish

- **ORIGINS** Ranges from East Africa, via the Indo-Pacific region, into the Pacific Ocean.
- **SIZE** 6¾ in (17 cm).
- **DIET** Should take meaty foods.
- **WATER** Temperature 79–82°F (26–28°C); alkaline (pH 8.2–8.3) with SG 1.021–1.024.
- **TEMPERAMENT** Territorial.

Cross-hatched patterning, a black band down the face, and a blue area between the eyes all help to identify this species. In addition, the caudal fin displays a vertical black bar and is edged with yellow. Like other members of the group, Latticed Butterflyfish cannot be kept safely with corals or other invertebrates, since they will destroy them.

Chaetodon ulietensis

Double-Saddle Butterflyfish

- **ORIGINS** Extends from Australia's Great Barrier Reef northward and eastward through the Pacific.
- **SIZE** 6¾ in (17 cm).
- **DIET** Should take meaty foods.
- **WATER** Temperature 79–82°F (26–28°C); alkaline (pH 8.2–8.3) with SG 1.021–1.024.
- **TEMPERAMENT** Territorial.

The prominent black spot on the caudal peduncle, which is otherwise yellowish, is characteristic of the Double-Saddle, but the key feature of this species is the white, saddlelike area extending down the sides of the body from the center of the back. Butterflyfish can be distinguished from angelfish because they do not have a spine on the gill cover. Butterflyfish are diurnal by nature.

Chaetodon melannotus

Black-Back Butterflyfish

- **ORIGINS** Ranges over a wide area, from the Red Sea through the Indian Ocean and into the Pacific.
- **SIZE** 6¾ in (17 cm).
- **DIET** Should take meaty foods.
- **WATER** Temperature 79–82°F (26–28°C); alkaline (pH 8.2–8.3) with SG 1.021–1.024.
- **TEMPERAMENT** Territorial.

The Black-Back has yellow edging around its body and a blackish area on its back. The black patch may be more extensive in some individuals than others, probably reflecting regional variations. It is best to wait for several months before introducing Black-Backs or any other butterflyfish to a new tank, because they will adapt much better to a mature tank system with a stable water chemistry.

Chaetodon auriga

Threadfin Butterflyfish

- **ORIGINS** Ranges from East Africa through the Indian Ocean and into the Pacific.
- **SIZE** 6³/₄ in (17 cm).
- **DIET** Should take meaty foods.
- **WATER** Temperature 79–82°F (26–28°C); alkaline (pH 8.2–8.3) with SG 1.021–1.024.
- **TEMPERAMENT** Territorial.

This fish often has a threadlike extension at the back of the dorsal fin. The front of the body is white, but the rear part is yellow, including the caudal fin. There are false eye-spots on the dorsal fin, while the real eyes are hidden by a black band. Butterflyfish have not yet been bred successfully in aquariums. In the wild, they spawn toward dusk near the surface, and their young feed on plankton.

Chaetodon ephippium

Saddleback Butterflyfish

- **ORIGINS** Ranges from East Africa through the Indian ocean and into the Pacific.
- **SIZE** 6³/₄ in (17 cm).
- **DIET** Should take meaty foods.
- **WATER** Temperature 79–82°F (26–28°C); alkaline (pH 8.2–8.3) with SG 1.021–1.024.
- **TEMPERAMENT** Territorial.

The prominent black area on the rear upperparts of this fish is separated from the rest of the body by a white band that curves down to the caudal peduncle. Orange-yellow markings occur on the face and throat and also along the edges of the caudal fin. As with other species, Saddlebacks are best housed singly; otherwise, disputes will occur.

Chaetodon collare

Red-Tail Butterflyfish

- **ORIGINS** From East Africa, via Indonesia, north to Japan and south to Australia's eastern coast.
- **SIZE** 7 in (18 cm).
- **DIET** Should take meaty foods.
- **WATER** Temperature 79–82°F (26–28°C); alkaline (pH 8.2–8.3) with SG 1.021–1.024.
- **TEMPERAMENT** Territorial.

The most obvious feature of this species of butterflyfish is the red tail, which has a narrow black-and-blue border at the rear. A striking white band extends down each side of the head behind the eyes. The overall coloration of the Red-Tail Butterflyfish is unusually dark for a member of this group, although there is an attractive diamond-shaped pattern stretching across most of the body.

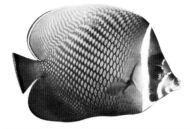

NIBBLERS OF THE REEF

The natural diet of butterflyfish, such as this Spotband Butterflyfish (*Chaetodon punctatofasciatus*), is rich in coral polyps and sponges, which the fish nibble with their extended jaws. As youngsters, however, some butterflyfish act as cleaners, using their mouthparts to remove parasites from other reef residents. The delicate jaws are easily damaged, so these fish need to be transported in large bags. Any abrasion to the mouth may cause them to refuse to eat after a move. Brine shrimp and worms can often be used to wean butterflyfish onto artificial diets.

Chaetodon lunula

Raccoon Butterflyfish

- **ORIGINS** Extends from the Red Sea eastward past Indonesia, reaching as far as the Pacific Ocean.
- **SIZE** 8¹/₄ in (21 cm).
- **DIET** Should take meaty foods.
- **WATER** Temperature 79–82°F (26–28°C); alkaline (pH 8.2–8.3) with SG 1.021–1.024.
- **TEMPERAMENT** Territorial.

One of the larger butterflyfish, the Raccoon has a distinguishing broad, black, sloping band that extends in a curve from the dorsal fin to the gill covers. Yellow-orange lines run along the edges of this band, while a white stripe across the top of the head separates it from the dark patch around the eyes, which resembles a mask. This gives the face a vaguely raccoonlike appearance. There is a large, black spot at the base of the tail.

Chaetodon semilarvatus

Red-Lined Butterflyfish

- 🌐 **ORIGINS** The distribution of this butterflyfish is restricted to the Red Sea.
- 🐚 **SIZE** 7¹/₂ in (19 cm).
- 🍽 **DIET** Should take meaty foods.
- 🌊 **WATER** Temperature 79–82°F (26–28°C); alkaline (pH 8.2–8.3) with SG 1.021–1.024.
- 😊 **TEMPERAMENT** Territorial.

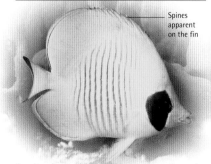

Spines apparent on the fin

A series of red, slightly curving lines run down the body of this fish, contrasting with its overall yellow coloration. On the face, there is an irregular black spot that extends around the eyes. When choosing a fish, check for signs of protozoan infections, since butterflyfish are prone to such diseases.

Chaetodon vagabundus

Vagabond Butterflyfish

- 🌐 **ORIGINS** From the Red Sea through the Indian Ocean and past Indonesia into the Pacific.
- 🐚 **SIZE** 6³/₄ in (17 cm).
- 🍽 **DIET** Should take meaty foods.
- 🌊 **WATER** Temperature 79–82°F (26–28°C); alkaline (pH 8.2–8.3) with SG 1.021–1.024.
- 😊 **TEMPERAMENT** Territorial.

Although the pattern of body markings is similar to that of the Threadfin Butterflyfish *(see p.257)*, the Vagabond can be identified at a glance by the black stripe extending down from the dorsal fin and a smaller black band across the tail. Vagabonds are considered harder to keep than Threadfins.

FEEDING ADVANTAGE

The long snout seen in many butterflyfish, especially the Long-Nosed Butterflyfish (*Forcipiger longirostris*), is an effective tool that enables these fish to probe the reef for edible items. The length of the Black Long-Nosed's snout is roughly a third of the fish's total length. This means that it can reach farther into the coral than its relatives, giving it an inherent advantage in the quest for food. Being adaptable in its feeding habits, this butterflyfish is one of the easier species to wean onto substitute foods in aquariums.

Chaetodon mertensii

Mertens's Butterflyfish

- 🌐 **ORIGINS** Range extends from the Great Barrier Reef, off Australia's eastern coast, into the Pacific Ocean.
- 🐚 **SIZE** 5¹/₂ in (14 cm).
- 🍽 **DIET** Should take meaty foods.
- 🌊 **WATER** Temperature 79–82°F (26–28°C); alkaline (pH 8.2–8.3) with SG 1.021–1.024.
- 😊 **TEMPERAMENT** Territorial.

This striking butterflyfish has variable black-and-white patterning on its flanks. The rear of the body varies from yellow to orange and is bordered by a narrow black and white band; the caudal fin is similar in color. There is a distinctive black spot just in front of and below the dorsal fin, and a vertical black stripe runs through each eye. The coloration of this species, as with other butterflyfish, typically changes at night. This helps to provide additional camouflage at a time when the fish would naturally retreat among the coral polyps for protection. When raised, the dorsal fin has a distinct serrated appearance. Mertens's is one of the smaller butterflyfish species. When buying, it is best to avoid exceptionally small individuals, because it can be difficult to acclimatize them successfully to home aquarium surroundings.

Chaetodon unimaculatus

Teardrop Butterflyfish

- **ORIGINS** The coast of East Africa eastward, via the Indian Ocean and Indonesia, to the Pacific.
- **SIZE** 8 in (20 cm).
- **DIET** Should take meaty foods.
- **WATER** Temperature 79–82°F (26–28°C); alkaline (pH 8.2–8.3) with SG 1.021–1.024.
- **TEMPERAMENT** Territorial.

The smudged appearance of the large black spot in the center of the upper body is responsible for the Teardrop Butterflyfish's name. As with most butterflyfish, there is also a black stripe running through the eyes. Narrow black edging extends along the rear of the body, crossing the caudal peduncle. The caudal fin is unusual in that it is essentially unpigmented, with a clear appearance.

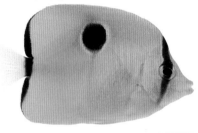

Chaetodon tinkeri

Tinker's Butterflyfish

- **ORIGINS** This butterflyfish is another of the striking species to be found in the Red Sea.
- **SIZE** 6¼ in (16 cm).
- **DIET** Should take meaty foods.
- **WATER** Temperature 79–82°F (26–28°C); alkaline (pH 8.2–8.3) with SG 1.021–1.024.
- **TEMPERAMENT** Territorial.

The distinctive appearance of Tinker's Butterflyfish means that it is easy to identify. The typical eye stripe is yellow rather than black, while the majority of the body is white and patterned with spots. An oblique dark patch extends from near the front of the dorsal fin to below the caudal peduncle. Unfortunately, Tinker's Butterflyfish cannot be sexed visually. Like all butterflyfish, this species needs well-oxygenated water to thrive.

Heniochus acuminatus

Bannerfish

- **ORIGINS** Extends across the Red Sea through the Indian Ocean and out into the Pacific.
- **SIZE** 10 in (25 cm).
- **DIET** Should take meaty foods.
- **WATER** Temperature 79–82°F (26–28°C); alkaline (pH 8.2–8.3) with SG 1.021–1.024.
- **TEMPERAMENT** Tolerant if kept in a shoal.

These butterflyfish are particularly eye-catching, thanks to the stunning elongation of the dorsal fin. The body bears two black bands separated by a silvery-white area, with the base of the dorsal fin and the caudal fin varying from yellow to yellow-orange. This latter feature helps to distinguish the Bannerfish from the Moorish Idol (see p.281).

— Face has black-and-white coloration

Forcipiger longirostris

Long-Nosed Butterflyfish

- **ORIGINS** Northern and western parts of the Indian Ocean, off East Africa, and in the Pacific region.
- **SIZE** 10¾ in (27 cm).
- **DIET** Should take meaty foods.
- **WATER** Temperature 79–82°F (26–28°C); alkaline (pH 8.2–8.3) with SG 1.021–1.024.
- **TEMPERAMENT** Territorial.

The black markings on this fish are confined to a triangular area extending down from the base of the dorsal fin to the level of the eye and then from here in a straight line to the end of the snout. The black dot just below the caudal peduncle is a false eye-spot. These and other butterflyfish must not be mixed with aggressive species, such as triggerfish and lionfish.

Chelmon rostratus

Copper-Band Butterflyfish

- **ORIGINS** From East Africa, via the Indian Ocean and Indonesia, into the Pacific.
- **SIZE** 8 in (20 cm).
- **DIET** Should take meaty foods.
- **WATER** Temperature 79–82°F (26–28°C); alkaline (pH 8.2–8.3) with SG 1.021–1.024.
- **TEMPERAMENT** Territorial.

This species has four coppery-orange bands edged with black running down the side of the body. There is a black circle with a white surround high up at the rear of the body and black markings on the tail. Butterflyfish may be crowded out by other fish at feeding time. Try offering food at several locations at once to make sure they get their share.

Bands evenly spaced

GOBIES

Gobies are an adaptable group of fish. Representatives are found in freshwater, marine, and brackish surroundings, in both tropical and temperate regions. The small size of most species and the fact that they are relatively easy to keep make these fish a good choice for the home aquarium, where some species have even been bred successfully. Males often guard the eggs, which may be laid in the crevices of rocks, and the young fish grow rapidly. Unfortunately, many species of gobies are naturally short-lived, so it is advisable to start out with young individuals.

Stunning patterning and bright coloration are features of many gobies and dartfish, as illustrated by these **Zebra Gobies** (*Ptereleotris zebra*).

Elacatinus oceanops

Neon Goby

- **ORIGINS** From the coast of Florida in the western Atlantic and down through the Caribbean region.
- **SIZE** 2 in (5 cm).
- **DIET** Marine flake and other small foods.
- **WATER** Temperature 77–79°F (25–26°C); alkaline (pH 8.1–8.3) with SG 1.020–1.024.
- **TEMPERAMENT** Relatively tolerant.

Neon Gobies have a brilliant blue stripe, edged with black, running from nose to tail. They belong to a group known as cleaner gobies, so called because part of their food intake in the wild comes from cleaning skin parasites off other fish. These gobies may adopt a favorite vantage point in the tank where water currents will carry suspended food particles to them. Neon Gobies rank among the most easily bred of all marine aquarium fish.

Cryptocentrus cinctus

Yellow Prawn Goby

- **ORIGINS** East Africa through the Indo-Pacific region, to Japan in the north and Australia in the south.
- **SIZE** 3 in (7.5 cm).
- **DIET** Small invertebrates and similar foods.
- **WATER** Temperature 77–79°F (25–26°C); alkaline (pH 8.1–8.3) with SG 1.020–1.024.
- **TEMPERAMENT** Tolerant but often territorial.

Bright sulfur-yellow coloration is characteristic of these fish. The pale bluish-white spots on the head also extend to the fins, while the body may display faint barring. The eyes are set high on the head, ensuring that the fish can still see when they are partially concealed in the substrate. These gobies need to be housed with symbiotic *Alpheus* pistol shrimp. The shrimp dig a burrow home which they share with the fish; in return, the gobies defend the burrow against predators.

Lythrypnus dalli

Catalina Goby

- **ORIGINS** Restricted to the eastern Pacific, where it occurs from southern California to the Gulf of California.
- **SIZE** 2¼ in (6 cm).
- **DIET** Small invertebrates.
- **WATER** Temperature 64–72°F (18–22°C); alkaline (pH 8.1–8.3) with SG 1.020–1.024.
- **TEMPERAMENT** Tolerant but often territorial.

The number of blue bands on the body of this vibrantly colored species can vary from three to six. The male, shown here, has longer front rays on its dorsal fin than the female. Originating from more temperate waters than most popular marine fish, Catalina Gobies should ideally be kept at a lower water temperature. They may still thrive in warmer water, but they will not live as long. These gobies can be bred successfully in the aquarium.

Nemateleotris magnifica

Firefish

- **ORIGINS** From East Africa eastward through the Indo-Pacific region, north to Japan and south to Australia.
- **SIZE** 2³/₄ in (7 cm).
- **DIET** Small crustaceans.
- **WATER** Temperature 77–79°F (25–26°C); alkaline (pH 8.1–8.3) with SG 1.020–1.024.
- **TEMPERAMENT** Social.

The most obvious feature of the Firefish is its long, narrow dorsal fin. The pelvic fins, which are not fused, are also long. A Firefish's creamy-white body becomes yellower toward the rounded head and redder toward the tail. These fish usually occupy the middle layer of the aquarium. When danger threatens, they retreat into a burrow excavated in the substrate, or sometimes into a rocky crevice.

Valenciennea strigata

Yellowhead Sleeper Goby

- **ORIGINS** From East Africa eastward through the Indo-Pacific region, north to Japan and south to Australia.
- **SIZE** 7 in (18 cm).
- **DIET** Small invertebrates.
- **WATER** Temperature 77–79°F (25–26°C); alkaline (pH 8.1–8.3) with SG 1.020–1.024.
- **TEMPERAMENT** Social.

This goby is easily identified by the bluish streaks on the sides of its yellow face. Yellowheads associate in small groups and communicate through movements of the mouth. Their tank should have a dense layer of sand, which the fish will excavate with their mouths in search of small worms and other edible items. The debris passes out through the gills and may cause harm if it is deposited on corals and similar invertebrates.

Valenciennea puellaris

Orange-Spotted Sleeper Goby

- **ORIGINS** The western Pacific, north to the Ryukyu Islands and south to Australia's Great Barrier Reef.
- **SIZE** 3 in (7.5 cm).
- **DIET** Small invertebrates.
- **WATER** Temperature 77–79°F (25–26°C); alkaline (pH 8.1–8.3) with SG 1.020–1.024.
- **TEMPERAMENT** Reasonably social.

Like other Prawn Gobies, this fish has formed a remarkable association with pistol shrimp (*Alpheus* species), whose burrows it shares. When the goby is looking out of the burrow, the shrimp knows that it is safe to come out. Keep these gobies in pairs. The female lays up to 1,000 eggs in a cave. She stays with them, and when they are due to hatch, the male piles up the substrate and seals her inside. She emerges with her brood when they are free-swimming.

Females lack the long, threadlike extensions to the first dorsal fin

UNMOVED BY TURBULENCE

Many gobies naturally inhabit turbulent areas of water, where strong currents could easily sweep them away. Some gobies have developed a unique mechanism that enables them to thrive in such potentially troublesome surroundings. In these fish, the pelvic fins—which are present on each side of the body, just below the gill covers—have become fused. This adaptation has created a suckerlike device that allows the fish to anchor itself firmly to rocks. The suction provided by the fins is so powerful that the fish can even cling to vertical surfaces.

Gobiodon citrinus

Citron Goby

- **ORIGINS** Red Sea eastward to Samoa, north to the Ryukyu Islands, and south to the Great Barrier Reef.
- **SIZE** 2¹/₂ in (6.25 cm).
- **DIET** Marine flake and other small foods.
- **WATER** Temperature 77–79°F (25–26°C); alkaline (pH 8.1–8.3) with SG 1.020–1.024.
- **TEMPERAMENT** Relatively tolerant.

The orange-yellow coloration of the Citron Goby is broken by light bluish stripes on the head and at the base of the anal and dorsal fins. Like other gobies, it can be distinguished from blennies of a similar size by the fact that the dorsal fin is clearly divided into two parts. Citrons are more confident than other gobies, swimming freely, perhaps because they are protected by a foul-tasting body mucus that deters most predators.

SQUIRRELFISH

The red color, large eyes, and fitful movements of these attractive fish are reminiscent of terrestrial squirrels. The eye size is an adaptation for nocturnal living in sheltered areas near the base of coral reefs, where it is vital to gather as much light as possible from the gloomy surroundings. Certain species are known as soldierfish, and this name is often used interchangeably with squirrelfish, but in zoological terms, only the subfamily Myripristinae can be correctly described as soldierfish.

Squirrelfish spend the daylight hours sheltering in crevices then emerge at night to patrol their territory. This is a group of **Bigeye Soldierfish** (*Myripristis bernelti*).

Myripristis bernelti

Bigeye Soldierfish

- ⊕ **ORIGINS** Found on reefs over a wide area from East Africa across the Pacific to Tahiti and Hawaii.
- ◑ **SIZE** 10 in (25 cm).
- ◐ **DIET** Animal-based foods.
- ◒ **WATER** Temperature 72–79°F (22–26°C); alkaline (pH 8.0–8.3) with SG 1.022–1.025.
- ⊕ **TEMPERAMENT** Social with its own kind.

The large eyes and the mottled appearance caused by the dark-edged scales are the most distinctive features of this species. It will adapt to daylight feeding. Soldierfish are fairly easy to keep, but they should be housed in groups; otherwise, they remain nervous in aquarium surroundings. This means that a relatively large tank is required.

Dark vertical streak at the edge of the gill covers

Myripristis vittata

Whitetip Soldierfish

- ⊕ **ORIGINS** Indo-Pacific; range includes the Maldives, Australia's Great Barrier Reef, and Samoa.
- ◑ **SIZE** 15 cm (6 in).
- ◐ **DIET** Fresh or thawed invertebrates, such as mussels.
- ◒ **WATER** Temperature 72–79°F (22–26°C); alkaline (pH 8.0–8.3) with SG 1.022–1.025.
- ⊕ **TEMPERAMENT** Social with its own kind.

White leading edge to fins

The white tips on the front dorsal fin help to identify the Whitetip Soldierfish. Like other soldierfish, it can vocalize, making a grunting sound by contracting the muscles surrounding its swim bladder and grinding its pharyngeal teeth (located at the back of the throat). Grunting is most likely to be heard if the fish feels threatened. This species usually swims in the lower part of the aquarium, although, like other soldierfish, it prefers to feed in the upper levels, rather than from the substrate.

Sargocentron caudimaculatum

Tailspot Squirrelfish

- ⊕ **ORIGINS** Ranges from East Africa, through the Indo-Pacific, to the Maldives, Vanuatu, and Samoa.
- ◑ **SIZE** Up to 10 in (25 cm).
- ◐ **DIET** Fresh or prepared meat-based foods.
- ◒ **WATER** Temperature 72–79°F (22–26°C); alkaline (pH 8.0–8.3) with SG 1.022–1.025.
- ⊕ **TEMPERAMENT** Shy, but social with its own kind.

The large scales that characterize this group of fish are clearly visible in the Tailspot Squirrelfish, where they are emphasized by their silvery outlines. These squirrelfish are shy by nature, so the aquarium needs to incorporate the equivalent of rocky ledges, where the fish can hide away. In addition, the lighting must not be too bright, or it may deter the fish from feeding.

Plain silvery white area

Sargocentron violaceum

Violet Squirrelfish

- **ORIGINS** Reefs throughout the Indo-Pacific region, including the Maldives, Vanuatu, and Samoa.
- **SIZE** 9 in (23 cm).
- **DIET** Likes crustaceans, such as shrimp.
- **WATER** Temperature 72–79°F (22–26°C); alkaline (pH 8.0–8.3) with SG 1.022–1.025.
- **TEMPERAMENT** Not suitable for a reef aquarium.

Both dorsal fins erect

Squirrelfish generally have a divided dorsal fin, with the taller part located close to the caudal fin. In this species, the fins as well as the body are suffused with violet. Found at greater depths than many other reef fish, Violet Squirrelfish will tolerate lower water temperatures. They will not thrive if the water in the aquarium is too warm, and often become reluctant to feed.

Sargocentron diadema

Crown Squirrelfish

- **ORIGINS** Ranges widely from the Red Sea to the Hawaiian islands.
- **SIZE** 8 in (20 cm).
- **DIET** Shellfish and fish.
- **WATER** Temperature 72–79°F (22–26°C); alkaline (pH 8.0–8.3) with SG 1.022–1.025.
- **TEMPERAMENT** Not suitable for mixing with smaller fish.

The white stripes running along the sides of the body help to distinguish the Crown Squirrelfish from related species. The stripes extend across the gill covers and around the mouth. These are lively, active fish and, like other members of the group, are best kept as a small shoal in a single-species setup. Relatively subdued lighting above the aquarium is recommended. A blue fluorescent night-light will allow you to watch the fish as they become active after dark.

Neoniphon opercularis

Blackfin Squirrelfish

- **ORIGINS** Found throughout the Indo-Pacific region, including the Maldives and Samoa.
- **SIZE** Up to 14 in (36 cm).
- **DIET** Fresh and thawed invertebrates.
- **WATER** Temperature 72–79°F (22–26°C); alkaline (pH 8.0–8.3) with SG 1.022–1.025.
- **TEMPERAMENT** Highly social with its own kind.

Front black dorsal fin folded down

Like other squirrelfish, the Blackfin lives in shoals. Although a group of squirrelfish in a tank is likely to contain both males and females, it has not yet been possible to breed squirrelfish in home aquariums. They probably need the stimulus of large numbers of their own kind to trigger spawning. Space is likely to be restricted in the tank, so there will almost certainly be too few fish in the shoal. Only when the females swell with eggs is any difference evident between the sexes.

Holocentrus rufus

Longspine Squirrelfish

- **ORIGINS** Tropical western Atlantic, including the vicinity of the Bahamas.
- **SIZE** 6 in (15 cm).
- **DIET** Meat-based foods.
- **WATER** Temperature 72–79°F (22–26°C); alkaline (pH 8.0–8.3) with SG 1.022–1.025.
- **TEMPERAMENT** Kills invertebrates and small companions.

A distinctive feature of this species is the long rear part of the dorsal fin. All squirrelfish are well protected against would-be predators by the sharp spines incorporated into their fins. In this particular genus, additional protection is afforded by similar projections on the gill covers. This means that they need to be caught with care. It is preferable to steer them into a suitable container, rather than using a net and risking injuring the fish.

WHEN RED BECOMES BLACK

The vivid red colors of squirrelfish make them highly sought after by aquarists, but the real function of the intense color is to camouflage the fish in their natural habitat. Red light does not penetrate well through water, while blue light passes down to the greater depths inhabited by squirrelfish. The absence of red light results in the red fish appearing black, which makes them difficult to spot in the dim water around the reef base. Squirrelfish and soldierfish tend to be more active after dark, so they will not thrive in a brightly lit tank.

LIONFISH AND SCORPIONFISH

Beautiful but potentially deadly, the members of this family are equipped with a painful venom. Their tank needs to be serviced with great care, because it is easy to catch a hand on one of their stinging spines. Transferring fish between tanks also needs to be carried out with caution. Never be tempted to use your hand to free a fish that becomes enmeshed in the material of the net, since this can result in a painful sting. Make sure you use a net that is large enough to accommodate the entire fish, and then, having caught it, invert the net carefully to let the fish swim out on its own. It will soon free itself, even if it is initially caught up in the netting.

The colorful, boldly marked **Zebra Lionfish** (*Dendrochirus zebra*) can be recognized by the dark spot at the base of the gill cover.

Rhinopias frondosa

Goose Scorpionfish

- 🌐 **ORIGINS** Ranges from East Africa across the Pacific to Indonesia, southern Japan, and the Caroline Islands.
- ♻ **SIZE** 9 in (23 cm).
- 🐟 **DIET** Fish-based foods.
- 💧 **WATER** Temperature 77–79°F (25–26°C); alkaline (pH 8.0–8.3) with SG 1.021–1.024.
- 😐 **TEMPERAMENT** Reasonably compatible with its own kind.

The color of these fish varies across their range and can also be influenced by their surroundings. Predatory by nature, they are not suitable for keeping with smaller fish or invertebrates. They ambush their prey, since they are not powerful swimmers. Like lionfish, their eyes are located on the top of the head, ensuring good visibility.

Pterois volitans

Volitans Lionfish

- 🌐 **ORIGINS** The Pacific region, from the Malay Peninsula to Japan, the eastern coast of Australia, and Pitcairn Island.
- ♻ **SIZE** 15 in (38 cm).
- 🐟 **DIET** All meat-based foods.
- 💧 **WATER** Temperature 77–79°F (25–26°C); alkaline (pH 8.0–8.3) with SG 1.021–1.024.
- 😐 **TEMPERAMENT** Smaller companions will be eaten.

Slow-moving by nature, the Volitans Lionfish is one of the larger members of its group. It spends more time swimming in the middle and upper layers of the tank than related species. The rays forming both dorsals fin are separate; the pectoral fins on either side of the body are partly divided.

Pectoral fins resemble feathers

Venom is present in the fins

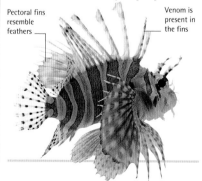

Pterois radiata

Clearfin Lionfish

- 🌐 **ORIGINS** Occurs over a very wide area, from the Red Sea eastward across the Indo-Pacific to Oceania.
- ♻ **SIZE** 8 in (20 cm).
- 🐟 **DIET** Fish-based foods.
- 💧 **WATER** Temperature 77–79°F (25–26°C); alkaline (pH 8.0–8.3) with SG 1.021–1.024.
- 😐 **TEMPERAMENT** Will prey on invertebrates.

The white banding on the tail and the white tips to the dorsal and pectoral fins are characteristic of the Clearfin. If you are stung by any lionfish, put your hand in hot water to coagulate the venom, and pour vinegar on the wound to ease the pain. You should seek prompt medical advice.

White-tipped, not mottled, spines

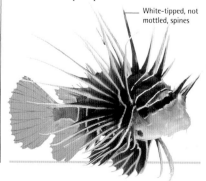

Pterois antennata

Spotfin Lionfish

- **ORIGINS** From East Africa and the Red Sea through the Indo-Pacific to eastern Asia and Australia.
- **SIZE** 8 in (20 cm).
- **DIET** Fish-based foods.
- **WATER** Temperature 77–79°F (25–26°C); alkaline (pH 8.0–8.3) with SG 1.021–1.024.
- **TEMPERAMENT** Tends to prey on smaller companions.

The dark spots with white edging running across the pectoral fins are clearly apparent when the fins are extended. If the fin rays are damaged, they will usually regrow over the course of several months, but perhaps not to their previous length. Lionfish may live for 12 years or more in an aquarium.

Shape of fish resembles seaweed

SEDENTARY SCORPIONS

Scorpionfish, such as the Merlet Scorpionfish (*Rhinopias aphanes*) pictured below, are sedentary, making them difficult to spot against the background of the reef. Species that have elaborate fins use them like fans, moving them back and forth to steer prey into a position—such as up against a rock—where it can be engulfed by the surprisingly large mouth. Lionfish sometimes shake and twitch unexpectedly: this is normal behavior, designed to shed dead skin and displace algae or even parasites from their bodies.

Pterois miles

Devil Lionfish

- **ORIGINS** Distribution restricted to the vicinity of the Red Sea and East Africa.
- **SIZE** 12 in (30 cm).
- **DIET** Invertebrates and fish-based foods.
- **WATER** Temperature 77–79°F (25–26°C); alkaline (pH 8.0–8.3) with SG 1.021–1.024.
- **TEMPERAMENT** Tolerant of its own kind.

The fins of this lionfish are very broad and banded along their length, rather than tapering to a point. Lionfish are initially reluctant to take inert foods. Offering them goldfish as a substitute diet is not recommended: quite apart from welfare concerns, the lionfish are likely to overeat and may die from gut impaction and liver failure. Weaning a lionfish onto prepared foods can be achieved by using feeding tongs to dangle food near the mouth.

Dendrochirus biocellatus

Fumanchu Lionfish

- **ORIGINS** The Pacific region, from the coast of Southeast Asia northward toward Japan and south to Australia.
- **SIZE** 5 in (12.5 cm).
- **DIET** Invertebrates and fish-based foods.
- **WATER** Temperature 77–79°F (25–26°C); alkaline (pH 8.0–8.3) with SG 1.021–1.024.
- **TEMPERAMENT** Safe with placid, similar-sized fish.

The Fumanchu, one of a number of dwarf lionfish, has a prominent pair of eyespots at the rear of its body. Dwarf lionfish mix better with other species than their larger relatives, but they are unlikely to breed in the typical home aquarium. Some species give birth to live young, but most are egg-layers that spawn near the surface.

Dendrochirus brachypterus

Shortfin Lionfish

- **ORIGINS** Range extends from the Red Sea and the Indo-Pacific to islands in Oceania.
- **SIZE** Up to 7 in (18 cm).
- **DIET** Invertebrates and fish-based foods.
- **WATER** Temperature 77–79°F (25–26°C); alkaline (pH 8.0–8.3) with SG 1.021–1.024.
- **TEMPERAMENT** Safe with larger, nonaggressive fish.

The Shortfin Lionfish's fins are shorter but no less deadly than those of other lionfish. The males of this species become darker when in breeding condition. As with other members of this group, this fish must not be exposed to bright lighting above its tank, which could damage its eyes.

WRASSE

Wrasse are lively, active fish found in most warm seas, though the majority come from tropical waters. They tend to be colorful, displaying elaborate patterning on their bodies. Their shape varies from elongated to deep and compressed, and there is a significant range in size among members of the group; some of the larger species are too big for the home aquarium. They are not hard to maintain, feeding readily, but even tame specimens can inflict a painful bite if offered food by hand. Their fanglike teeth are more usually applied to cracking into invertebrates, such as sea urchins, which form part of their natural diet.

The orange markings of these young **Orange-Spot Wrasse** (*Coris aygula*) will disappear as they grow. They can attain lengths of up to 4 ft (1.2 m).

Cirrhilabrus rubriventralis

Fairy Wrasse

- **ORIGINS** Northern stretches of the Red Sea; also reported in the waters around Sri Lanka.
- **SIZE** 4 in (10 cm).
- **DIET** Fresh and thawed marine foods; also algae.
- **WATER** Temperature 77–79°F (25–26°C); alkaline (pH 8.0–8.3) with SG 1.020–1.025.
- **TEMPERAMENT** Can be kept safely with other small fish.

Do not mix this wrasse species with invertebrates such as crustaceans, because it is likely to prey on them. In the wild, a solitary male, recognizable by the blue longitudinal stripes along its body and the prominent black spot at the top of the caudal peduncle, lives in association with a group of females. Fairy Wrasse stay close to the reef, rarely being observed swimming in open water. Nothing is known about their breeding habits.

Coris formosa

Formosa Wrasse

- **ORIGINS** From East Africa through the Indo-Pacific region, extending as far as Australia.
- **SIZE** Up to 20 in (50 cm).
- **DIET** Fresh and thawed marine foods.
- **WATER** Temperature 77–79°F (25–26°C); alkaline (pH 8.0–8.3) with SG 1.020–1.025.
- **TEMPERAMENT** Will prey on invertebrates.

Formosa Wrasse change dramatically as they mature. Juveniles are not dissimilar to some Clownfish in their coloration, being orange and white with black borders. Adult Formosa Wrasse have a predominantly green body dotted with dark spots, and blue stripes on the head. These wrasse bury themselves in the substrate of their aquarium and become solitary by nature as they grow older.

Coris gaimard

Clown Wrasse

- **ORIGINS** Extends from the Red Sea through much of the Pacific region, reaching Japan and Hawaii.
- **SIZE** 16 in (40 cm).
- **DIET** Meat-based foods, such as invertebrates.
- **WATER** Temperature 77–79°F (25–26°C); alkaline (pH 8.0–8.3) with SG 1.020–1.025.
- **TEMPERAMENT** Do not mix with small companions.

Adult coloration

Young Clown and Formosa Wrasse are very similar, but a close examination of the Clown's head shows that the black-bordered stripe does not extend below the top of the eye (as it does on the Formosa Wrasse), and the black spot is absent from the dorsal fin. Mature Clown Wrasse can be sexed by the green stripe above the male's anal fin. There are also regional color differences; Pacific fish, for example, have a bright yellow caudal fin.

Gomphosus varius

Bird Wrasse

- ⬡ **ORIGINS** Extends over a wide area, from East Africa across the Pacific Ocean as far as Hawaii.
- ⬡ **SIZE** 12 in (30 cm).
- ⬡ **DIET** Omnivorous, but prefers meat-based foods.
- ⬡ **WATER** Temperature 77–79°F (25–26°C); alkaline (pH 8.0–8.3) with SG 1.020–1.025.
- ⬡ **TEMPERAMENT** Not social with its own kind.

Female

Male Bird Wrasse are greenish, while females are browner on the flanks with an orange stripe running from the snout to the eyes. The pronounced snout enables this wrasse to forage for food in crevices. The snout will not be apparent until the fish is about 4 in (10 cm) long. Up until this stage, the fish is small enough to swim directly into crevices on the reef in search of edible items swept there by the currents. It is best to obtain young fish, which will be more adaptable in their feeding habits. Older individuals will not forage for their food in the open, although they are likely to graze on algal growth in the aquarium.

Halichoeres trispilus

Four-Spot Wrasse

- ⬡ **ORIGINS** Found in the vicinity of East Africa; replaced by a similar species (*H. chrysus*) farther east.
- ⬡ **SIZE** 4 in (10 cm).
- ⬡ **DIET** Prepared foods.
- ⬡ **WATER** Temperature 77–79°F (25–26°C); alkaline (pH 8.0–8.3) with SG 1.020–1.025.
- ⬡ **TEMPERAMENT** Very active; intolerant of their own kind.

The common name of this dwarf wrasse stems from the presence of three spots on the dorsal fin and another on the caudal peduncle, although it is not the only species with this type of patterning. In the most common coloration, the top half of the body is yellow, while the underparts are white. Dominant males, like the one shown below, are more brightly colored. Although small, Four-Spots are still aggressive toward one another.

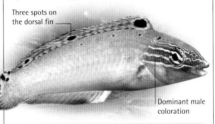

Three spots on the dorsal fin

Dominant male coloration

SPECIAL RELATIONSHIP

The interaction between the Blue Streak Cleaner Wrasse shown below *(see p.268)* and other reef fish is a good example of interspecies communication. The wrasse position themselves at particular areas on the reef, and other species come to be cleaned. The fish adopt different postures to indicate to the wrasse which parts of their bodies need grooming. The normal relationship between predator and prey is suspended, enabling the wrasse to venture into the jaws of some larger fish without the risk of being snapped up.

Bodianus rufus

Spanish Hogfish

- ⬡ **ORIGINS** Found throughout the Caribbean region as far as Bermuda, and southward to Brazil.
- ⬡ **SIZE** 12 in (30 cm).
- ⬡ **DIET** Prefers to feed on invertebrates.
- ⬡ **WATER** Temperature 77–79°F (25–26°C); alkaline (pH 8.0–8.3) with SG 1.020–1.025.
- ⬡ **TEMPERAMENT** Males will quarrel with each other.

The striking coloration of the Spanish Hogfish is consistent between individuals, and the patterning of these fish does not alter significantly during their lives. The adults can be readily sexed, because the heads of males turn from blue to yellow when they mature. However, the depth of water in which these hogfish occur may affect their appearance; those found in deeper areas tend to be redder along their backs. They need a large aquarium, and must be kept with companions of a similar size. Take care when handling these fish, since they have razor-sharp teeth.

Male

Lienardella fasciatus

Harlequin Tuskfish

- ⬡ **ORIGINS** The Pacific region, close to Asia and Australia, where it is present on the Great Barrier Reef.
- ⬡ **SIZE** 10 in (25 cm).
- ⬡ **DIET** Meat-based foods.
- ⬡ **WATER** Temperature 77–79°F (25–26°C); alkaline (pH 8.0–8.3) with SG 1.020–1.025.
- ⬡ **TEMPERAMENT** Solitary by nature.

In spite of its name, the Harlequin Tuskfish is a wrasse, albeit with a relatively broad body and distinctive blue teeth. The flexible jaws allow the teeth to protrude, when they resemble tusks. This enables the fish to turn over stones and grab invertebrates. The young display dark eye-spots on their pelvic, dorsal, and anal fins, which disappear as they mature. Tuskfish will burrow into the substrate, like other wrasse.

Bodianus anthioides

Lyretail Hogfish

🌐 **ORIGINS** From the Red Sea eastward through much of the Pacific to Japan and New Caledonia.

🔄 **SIZE** 10 in (25 cm).

🍽 **DIET** Primarily invertebrates.

💧 **WATER** Temperature 77–79°F (25–26°C); alkaline (pH 8.0–8.3) with SG 1.020–1.025.

😐 **TEMPERAMENT** Do not mix with invertebrates.

These stunningly attractive wrasse undergo a progressive color change. In young fish, the area around the lips is white, and the front of the body is more orange than the red seen in adults. The black areas on the body become more evident as the fish age. The Lyretail Hogfish has an active nature and will spend long periods swimming around the tank. It will not burrow into the substrate, like some wrasse, but prefers to have a suitable area in among the rockwork of the aquarium where it can rest. Avoid mixing Lyretails with invertebrates such as crustaceans, which are their natural prey in the wild. Companion fish should be of a similar size, since Lyretail Hogfish are territorial. Unfortunately, it does not appear possible to sex these wrasse visually. Their reproductive behavior has not been documented to date.

Labroides dimidiatus

Blue Streak Cleaner Wrasse

🌐 **ORIGINS** From the Red Sea through the Indo-Pacific to Japan and eastern Australia, extending to Oceania.

🔄 **SIZE** 4 in (10 cm).

🍽 **DIET** Marine flake and freeze-dried foods.

💧 **WATER** Temperature 77–79°F (25–26°C); alkaline (pH 8.0–8.3) with SG 1.020–1.025.

😐 **TEMPERAMENT** Can be kept together.

Like other cleaners, Blue Streaks naturally feed on the sides of other reef fish, removing parasites and eating mucus and loose skin. They must not be housed with delicate-skinned fish, since their attentions may cause damage, but they normally adapt readily to an alternative diet of prepared foods. Keep only one in the tank if other species are present; otherwise, the cleaners may harass the rest of the occupants. Pairs in good condition may spawn in the aquarium, but attempts to rear the fry almost always fail. At night, Blue Streak Wrasse retreat into dark crevices.

Long, narrow body, with obvious striping

Larabicus quadrilineatus

Four-Line Wrasse

🌐 **ORIGINS** The distribution of this particular wrasse appears to be restricted to the Red Sea.

🔄 **SIZE** 5 in (12.5 cm).

🍽 **DIET** Prepared diets, including thawed live foods.

💧 **WATER** Temperature 77–79°F (25–26°C); alkaline (pH 8.0–8.3) with SG 1.020–1.025.

😐 **TEMPERAMENT** Adults more aggressive than young.

Low dorsal fin

Known only since 1973, Four-Line Wrasse undergo a dramatic change in feeding habits as they mature. While young, they obtain food by cleaning other fish, but when they become adults, they switch to feeding on coral polyps. As a result, they are not a good choice for a reef aquarium, despite their small size. Males also change in color, losing the striped body patterning but developing a small angular blue stripe under the eye.

COLOR CHANGES

There is often such a difference in the color of young and adult wrasse that they look like separate species. The fish below is a juvenile Diana's Hogfish (*Bodianus diana*), but when it matures, its body will turn red and it will lose the white spots. Males may also change color if their status alters. In each population, there is a dominant male—the largest and most brightly colored individual—plus a number of females and subordinate males. If the dominant male dies, one of the subordinate males will change color and assume the role.

Thalassoma jansenii

Jansen's Wrasse

- **ORIGINS** From the Maldives eastward to Fiji, northward to Japan, and as far south as Australia's east coast.
- **SIZE** 7 in (18 cm).
- **DIET** Krill, mussels, and similar foods.
- **WATER** Temperature 77–79°F (25–26°C); alkaline (pH 8.0–8.3) with SG 1.020–1.025.
- **TEMPERAMENT** Do not mix with other wrasse.

Jansen's Wrasse is active by nature, so its tank should include plenty of open space for swimming, as well as suitable retreats where the fish can hide away, especially at night. It cannot be housed safely in a reef aquarium with invertebrates. Males can usually be recognized by their brighter coloration, particularly on the head. Good water quality, along with currents that mimic those of the reef, is important to ensure good health.

Thalassoma duperrey

Saddle Wrasse

- **ORIGINS** Found in the Pacific Ocean in the vicinity of the Hawaiian islands, where it is very common.
- **SIZE** 12 in (30 cm).
- **DIET** Various crustacean-based foods.
- **WATER** Temperature 77–79°F (25–26°C); alkaline (pH 8.0–8.3) with SG 1.020–1.025.
- **TEMPERAMENT** Becomes more territorial with age.

As the Saddle Wrasse becomes older, it develops the characteristic light band of color behind the head. This varies from orange to yellow, depending on the individual. Dominant males display the brightest coloration. Young fish have a dark upper body, with paler underparts. Saddle Wrasse have good appetites and grow rapidly, so the aquarium's filtration system must be highly efficient. Saddle Wrasse should generally be housed individually.

Thalassoma hardwicke

Six-Bar Wrasse

- **ORIGINS** From East Africa to Japan, southward to Australia, and extending as far as Tuamotu in Oceania.
- **SIZE** 8 in (20 cm).
- **DIET** Will eat prepared foods, even marine flake.
- **WATER** Temperature 77–79°F (25–26°C); alkaline (pH 8.0–8.3) with SG 1.020–1.025.
- **TEMPERAMENT** Territorial.

Vertical black barring across the body

As with other *Thalassoma* wrasse, the Six-Bar needs a sandy base to its aquarium so that it can burrow, but it will also colonize rocky retreats. The feeding habits of Six-Bars are such that they cannot be trusted with invertebrates. They are very lively, so choose tankmates with a similar nature. Avoid slow swimmers, opting instead for active fish such as tangs or even triggerfish.

Pseudocheilinus hexataenia

Six-Line Wrasse

- **ORIGINS** The Red Sea eastward through the Indo-Pacific to Oceania. Extends to Japan and Australia's east coast.
- **SIZE** 4 in (10 cm).
- **DIET** Eats a variety of prepared foods.
- **WATER** Temperature 77–79°F (25–26°C); alkaline (pH 8.0–8.3) with SG 1.020–1.025.
- **TEMPERAMENT** Relatively shy.

This colorful wrasse can be distinguished from other similar species by a black spot on the caudal peduncle and six pairs of alternating blue and yellow horizontal body stripes. Six-Line Wrasse will eat a variety of foods, but they initially prefer thawed items, such as lobster eggs, to marine flake. Encourage the fish to sample as wide a range of food items as possible. Only youngsters will get along well together in the same surroundings.

Novaculichthys taeniourus

Dragon Wrasse

- **ORIGINS** Extends from the Red Sea across the Pacific Ocean to the coast of Panama.
- **SIZE** 10 in (25 cm).
- **DIET** Small pieces of meat-based foods.
- **WATER** Temperature 77–79°F (25–26°C); alkaline (pH 8.0–8.3) with SG 1.020–1.025.
- **TEMPERAMENT** Adults are territorial.

As Dragon Wrasse mature, they lose the rays at the front of the dorsal fin, and the body also becomes less colorful as its greenish hue disappears. These fish excavate the substrate in search of edible items and may burrow into the sand to avoid danger. They grow more territorial with age, so keep them with nonaggressive fish that attain a similar size. Dragon Wrasse will devour any invertebrates in their tank.

BATFISH

These unusually deep-bodied fish need an aquarium at least 24 in (60 cm) deep and wide. Young batfish are far more colorful than adults, and, once established in their quarters, they grow fast. Despite their size, they are not aggressive toward unrelated fish, but they should be kept apart from species that might nip their elaborate fins. Large batfish can prove very destructive in a reef aquarium.

Although batfish, such as these **Orbicularis Batfish** (*Platax orbicularis*), occur in shoals in the wild, they are not suited to group living in the aquarium and need to be kept singly.

Platax batavianus

Humpback Batfish

- **ORIGINS** Ranges from the vicinity of southern Japan down through Indonesia to Australia's Great Barrier Reef.
- **SIZE** 26 in (65 cm).
- **DIET** Fresh or thawed meat-based foods.
- **WATER** Temperature 77–79°F (25–26°C); alkaline (pH 8.1–8.3) with SG 1.021–1.024.
- **TEMPERAMENT** Intolerant.

Although the Humpback is less commonly available than other batfish, its care does not pose any particular problems. It may be necessary to feed these fish on live brine shrimp at first, until they adapt to a more varied diet, so be sure to have sufficient cultures set up for this purpose. Their mouths are small for their size, and this should be reflected in the type of food offered.

Fins have a tassled appearance

Platax orbicularis

Orbicularis Batfish

- **ORIGINS** Ranges from the Red Sea to the eastern Pacific, as far as Papua New Guinea.
- **SIZE** 20 in (51 cm).
- **DIET** Thawed marine foods, such as lobster eggs.
- **WATER** Temperature 77–79°F (25–26°C); alkaline (pH 8.1–8.3) with SG 1.021–1.024.
- **TEMPERAMENT** Best kept apart from other batfish.

The dorsal, anal, and ventral fins of Orbicularis Batfish become shorter as they grow older. As juveniles, these fish inhabit mangrove swamps, where their unusual elongated appearance, complete with brown barring on the body, helps to disguise their presence among the mangrove roots. As they grow older, they move to a reef habitat. They become more grayish in appearance and change significantly in shape to a more disklike outline. Good water quality is important to ensure healthy fins in these fish.

Juvenile

Platax pinnatus

Pinnate Batfish

- **ORIGINS** Ranges from East Africa and the Red Sea to the central area of the Indo-Pacific region.
- **SIZE** 16 in (41 cm).
- **DIET** Live brine shrimp; thawed marine foods.
- **WATER** Temperature 77–79°F (25–26°C); alkaline (pH 8.1–8.3) with SG 1.021–1.024.
- **TEMPERAMENT** Safe with nonaggressive, larger fish.

Pinnate Batfish are harder to establish in the aquarium than others, which is unfortunate in view of their spectacular appearance. It is difficult to wean them onto a varied diet of inert foods. Keeping them on their own should help, and they may eventually become tame enough to take food from the fingers. The red edging around their bodies disappears with age.

The red edging makes the fish resemble an inedible flatworm (Platyhelminth) when lying horizontally

---

Platax teira

Tiera Batfish

- **ORIGINS** From the Red Sea through the Indo-Pacific, and beyond Indonesia to the east coast of Australia.
- **SIZE** 25 in (63 cm).
- **DIET** Varied marine foods.
- **WATER** Temperature 77–79°F (25–26°C); alkaline (pH 8.1–8.3) with SG 1.021–1.024.
- **TEMPERAMENT** Do not keep with invertebrates.

The Tiera Batfish has a distinctive rounded face. Like other batfish, it relies on its appearance for camouflage when it is young, looking much like a mangrove leaf drifting in the water. It "plays dead" and floats on its side if danger threatens and will often behave in this fashion after being transferred to new aquarium surroundings. This is a normal reaction, and should not be taken as a cause for concern.

GROUPERS AND GRUNTS

These colorful fish are not especially difficult to maintain in aquarium surroundings, but they will grow to a large size and so will eventually require a spacious setup. This needs to be taken into consideration from the outset, since it is not easy to find new homes for such fish once they have outgrown the average-sized home aquarium. While some groupers and grunts are shoaling fish, others tend to seek out suitable retreats on the reef where they can lurk. Sexing can be difficult, since many species show hermaphrodite characteristics and are able to change gender to suit their environment.

The markings of groupers, such as this **Blacktip Grouper** (*Epinephelus fasciatus*), can be quite variable.

Porkfish

Anisotremus virginicus

- **ORIGINS** Occurs throughout the Caribbean region, from Florida down to the northern coast of South America.
- **SIZE** 12 in (30 cm).
- **DIET** Fresh or thawed meat-based foods.
- **WATER** Temperature 77–79°F (25–26°C); alkaline (pH 8.1–8.3) with SG 1.021–1.024.
- **TEMPERAMENT** Preys on invertebrates.

Porkfish are members of the grunt family— a name derived from their ability to produce sounds that resemble the grunts of pigs. They have a long, steep forehead and large eyes. Juveniles differ markedly from the adult seen above, having black stripes running the length of their bodies and a distinctive black blotch on the caudal peduncle.

Panther Grouper

Cromileptes altivelis

- **ORIGINS** East Africa across the Pacific to Japan, Australia's Great Barrier Reef, and Vanuatu.
- **SIZE** 27 in (70 cm).
- **DIET** Dried and thawed marine foods.
- **WATER** Temperature 77–79°F (25–26°C); alkaline (pH 8.1–8.3) with SG 1.021–1.024.
- **TEMPERAMENT** Will prey on smaller companions.

The spotted appearance of these distinctive fish becomes more pronounced with age. Panther Groupers have prodigious appetites and grow fast, so it is important that the filtration system in their aquarium is effective enough to handle the resulting volume of waste. Loss of appetite in this species often signals a deterioration of water quality, in which pH can drop markedly.

Raised dorsal fin

Swiss Guard Basslet

Liopropoma rubre

- **ORIGINS** Caribbean region, from Florida down to the Venezuelan coast.
- **SIZE** 4 in (10 cm).
- **DIET** Fresh and thawed marine foods, as well as flake.
- **WATER** Temperature 77–79°F (25–26°C); alkaline (pH 8.1–8.3) with SG 1.021–1.024.
- **TEMPERAMENT** Mixes well.

These small, attractive fish are ideal for a mixed-species aquarium, where they tend to occupy the area from midwater downward. They are adept at hiding away in nooks and crevices, proving to be rather shy by nature. Their unusual common name reflects the fact that their coloration resembles the uniform of the Papal Swiss Guard based in Vatican City.

Epinephelus fasciatus

Blacktip Grouper

- **ORIGINS** Ranges eastward from the Red Sea, via the Indo-Pacific region, to Oceania.
- **SIZE** 14 in (35 cm).
- **DIET** Thawed and dry marine foods.
- **WATER** Temperature 77–79°F (25–26°C); alkaline (pH 8.0–8.3) with SG 1.021–1.024.
- **TEMPERAMENT** Do not house with invertebrates.

The appearance of these groupers, or Rock Cod as they are also known, can be very variable. Some individuals show more numerous or extensive reddish-orange body markings and smaller white patches than others. However, the presence of black tipping along the dorsal fin is a consistent feature. Blacktip Groupers spend most of their time on or near the base of the aquarium, where they also feed. They are solitary by nature.

Hypoplectrus guttavarius

Shy Hamlet

- **ORIGINS** Found in the Caribbean region, from Florida down to the coast of South America.
- **SIZE** 6 in (15 cm).
- **DIET** Fresh or thawed marine foods.
- **WATER** Temperature 77–79°F (25–26°C); alkaline (pH 8.0–8.3) with SG 1.021–1.024.
- **TEMPERAMENT** Retiring and peaceful.

The back of this hamlet is dark in some specimens but a much bluer shade in others. These fish need nooks and crannies in the aquarium where they can hide, although they do become bolder once they are established in their quarters. It is possible to breed Shy Hamlets in home aquariums. Amazingly, each fish is a hermaphrodite, possessing both male and female sex organs, so any two Shy Hamlets should be able to mate successfully.

Hypoplectrus gemma

Blue Hamlet

- **ORIGINS** The Caribbean region, from the coast of Florida down to northern South America.
- **SIZE** 5 in (13 cm).
- **DIET** Thawed marine foods.
- **WATER** Temperature 77–79°F (25–26°C); alkaline (pH 8.0–8.3) with SG 1.021–1.024.
- **TEMPERAMENT** Predatory.

The blackish upper and lower edges on the caudal fin distinguish this species from the much rarer Indigo Hamlet (*H. indigo*). Blue Hamlets are predatory by nature and cannot be kept safely with smaller companions. They are more sensitive to water quality than their larger relatives; the total nitrate reading should not be allowed to rise above 10 ppm. For this reason, the tank's filtration system should include a protein skimmer.

Cephalopholis argus

Blue-Spotted Grouper

- **ORIGINS** Ranges from the Red Sea throughout the Pacific Ocean.
- **SIZE** 20 in (50 cm).
- **DIET** Thawed marine foods.
- **WATER** Temperature 77–79°F (25–26°C); alkaline (pH 8.0–8.3) with SG 1.021–1.024.
- **TEMPERAMENT** Will prey on smaller fish.

The striking coloration of these groupers has made them popular with marine aquarists, but as in the case of similar species, they are likely to grow rapidly and can reach a large size. This demands both a correspondingly large aquarium and an efficient filtration system. Any loss of color is usually a sign of deteriorating water quality; this needs to be monitored closely. Blue-Spotted Groupers are not particularly active fish, seeking out suitable retreats near the bottom of the aquarium, where they can lurk and wait for food. Their attractive patterning of black-ringed blue spots extends over the fins as well as the body, but the distribution of spotted markings differs between individuals. The spots appear to glitter when caught by the light.

INTIMIDATING GAPE

Groupers are not the most active fish on the reef, but they are still effective predators, thanks to their quick reflexes, large mouths, and fearsome array of teeth, as visible in this Marbled Grouper (*Epinephelus polyphekadion*). The spotted patterning helps to conceal them as they lie in wait for small fish or invertebrates to swim within reach. A grouper will also open its mouth wide to intimidate rivals. Some groupers reach a huge size on the reef, with relatives of the Blacktip Grouper *(see top left)* weighing 1,000 lb (450 kg) and measuring nearly 10 ft (3 m).

GRAMMAS, DOTTYBACKS, AND BASSLETS

Not only do these rank as some of the most beautifully colored marine fish, but their small size also means that they can be accommodated easily. Furthermore, successful aquarium breeding is becoming more frequent, but compatibility can be a serious issue in smaller tanks, because these fish are very territorial by nature. Members of this group can be incorporated successfully into an invertebrate tank. Rockwork with suitable retreats must be included in the aquarium to replicate the numerous hiding places they frequent on the reef.

The colorful **Royal Gramma** (*Gramma loreto*) was one of the first marine fish to be bred in aquarium surroundings during the 1960s.

Pictichromis porphyrea

Magenta Dottyback

- **ORIGINS** From Japan, via Indonesia, to the east coast of Australia.
- **SIZE** 2 in (5 cm).
- **DIET** Small thawed and freeze-dried foods.
- **WATER** Temperature 77–79°F (25–26°C); alkaline (pH 8.1–8.3) with SG 1.020–1.024.
- **TEMPERAMENT** Aggressive to its own kind.

These small fish inhabit holes in the reef, through which they move easily, thanks to their narrow, elongated body shape. They are inconspicuous, remaining hidden for long periods and then darting out to seize a morsel of floating food. Like related species, the Magenta Dottyback will eat food that falls to the tank floor. Dottybacks have sharp teeth and are capable of inflicting a painful bite, irrespective of their size.

Pictichromis diadema

Diadem Basslet

- **ORIGINS** The western Pacific region, ranging from northern Australia and Indonesia to Japan.
- **SIZE** 3 in (7.5 cm).
- **DIET** Fresh and thawed animal foods.
- **WATER** Temperature 77–79°F (25–26°C); alkaline (pH 8.1–8.3) with SG 1.020–1.024.
- **TEMPERAMENT** Highly territorial.

A broad purple streak extends from the upper lip of the Diadem Basslet and narrows to a point near the rear of the dorsal fin. The rest of the body varies from yellow to yellowish-orange. Like other members of the group, these basslets may hang at strange angles in the water. This is normal and not generally a cause for concern, as it would be with many other fish. Loss of color is more significant and can indicate a decline in water quality.

Pseudochromis dutoiti

Dutoiti Dottyback

- ⊕ **ORIGINS** Found off the east coast of Africa, in northern and western parts of the Indian Ocean.
- ⟳ **SIZE** 3¹/₂ in (9 cm).
- ⟳ **DIET** Prepared live foods of suitable size.
- ⟳ **WATER** Temperature 77–79°F (25–26°C); alkaline (pH 8.1–8.3) with SG 1.020–1.024.
- ⟳ **TEMPERAMENT** May quarrel with related fish.

Dutoiti Dottybacks can be bred in the home aquarium. To maximize breeding success, three individuals should be introduced into a large reef aquarium that has plenty of well-spaced retreats. After mating occurs, the spawn is guarded by the male in a safe locality, such as a hole in a rock. The eggs hatch approximately six days after spawning. Rotifers are a suitable food for feeding the fry. An adult pair of Dutoiti Dottybacks may spawn over 20 times during the course of a year.

Neon-blue streaks along back and on face

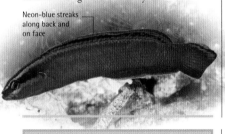

STAYING ALIVE

The members of this group are typically small in size, making them vulnerable to a wide range of predators. The danger is compounded because they are not able to swim strongly to escape danger. To protect themselves, these fish use their small size to retreat into crevices in the reef and tend not to venture far from these relatively safe havens. Each individual learns to recognize its own territory and uses interconnecting holes in its reef domain like escape tunnels, darting through them in order to avoid any attempted pursuit or ambush.

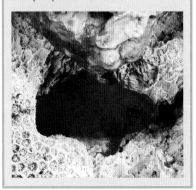

Pictichromis paccagnellae

Royal Dottyback

- ⊕ **ORIGINS** Pacific region, occurring from Indonesia southward to Australia, and to Vanuatu in the east.
- ⟳ **SIZE** 2³/₄ in (7 cm).
- ⟳ **DIET** Fresh and thawed animal foods.
- ⟳ **WATER** Temperature 77–79°F (25–26°C); alkaline (pH 8.1–8.3) with SG 1.020–1.024.
- ⟳ **TEMPERAMENT** Highly territorial.

The Royal Dottyback has coloration similar to the Royal Gramma (*see below*), a Caribbean rather than a Pacific species. It can be distinguished by the more distinct boundary between the purple and yellowish areas of the body. When buying these fish, select individuals with a strong coloration, since these are most likely to be in good health. Subsequent loss of color may be a sign of poor water quality or territorial conflict.

Gramma loreto

Royal Gramma

- ⊕ **ORIGINS** The Caribbean region, from Florida to the northern coast of South America.
- ⟳ **SIZE** 3 in (7.5 cm).
- ⟳ **DIET** Fresh and thawed animal foods.
- ⟳ **WATER** Temperature 77–79°F (25–26°C); alkaline (pH 8.1–8.3) with SG 1.020–1.024.
- ⟳ **TEMPERAMENT** Highly territorial.

Black spot most evident when dorsal fin raised

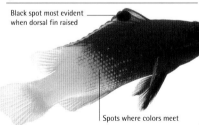

Spots where colors meet

One of the most beautiful members of a colorful group, the Royal Gramma has a purplish front half to its body, with the rear part varying from yellow to orange. A black stripe runs through the eye, and the dorsal fin has a black spot. There is a good chance of successful spawning, but for breeding purposes it is vital to introduce the fish to the tank at the same time, since a newcomer is likely to be persecuted.

Pseudochromis fuscus

Golden Dottyback

- ⊕ **ORIGINS** From Australia's Great Barrier Reef northward through Indonesia and the Philippines.
- ⟳ **SIZE** 4 in (10 cm).
- ⟳ **DIET** Fresh and thawed animal foods.
- ⟳ **WATER** Temperature 77–79°F (25–26°C); alkaline (pH 8.1–8.3) with SG 1.020–1.024.
- ⟳ **TEMPERAMENT** Shy but highly territorial.

Rich canary-yellow coloration over the entire body characterizes the Golden Dottyback. Being shy, it lives close to the floor of the aquarium, where it seeks out retreats. Despite spending much of its time near the substrate, this gramma will not seek its food there, instinctively feeding instead on items in suspension. It will, however, prey on small worms or tiny crustaceans lurking in the vicinity of the rockwork.

Gramma melacara

Blackcap Basslet

- ⊕ **ORIGINS** The Caribbean region, occurring in the area between Florida and northern South America.
- ⟳ **SIZE** 4 in (10 cm).
- ⟳ **DIET** Fresh and thawed animal foods.
- ⟳ **WATER** Temperature 77–79°F (25–26°C); alkaline (pH 8.1–8.3) with SG 1.020–1.024.
- ⟳ **TEMPERAMENT** Highly territorial.

Stunning shades of mauve and purple extending to the fins, and a dark area on the head, are the key features of this basslet, which is among the more territorial members of the group. If you intend to keep more than one Blackcap Basslet, ensure that the base of the tank is divided by the decor into different areas where the fish can establish themselves. They are less likely to quarrel if kept in groups of three, rather than pairs.

Long dorsal fin

BLENNIES AND MANDARINFISH

The members of these two families have similar care requirements, and they can even be kept together in the same tank, alongside invertebrates and placid fish, such as seahorses and pipefish. It is important that blennies and mandarinfish are not harried by their tankmates, because they will produce an unpleasant, protective slime from their bodies to deter assailants. Some blennies have evolved the predatory trick of copying the appearance of cleaner wrasse *(see p.268)* and then biting chunks out of fish expecting to be cleaned.

The striking markings of the **Mandarin Fish** (*Pterosynchiropus splendidus*) extend to its elaborate fins. Because of their patterning, members of this group are also called psychedelic fish.

Mandarin Fish
Pterosynchiropus splendidus

- **ORIGINS** The western Pacific region, off the coast of Southeast Asia and China, extending up to Japan.
- **SIZE** 2½ in (6 cm).
- **DIET** Live brine shrimp and thawed foods.
- **WATER** Temperature 77–79°F (25–26°C); alkaline (pH 8.1–8.3) with SG 1.020–1.024.
- **TEMPERAMENT** Males are aggressive.

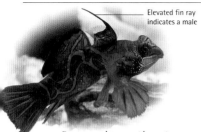

Elevated fin ray indicates a male

Be prepared to provide copious amounts of live foods if you choose Mandarin Fish. They really need to be kept in well-established reef tanks so that they can forage for their natural prey of small invertebrates. If housed together, male Mandarin Fish are likely to fight, but males can be identified easily, as they have an extended fin ray at the front of the dorsal fin. No two individuals have matching body patterning.

Spotted Mandarin
Synchiropus picturatus

- **ORIGINS** The western Pacific region, extending down to the northern coast of Australia.
- **SIZE** 2¾ in (7 cm).
- **DIET** Live brine shrimp and thawed foods.
- **WATER** Temperature 77–79°F (25–26°C); alkaline (pH 8.1–8.3) with SG 1.020–1.024.
- **TEMPERAMENT** Males are aggressive.

Wildly spotted body patterning distinguishes this species. The spots are made up of concentric rings, set against a greenish background. As they comb the algae-covered rocks, these fish suck in tiny microbes, expelling particles of mud via their gills. Keep the tank covered, because it is not unknown for Mandarins to leap out of the water. Aquarium spawnings are very rare. The eggs develop near the water's surface.

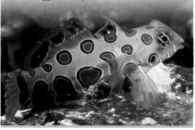

Bicolor Blenny
Ecsenius bicolor

- **ORIGINS** From the Maldives, in the Indo-Pacific, and eastward to Australia's Great Barrier Reef.
- **SIZE** 4 in (10 cm).
- **DIET** Small live foods and algae.
- **WATER** Temperature 77–79°F (25–26°C); alkaline (pH 8.1–8.3) with SG 1.020–1.024.
- **TEMPERAMENT** Not aggressive to unrelated fish.

Taller rear section to dorsal fin

Golden-yellow near the tail

The Bicolor Blenny is difficult to describe because its appearance differs not only between populations but also during the breeding period, when the males display a red-and-white barred patterning before turning blue with white flank markings. The females are yellow. These blennies can be housed in a reef tank, where they will browse on algae and will often be seen resting on top of a favored rocky outcrop.

Meiacanthus atrodorsalis

Forktail Blenny

- **ORIGINS** The coast of East Africa through the Indo-Pacific, north to Japan and south to northern Australia.
- **SIZE** 3 in (7.5 cm).
- **DIET** Small live foods and algae.
- **WATER** Temperature 77–79°F (25–26°C); alkaline (pH 8.1–8.3) with SG 1.020–1.024.
- **TEMPERAMENT** Not aggressive to unrelated fish.

This blenny is bluish-green on the head and becomes yellowish toward the tail, often with some black striping. Its small size makes it look harmless, but the Forktail has specialized teeth that enable it to inject venom when it bites. Most predatory fish recognize the Forktail's coloration and tend to leave it alone. Some other species mimic this blenny's appearance in order to gain protection for themselves. Beware: the Forktail's venom is also painful to people.

Meiacanthus smithi

Smith's Sawtail Blenny

- **ORIGINS** Restricted to northern and western parts of the Indian Ocean off the coast of East Africa.
- **SIZE** 3½ in (9 cm).
- **DIET** Small live foods and algae.
- **WATER** Temperature 77–79°F (25–26°C); alkaline (pH 8.1–8.3) with SG 1.020–1.024.
- **TEMPERAMENT** Not aggressive to unrelated fish.

| Streaked caudal fin

Smith's Sawtail has a pale grayish-white coloration, with a pinkish hue on the underparts and a prominent black stripe running along the top of the body down to the eye. The sawtail effect is produced by the darker markings in the caudal fin. As with other *Meiacanthus* species, Smith's Sawtail has a functional swim bladder, so it can control its buoyancy effectively. This blenny is therefore quite active by nature, but it still prefers not to venture far from the security of its rocky habitat.

Ophioblennius atlanticus

Atlantic Fanged Blenny

- **ORIGINS** The Caribbean region, from Florida down to the northern coast of South America.
- **SIZE** 4¾ in (12 cm).
- **DIET** Algae plus small live foods.
- **WATER** Temperature 77–79°F (25–26°C); alkaline (pH 8.1–8.3) with SG 1.020–1.024.
- **TEMPERAMENT** Highly territorial.

The coloration of these blennies is variable and is influenced in part by the background. The lips are invariably red, as are the edges of the dorsal fin. While the pelvic fins are yellowish, the body itself tends to be quite dark. These fish lack scales on their bodies. The Atlantic Fanged Blenny belongs to a group called the combtooth blennies, so named because their teeth resemble those of a comb. They establish territories, which they will defend against all fish.

Salarias fasciatus

Jeweled Rockskipper

- **ORIGINS** From East Africa throughout the Indo-Pacific region, including Australia and Japan.
- **SIZE** 4 in (10 cm).
- **DIET** Small live foods and algae.
- **WATER** Temperature 77–79°F (25–26°C); alkaline (pH 8.1–8.3) with SG 1.020–1.024.
- **TEMPERAMENT** Not aggressive to unrelated fish.

A combination of mottled light and brown bands running down the sides of the body help to break up the Jeweled Rockskipper's outline. The long dorsal fin is similarly patterned, while the outer part of the eye has spokelike markings. As their name suggests, Jeweled Rockskippers inhabit the lower reaches of the aquarium, where they blend in well against rockwork. They will dart back quickly into a nearby crevice if danger threatens.

TESTING THE WATER

One of the characteristics of many blennies, including this Orange-Spot Blenny (*Blenniella chrysospilos*), is the presence of sensory feelers, known as cirri, on the top of the head. The branched structure of the cirri may help these blennies detect local currents, or water movements that may indicate the approach of a predator. Looking much like part of the coral reef, these feelers probably also help to disguise the fish when they are at rest. The shape of the cirri is identical between members of the same species, but it is not consistent throughout the group as a whole.

RABBITFISH

The name of these fish derives partly from the rabbitlike way in which they browse on marine algae and also partly from their harelike upper lip. They are related to tangs and surgeonfish (*see pp. 236–239*) and need similar care. There is much debate over the relationships between the species in this group. Sexing is not usually possible, although females are often larger than males.

The alternative name, "spinefoot," for a number of these fish derives from the venomous defensive spikes at the front of the dorsal fin. Pictured here is the attractively marked **Golden-Spotted Spinefoot** (*Siganus punctatus*).

Lo magnificus
Magnificent Foxface

- **ORIGINS** Restricted to northern and western parts of the Indian Ocean, off the coast of East Africa.
- **SIZE** 8 in (20 cm).
- **DIET** Plant matter and small live foods.
- **WATER** Temperature 79–82°F (26–28°C); alkaline (pH 8.1–8.3) with SG 1.021–1.024.
- **TEMPERAMENT** Do not mix with other foxfaces.

A single broad, black band running down each side of the face distinguishes the Magnificent Foxface. The "saddle" region on the upper back is invariably dark, while the underparts are usually whiter, and yellow is often evident on the fins. The Magnificent Foxface may feed on organpipe corals as well as algae if housed in a reef aquarium.

Exact coloration varies between individuals

Siganus virgatus
Double-Barred Spinefoot

- **ORIGINS** Extends from Australia's Great Barrier Reef northward to Indonesia and the Philippines.
- **SIZE** 12 in (30 cm).
- **DIET** Plant matter and small live foods.
- **WATER** Temperature 79–82°F (26–28°C); alkaline (pH 8.1–8.3) with SG 1.021–1.024.
- **TEMPERAMENT** Usually placid.

The color of the Double-Barred Spinefoot is much brighter during the day, when the fish bears some resemblance to the Foxface. It displays yellow coloration along the upper back, extending to the caudal fin, and black bars across the head. These spinefoots are also able to change their color to merge in with their background. Like other rabbitfish, they can be weaned easily onto artificial diets.

Siganus vulpinus
Foxface

- **ORIGINS** Extends from the eastern coast of southeast Asia north to Japan and across the Pacific.
- **SIZE** 10 in (25 cm).
- **DIET** Plant matter and small live foods.
- **WATER** Temperature 79–82°F (26–28°C); alkaline (pH 8.1–8.3) with SG 1.021–1.024.
- **TEMPERAMENT** Intolerant of its own kind.

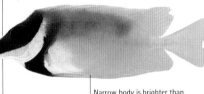

Narrow body is brighter than that of the Magnificent Foxface

The Foxface has a prominent black band running through the eyes to the jaws, with a white area beneath, plus black edging extending down from the gill covers. Foxfaces vary so much throughout their wide range that different populations are sometimes regarded as separate species. These fish need to be handled with particular care, because it is very easy to catch your hands on the dorsal fin's defensive spines, which can inflict painful wounds.

Siganus unimaculatus
One-Spot Foxface

- **ORIGINS** Restricted to the northwestern Pacific, specifically between Japan and the Philippines.
- **SIZE** 8 in (20 cm).
- **DIET** Plant matter and small live foods.
- **WATER** Temperature 79–82°F (26–28°C); alkaline (pH 8.1–8.3) with SG 1.021–1.024.
- **TEMPERAMENT** Intolerant of its own kind.

The One-Spot is very similar in appearance to the Foxface, showing the same characteristic black-and-white facial coloration and yellow body. What sets this fish apart, however, is the black spot just below the dorsal fin. The One-Spot and the Foxface are sometimes regarded as different forms of the same species, with different areas of distribution. Their mutual intolerance means that these two fish should never be housed together.

Black area continues along the underparts

HAWKFISH

The predatory hawkfish are so called because of their habit of swooping down from above on to their victims, in a similar way to birds of prey. Long pectoral fins enable these fish to rest securely on a rocky perch so that, although they may sway in the current, they will not be swept away by the swell. This is very important, because hawkfish lack a swim bladder and so have difficulty in maintaining their buoyancy in the water.

Like other hawkfish, the **Pixy Hawkfish** (*Cirrhitichthys oxycephalus*) is not an active swimmer, preferring instead to ambush its prey.

Cirrhitichthys oxycephalus

Pixy Hawkfish

- ⊕ **ORIGINS** The Red Sea through the Indo-Pacific region to the western coast of Central America.
- ⬭ **SIZE** 3 in (8 cm).
- ⬗ **DIET** Marine flake and thawed livefoods.
- ⬯ **WATER** Temperature 77–79°F (25–26°C); alkaline (pH 8.1–8.3) with SG 1.020–1.024.
- ⬤ **TEMPERAMENT** Keep separate from other hawkfish.

The Pixy Hawkfish's dominant reddish coloration becomes blotched on the lower part of its body. It makes an interesting aquarium occupant, but like other hawkfish, it should not be housed with crustaceans in a reef aquarium, since its sharp teeth can make easy work of even a crab's shell. Hawkfish will also eat worms and smaller fish.

Paracirrhites arcatus

Arc-Eyed Hawkfish

- ⊕ **ORIGINS** East Africa through the Indian Ocean to Hawaii and other areas of the Pacific.
- ⬭ **SIZE** 5½ in (14 cm).
- ⬗ **DIET** Marine flake and thawed live foods.
- ⬯ **WATER** Temperature 77–79°F (25–26°C); alkaline (pH 8.1–8.3) with SG 1.020–1.024.
- ⬤ **TEMPERAMENT** Sedentary by nature.

This hawkfish is characterized by delicate markings of orange and pale blue on the head, which form an arc around the eye, and a horizontal white stripe along the rear of the body. The Arc-Eyed Hawkfish can move very swiftly when an edible item catches its eye. It will readily learn to take prepared foods and can even be persuaded to feed from the hand.

Prominent pectoral fin

Oxycirrhites typus

Longnose Hawkfish

- ⊕ **ORIGINS** From the Red Sea through the Indian Ocean to the eastern Pacific seaboard.
- ⬭ **SIZE** 5 in (13 cm).
- ⬗ **DIET** Marine flake and thawed live foods.
- ⬯ **WATER** Temperature 77–79°F (25–26°C); alkaline (pH 8.1–8.3) with SG 1.020–1.024.
- ⬤ **TEMPERAMENT** Choose companions carefully.

Serrated dorsal fin

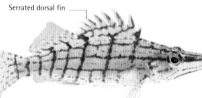

The elongated mouthparts of the Longnose Hawkfish, combined with its red-banded body, give this species an unmistakable appearance. Arrange the aquarium decor securely so that the fish has several vantage points near the surface from which it can watch over the rest of the tank. The Longnose Hawkfish can jump well, and it should therefore be kept in a covered aquarium. Hawkfish cannot be sexed visually. Keep these fish singly to avoid fighting.

Neocirrhites armatus

Flame Hawkfish

- ⊕ **ORIGINS** The Great Barrier Reef, off Australia's eastern coast, and throughout the Pacific region.
- ⬭ **SIZE** 3 in (7.5 cm).
- ⬗ **DIET** Marine flake and thawed live foods.
- ⬯ **WATER** Temperature 77–79°F (25–26°C); alkaline (pH 8.1–8.3) with SG 1.020–1.024.
- ⬤ **TEMPERAMENT** Can be predatory.

This is one of the most colorful hawkfish, its stunning red coloration augmented by black spectacles and a black area running along the top of the body onto the dorsal fin. It shows to best effect if given a perch, such as a cured sea fan, which it can adopt as a vantage point. Breeding in the aquarium is highly unlikely, since this fish spawns in harems.

SEA HORSES

Sea horses are among the most instantly recognizable and fascinating of all marine fish. Less well known, however, are their elongated relatives called pipefish. Found in temperate and tropical waters, both groups share unusual breeding habits, with the male caring for the eggs and often carrying them in a pouch on the front of his body. These fish need to be fed up to four times a day, since they eat almost constantly in the wild.

Sea horses adopt a vertical posture when resting but swim with their body tilting forward. These are **Barbour's Sea horses** (*Hippocampus barbouri*).

Hippocampus kuda
Common Sea horse

- **ORIGINS** The Red Sea through the Indo-Pacific region to the east coast of Asia and the north of Japan.
- **SIZE** 10 in (25 cm).
- **DIET** Primarily live brine shrimp.
- **WATER** Temperature 75–77°F (24–25°C); alkaline (pH 8.1–8.3) with SG 1.021–1.024.
- **TEMPERAMENT** Inoffensive.

The Common Sea horse is often, but not always, yellow in color. As with other members of this group, it is an expert at camouflage, changing its color to blend in with the surroundings. This makes it difficult to distinguish between different species with certainty. All sea horses lack a caudal fin, which is replaced instead by a prehensile tail. This allows the fish to anchor themselves to items such as seaweed fronds. Sea horses can be incorporated as part of a reef aquarium and kept in the company of other very gentle fish. Their propulsive power is provided by the dorsal fin.

Tapering body

Curled tail used for anchorage

Hippocampus erectus
Lined Sea horse

- **ORIGINS** From Eastern North America down through the Caribbean to northern South America.
- **SIZE** 6 in (15 cm).
- **DIET** Mainly live brine shrimp.
- **WATER** Temperature 75–77°F (24–25°C); alkaline (pH 8.1–8.3) with SG 1.020–1.024.
- **TEMPERAMENT** Placid and sedentary.

The Lined Sea horse is one of the smaller species, displaying the typical protective bony rings around its body. Lined Sea horses tend to have a shorter life span than their larger relatives—about two years compared to five. Breeding results in aquariums have improved significantly over recent years. The female lays her eggs directly in the male's brood pouch, and the young emerge into the aquarium about a month later.

Prehensile tail

Dunckerocampus dactyliophorus
Banded Pipefish

- **ORIGINS** From the Red Sea and the East African coast through the Indian Ocean to the Pacific.
- **SIZE** 7 in (18 cm).
- **DIET** Brine shrimp and other small live foods.
- **WATER** Temperature 75–77°F (24–25°C); alkaline (pH 8.1–8.3) with SG 1.021–1.024.
- **TEMPERAMENT** Placid.

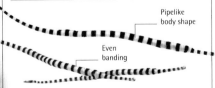

Pipelike body shape

Even banding

This pipefish has alternating bands of pale yellow and reddish-brown along the length of its body. Its predominantly red caudal fin is edged with white. The other fins on its body are very small and inconspicuous, and the tail provides the main propulsive thrust for swimming. Pipefish need a quiet aquarium where they will not be harried by the other occupants. The elongated snout is used to suck small invertebrates into the mouth. The diet of larger pipefish can be supplemented with the fry of livebearers such as guppies.

Syngnathoides biaculeatus
Alligator Pipefish

- **ORIGINS** From the Red Sea, via the Indo-Pacific region, as far as southern Japan and northeastern Australia.
- **SIZE** 12 in (30 cm).
- **DIET** Mainly live brine shrimp.
- **WATER** Temperature 75–77°F (24–25°C); alkaline (pH 8.1–8.3) with SG 1.021–1.024.
- **TEMPERAMENT** Very placid.

This pipefish has filamentous areas on its head, which may help to conceal its presence. It is light green, with a yellower tone to its underparts. The tail is prehensile, allowing the Alligator Pipefish to anchor itself to coral. The dorsal fin is inconspicuous, while the caudal, anal, and pelvic fins are absent. These pipefish are weak swimmers, so they rely on camouflage to evade predators. The eggs are carried stuck on the male's abdomen, not in a brood pouch.

Long, broad-ended snout

OTHER POPULAR MARINE FISH

A number of marine fish from other diverse groups are occasionally available to aquarists. Their requirements differ widely; some predatory species need a fish-only setup, while others can be housed safely in a reef aquarium. Never be tempted to choose a fish simply by its appearance. Make sure you can identify it with certainty, not only so that you can be sure of the size it is likely to reach as an adult, but also so that you can find out about its environmental needs and whether it will be compatible with other aquarium occupants. Bear in mind that juveniles are more commonly offered than adults.

The Blue Ribbon Eel (*Rhinomuraena quaesita*) requires rocky retreats to be built into its tank to give it places to hide.

Aulostomus maculatus

Atlantic Trumpetfish

- **ORIGINS** Widely distributed around the Caribbean region, from Florida down to the South American coast.
- **SIZE** Up to 35 in (90 cm).
- **DIET** Fresh and thawed meat-based foods.
- **WATER** Temperature 77–79°F (25–26°C); alkaline (pH 8.1–8.3) with SG 1.021–1.024.
- **TEMPERAMENT** Do not mix with smaller companions.

These long, narrow-bodied fish hunt a variety of crustaceans and small fish in the wild. They can be tamed to feed from the hand, but it is difficult to wean them off live foods at first, so a supply of prepared live foods may be required. Sexing is easy, since males have a longer ray at the front of the dorsal fin. A spacious tank is needed to accommodate this species. Its smaller Indo-Pacific counterpart, *A. chinensis*, attains a length of only about 24 in (60 cm).

Aeoliscus strigatus

Coral Shrimpfish

- **ORIGINS** From the Red Sea all the way across the Pacific Ocean, as far east as Hawaii.
- **SIZE** 6 in (15 cm).
- **DIET** Prefers crustaceans.
- **WATER** Temperature 77–79°F (25–26°C); alkaline (pH 8.1–8.3) with SG 1.020–1.024.
- **TEMPERAMENT** Placid, suitable for a reef aquarium.

Shrimpfish always swim vertically, usually with their heads pointing downward, and adopt a horizontal posture only when they feed. The body is protected by bony plates, and the mouth is small. Shrimpfish should be kept together in groups of four to six individuals. Male fish build a nest in which several females will lay their eggs, but successful spawning in an aquarium is unlikely.

Monocentris japonica

Pine-Cone Fish

- **ORIGINS** Ranges from the Red Sea through the Indian Ocean to southern Japan and east of Papua New Guinea.
- **SIZE** 6 in (15 cm).
- **DIET** Fresh and thawed marine foods.
- **WATER** Temperature 77–79°F (25–26°C); alkaline (pH 8.1–8.3) with SG 1.020–1.024.
- **TEMPERAMENT** Nonaggressive.

These fish represent a very ancient lineage that has altered little over millions of years. Pine-Cone Fish have a light-emitting organ under each eye, which may assist when hunting invertebrates at night. In the aquarium, they prefer low light levels. You may need to wean newly acquired individuals off foods such as live brine shrimp and onto similar prepared diets.

Zanclus cornutus

Moorish Idol

- **ORIGINS** Widely distributed throughout the Indo-Pacific region.
- **SIZE** Up to 10 in (25 cm).
- **DIET** Plant matter and meat-based foods, such as squid.
- **WATER** Temperature 77–79°F (25–26°C); alkaline (pH 8.1–8.3) with SG 1.020–1.024.
- **TEMPERAMENT** Generally shy but sometimes a bully.

This close relative of the Rabbitfish is difficult to establish in a new aquarium. Plenty of plant matter will initially be needed to replace the algae that forms much of its normal diet. It is best to specialize with this species, keeping just one fish with no other companions. In the wild, Moorish Idols live in shoals, but in aquariums they tend to quarrel if housed in a group. As they mature, adults develop hornlike swellings in front of their eyes.

Dorsal fin extends beyond the caudal

Yellow banding

Beaklike jaws with thick lips for feeding on algae

Equetus lanceolatus

Jack-Knife Fish

- **ORIGINS** Occurs in the Caribbean, from Florida down to the coast of South America.
- **SIZE** Up to 10 in (25 cm).
- **DIET** Fresh and thawed meat-based foods.
- **WATER** Temperature 77–79°F (25–26°C); alkaline (pH 8.1–8.3) with SG 1.020–1.024.
- **TEMPERAMENT** Becomes more aggressive with age.

The striking appearance of these fish results from the upright position of the first of the two dorsal fins, with its dark stripe curving down the body to the tip of the caudal fin. This fin arrangement means that it is not a fast-swimming species. Choose companions carefully, because the elaborate fins are easily damaged.

Pterapogon kauderni

Bangaii Cardinalfish

- **ORIGINS** Occurs around Indonesia's Bangaii Islands, close to Sulawesi (formerly the Celebes).
- **SIZE** 3¼ in (8 cm).
- **DIET** Fresh and thawed meat-based diets.
- **WATER** Temperature 77–79°F (25–26°C); alkaline (pH 8.1–8.3) with SG 1.020–1.024.
- **TEMPERAMENT** Relatively placid.

Although they have a restricted range in the wild, the breeding habits of Bangaii Cardinalfish have allowed relatively easy creation of aquarium strains. These fish are mouth-brooders, with the male carrying the eggs until they hatch. The fry are large enough to be fed brine shrimp. Bangaiis can be kept in small groups, alongside other nonaggressive species.

Deeply forked caudal fin

White-spotted pattern

Plotosus lineatus

Saltwater Catfish

- **ORIGINS** Ranges from the Red Sea throughout the Indo-Pacific region.
- **SIZE** Up to 16 in (40 cm).
- **DIET** Thawed or fresh meat-based foods.
- **WATER** Temperature 77–79°F (25–26°C); alkaline (pH 8.0–8.3) with SG 1.020–1.024.
- **TEMPERAMENT** Highly social only when young.

Think carefully before choosing this species for an aquarium. Young Saltwater Catfish, like those shown below, are social fish that must always be kept in groups. When they reach 6 in (15 cm) in length, however, they lose the distinctive white body stripes and their behavior changes; they start to prefer a more solitary lifestyle and should be kept singly. In addition, care must be taken to avoid their potentially lethal venomous fin spines.

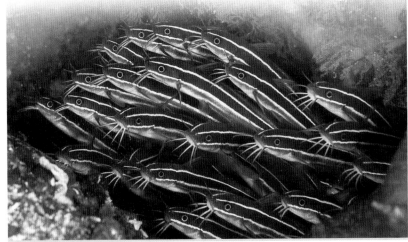

HIDING PLACES

On the reef, just as in the aquarium, fish will make the most of whatever retreats they can find. Reef shipwrecks, for example, are rapidly colonized by a variety of species. This Giant Moray Eel (*Gymnothorax javanicus*) has taken up residence in a piece of piping. This unusual hideaway allows it to lurk unseen and surge out to seize passing prey, just as it would from a reef crevice. However, not all fish that approach the eel's lair will be eaten. Some small species come in search of leftover food scraps, darting boldly close to the eel's mouth to snatch floating morsels.

Ptereleotris zebra

Zebra Goby

- **ORIGINS** From the Red Sea across the Pacific, to the Ryukyu Islands; south to Australia's Great Barrier Reef.
- **SIZE** 4 in (10 cm).
- **DIET** Small invertebrates favored.
- **WATER** Temperature 77–79°F (25–26°C); alkaline (pH 8.1–8.3) with SG 1.020–1.024.
- **TEMPERAMENT** Usually placid; adults more territorial.

A small group of Zebra Gobies can be kept in a reef aquarium, but their food needs to be carried on the current rather than lying on the bottom. These fish seek small caves where they can retreat, often sharing holes. A pair will spawn in aquarium surroundings, with the female guarding the eggs until they hatch.

Opistognathus aurifrons

Yellow-Headed Jawfish

- **ORIGINS** The Caribbean, from Florida and the Bahamas down to the coast of Venezuela.
- **SIZE** 4 in (10 cm).
- **DIET** Small fresh and thawed meat-based foods.
- **WATER** Temperature 77–79°F (25–26°C); alkaline (pH 8.1–8.3) with SG 1.020–1.024.
- **TEMPERAMENT** Shy and nonaggressive.

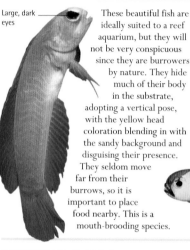

Large, dark eyes

These beautiful fish are ideally suited to a reef aquarium, but they will not be very conspicuous since they are burrowers by nature. They hide much of their body in the substrate, adopting a vertical pose, with the yellow head coloration blending in with the sandy background and disguising their presence. They seldom move far from their burrows, so it is important to place food nearby. This is a mouth-brooding species.

Nemateleotris decora

Purple Fire Goby

- **ORIGINS** The western Pacific, north to the Ryukyu Islands and south to Australia's Great Barrier Reef.
- **SIZE** 3 in (7.5 cm).
- **DIET** Small crustaceans.
- **WATER** Temperature 77–79°F (25–26°C); alkaline (pH 8.1–8.3) with SG 1.020–1.024.
- **TEMPERAMENT** Reasonably social.

The dorsal fin of these small fish has two parts, and they may raise the taller front portion, almost as if in a threatening gesture. They are suitable for a reef aquarium, where they will use holes in the rockwork as retreats. Do not house these fish with invertebrate predators, such as hermit crabs and bristleworms, which will prey on them at night. Feed them on brine shrimp at first, but later they can be weaned onto frozen planktonic foods.

Front section of dorsal fin held flat

Rear portion of dorsal fin

Synanceia horrida

Horrid Stonefish

- **ORIGINS** Ranges from the Red Sea throughout the Indo-Pacific region, extending east to Oceania.
- **SIZE** 12 in (30 cm).
- **DIET** Animal-based foods.
- **WATER** Temperature 77–79°F (25–26°C); alkaline (pH 8.1–8.3) with SG 1.020–1.024.
- **TEMPERAMENT** Predatory.

The Horrid Stonefish's appeal lies in its amazing camouflage. As an aquarium occupant, stonefish normally have to be kept on their own because of their highly predatory natures, although feeding them is quite straightforward. Great care needs to be taken when catching one of these fish or servicing its tank to avoid being injured by the stonefish's venomous spines. The spines contain a toxin that can cause severe tissue damage.

THE PATIENT PREDATOR

The ultimate ambush experts in the marine world are stonefish, such as this Popeyed Sea Goblin (*Inimicus didactylus*). They spend their time lying camouflaged on the seabed, often partially buried or concealed among seaweed. The eyes, which are positioned on the top of the head to give all-around visibility, alert the stonefish to the approach of potential prey, which is snapped up by the cavernous mouth. A stonefish can swim but typically prefers to remain hidden from view. Barefoot swimmers risk being impaled on the fish's venomous spines when wading in shallow water.

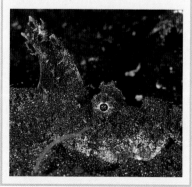

Sphaeramia nematoptera

Pajama Cardinalfish

- **ORIGINS** Eastern Pacific, from Java up to the Ryukyu Islands and south to Australia's Great Barrier Reef.
- **SIZE** 3¼ in (8 cm).
- **DIET** Fresh and thawed meat-based marine foods.
- **WATER** Temperature 77–79°F (25–26°C); alkaline (pH 8.1–8.3) with SG 1.020–1.024.
- **TEMPERAMENT** Generally placid.

These unusually patterned members of the cardinalfish family are ideal for a community marine setup or a reef aquarium, since they are rarely aggressive, even toward invertebrates. However, they may occasionally disagree among themselves if they are kept in a small group. To avoid overcrowding, allow 12 in (30 cm) of tank length for every one of these fish. By introducing them all to the aquarium at the same time, you can minimize the risk of territorial disputes. Pajamas have large eyes, indicating that they tend to be most active after dark, but they will also prove lively in an aquarium during the day. Transferring them to artificial diets is usually not difficult, although they will typically refuse marine flake foods. Brine shrimp are very popular with this species, and these can also be used to rear any young that are produced. Pajama Cardinalfish are mouth-brooders.

Echidna nebulosa

Snowflake Moray Eel

- **ORIGINS** Extends from the coast of East Africa and the Red Sea across the Indo-Pacific region to Oceania.
- **SIZE** 39 in (100 cm); 12 in (30 cm) in aquariums.
- **DIET** Meat-based foods, fresh and thawed.
- **WATER** Temperature 75–79°F (24–26°C); alkaline (pH 8.0–8.3) with SG 1.019–1.026.
- **TEMPERAMENT** Aggressive; do not mix with other fish.

These predatory denizens of the reef will settle well in a marine aquarium. The aquarium must be securely covered to prevent escape, and a large crevice in which the eel can hide is essential. Moray eels locate food by its waterborne scent, the small eyes being an indication of their poor vision. Do not try to hand-feed these fish—they can inflict serious bites. Instead, use special tongs usually sold for offering food to snakes.

Taeniura lymma

Blue-Spotted Ray

- **ORIGINS** Ranges from the Red Sea to the vicinity of southern Japan and Australia's eastern coast.
- **SIZE** 10 in (25 cm).
- **DIET** Mainly crustaceans and mollusks.
- **WATER** Temperature 75–77°F (24–25°C); alkaline (pH 8.1–8.3) with SG 1.021–1.024.
- **TEMPERAMENT** Keep separate.

Long tail

Body undulates when swimming

These fish are potentially dangerous because they have a toxic spine at the base of the tail. During any maintenance work, such as cleaning the filter, it is important to partition the aquarium in order to avoid any risk of being stung by the spine. Be sure that there are extensive open areas in the tank for swimming. These rays will spend much of their time close to the floor of the aquarium, which is where they search for food. Rays have a cartilaginous rather than a bony skeleton and no swim bladder, relying instead on their large, fatty liver to provide buoyancy.

Rhinomuraena quaesita

Blue Ribbon Eel

- **ORIGINS** Seas around southeast Asia, extending south to Australia and north almost to Japan.
- **SIZE** 48 in (120 cm); 15 in (38 cm) in aquariums.
- **DIET** Live invertebrates may be needed.
- **WATER** Temperature 75–79°F (24–26°C); alkaline (pH 8.0–8.3) with SG 1.019–1.026.
- **TEMPERAMENT** Will escape if aquarium is not covered.

The color of these ribbon eels alters with age and may also reflect a change in gender. They start off as black, then change to the blue form (which used to be considered as a separate species), and later become more yellow. All females change from males and are yellower overall. A ribbon eel needs rocky crevices where it can lurk. It prefers live prey, so weaning it onto prepared foods can be difficult. Initially, try waving inert foods on tongs near the eel's lair.

DIRECTORY OF
MARINE
INVERTEBRATES

SPONGES AND ANEMONES

These invertebrates are an integral part of the reef ecosystem. Anemones are soft-bodied creatures with flowing, stinging tentacles for catching prey. Sponges have a more rigid body structure and are filter-feeders. Both are sometimes known as sessile invertebrates because, like corals, they anchor themselves to the reef. Sponges build up their body casing from minerals, such as calcium and silica, combined with a jellylike substance called spongin. As a sponge grows, it is possible to take off pieces and establish these "cuttings" elsewhere. Anemones reproduce either sexually or asexually, depending on the species.

The **Four-Colored Anemone** (*Entacmaea quadricolor*) is used as a retreat by clownfish (see pp.244–245).

Pachycerianthus species

Cerianthus

- **ORIGINS** The western Pacific, notably from Singapore north to the Philippines.
- **SIZE** 12 in (30 cm).
- **DIET** Meat-based foods. Drop food on the anemone once or twice a week.
- **WATER** Temperature 77–79°F (25–26°C); alkaline (pH 8.1–8.3) with SG 1.020–1.024.

This anemone constructs a protective tube from mucus and sand. During the day, the anemone retreats into the tube, anchored in the substrate. At night, it preys on crustaceans and slow-swimming fish. In the aquarium, avoid housing it with creatures such as seahorses, which it will eat. Site it away from corals and other anemones, and take care not to touch its stinging tentacles.

Axinella species

Orange Cup Sponge

- **ORIGINS** Widely distributed throughout the Indo-Pacific region, including off Sri Lanka and Indonesia.
- **SIZE** 6 in (15 cm).
- **DIET** Invertebrate food. Will also take puréed shrimp and mussels.
- **WATER** Temperature 77–79°F (25–26°C); alkaline (pH 8.1–8.3) with SG 1.020–1.024.

Orange Cups will not thrive in silt, and their tank needs a relatively strong current to prevent debris from building up in the cup. Their shape also makes them vulnerable to being choked by algae. If the cup's rim is paler than the rest of the sponge, it has dried out at some point previously. This will prove fatal in the long run, so avoid such specimens.

Condylactis gigantea

Pink-Tipped Anemone

- **ORIGINS** Ranges widely throughout the Caribbean and western Atlantic, reaching Bermuda and Brazil.
- **SIZE** 16 in (40 cm).
- **DIET** Meat-based foods. Typically requires feeding every two days or so.
- **WATER** Temperature 77–79°F (25–26°C); alkaline (pH 8.1–8.3) with SG 1.020–1.024.

These anemones vary naturally in color; those with purplish tentacles tend to be more popular than pale-tentacled specimens. Preferring bright light, they are ideal for a reef tank, where they will anchor themselves in rocky crevices. They fare well in dimmer light, too, so they can be included in a setup intended primarily for fish. Pale dots on the tentacles indicate the positions of stinging cells.

Rhodactis species

Elephant Ears

- **ORIGINS** Distribution extends from the central part of the Indo-Pacific as far south as Australia.
- **SIZE** 8 in (20 cm).
- **DIET** May prey on brine shrimp. Will also consume some meat-based foods.
- **WATER** Temperature 77–79°F (25–26°C); alkaline (pH 8.1–8.3) with SG 1.020–1.024.

This flattened, disklike anemone has very short tentacles and resembles a coral. Unlike true stony corals, it lacks a hard body casing. Also, each anemone body is separate, while coral polyps are joined. Elephant Ears contains symbiotic bacteria that provide it with food when they photosynthesize. It needs strong light to thrive and benefits from water movement in the tank.

Amphimedon compressa

Red Tree Sponge

- **ORIGINS** Widely distributed throughout the waters of the Caribbean.
- **SIZE** 8 in (20 cm).
- **DIET** Plankton or puréed food. Will need feeding every day or two.
- **WATER** Temperature 77–79°F (25–26°C); alkaline (pH 8.1–8.3) with SG 1.020–1.024.

Despite its name, the Red Tree Sponge is often pinkish rather than red. This paler hue is not a reflection of poor health, because this species varies naturally in coloration. What is likely to be a sign of ill-health, however, is the appearance of white patches on the body—the coloration should be uniform. Red Tree Sponges require dimly lit surroundings. They feed on microscopic plankton, so they must be placed in a current to ensure that food is wafted to them. Under no circumstances should these sponges be allowed to dry out when they are being moved, since this can prove fatal. It is also not advisable to introduce these or other sponges to a recently established reef aquarium, which is unlikely to contain enough natural food for them. When purchasing a Red Tree Sponge, especially a large specimen, check that there are no tiny crustaceans lurking among its branches.

COLONIAL LIFE

Some anemones, such as these Yellow Indonesian Polyps (*Parazoanthus* sp.), live in colonies. They are vulnerable to predators, such as marine angelfish *(see pp.252–255),* when their tentacles are exposed (below right). If danger approaches, an anemone will pull its tentacles into its body (below left). One advantage of communal living is that when one anemone withdraws its tentacles, all its neighbors are instantly alerted to the threat. The length of the extended tentacles gives an insight into the health of the colony. Shortened tentacles suggest poor water quality, ill-health, or individuals that have recently inflicted a sting. All the anemones in a colony are likely to be clones of one another. They reproduce asexually, sending out runners that develop into new anemones, enabling the colony to grow in size.

Heteractis magnifica

Magnificent Anemone

- **ORIGINS** Ranges throughout the Indo-Pacific region, from the Red Sea eastward as far as Samoa.
- **SIZE** 40 in (100 cm).
- **DIET** Meat-based foods. Avoid overfeeding, which will impair water quality.
- **WATER** Temperature 77–79°F (25–26°C); alkaline (pH 8.1–8.3) with SG 1.020–1.024.

The body of this anemone ranges from purplish-pink to white and even avocado green, although much of it is hidden by the tentacles, which are typically over 3 in (7.5 cm) long. If the tentacles are largely retracted, the water quality is likely to have deteriorated. This anemone sometimes occurs in groups on the reef, often close to the surface.

CORALS

Corals form the centerpiece of any reef aquarium.
A coral is a colony of linked organisms called polyps.
In stony corals, which form the foundation of the reef,
the polyps have a hard body casing of calcium carbonate,
while the polyps of soft corals are supported by a less
rigid calcareous structure. Corals need plenty of space,
and overcrowding will hinder their growth. Many
corals contain symbiotic zooxanthellae (algae). When
the algae photosynthesize, they provide food for both
themselves and the coral, so good lighting in the tank
is vital. If a piece of coral breaks off, it can be used to
establish a new colony elsewhere in the tank.

The polyps of the **Cauliflower Coral** (*Pocillopora damicornis*) are
shown here in close-up. Corals can be identified by their polyp
shape, which is a relatively consistent feature within each species.
Coloration, which can be much more variable, is a less reliable guide.

Plerogyra sinuosa

Bubble Coral

- **ORIGINS** Extends from the Red Sea eastward through the entire Indo-Pacific region to Samoa.
- **SIZE** 39 in (100 cm).
- **DIET** Symbiotic, with internal algae providing food. Will also feed on plankton and brine shrimp.
- **WATER** Temperature 77–79°F (25–26°C); alkaline (pH 8.1–8.3) with SG 1.020–1.024.

Inhabiting fairly exposed areas of the reef, these
corals have a relatively low, compact shape that
helps to protect them from damage. Bubble Corals,
which sometimes form massive colonies, are
nocturnal creatures, only putting out their polyps
in search of small prey after dark. White stripes
across the individual bubblelike swellings indicate
the location of the stinging cells, or nematocysts.

Tubastrea aurea

Orange Polyp Coral

- **ORIGINS** Widely distributed on reefs throughout the Indo-Pacific region.
- **SIZE** 4 in (10 cm).
- **DIET** Shrimp and other meat-based foods. Drop tiny pieces of food into the open coral heads.
- **WATER** Temperature 77–79°F (25–26°C); alkaline (pH 8.1–8.3) with SG 1.020–1.024.

A stunning appearance and simple care needs make
this orange coral an ideal choice for home aquariums.
After transfer to a new tank, the polyps may
remain closed for a week. Because Orange Polyp
Corals inhabit shady areas, their bodies lack
symbiotic algae, so they feed by catching food with
their tentacles. Reproduction is asexual, with new
polyps budding off from the base of existing ones.

Lobophyllia hemprichii

Lobed Brain Coral

- **ORIGINS** From East Africa and the Red Sea, via the Indo-Pacific, to the Marshall Islands in the east.
- **SIZE** 16 in (40 cm).
- **DIET** Symbiotic, with internal algae producing nutrients. Will take small amounts of a proprietary food.
- **WATER** Temperature 77–79°F (25–26°C); alkaline (pH 8.1–8.3) with SG 1.020–1.024.

A twisting, involuted appearance characterizes
Lobed Brain Corals, which can grow to a large
size. Their coloration ranges from grayish-blue to
green to deep red. These stony corals are found
on deeper parts of the reef and are nocturnal in
habit. When feeding Lobed Brain Corals, it helps
to use a pipette so that the food can be placed
directly in the vicinity of the coral.

Goniopora species

Flowerpot Coral

- **ORIGINS** Occurs widely throughout the Indo-Pacific, from the Red Sea and East Africa to Fiji and Samoa.
- **SIZE** 8 in (20 cm).
- **DIET** Plankton and symbiosis; good lighting conditions are vital to ensure that the internal algae are healthy.
- **WATER** Temperature 77–79°F (25–26°C); alkaline (pH 8.1–8.3) with SG 1.020–1.024.

The color of this stony coral varies, depending on the color of the zooxanthellae in its polyps. If gray blotches appear, however, this may signify the onset of a serious illness. The polyps, which are permanently extended, are long, delicate, and featherlike. This coral likes strong currents and bright lighting, but there is a risk that it may be attacked by external algae under such conditions.

NIGHT BLOOMING

Corals may reproduce either asexually, by a process known as budding, or by sexual means, which enables them to spread farther afield. The problem with sexual reproduction is that when a female coral releases an egg (as shown here), the currents around the reef make the chances of the egg's coming into contact with sperm relatively slight. To improve the chances of fertilization, entire coral populations simultaneously release their gametes, using the lunar cycle to coordinate this mass reproduction. In fact, they are so prolific that the sea temporarily turns white with eggs and sperm, greatly increasing the likelihood of fertilization. This phenomenon is called night blooming. The young coral larvae drift away on the current to colonize new reef areas.

Xenia species

Pulsing Coral

- **ORIGINS** Ranges from the Red Sea eastward to the Indo-Pacific region, including the Philippines.
- **SIZE** 3½ in (8 cm).
- **DIET** Mainly symbiotic; good lighting conditions are vital to ensure that the internal algae are healthy.
- **WATER** Temperature 77–79°F (25–26°C); alkaline (pH 8.3) with SG 1.020–1.024.

A delicately branched appearance typifies this treelike coral. The permanently extended polyps move continuously during the daytime, not in search of food but to create water currents that will bring oxygen to the coral. As with other corals that live by symbiosis, the Pulsing Coral needs bright lighting and good water circulation. Any nitrate in the water will harm this coral.

Pocillopora damicornis

Cauliflower Coral

- **ORIGINS** Extends from East Africa and the Red Sea throughout the Indo-Pacific region.
- **SIZE** 4 in (10 cm).
- **DIET** Filter-feeder, requiring very fine particles of food that it can sift from the water.
- **WATER** Temperature 77–79°F (25–26°C); alkaline (pH 8.1–8.3) with SG 1.020–1.024.

This attractive coral is named after its cauliflower-like growth pattern. The Cauliflower Coral varies from pinkish-blue to pure blue. Site this coral in the upper levels of the tank; on the reef, it normally occurs close to the surface. As with all stony corals, a special supplement containing minerals, such as calcium, and trace elements, including strontium, should be added to the water.

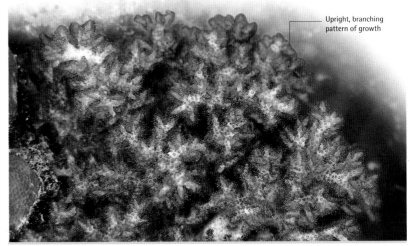

Upright, branching pattern of growth

CRUSTACEANS

Crustaceans are justifiably popular in reef aquariums because they provide both color and movement. However, they must be chosen carefully because they can be aggressive and predatory. These invertebrates have a hard body casing called an exoskeleton. They also possess jointed limbs, with the front pair often modified into claws for digging, grasping, and cutting. Crustaceans grow by a series of molts, inflating themselves with hemolymph so that their exoskeleton splits open. The new exoskeleton hardens soon after the crustacean has emerged from the old one. Any lost limbs may be regenerated during the molting process.

Some remarkable relationships involving crustaceans have formed on the reef. The **Swimming Crabs** (*Lissocarcinus* species) shown here do not burrow into the sand for protection but retreat into the stinging tentacles of tube anemones to escape from danger.

Calappa flammea

Shame-Faced Crab

- **ORIGINS** Widely distributed through the entire Caribbean region.
- **SIZE** Body is 8 in (20 cm) across.
- **DIET** Live foods. Will scavenge for pieces of fish and shellfish, but be careful not to overfeed.
- **WATER** Temperature 77–79°F (25–26°C); alkaline (pH 8.1–8.3) with SG 1.020–1.024.

Shame-Faced Crabs are so called because of the way in which they seem to hide their eyes behind their greatly enlarged claws. Although they often lie concealed under the substrate, these crabs can be rather disruptive in a typical reef tank, because they scavenge aggressively and prey on mollusks and other invertebrates. Shame-Faced Crabs do, however, get along well with fish such as gobies.

Neopetrolisthes ohshimai

Porcelain Crab

- **ORIGINS** The Indo-Pacific region, from the Asian coast south to Australia and east to the central Pacific.
- **SIZE** Body is 1 in (2.5 cm) across.
- **DIET** Will take freeze-dried and tablet foods. Try to direct food to the vicinity of the crab.
- **WATER** Temperature 77–79°F (25–26°C); alkaline (pH 8.1–8.3) with SG 1.020–1.024.

These tiny crabs seek sanctuary from would-be predators by hiding among the stinging tentacles of sea anemones, to which they appear to be immune. Note that Porcelain Crabs will attack any clownfish introduced into the tank. They feed by trapping tiny particles of food with feathery projections on their jaws. Unfortunately, there is no way of sexing these crabs visually.

Variable red-and-white patterning

Enoplometopus occidentalis

Red Lobster

- **ORIGINS** Found throughout the Indo-Pacific region, ranging as far east as the Hawaiian islands.
- **SIZE** Body is 5 in (12.5 cm) long.
- **DIET** Shrimp, fish, and food tablets. Try to ensure that the food is placed within the lobster's reach.
- **WATER** Temperature 77–79°F (25–26°C); alkaline (pH 8.1–8.3) with SG 1.020–1.024.

These strikingly colored lobsters are less conspicuous in the aquarium than their color would suggest. This is because they are nocturnal and usually hide during the daytime. Red Lobsters are territorial and will use their powerful claws to fight ferociously if they are housed together. These lobsters will prey on small fish, but they will also take inanimate foods.

Dardanus lagopodes

Blade-Eyed Hermit Crab

- **ORIGINS** From the Red Sea eastward throughout the entire Indo-Pacific region.
- **SIZE** Body is 2½ in (6 cm) long.
- **DIET** Proprietary hermit crab food or meat-based foods, including pieces of fish and shellfish.
- **WATER** Temperature 77–79°F (25–26°C); alkaline (pH 8.1–8.3) with SG 1.020–1.024.

Hermit crabs are unsuited to reef aquariums because they will prey on the occupants. They are best kept in a tank with nonaggressive fish, where they may prove useful in finishing off uneaten food. Hermit crabs do not have their own shells but take over those of mollusks, swapping to larger shells as they grow. Make sure there is a series of larger shells available in the tank for this purpose.

Allogalathea elegans

Feather Star Squat Lobster

- **ORIGINS** Widely distributed throughout the entire Indo-Pacific region.
- **SIZE** Body is 1¾ in (2 cm) long.
- **DIET** Prefers thawed foods, but will also take small freeze-dried items.
- **WATER** Temperature 77–79°F (25–26°C); alkaline (pH 8.1–8.3) with SG 1.020–1.024.

These tiny lobsters have a striped, egg-shaped body. They should not be mixed with larger predatory species of any kind. The tank must include Feather Starfish (see p.297), since the lobsters live among their arms, avoiding detection by modifying their coloration so that they blend in with their hosts. Small crevices in nearby rockwork will serve as hiding places for these shy, retiring lobsters.

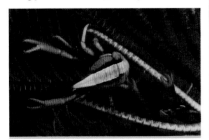

Panulirus versicolor

Purple Spiny Lobster

- **ORIGINS** The Pacific region, especially around Singapore and parts of Indonesia.
- **SIZE** Body is 9 in (22.5 cm) long.
- **DIET** Fish, shrimp, and tablet food. Should be fed a relatively small quantity once a day.
- **WATER** Temperature 77–79°F (25–26°C); alkaline (pH 8.1–8.3) with SG 1.020–1.024.

This large lobster can be identified by its banded body patterning, striped legs, and the blue area on its tail. The antennae, which help the lobster to find food, are often longer than the body, so a spacious tank is required. If an antenna breaks off, it should regrow over successive molts, but not necessarily to its original length. The Purple Spiny Lobster can be housed safely with large companions.

Lissocarcinus laevis

Harlequin Swimming Crab

- **ORIGINS** Widely distributed throughout the entire Indo-Pacific region.
- **SIZE** Body is 1¼ in (3 cm) across.
- **DIET** Animal-based foods. Provide relatively small pieces that can be consumed easily.
- **WATER** Temperature 77–79°F (25–26°C); alkaline (pH 8.1–8.3) with SG 1.020–1.024.

This red-and-white crab must always be housed alongside tube anemones, with which it forms a close association. Like other types of swimming crab, it has paddlelike hind legs that help it to swim efficiently. It is possible to sex this species by examining the underside of the body, since males have a narrower, more triangular abdominal region than females. Breeding in aquariums has yet to be achieved with this rather solitary crab.

BOXING CHAMP

Some crustaceans evade danger by hiding under the sand, or among the stinging tentacles of anemones, while others rely on camouflage to conceal their presence. The Common Boxing Crab (*Lybia tessellata*), shown here, has a more direct way of protecting itself: it carries a small anemone in each claw and uses them like weapons. If threatened, the crab thrusts one claw forward, followed by the other, like a boxer delivering punches. This is enough to persuade most predators to back off. The Boxing Crab lets go of its anemones only when it molts, picking them up again when its new exoskeleton hardens. If the crab is challenged before this, it responds as though it were still holding the anemones.

Periclimenes yucatanicus

Spotted Cleaner Shrimp

- **ORIGINS** Occurs throughout the Caribbean region and in the western part of the Atlantic.
- **SIZE** Body is ³/₄ in (2 cm) long.
- **DIET** Marine flake and small live foods. Often scavenges food given to its host anemone.
- **WATER** Temperature 77–79°F (25–26°C); alkaline (pH 8.1–8.3) with SG 1.02–1.024.

These shrimp have a yellow-and-white body, dark bands of red and white on their limbs, and white antennae. They need to be housed with a suitable host anemone, such as the Pink-Tipped Anemone (*see p.286*). Breeding is possible in aquariums; the female carries the green eggs under her abdomen. However, predation by other tank occupants makes it highly unlikely that any of the young will survive.

Lysmata debelius

Scarlet Cleaner Shrimp

- **ORIGINS** The Indo-Pacific region, ranging from the Maldives to Japan, Indonesia, and the Society Islands.
- **SIZE** Body is 1¼ in (3 cm) long.
- **DIET** Marine flake and small live foods. Try to ensure that food falls within reach of the shrimp.
- **WATER** Temperature 77–79°F (25–26°C); alkaline (pH 8.1–8.3) with SG 1.020–1.024.

Scarlet-red coloration is a feature of these shrimp, which also display white spots on the body, white lower limbs, and white antennae. In reef aquariums, they are less shy if kept as a small group rather than on their own.
Each maintains a small area of territory that includes a number of snug retreats.

Antennae are used as feelers in murky surroundings

Rhynchocinetes uritai

Dancing Shrimp

- **ORIGINS** Widely distributed throughout the entire Indo-Pacific region.
- **SIZE** Body is 1½ in (4 cm) long.
- **DIET** Prepared foods, which can include marine flake. Will also eat small pieces of fish.
- **WATER** Temperature 77–79°F (25–26°C); alkaline (pH 8.1–8.3) with SG 1.020–1.024.

Prominent eyes and a hump-backed appearance help to identify this crustacean. Males have larger claws than females. Dancing Shrimp are social by nature, but they will attack anemones that are not protected by stinging tentacles, and they will also eat coral polyps. Their movements resemble those of a tango dancer, advancing in a very deliberate fashion and then briefly pausing.

LIVING DANGEROUSLY

Cleaner Shrimp appear to have a death wish, actively seeking out and climbing all over fish that could easily snap them up. The potential predators refrain from devouring them because the shrimp use their powerful claws to remove parasites and skin debris from the bodies of the fish. The shrimp below (*Lysmata amboinensis*) is "cleaning" a Moray Eel (*Gymnothorax* species).

This arrangement helps to keep the eel healthy, while the shrimp gets to eat whatever it can remove. The shrimp tend to clean relatively sedentary fish species so that they are not carried off into the depths. How these relationships come about is unclear, because in aquariums the shrimp will even perform this service for fish that they do not encounter in the wild.

Saron species

Marble Shrimp

- **ORIGINS** Throughout the Indo-Pacific region, from the Red Sea, via Indonesia, to Hawaii.
- **SIZE** Body is 1¼ in (3 cm) long.
- **DIET** Eats small live foods, and will also scavenge in the tank for items such as marine flake.
- **WATER** Temperature 77–79°F (25–26°C); alkaline (pH 8.1–8.3) with SG 1.020–1.024.

These sociable shrimp have marbled patterning—often a whitish-green with darker markings. They are shy and are unlikely to be seen during the day. When they do emerge, it quickly becomes obvious that their daytime coloration can be very different from their appearance at night. It is best to offer them food only after dark. Marble Shrimp should not be kept in a tank with live corals.

Stenopus hispidus

Coral Banded Shrimp

- **ORIGINS** Found on tropical reefs throughout the world's oceans, particularly the Pacific.
- **SIZE** Body is 2½ in (6 cm) long.
- **DIET** Meat-based foods. If frozen supplies are used, make sure that the food is thawed completely before use.
- **WATER** Temperature 77–79°F (25–26°C); alkaline (pH 8.1–8.3) with SG 1.020–1.024.

The ancestors of these shrimp were present in the oceans more than 10 million years ago, when North and South America were still unattached continents. As the land bridge formed, the populations in the tropical regions of the Pacific and Atlantic became separated, resulting in different color forms and new species in different locations. Female Coral Banded Shrimp are larger than males. They also have bright red banding on their upperparts, which appears brownish in males. Keep these shrimp in true pairs, if not singly, because otherwise they will fight to the death. Tankmates need to be chosen carefully because these invertebrates will prey on other shrimp and smaller fish. The male Coral Banded Shrimp is unusual in that he collects food for the female and actively feeds her in a cave. She signals by clicking her pincers, and he responds by passing over food.

Synalpheus species

Pistol Shrimp

- **ORIGINS** Present in various forms in the Indo-Pacific, from the Red Sea to Japan and Hawaii.
- **SIZE** Body is 1½ in (4 cm) long.
- **DIET** Small live foods and food tablets. As always, avoid overfeeding, since uneaten food will pollute the water.
- **WATER** Temperature 77–79°F (25–26°C); alkaline (pH 8.1–8.3) with SG 1.020–1.024.

Pistol Shrimp have one enlarged claw that can produce a loud noise like a gunshot when it snaps together. The "shot" is so loud that it may sound as if the tank has shattered. Pistol Shrimp use these sound blasts to stun small prey. If the pistol claw is lost, the other claw will enlarge to compensate while the damaged limb regrows. Another key feature of Pistol Shrimp is their small eyes, which are partially obscured by the protective carapace.

Thor amboinensis

Broken-Back Shrimp

- **ORIGINS** Widely distributed throughout the Indo-Pacific region, including off the island of Amboina, Indonesia.
- **SIZE** Body is ¾ in (2 cm) long.
- **DIET** Thawed or freeze-dried live foods. Offer food in proximity to the host anemone.
- **WATER** Temperature 77–79°F (25–26°C); alkaline (pH 8.1–8.3) with SG 1.020–1.024.

This crustacean can be distinguished by its bold markings of dull orange and white and by the way in which it keeps its hindquarters raised above the rear of its body. It typically lives in association with *Heteractis* sea anemones, among which it scavenges for food. The Broken-Back Shrimp usually lies on top of its host, rather than retreating into its tentacles.

MOLLUSKS AND ANNELIDS

There are few more diverse invertebrate groups than the mollusks. Marine gastropod mollusks possess a protective shell and a "foot" for locomotion, just like the land snails. Bivalves, such as clams, have a two-part, hinged shell and a more sedentary lifestyle. The shell-less sea slugs can be difficult to maintain in aquariums, because their dietary requirements are so specific. Cephalopods, which include squid and octopuses, can also be problematic, since they require large tanks and are extremely sensitive to water conditions. Annelids, in contrast, are a group of segmented worms.

The **Spanish Dancer** (*Hexabranchus imperialis*) is the largest and also one of the most attractive of all sea slugs, thanks to its vivid red-and-white coloration. Originating from the Indo-Pacific, it has a graceful swimming motion.

Limaria scabra

Flame Scallop

- **ORIGINS** Widely distributed throughout the Caribbean region, and in parts of the western Atlantic.
- **SIZE** 2½ in (6 cm) in diameter.
- **DIET** Filter-feeding. Provide a prepared food or a blend of puréed shellfish and seawater.
- **WATER** Temperature 77–79°F (25–26°C); alkaline (pH 8.1–8.3) with SG 1.020–1.024.

The true beauty of this scallop is only apparent when it opens its shell and exposes the scarlet-red interior and tentacles. (One closely related form has off-white tentacles.) Flame Scallops anchor themselves to rockwork, so place them near the front of the tank where they will be clearly visible. They may breed successfully in the tank, giving rise to small groups of young.

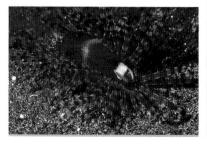

Tridacna crocea

Crocus Clam

- **ORIGINS** Southeast Asia, originating from the coastal waters around Singapore and parts of Indonesia.
- **SIZE** 8 in (20 cm).
- **DIET** Filter-feeding and symbiosis. Needs fine particulate food and good lighting to maintain its internal algae.
- **WATER** Temperature 77–79°F (25–26°C); alkaline (pH 8.1–8.3) with SG 1.020–1.024.

Its stunning blue interior, emphasized by the external fluting, makes this small, slow-growing bivalve a popular choice for the invertebrate tank.

High-intensity lighting is vital for the well-being of this clam. The blue coloring is produced by algae living in the mantle (the inner lining of the shell), and these microorganisms need bright light in order to be able to carry out photosynthesis, which provides the clam with most of its nutrients. A healthy clam will not only be well colored, but will also be able to close rapidly when touched gently with a finger. Crocus Clams obtain some of their food by filtering tiny creatures from seawater, which is drawn into the body via an opening called an inlet siphon. Prepared foods for filter-feeders enable Crocus Clams to feed well in aquariums.

Sabellastarte magnifica

Featherduster Worm

- ⊕ **ORIGINS** Occurs widely throughout the tropical western Atlantic and the Caribbean region.
- ⊕ **SIZE** 4 in (10 cm).
- ⊕ **DIET** Filter-feeding. Use prepared food, or a blend of puréed shellfish and seawater.
- ⊕ **WATER** Temperature 77–79°F (25–26°C); alkaline (pH 8.1–8.3) with SG 1.020–1.024.

This annelid worm's body is housed in a protective tube anchored to rockwork. The feathery tentacles projecting from the top of the tube collect floating food particles. Featherduster Worms occur in a wide range of colors. Keeping them in groups will encourage breeding. After reproduction, adult Featherduster Worms shed their feeding tentacles to prevent them from taking their own larvae. The tentacles start to regrow about two weeks later.

Octopus cyanea

Common Tropical Octopus

- ⊕ **ORIGINS** Coral reefs off Asia and throughout the Indo-Pacific region, especially off Indonesia.
- ⊕ **SIZE** 12 in (30 cm).
- ⊕ **DIET** Fish and crustaceans, which must be fully thawed if frozen. Feed according to appetite each day.
- ⊕ **WATER** Temperature 77–79°F (25–26°C); alkaline (pH 8.1–8.3) with SG 1.020–1.024.

This octopus should be housed singly, alongside coral and sponge species that it will not harm. The tank must be firmly covered, since it can escape through even a small gap in the hood. When introducing an octopus to new surroundings, leave it undisturbed and without lighting for a day or so, or it may eject its ink, with potentially fatal consequences in this restricted environment.

Lobatus gigas

Queen Conch

- ⊕ **ORIGINS** The Caribbean region, including the Florida coast and around the Bahamas.
- ⊕ **SIZE** 10 in (25 cm).
- ⊕ **DIET** Marine algae and scavenged waste matter. Place food within easy reach of the conch.
- ⊕ **WATER** Temperature 77–79°F (25–26°C); alkaline (pH 8.1–8.3) with SG 1.020–1.024.

These slow-moving, spectacular-looking gastropods grow to a large size and require a suitably spacious aquarium. Fortunately, care of the Queen Conch is relatively straightforward, and feeding presents no particular difficulties. When threatened, this conch will simply withdraw into the safety of its heavy-weight shell, which is more than a match for most would-be predators. The shell, which has a light-colored exterior, is pinkish-white inside.

Cypraea tigris

Tiger Cowrie

- ⊕ **ORIGINS** From the Red Sea and the coast of East Africa eastward through the Indo-Pacific to Hawaii.
- ⊕ **SIZE** 6 in (15 cm).
- ⊕ **DIET** Caulerpa (a marine alga) and meat-based foods. Will tend to scavenge for food around the aquarium.
- ⊕ **WATER** Temperature 77–79°F (25–26°C); alkaline (pH 8.1–8.3) with SG 1.020–1.024.

The smooth, oval shell of this gastropod is covered with dark spots. It is partially hidden by the mantle, which extends around its sides. These algal browsers are not difficult to keep and are easy to feed. However, their size means that they may damage the structure of the reef by dislodging corals and other sessile invertebrates, especially at night, when they are most active.

THE REEF'S VITAL BROWSER

Although they could not be said to be the most striking occupants of a reef tank, Turbo Snails (*Haliotis varia*) are certainly among the most significant. Often called Turban Snails because of their conical shape, these mollusks are useful because they browse almost exclusively on algae, keeping its growth in the aquarium under control. They perform a similar function on the reef, where unchecked algal growth could choke the surfaces of the corals and threaten their survival. Small Turbo Snails may be inadvertently introduced to the tank on live rock. They can largely be left to fend for themselves.

ECHINODERMS

Starfish are the best-known echinoderms, but the group also features feather and brittle stars, as well as sea urchins and sea cucumbers. While they vary in shape, all these creatures have an internal body structure formed from calcium and tube feet that enable them to move over the seabed. Some echinoderms, including many starfish, are easy to maintain in aquariums, but sea cucumbers have a devastating habit of eviscerating themselves when they become distressed, fatally polluting the water for the other tank occupants.

Blue Starfish (*Linckia laevigata*), which are of Asiatic origin, are a good introduction to this group of marine invertebrates. However, they are particularly vulnerable to a species of bivalve mollusk that burrows into their arms, causing paralysis.

Fromia elegans

Little Red Starfish

- **ORIGINS** Occurs throughout the Indo-Pacific region, especially off Indonesia.
- **SIZE** 3¼ in (8 cm).
- **DIET** Mussel, clam, and shrimp meat. Beware of overfeeding these and other echinoderms.
- **WATER** Temperature 77–79°F (25–26°C); alkaline (pH 8.1–8.3) with SG 1.020–1.024.

With their bright red coloration, Red Starfish make small, attractive additions to a reef tank. Juvenile Red Starfish can be recognized by the black tips of their arms (as shown below). Unlike some of their relatives, these starfish can be kept safely with other invertebrates. However, avoid housing Red Starfish with larger predatory starfish species, which will eat them. Some crustaceans may also prey on them.

Protoreaster lincki

Red-Knobbed Starfish

- **ORIGINS** Ranges from the Red Sea to Indonesia and islands of the Pacific.
- **SIZE** 12 in (30 cm).
- **DIET** Mussel, clam, and shrimp meat. Place a small amount of food under the starfish each day.
- **WATER** Temperature 77–79°F (25–26°C); alkaline (pH 8.1–8.3) with SG 1.020–1.024.

Swollen red areas on the body and arms help to identify these predatory starfish; the red areas form a meshlike pattern against a whitish background. Red-Knobbed Starfish are relatively easy to keep, but they will attack and feed on sessile invertebrates sharing their quarters, as well as any mollusks that they can ambush, so their companions need to be chosen very carefully.

Ophiomastix species

Brittle Starfish

- **ORIGINS** Tropical areas, notably off the coast of Florida and throughout the Caribbean region.
- **SIZE** 6 in (15 cm).
- **DIET** Mussels, shrimp, and other meaty foods. Try to vary the type of food offered.
- **WATER** Temperature 77–79°F (25–26°C); alkaline (pH 8.1–8.3) with SG 1.020–1.024.

Brittle Starfish have a small central disk and long, thin arms that are fringed along their entire length. They are ideal starfish for a reef tank, because they will scavenge for food without harming the other invertebrates that share their quarters. They can locate food by smell and are able to reach into nooks and crannies to seek out food particles that are beyond the reach of other tank occupants.

Himerometra robustipinna

Feather Starfish

- **ORIGINS** Off the coast of Southeast Asia, notably in the vicinity of Singapore and parts of Indonesia.
- **SIZE** 7 in (18 cm).
- **DIET** Brine shrimp and other small foods. Try to place the food close to the starfish to encourage it to feed.
- **WATER** Temperature 77–79°F (25–26°C); alkaline (pH 8.1–8.3) with SG 1.020–1.024.

This placid, red starfish usually feeds after dark, waving its highly mobile, featherlike arms in the current to collect fine particles of food. Feather Starfish can be kept in groups, providing retreats for small, nonaggressive fish, such as gobies. As with other starfish species, if an arm breaks off, it will usually regenerate. Good water conditions are vital for the overall well-being of Feather Starfish.

Culcita novaeguineae

Bun Starfish

- **ORIGINS** Widely distributed throughout the entire Indo-Pacific region.
- **SIZE** 10 in (25 cm).
- **DIET** Mussel, clam, and shrimp meat. Offer small amounts each day, removing any uneaten scraps.
- **WATER** Temperature 77–79°F (25–26°C); alkaline (pH 8.1–8.3) with SG 1.020–1.024.

When Bun Starfish are mature, their bodies fill out so much that their five legs are no longer visible. One of the heaviest of all starfish, they have spotted upperparts, but their markings and coloration are highly variable. These predatory starfish are best housed alongside nonaggressive fish, rather than in a reef setup, because they will attack sessile invertebrates, such as corals.

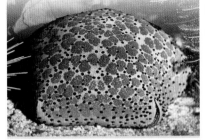

Pseudocolochirus axiologus

Sea Apple

- **ORIGINS** Found on coral reefs off Indonesia, and also on Australia's Great Barrier Reef.
- **SIZE** 6 in (15 cm).
- **DIET** Food particles in suspension. Use liquid foods. May also take brine shrimp.
- **WATER** Temperature 77–79°F (25–26°C); alkaline (pH 8.1–8.3) with SG 1.020–1.024.

The purplish-blue form of this echinoderm, which occurs on the Great Barrier Reef, is exceptionally beautiful. Sea Apples of Indonesian origin are smaller, with grayish-pink bodies. The tentacles projecting from the body are used to extract fine food particles from the water. Tankmates for Sea Apples must be chosen carefully, because some fish will try to bite off their feeding tentacles.

SPINELESS VERTEBRATES

Sea squirts, such as the striking blue variety seen here (*Rhopalaea crassa*), consist of little more than a baglike body known as a tunic. These reef animals feed by drawing water in through the large hole on the top of the body, then filtering it for edible particles, and finally passing it out through a smaller orifice called the exhalant siphon. Like the other species featured in this section, sea squirts do not have a backbone, yet these creatures are actually classified in the same phylum as vertebrates. This apparent oddity of classification comes about because when they are free-swimming larvae, sea squirts possess a well-developed nervous system. This includes a stiff central support, called a notochord, which is similar to the spinal cord. Once they have matured, however, sea squirts adopt a wholly sedentary lifestyle and lose these features.

Sea squirts can grow to a relatively large size, with some species measuring up to 20 in (50 cm) in length. Being highly vulnerable to predators, they inhabit inconspicuous areas of the reef, where they often associate in groups. Like Turbo Snails (*see p.295*), sea squirts may turn up unexpectedly in a marine aquarium, being introduced accidentally when pieces of live rock are added to the tank.

INTRODUCTION TO
POND FISH

What to consider

A carefully chosen and well-designed pond can be the focal point of a garden, but finding the right style can be a daunting task. The choice of design is partly personal but should also take into account the requirements of the fish, as well as the amount of space you have available, the existing landscaping in your garden, your level of building expertise, and the amount of time and money you have to spend.

Although the basic construction techniques used to create ponds of any type are similar, the resulting effect created by the landscaping and planting can be strikingly different. Ponds can be designed in an informal, naturalistic way or in a formal style, based on stark, geometric patterns, such as rectangles.

A good starting point when deciding on the type of pond to build is to consider the species of fish you would like to keep. Goldfish are suitable for most types of ponds, but large pond fish, such as koi, require a considerable volume of water if they are to thrive. A pond for these fish must have a minimum surface area of at least 100 ft^2 (10 m^2). The constraints created by keeping koi mean that they are

This modern pond design blends seamlessly with the style of the house. It includes a waterfall feature and decking but no plants.

A naturalistic style of pond (below), with native plants in and around the edges, helps to create retreats for neighborhood wildlife.

Formal ponds are frequently seen in the grounds of European stately homes, reflecting a gardening trend that began in North Africa. Such ponds generally have few plants present, other than water lilies.

invariably kept in a larger formal design of pond. Such ponds are costly to construct, however, not just because they need to be large but also because they require an effective filtration system to maintain water quality and clarity.

Naturalistic ponds are simple and inexpensive to create using flexible pond liner (see p. 302). At certain times of year, however, the fish may not be particularly conspicuous in this type of pond; during early summer, for example, the water may be green with algae and the pond partially covered with profuse plant growth. A filtration system is less essential in a pond of this type, but regular maintenance is required to remove dead plant material and clean out silt (see pp. 318–320).

Depth is an important consideration, especially in temperate areas. Fish instinctively retreat to the bottom of a pond as the temperature drops toward freezing. Ponds should therefore include an area that is at least 4 ft (1.2 m) deep, to ensure that it will not freeze to the bottom in even the most severe winters.

This koi pond (below) incorporates a gravel border, a low planting plan, and large rocks, reflecting the Oriental origins of the fish kept there.

SITING GUIDELINES

- Ponds should be constructed only in areas where the ground is relatively level.

- Choose a location that allows you to view the pond from inside the house.

- Consider the availability of a power supply for running filters and fountains.

- Avoid building a pond close to trees, because the growth of their roots may damage the foundations or puncture the pond liner.

- Avoid areas that are naturally prone to becoming waterlogged, because excess water can collect under the pond liner.

- Site the pond in a sheltered spot, to stop leaves from blowing into it.

- Choose a site that is not in direct sunlight during the hottest part of the day.

Construction choices

The availability of modern pond construction materials provides a wide choice for the hobbyist. Creating a pond using flexible pond liner or installing a preformed pond unit *(see box, below)* does not require advanced construction skills and can be relatively inexpensive. Large concrete ponds, by contrast, are considerably more expensive and may even require the services of a professional installer.

A wide range of flexible liners and preformed pond units are available to suit most budgets, but it can be a false economy to purchase the cheapest option. Less expensive liners, such as polyethene, may not last as long as higher-quality flexible liners, such as PVC and butyl rubber, which are more resistant to attack by the ultraviolet component of sunlight. Butyl rubber is probably the best material to choose, partly because it is very elastic and so will not crease as much during installation as other materials. A PVC pond liner is a somewhat cheaper option, with a correspondingly shorter life span; if choosing PVC, select a thicker grade, typically 1 mm, with a reinforcing nylon weave providing extra strength and durability. Preformed pond units are also available in a range of materials, of which rubberized versions are generally the most durable.

Flexible pond liner is sold by the square foot, in rolls of various widths, so careful planning is required to be sure you purchase sheeting of the correct dimensions. The amount of liner required is easily determined for any shape of pond using the following method. First, determine the length of the pond at its longest point and the width at its widest point. Next, adjust these dimensions to allow for sufficient liner to fit into the deepest part of the pond: to do this, multiply the maximum depth of the pond by two, and add this to both the length and width figures respectively. Finally, add a further 18 in (45 cm) to each dimension to provide extra liner to overlap the edge of the pond; the final figures give you the overall width and length of liner required.

FLEXIBLE LINER OR PREFORMED POND UNIT?

FLEXIBLE POND LINERS

Building a pond using flexible liner is relatively straightforward and allows you to exactly tailor the design to your needs. All types of liners must be used with a suitable underlay material, which cushions it from sharp objects; commercial products are available, but large pieces of old carpet or a layer of sand can work just as well.

PREFORMED POND UNIT

Preformed units made from plastic, fiberglass, or rubber are quick and easy to install and are manufactured in a wide range of shapes and sizes. Many designs include a shelf around the edge for cultivating marginal plants. Pale-colored units should usually be avoided, because they can look artificial.

PROS

● Can be used to create any size or shape of pond and is ideal for a more natural look.

● Durable; top-quality liners can last for up to 50 years before needing to be replaced.

● Suitable for use either in the ground or raised above it within a brick support.

CONS

● Can be punctured quite easily by the roots of some plants, requiring repair.

● Cheap liners can have a short life span and are a poor choice.

● Requires underlay, which can add to the cost.

● Silt can build up in folds or creases in the liner and can be difficult to clean out.

PROS

● Very easy to install, once the area has been prepared.

● A wide range of shapes and sizes are available, including units that are large enough to accommodate koi.

● Can be used both in the ground and for creating a raised or semi-raised pond.

CONS

● Units are molded to fixed shapes, which can constrain pond design.

● Some styles may not be deep enough for safe overwintering of fish.

● Fairly durable, but if split or otherwise damaged by invasive plant roots, they are not easily repaired.

POND FISH
SETTING UP
THE POND

Building a pond

Thorough planning is the key to a successful result when building a pond, whether using a preformed pond unit or flexible liner. Think through all aspects of the project before you start; draw up detailed plans of the design and planting, and consider practical matters, such as which materials and tools will be required, and even how to dispose of the large volumes of excavated earth that will be generated.

Excavation of a pond site can be physically demanding. Although small ponds can be excavated manually, it may be worthwhile to rent a small excavator for more extensive works. Alternatively, some pond suppliers may build the pond for you. Soil conditions can have a considerable effect on the ease of construction. In sandy areas, for example, the soil is loose and easy to dig, but it can be difficult to cut clean boundaries; clay soils, by contrast, are easily shaped but present much more strenuous digging conditions.

Edging materials, such as the stone slabs bordering this mature garden pond, are very effective at concealing the edges of the liner or preformed unit and give the pond a more natural look.

FLEXIBLE-LINER PONDS

Flexible liners are an extremely versatile material for pond construction and are ideal for informal ponds and unusual shapes. Detailed planning of the exact dimensions of the pond to be built is crucial, to avoid costly miscalculations of the amount of liner and underlay required *(see p.302).*

Mark out the second level with sand before starting to dig

❸ Reinforce the outer edge
Make sure the site is level. Define the edge of the pond with one layer of bricks set in mortar, to reduce the risk that the edges will collapse.

❶ Mark the outer edge
Use soft sand to define the outer edge of the pond, and use this as a guide to where to start digging.

Trail of sand

Spread mortar

❷ Dig the first level
Start by digging down around the perimeter to what will ultimately become the level of the marginal shelf, before going deeper.

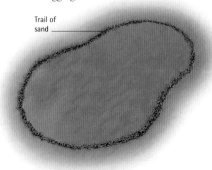

Rake the base to remove stones and any other sharp objects that could penetrate the liner.

Line of bricks follows contours of pond

PREFORMED POND UNITS

Creating a pond using a preformed unit is relatively straightforward. For a successful and lasting result, however, make sure the site is well prepared and free from stones or debris and that the unit is level when installed.

1 Mark out the perimeter
Balance the unit on blocks and place stakes at regular intervals around it. Use string laid around the stakes to outline the area to be excavated.

Stake

Blocks raise the unit from the ground

String marks out where to dig

Take measurements to check the depth at several points across the hole

The base must be prepared carefully. Ensure sure that any sharp debris is removed and the area is level.

2 Excavate the pond area
Shape the hole to match the pond unit as closely as possible, including the marginal shelves. Rake and firm the base then seat the unit on a layer of sand so that it fits exactly and is completely level.

Make sure the unit is not sloping using a level and a piece of lumber.

3 Stabilize the unit
Slowly fill the pond with water, pausing at intervals to check that the unit is still level. Backfill around the edges with sand until the unit is secure.

Use sand to fill the gap between the unit and the wall of the hole

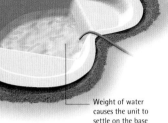

Weight of water causes the unit to settle on the base

Contour liner over marginal shelf

Using underlay, such as carpet, commercial matting, or a layer of soft sand, helps to prolong the life of the pond liner.

Allow an overhang to conceal the liner

4 Lay the liner
Flexible pond liner should be laid over a suitable underlay material (above). Position the liner so that the overlap around the perimeter is even.

5 Partially fill the pond
As you add the water, creases in the liner are evened out. Hold the edges of the liner or secure them with bricks.

Weight of water pulls the liner down

Check that each stone slab is seated evenly using a level.

6 Lay the hard edging
An informal surround can be created using stone slabs laid in mortar. Be careful not to drop mortar into the pond, however, because it is toxic to fish.

Cut back excess liner around the perimeter evenly using a pair of sharp scissors.

Excess liner can be trimmed away once pond is full

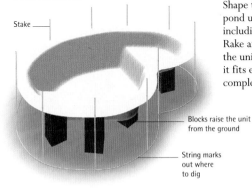

Pumps and filters

Garden ponds are often stocked with many more fish than would be found in a natural pond and so benefit from the addition of a filtration system to improve water quality. This is especially true for koi ponds, where crystal-clear water is desirable to give the best view of the colors and patterns of the fish. Filtration systems are driven by electric pumps, which can also be used to create fountains and water features.

Pond pumps fall into two main categories: submersible units that function underwater, and powerful surface pumps for use only in a dry location. Submersible pumps are generally smaller and often run on a low voltage so are most suitable for smaller ponds, while external pumps for larger ponds and extensive water features usually run off household electricity. To work effectively, a pump must have sufficient capacity to cycle all of the water in the pond in two hours, so calculate the volume of water in the pond before choosing a pump. Preformed pond units often have their volume marked on the base; alternatively, when filling the pond, measure the volume by attaching a flow meter to the hose.

The choice of pump also depends on its intended function. For example, when operating a fountain, the water needs to

Fountains are not simply decorative features but also improve oxygenation of the water. The cascade should be chosen carefully, to ensure that it does not spray outside the pond.

be pumped under high pressure but at a low output, whereas for the operation of a waterfall—where the water needs to be pumped uphill—a high output from the pump is required. Refer to the manufacturer's instructions and get advice from reputable retailers to be sure that the pump you choose is suitable for your intended use.

Maintaining water quality

Most filtration systems use separate pumps to draw water from the pond and through the filter, although pumps with an integral filter are also available. There are two main types of pond filter: internal units, which sit underwater in the pond, and external filters, which are sited outside the pond. All filtration systems function in a similar manner (see box, opposite), but the filtration media they house varies. When installing an internal filter, locate the pump as close as possible to the filter unit, to maintain the flow rate. By contrast, if the system uses an external filter chamber, locate the pump as far as possible from the filter outflow, to ensure that clean water is not simply pumped straight back through the system.

Ultraviolet systems, located between the pump and filter, are a further refinement that can be used to achieve very high water quality. Ultraviolet radiation emitted by the unit causes algae in suspension to form into clumps, which are then easily strained out of solution by the filter.

OUTDOOR ELECTRICITY

If you need to run an electrical supply from the house to the pond, hire a professional electrician, who will install cables in accordance with local building safety codes. Devices running directly off the main supply must always be connected via a ground-fault circuit interrupter (GFI). Low-voltage pumps, suitable for small ponds, do not present a hazard and need not be installed by an electrician.

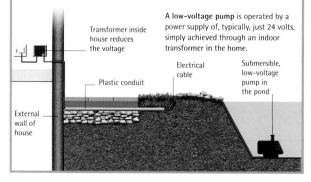

Transformer inside house reduces the voltage

A low-voltage pump is operated by a power supply of, typically, just 24 volts, simply achieved through an indoor transformer in the home.

Electrical cable

Plastic conduit

Submersible, low-voltage pump in the pond

External wall of house

POND FILTRATION EQUIPMENT

The pump is a critical component of a filtration system, providing the power to force the water through the filter unit. All designs of pumps, from submersible units to more powerful surface models, work on a similar principle: an electric motor draws water through the unit and expels it through an outlet. Most pumps are equipped with a pre-filter—ranging from sieve-type fixtures to foam-sponge attachments—which helps to prevent debris in the pond from entering the pump unit and causing a blockage.

Submersible pumps are designed to operate underwater and will not function outside the pond. An electric motor within the pump unit drives an impeller, which forces water out of the pump through the outlet. This creates a flow through the unit that draws water into the pump. The outlet can be connected to a pipe that leads to a filtration system or can supply a fountain or a waterfall.

Pond filtration systems work on the same principles as aquarium filters (see pp.34–35), albeit on a much larger scale. Water from the pond is passed through an internal or external filter unit that contains layers of filtration media. These can include brushes or layers of filter foam of different grades that sieve particulate matter from the water and also serve as biological filtration media, on which beneficial bacteria can grow. The bacteria break down ammonia produced by the fish, as well as other waste matter. Having passed through the media, the clean water then returns to the pond via the outflow or can be run through a further series of filters to maximize the water quality.

Check the filter flow rate and water quality regularly; any drop suggests that the filter needs cleaning. When cleaning a pond filter, always wash the media in dechlorinated or pond water, to safeguard the bacterial population. In a new pond, it will take time for beneficial bacteria to develop in the filter. To make sure ammonia does not build up to dangerous levels, add zeolite—a compound that removes ammonia from water—to the filter.

Outlet to fountain

T-piece allows pump to be attached to filtration system

Flow rate adjuster, held in place with a hose clip

Water is drawn in through slits in the outer casing

Electric motor with attached impeller

Impeller helps to drive water through the unit

Strainer traps material that could clog the impeller

The main chamber of an external filter unit is housed outside the pond. Water is pumped in at the top, where it is delivered, as a fine spray, onto the filtration media. It first passes down through a number of foam layers—from coarse to fine—where particulate matter is trapped, and then through a layer of biological media, before being returned to the pond.

Overflow pipe, which carries water back to the pond if the filter is blocked

Spray

Water pumped out of the pond enters the filter through the inlet pipe

Layers of foam help to sieve the water

Plastic biomedia in the base of the filter encourages growth of beneficial bacteria

Unit can be hidden behind suitable vegetation but must be accessible for maintenance

Filter unit

Return pipe concealed under rocks

Plastic tubing through which water exits the filter

Clean water flows back into the pond

Water is pumped out of the pond to the filter unit

Pond water is drawn into the pump

Pump positioned on a solid base

A typical pond setup uses a submersible pump to supply an external filter by the waterside. Pumps that are used for passing water through a filtration system should be robust enough not to become choked by debris from the pond. If sufficiently powerful, a pump can be employed to operate a waterfall or fountain in addition to supplying the filtration system.

Plants and landscaping

In addition to enhancing the look of a pond, plants help to maintain water quality, providing a healthy environment for the fish. The choice of plants will partly depend on the style of pond—a naturalistic pond looks best when heavily planted around the edges so that it blends seamlessly into its environment, while a contemporary look may be best achieved with more minimalist planting.

A well-balanced, healthy pond must contain two types of plants: oxygenators *(see pp.370–371)*, which release oxygen into the water, and floating plants *(see pp.372–373)*, which provide shelter from sunlight. Without these, or an efficient filtration system, the water in the pond can become overgrown with algae, which not only turns the water green but can also affect the health of some fish species, such as Sterlets *(see p.359)*. Plants in the body of the pond also absorb nitrate— the product of the breakdown of fish waste—which lessens the burden on the filtration system.

Incorporating plants into a koi pond is not straightforward, partly because of the depth of water and also because koi have a habit of digging up plants and browsing on the growing shoots. Most koi ponds, therefore, simply incorporate a few tall marginals and perhaps some water lilies, whose leaves help to protect the fish from sunburn in the clear water.

Planting

In a new pond, wait several days after filling before putting the plants in place, to allow the water temperature to rise to that of the environment. Pot plants as necessary *(see opposite)*, first inspecting them closely for any signs of disease or pests. In temperate areas, spring is the best time to introduce new plants into an existing pond, because aquatic plants start to grow rapidly at this time. If the pond is large, you may need waders to put plants in place, and special pond gloves should always be worn. These reach up to your shoulders and provide protection against waterborne diseases, such as Weil's disease *(see p.323)*—a potentially serious condition, spread by rodents, which causes jaundice.

TIPS FOR CHOOSING PLANTS

● Avoid buying plants in the winter when they are dormant, because it is impossible to tell how healthy they are.

● Examine plants carefully for potential pests, such as aquatic snails.

● Plants already set in containers will grow faster than bare-rooted plants, assuming they do not need repotting.

Plant has relatively few flowers developing

A number of flower stalks and buds developing

Healthy Marsh Marigold

Plant not thriving

Weeds in the container suggest unfavorable growing conditions

Healthy, green coloration to the leaves

Poor example

TYPES OF POND PLANTS

Plants for the pond can be divided into four categories, based on their growing habits and where in the pond they are to be found. Oxygenating plants, water lilies, and floating plants are truly aquatic, growing in or under the water. Marginal plants are a useful addition to the pond, not only as a decorative element but also to provide an excellent habitat for insects.

Oxygenating plants grow largely underwater, releasing oxygen during daylight hours.

Water lilies have attractive flowers and large leaves, which provide shade and protection for the fish.

Floating plants can rapidly spread across a pond, and their growth may need to be restricted.

Marginal plants can be cultivated in shallow water or boggy conditions around a pond's edge.

PLANTING POND PLANTS

Marginal plants and water lilies will grow readily in a layer of soil at the bottom of the pond but are most easily managed if they are grown in special planting baskets; this allows the plants to be moved as required, keeps a check on the growth of faster-growing species, and also minimizes the risk of liner damage by invasive plant roots. Choose a relatively large basket, to allow a good amount of space for growth, and fill it with special aquatic potting mix, which creates ideal conditions for pond plants. When planting, never bury the crown of the plant below the surface of the soil, because this will cause it to rot in the water. Oxygenating plants can also be planted in baskets, to contain their growth, while floating plants can simply be placed on the surface of the water.

Large stones can be added to planting baskets to stop them from tipping over; this is especially useful for tall plants.

Spreading gravel over the top of the soil helps to weigh down the plant until it has taken root within its planting basket.

Marginals should be planted in plastic baskets

Plant pots may need to be weighted

Young water lilies should be placed on bricks so that their leaves sit on the surface of the water

Marginal plants grow well in shallow water

Do not allow plants to dry out before planting

Marginal plants should be placed on the marginal shelf, with the top of the planting container positioned beneath the water level. Raise young water lilies on bricks at first, gradually lowering them as the plants grow larger.

PLANTING STYLES

The plants in and around a pond have a great effect on the overall impression created. Traditional, formal ponds often incorporate low-growing plants, such as water lilies, which do not mask the crisp, neat edges of the pond. Small ponds often benefit from the inclusion of taller, more architectural plants, such as reeds and grasses, which lift the eye, making the pond appear larger.

Three varieties of water lily (*Nymphaea* 'Escarboucle,' 'William Falconer,' and 'Marliacea Albida') adorn this large, formal pond, which is bordered by the tall, elegant spikes of *Iris laevigata* 'Variegata,' *Canna flaccida*, and *Schoenoplectus lacustris*. *Myriophyllum verticillatum* covers one corner of the pond.

The vertical emphasis of the planting in this courtyard pond, achieved through the use of tall marginals, such as irises and rushes, enhances the geometric lines of this modern style, while a single water lily (*Nymphaea* 'Gladstoneana') softens the look and provides cover for the fish.

Creative landscaping

Edging around a pond strengthens its perimeter and helps to disguise the edge of the pond liner. It can also prolong the life of the liner by shielding it from sunlight. Hard construction materials, such as paving slabs or bricks, laid around the edge of a pond give a more formal look, while natural stone or sod are ideal for a more informal pond. Another possibility is a wooden deck raised above water level, but the wood must first be treated with a nontoxic preservative to keep it from warping or rotting.

Consider the access to the pond: if this is across a lawn, regular foot traffic can quickly result in an unsightly muddy trail. If you do not want to construct a path, set paving slabs into the grass as an informal solution.

The planting and landscaping around the pond can be used to disguise pond equipment. An external filter, for example, can be hidden in vegetation in a flowerbed, although it must still be easily accessible for routine maintenance and servicing.

Moving water

A fountain is an attractive addition to any pond and also creates a healthier environment for the fish by improving the water's oxygen content. Water lilies prefer calm water, however, and will not thrive under the jet of a fountain, so they need to be located at the opposite end of the pond. Water currents created by the fountain can waft floating plants to one side of the pond; before adding plants, test the flow by floating a light plastic ball on the surface of the water while the fountain is operating. If the ball drifts away from where you want the plants to be, adjust the positioning of the fountain.

Oriental-style koi ponds often incorporate bridges and decorative features of Japanese life, such as bonsai trees and this popular style of bamboo water fountain (left). Japanese maples create a striking backdrop to the pond and can be grown in pots or in the ground.

Bridges not only provide an ideal vantage point from which to observe and feed the fish but can also be an attractive and decorative feature of the pond.

Decorative lighting allows you to enjoy your pond after dark and can also be mixed with other features, such as fountains, to create a striking effect.

Stepping stones can give a modern feel when made from decking raised on plinths, but make absolutely sure that any wood preservative used is not poisonous to fish.

Choosing and introducing fish

The availability of pond fish is seasonal, with the largest selection offered for sale during the spring and early summer months. This is a good time to purchase pond fish, because it gives hardy varieties that overwinter in the pond the opportunity to become well established before the onset of colder weather. Always obtain pond fish from a reputable supplier who allows you to inspect the fish before purchase.

Most aquarium stores offer a range of coldwater fish, but if you want exhibition-standard koi, you should seek out a specialist dealer. Pond fish are usually priced by size, with the largest individuals commanding the highest prices. It is best to start out with younger fish; this is not only less costly but also gives you the opportunity to tame them. Coldwater fish can grow rapidly under favorable conditions, so take this into account when considering stocking levels for the pond.

House newly acquired fish in an indoor aquarium or a small outdoor pond for a week or so before transferring them into the main pond, to be sure they are healthy. If you have only recently filled the pond, treat the water with the appropriate amount of dechlorinator before introducing the fish.

Stocking levels in a pond without filtration should not exceed 2 in (5 cm) of fish for every 1 ft² (0.09 m²) of pond surface area.

CHOOSING HEALTHY FISH

● Make sure your chosen fish is swimming without difficulty through the water before asking for it to be caught.

● View the fish from both sides, because a problem may be evident on one side only.

● Inspect the fish closely for signs of skin damage, including missing scales or reddened areas on the body.

● Look carefully for any signs of external parasites, which are hard to eliminate from a pond.

INTRODUCING NEW FISH

Most fish can be transported from the supplier to your pond in large plastic bags, but very large koi may have to be moved in vats. Care must be taken when transporting fish, to avoid subjecting them to unnecessary stress. If possible, use a local supplier to minimize the traveling time. Adding oxygen to the traveling bag lengthens the time that the fish can be kept in it, but never keep them confined for longer than necessary.

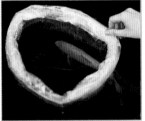

❶ Catching the fish
The supplier will usually catch the fish for you, transferring it into a bag filled with water.

❷ Oxygen is added
To sustain the fish during transport, the bag is inflated with oxygen and sealed tightly.

❸ Safe travel
A cardboard box protects the bag and keeps it upright. Never leave the bag in direct sunlight.

❹ Releasing the fish
Do not dump the water into the pond; allow the fish to swim out then discard the water.

Coldwater fish in the home

Many coldwater fish can be kept indoors in aquariums; indeed, some of the smaller, less colorful species are more visible there than in a pond. Even larger species, such as koi and sterlets, may be housed in a tank when small, but they must be moved outdoors when they outgrow their surroundings. Some coldwater fish cannot survive the winter in a pond, especially in temperate areas, and must be housed indoors until spring.

It is no coincidence that, numerically, goldfish (see pp.332–343) rank as the most popular pets in the world. Their care, whether indoors in aquarium surroundings or outdoors in a pond, is very straightforward. Not only are goldfish unfussy in their water chemistry requirements, but they will also live happily in unheated, dechlorinated tap water. Goldfish are generally not nervous or shy, especially once established in an aquarium, and it is usually possible to keep two or more together without difficulty. Different varieties can successfully be kept together in a single tank, but it is best to avoid mixing fancy varieties with goldfish with simple fins and tails; these attractive fish were bred for their looks rather than their

SETTING UP A COLDWATER AQUARIUM

The setup of a coldwater tank is very similar to that of a freshwater tank (see pp.38–42), except that it lacks a heater and can support a different range of plant species. Coldwater tanks benefit from a simple filtration system, such as an undergravel or power filter (see pp.34–35). Lighting is important; a light in the hood assists plant growth and enhances the visual impact of the fish. Pale-colored gravel is the best choice for a goldfish tank, providing a striking contrast to the fish's vivid colors; avoid red or blue gravel, because it makes their coloration appear dull. Goldfish excavate the gravel when searching for food, which can make it difficult to establish plants. Protect the plants by securing them in place with rocks.

swimming ability and may lose out to their more agile relatives in the daily competition to find food.

Although less commonly kept, small minnows and other similar-sized coldwater species will also thrive in planted aquariums indoors. Others, however, notably members of the sturgeon family, may not thrive at typical room temperature, especially when young; this is because they stop feeding at water temperatures of more than 68°F (20°C).

Seasonal accommodation

Some species of coldwater fish that are kept in a pond over the summer months must be moved indoors for the duration of the winter. This particularly applies to fancy goldfish varieties, which are not hardy enough to overwinter in cold, possibly freezing, water. Their corpulent body shape means that they are especially vulnerable to swim bladder problems linked with low water temperatures, which cause them to float at an abnormal angle in the water. Set up an aquarium in advance of the start of colder weather,

Shubunkins and other goldfish root around in the aquarium substrate, sucking in and spitting out pieces of gravel in their search for edible items.

so that the fish can be brought indoors before the temperature dips dangerously low. To transfer fish from the pond to a tank, simply net them out of the pond and transfer them to a suitable container, such as a clean plastic bowl. They can then be brought indoors and introduced into the aquarium.

Even hardy varieties of goldfish may benefit from spending their first winter in an aquarium. If left outdoors, the decline in temperature deters them from feeding and so slows their rate of growth. Moving them indoors also provides an opportunity to tame the young fish.

Goldfish that have been housed indoors during the winter months can be returned to their outdoor accommodation in the late spring, once the risk of cold weather has passed. Before transferring the fish, ensure that the temperature of the water in the pond is close to that of the aquarium, so as not to cause them undue stress.

An aquarium also provides an ideal environment to quarantine newly acquired fish, to ensure that they are healthy before releasing them into the pond. This is especially important for controlling parasites in a pond; these pests can be transferred into the pond on a new fish and may multiply unchecked and harm other pond occupants before their presence is identified. As a general guide, newcomers should be kept in an aquarium for a week or so, until it is clear that they are feeding well and appear to be generally healthy. Try to plan ahead if you are buying fish for a pond; set up the aquarium well in advance, and start the maturation of the filter by adding an existing fish, if possible. Larger coldwater fish, including bigger goldfish, will need a more powerful filtration system, such as a power filter, in addition to a simple undergravel filter.

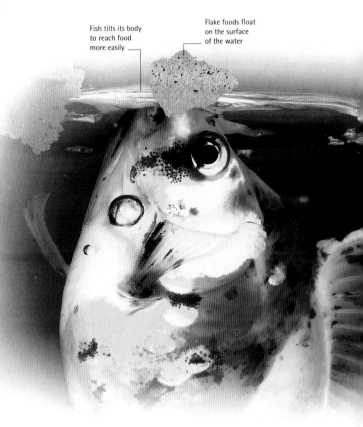

Fish tilts its body to reach food more easily

Flake foods float on the surface of the water

Goldfish feed readily on floating food, such as flake or pellets. Feeding fish like this helps to prevent food from being wasted and makes the tank easier to maintain.

COLDWATER TANK CHOICES

Always choose a tank that is larger than is needed at first, which will then allow goldfish to be accommodated without being overcrowded as they grow. Avoid goldfish bowls, which offer a limited surface area for gaseous exchange (although some have built-in aeration systems) and little swimming space. Pay attention to the type of goldfish that you are intending to keep in the tank, as some (below) with long fins will benefit from being kept in relatively deep water.

TANK MAINTENANCE

Regular partial water changes are important for the health of the occupants of a coldwater aquarium, because they reduce the levels of pollutants, such as nitrate, in the tank. Partial water changes should be carried out every two weeks or so, with up to a third of the tank's volume being replaced each time. The water is easily removed from the tank using a length of siphon tube *(see pp.50–51)*, and if a gravel cleaner is fitted to the tube, mulm that has built up on the surface of the gravel can be removed at the same time.

Pour water slowly into the tank, taking care not to disturb the plants

When carrying out a partial water change, the water to be used to fill the tank must be treated with water conditioner. It is a good idea to use water that has been allowed to stand overnight so that it is the same temperature as the water in the tank.

Indoor ponds

An indoor pond can be an attractive feature in a conservatory or inside the house, and, especially in temperate areas, it has a number of advantages over a similar pool outdoors. Under cover and in heated surroundings, it is possible to grow a much wider range of aquatic plants, including lotuses and tropical water lilies *(see pp. 378–379)*. A colorful backdrop can be created using houseplants that would not be hardy outdoors. The potential range of fish that can be kept is much wider, too, especially when the water is heated. A number of tropical species can be housed in an indoor pond; shoals of small fish such as Neon Tetras *(see p. 99)* can create a particularly striking display. Coldwater fish, such as koi, grow much faster in an indoor pond than they do outdoors. A typical Ogon *(see p. 352)*, for example—one of the more

rapidly growing varieties of koi—may grow to a length of 36 in (90 cm) by five years of age in this environment.

Constructing an indoor pond is very similar to creating a pond outside *(see pp. 304–305)*, but its location in the home must be carefully considered *(see box, below)*. Raised ponds are the most popular style indoors, especially if small fish are to be kept there, and broad edging provides an opportunity to sit and view the fish at close quarters. Always consider the safety of your design *(see p. 301)*, especially if young children are likely to visit.

It is important to have an efficient filtration system in an indoor pond, to maintain the quality of the water; partial water changes are also necessary, because, unlike an outdoor pond, it will not be periodically flushed through with rainwater. Evaporation will reduce the water level, so you will need to top off the pond regularly with fresh, dechlorinated water.

Achieving healthy plant growth in an indoor pond, such as this well-established conservatory pool, may require overhead lighting to maintain the plants through the winter, especially in temperate areas. Additional heating may also be necessary, depending on the plant species you choose to grow.

An indoor pond can be a striking focal point in a contemporary room design. This Asian-style koi pond (right), by architects Dransfield Owens de Silva, can be crossed via stepping stones to reach another room.

INDOOR POND CONSIDERATIONS

● Always seek professional advice if you wish to build a large indoor pond.

● Screen the windows in a conservatory to prevent large water temperature fluctuations.

● Provide adequate ventilation to prevent condensation problems and mold growth.

● Incorporate a means of draining the pond with no risk of flooding the area.

POND FISH
MAINTENANCE

Food and feeding

Coldwater fish, such as koi and goldfish, constantly forage for food—a behavioral characteristic that makes them seem hungry all the time, even though they may have no immediate need for sustenance. In the pond environment, they are able to browse on a range of natural foods, such as algae and plant matter, but they also need to be provided with a staple formulated food to achieve a balanced diet.

Most fishkeepers give their fish commercially prepared foods, such as pellets and flakes *(see box, opposite)*. These foods contain a scientifically formulated balance of ingredients, including proteins, fats, carbohydrates, vitamins, minerals, and fiber, all of which are essential for a healthy diet. These foods should form the basis of the diet that is offered to your fish.

It is important not to overfeed pond fish, because this can lead to obesity, which will shorten their life span. Also, excess uneaten food is difficult to remove from the pond, and any leftovers may rot, causing a deterioration in water quality. Offering only small portions of food will ensure that all of it is eaten within a few minutes.

Koi can be tamed if they are fed by hand at same place in the pond at regular times. Any sudden movements may cause them to dive back down into the depths of the pond.

A varied diet

Although pond fish will thrive on a diet of formulated sticks or pellets, it is good to provide them with some variety by supplementing their diet with other foods. Depending on the type and size of the food being offered, it should be chopped into suitable sizes so that all fish are able to eat it.

Most pond fish are omnivorous and will readily eat green leaf vegetables. These vegetables, such as fresh lettuce leaves, are a good source of fiber. Peas can also be given to the fish, once they have been removed from their shell. Shelled shrimp

SPECIALIST DIETS

Highly specific diets are now available with ingredients that are formulated to encourage growth, fertility, color, or a physical characteristic particular to one species. Some goldfish varieties, such as the Blue Oranda pictured here, can be given food to enhance the development of their hoods. These foods may look much the same as other pond fish foods, but the levels of protein, beta-carotene, and other components can vary significantly.

NUTRITIONAL BREAKDOWN

The protein levels in foods specifically designed to encourage growth are much higher than in general-purpose foods. High-growth foods should be given in summer, when the fish are most active.

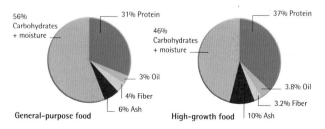

56% Carbohydrates + moisture — 31% Protein — 3% Oil — 4% Fiber — 6% Ash
General-purpose food

46% Carbohydrates + moisture — 37% Protein — 3.8% Oil — 3.2% Fiber — 10% Ash
High-growth food

Koi carp are well equipped to locate food. They rely on their eyesight and also on the sensory barbels around their mouths, which they use like feelers to search for food in the substrate.

TYPES OF FOOD

Prepared coldwater fish foods contain a cereal base, with additional ingredients to enhance color, aid digestion, or encourage growth. It is important to choose the most appropriate foods for all species of fish in the pond; for example, make sure there is food available that is suitable for the smallest species, to ensure that no fish will starve.

Frozen food tablets are prepackaged and should be defrosted before they are given to the fish.

Sticks are available in floating and sinking varieties and are suitable for koi and other large pond fish.

Pellets are suitable for medium-sized pond fish and are produced in varieties that either float or sink.

Flakes float on the surface of the pond and then sink. They are ideal for smaller coldwater species.

Automatic feeders dispense dry foods in predetermined quantities at regular intervals. The feeding interval can be altered easily, and the feeder will keep the food dry if it rains.

Special color foods help maintain and enhance the color of goldfish, koi, and other coldwater fish. These foods typically include significant amounts of shrimp meal and spirulina algae (*Arthrospira platensis*), which contain natural coloring agents, such as beta-carotene.

are a more expensive treat and can be given by hand. They must be defrosted before they are offered to the fish. Most pond fish will also eat brown bread rolled into small balls.

Koi eat a surprising range of foods. One of the most unusual is garlic, which may bring health benefits to the fish, since it has antiseptic qualities. It should be finely chopped before it is given to the fish, or it can be rubbed on the surface of unfamiliar foods to encourage koi to eat them. They will also feed on slices of orange; this provides a valuable source of vitamin C, which can boost the immune system of the fish. The koi will strip the flesh from the peel, which should then be removed from the pond.

Seasonal variations

Most fish are ectothermic, which means that their body temperature is governed by the temperature of the water around them, so that when the water temperature drops, a fish's body temperature also drops. Its ability to digest and assimilate food then decreases, which consequently lowers its appetite and activity level. In cold weather, fish should be offered only small quantities of foods that can be digested easily, since undigested food remaining in the gut could cause illness. The fish should not be fed at all once the water temperature falls below 50°F (10°C) because they are completely unable to digest food at this temperature.

To cater to this variability in requirements, foods are available with ingredients appropriate to the seasonal dietary needs of pond fish. For example, dried foods based on wheat germ, which is easy to digest, are ideal for use during spring and fall when the water temperature is lower.

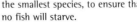

Pond management

Ponds need regular maintenance if they are to provide fish with a healthy and safe environment. Problems are most likely to arise in the first year, before the pond is established, and are typically caused by overfeeding or overstocking. Frequent monitoring and an awareness of seasonal changes will help avoid the major pitfalls.

A well-kept pond, that has healthy planting and good water quality is an attractive addition to any garden and will provide a healthy habitat for pond fish.

Spring checks

As fish begin to stir out of their period of winter dormancy, spent in the depths of the pond, they are very susceptible to minor illnesses, such as bacterial infections, which can rapidly overwhelm their weakened immune systems. Thorough inspection of individual fish will help identify and treat illnesses in their earliest stages *(see pp.322–324)*. Once the water temperature increases, fish regain their appetites and become better able to fight infections. This surge in appetite, and the resulting increase in waste products, causes a rise in ammonia levels in the water, so now is a good time to maintain and service filtration equipment. The beneficial bacteria in biological filters are inactive during cold weather, and such filters may need to be reseeded with bacterial cultures. Live and freeze-dried cultures are available from suppliers of pond equipment. The addition of zeolite, a chemical that absorbs ammonia directly from the water, may also be beneficial until the filter is fully functioning again.

Pond fish begin to show spawning behavior in late spring, when new plant growth provides surface cover for the fish and for any eggs and resulting fry in the pond. Check regularly in case any fish have become trapped in reeds or

SEASONAL PLANT CARE

Plants around the pond may benefit from a layer of leaf mulch to protect them in winter. Plants in the pond itself that are vulnerable to freezing weather must be transferred indoors before the first frost. Although the winter pond may look bare (below left), the plants can be returned to the pond in spring and will grow quickly over the summer (below right).

Removing leaves in the fall is easier if netting is placed over the surface of the pond.

other plants or any females have been driven out of the pond by overenergetic males. During the summer, the increased temperature of the water and greater activity levels of the fish result in lower oxygen levels. If not controlled, plant and algal growth also reduces oxygen levels in the water. To maintain oxygenation, fountains and other water features should be left on overnight, or special aeration equipment should be installed.

Preparing for winter

Pond plants begin to die back in the fall, and excess foliage should be removed. Falling leaves should not be allowed to accumulate on the surface of the pond, because they decompose in the water and can harm the fish. Covering the surface of the pond with netting keeps leaves out of the water and allows them to be collected easily.

If there are delicate fish in the pond, or any young from a late spawning, they should be caught and transferred to an aquarium for the winter to ensure their survival. When the water temperature falls below 43°F (6°C), the remaining fish may enter an almost completely motionless state and will not require feeding until spring. Below around 39°F (4°C), a warmer layer of water will develop in the deepest

Algal blooms—sudden flushes of algal growth—can be a problem in warm weather. Removing any dying or dead leaves from plants around the pond will help limit algal proliferation.

Blanketweed is a type of filamentous alga, which can trap fish. It should be removed regularly using a stick.

Duckweed grows rapidly and will entirely cover the surface of a pond. It can easily be controlled by scooping it off the surface.

REGULAR MAINTENANCE TASKS

DAILY

● Check to see if fish are showing signs of ill health or behaving strangely.

● Ensure that the filtration system, if present, is functioning correctly.

● Feed the fish, according to their appetite, several times during the day, except in the winter (see p.317) or in very hot weather.

● Note the water level in the pond; sudden falls indicate a leak in the pond liner.

● Monitor the ammonia and nitrite levels, especially in a newly established pond.

WEEKLY

● During the growing season, remove faded flowers of marginals, unless seed is required.

● Top off the water level in the pond if the evaporation level is high, using water treated with a dechlorinating product.

● Test the oxygen levels in the water, especially in hot weather.

● Check for any signs of plant pests, such as aphids, removing them from the vegetation where necessary.

MONTHLY

● Check the nitrate level of the pond water. It should not rise above 50 mg/l.

● Remove blanketweed so that it cannot choke other plants and pond fish.

● Prevent any buildup of algae on bridges or decking, which could make them slippery. Remove it by scrubbing the surface of the wood with a clean brush.

● Watch for any signs of moss growing on the surface of paving or stepping stones close to and surrounding the pond.

● Rather than stocking the pond to the maximum capacity at the outset, add further fish gradually over the spring and summer months.

part of the pond, where pond fish spend the winter. If there is a submersible pump installed, position it more than 6 in (15 cm) from the bottom of the pond, and switch off water features, such as waterfalls or fountains—otherwise, these will circulate and cool the water by mixing the colder surface layers with the warm layer below.

In mild areas, a pond heater can help to prevent the surface of the pond from freezing over in the winter. It will stop the area around the heater from freezing, allowing noxious gases produced by decomposing plant matter to escape from under the ice. If ice has formed on the surface of the pond, never try to smash it, because the shock waves will traumatize the fish and may even prove fatal. Instead, melt the ice slowly by carefully holding a hot saucepan on the surface of the pond.

During the spring, place a net over the surface of the pond or stretch it across a framework to protect exhausted fish from opportunistic predators. Decoys are available to deter birds, but they are unlikely to stop raccoons or cats.

CLEANING OUT A POND

Over time, sediment accumulates in the pond, and plant growth proliferates, inevitably reducing the area of water that is accessible to the fish. At intervals of a year or so, it is a good idea to unertake a major clearout. The best time is in early spring, because the pond will have time to reestablish itself in the warm summer months. If any cases of serious illness have occurred within the pond, it may require disinfection. Some preformed pond units can be lifted out of the ground to make this task easier.

● Before starting the clearout, catch the fish and move them to a location where they will be safe.

● Siphon or bail out the pond water, removing other aquatic life, such as snails or dragonfly larvae.

● Divide and repot water lilies and marginal plants.

● Remove the silt using a spade or scoop, and hose out the base of the pond. The used silt can be dumped on flowerbeds.

● Refill the pond, adding a suitable volume of water conditioner.

● Allow the water temperature to rise before returning the fish and plants to the pond.

The surface of a pond can become choked with aquatic vegetation (top). Clearing out the pond, by thinning or cutting back the plants and removing dead matter, provides the fish with a larger swimming space and makes the area neater, safer, and more attractive (bottom).

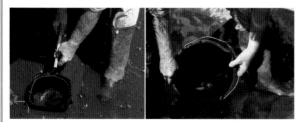

Remove the fish before cleaning, watching closely for small fry; transfer the fish to a safe container.

Reintroduce the fish only after any replanting is complete; allow the fish to settle without further disturbance.

ILLNESS AND TREATMENT

Health concerns

Regular pond maintenance and water-quality checks help keep diseases away from fish, but illnesses still occur, even in the best-kept ponds. The first sign of a problem may be a fish floating at the surface, by which time it is probably too late for effective treatment. For this reason, it is vital to set up a routine for examining fish; feeding time provides an ideal opportunity to check their appearance and behavior.

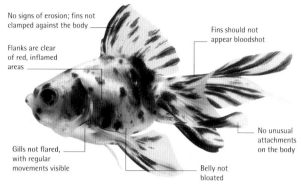

No signs of erosion; fins not clamped against the body

Fins should not appear bloodshot

Flanks are clear of red, inflamed areas

No unusual attachments on the body

Gills not flared, with regular movements visible

Belly not bloated

A healthy fish appears active and feeds well. This Fantail Goldfish (*Carassius auratus*) demonstrates what to look in a well-kept specimen.

Environmental problems

The health of pond fish is hugely influenced by environmental conditions. During spells of hot weather, for example, evaporation can significantly lower water levels, which has the effect of concentrating dissolved nitrogenous waste. At the same time, elevated temperatures drive oxygen out of the water; the combination of nitrate and oxygen stress can be fatal, especially for larger fish. Many of these problems can be avoided simply by topping off water levels regularly during the summer and incorporating a pump and filter; these improve water quality, break down waste, and increase oxygen content by creating water movement. Overstocking a pond, especially if it is not well established, places great stress on its occupants, and fish may succumb to usually benign bacteria that are present naturally in the water. Overfeeding is another common environmental problem, especially in temperate areas in the spring and fall; uneaten food decomposes in the water, encouraging populations of pathogens.

Dealing with disease

Disease-causing bacteria, fungi, and parasites may be introduced into the pond whenever it is stocked with fish or plants. Undesirable organisms can also be brought in on the bodies of animals, especially wading birds, that move from pond to pond. These can multiply and cause serious harm

Bloat, or dropsy, is a serious condition, often caused by a fungal or bacterial infection. Isolate affected fish as a precaution, but this does not normally prove highly infectious.

Fin rot begins in the inter-fin ray membranes but spreads down the fins until it reaches the body, when it can be fatal. In extreme cases, rotten parts of fins may need to be cut off under anesthesia.

Fish lice can resemble small patches of green algae on the body of a fish

Fish lice are crustacean parasites that feed on fish blood. They are obvious when attached to the body but also live free in the water for up to two weeks.

Dying or dead fish will often float to the surface of the pond. If not spotted or attended to quickly, they are likely to be taken by bird or animal scavengers.

TREATING SICK FISH

Treatments for pond fish can be delivered to an individual or to the whole pond. Individual treatment in an isolation tank (right) or by injection (below) is preferable where bacterial, fungal, or gill problems are suspected, while whole-pond treatments are more appropriate where there is a generalized parasite problem.

Antibiotic injections and simple surgery can be carried out by the experienced fishkeeper. Veterinarians who specialize in fish are most often consulted about koi, because of the high value of these fish.

Fish receiving treatment

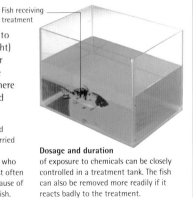

Dosage and duration of exposure to chemicals can be closely controlled in a treatment tank. The fish can also be removed more readily if it reacts badly to the treatment.

HUMAN HEALTH

● Always supervise children near a pond, even if you believe it to be childproof. Some poolside plants can be toxic if eaten.

● Use long rubber gloves when servicing the pond, and avoid dipping your hands in pondwater. Rats can contaminate the water with their urine, which may carry Weil's disease, a serious illness that resembles the flu in its early stages.

before their presence is detected, and eliminating them can be very difficult. A table of the most common conditions seen in pond fish, as well as treatment strategies, follows on page 324.

If your fish are affected, you are most likely to first notice changes in their behavior and feeding patterns; a sick fish may, for example, distance itself from others or take refuge behind a plant. If disease is suspected, affected fish should immediately be removed from the pond and kept in isolation, preferably in a large aquarium *(see above)*. Here you can inspect the body close up and check for symptoms of disease or parasite infestation. Fish lice will be visible in this environment, and you should also be able to detect gill flukes much earlier than would be possible in a pond. Treatments can be carried out in the tank itself, or in smaller baths, and the fish's progress can be readily monitored before reintroduction to the pond.

If a fish is affected with a disease or parasite, check other fish to determine whether there is a general problem in the pond or the disease is an isolated instance. Look out, too, for secondary infections. Sometimes the entire pond needs treatment with commercial chemicals, but often it is sufficient to treat individual fish. Check all water-quality parameters before reintroducing the fish; minimizing environmental stress will help prevent recurrence of the condition.

Certain diseases, such as the rapidly spreading koi herpesvirus (KHV), are untreatable, emphasizing the importance of isolating new fish before introducing them to a pond *(see p.311)* and seeking professional advice if many fish become ill.

Pale-colored fish, especially koi, living in clear water are at risk from sunburn. With such delicate specimens, it is best to incorporate some form of screening above the water to protect the fish.

POISONS IN THE BACKYARD

A number of garden chemicals can be harmful to fish and other aquatic life if they enter the pond. This can occur either as the result of runoff, caused by water draining into the pond, or by the chemicals wafting on to the surface. Be particularly careful if you are using any pesticides or herbicides near the pond—whether for the garden itself or when treating pets, such as dogs, against fleas— to make sure there is no risk of contamination. Otherwise, you can quickly find that you have lost all the fish.

Pond fish can die rapidly if exposed to harmful chemicals. Never use a hose to refill the pond without using a water conditioner—there is chlorine or chloramine in most city tap water.

BACTERIAL, FUNGAL, AND VIRAL DISEASES

CONDITION	AT RISK	SYMPTOMS	TREATMENT
Fin (tail) rot	Goldfish, especially long-finned variants, and koi	Ulceration and damage to the fins, with reddish streaking. Especially common in cold weather under poor environmental conditions.	Commercial antibacterial remedy to treat causal *Flexibacter* bacteria. Improve water quality.
Septicemia	All fish, especially koi	Often begins as superficial skin damage. This spreads, creating external ulceration, reddening of the body, and lethargic behavior.	Topical wound treatment. Antibiotics are needed to deal with the bacteria on the body surface to prevent them from affecting the vital organs. Improve water quality.
Spring viremia of carp (SVC)	Koi, goldfish, and other carp	Hemorrhaging under the skin and bloated body. Autopsy reveals liver and spleen enlarged and accumulation of fluid.	No treatment is available for this highly infectious viral disease. State/Provincial Veterinarian should be notified of a suspected outbreak. Improve water quality.
Carp pox	Koi, goldfish, and other carp	Whitish swellings develop over the surface of the body. These may fuse together to create larger areas of swelling.	It is not possible to treat this viral illness, but affected fish may recover, especially if kept in clean water to prevent secondary infection. Improve water quality.
Lymphocystis	Most prevalent in goldfish but also in koi	Isolated whitish swellings over the body surface may sometimes become enlarged and branched.	No treatment is available for this viral illness; though disfiguring, it rarely causes problems and does not spread rapidly. Improve water quality.
Fungus	All fish	Whitish areas resembling cotton fluff evident on the fins or on the body. The fungus typically gains access at the site of an injury.	Use a medicated bath for sick individuals. More general pond treatments are also available. A partial water change is likely to be beneficial.

PARASITES AND ENVIRONMENTAL CONDITIONS

CONDITION	AT RISK	SYMPTOMS	TREATMENT
Fish lice (*Argulus* species)	Any fish, especially koi and goldfish	Semitransparent lice, up to ½ in (1 cm) long, are seen close to the base of the fins. Fish frequently rub themselves to relieve irritation.	Commercial remedy. Lice can be removed with forceps, but mechanical removal may result in infection. Improve water quality.
Gill flukes (*Dactylogyrus* species)	Any fish	Gill covers flare open, and fish have obvious difficulty breathing. Parasites cause irritation and excess mucus production in the gills.	Commercial remedy. Be careful when handling affected fish because respiration is compromised by the presence of these flukes.
Anchor worm (*Lernaea* species)	Any fish	Crustacean parasites, just under 1 in (2.5 cm) long, hang down from the sides of the body. The parasites feed directly on the fish's body and can cause ulcers.	Adult worms can be removed with forceps; free-swimming *nauplii* should be destroyed using commercial treatments added to the water.
Leeches (annelids)	Any fish	The soft-bodied parasites attach to softer parts of the fish and suck up body fluids. Large numbers cause anemia and may spread other diseases.	Commercial remedy. Eggs are very resistant to treatment. The fish may need to be removed from the pond and the eggs destroyed using calcium hydroxide.
White spot (*Ichthyophthirius multifiliis*)	Any fish	Small white spots over the body, which ulcerate and are likely to become infected. Fish rub their sides against the pond; other symptoms include lethargy and appetite loss.	Treat in a salt bath or with a commercial remedy.
Suffocated fish	Goldfish and smaller koi	Occasionally occurs in ponds used by spawning frogs. The male frog mistakenly grabs a goldfish around the head, stopping its gill movements.	Check fish regularly. Net any individual that has a frog holding on to it; this should cause the amphibian to loosen its grip. Otherwise, gently separate the frog and the fish.

POND FISH

BREEDING

Reproductive cycle

The breeding triggers for coldwater fish are the rise in water temperature and increasing day length that occur in spring. Spring also sees an increase in insect life and aquatic crustaceans, providing food to bring adult fish into breeding condition and to sustain the fry. The breeding season is thus more prescribed than in many tropical species, for whom rainfall rather than temperature is the most significant factor.

You can estimate when your fish are likely to breed by taking regular measurements of the water temperature in the pond, since spawning typically occurs at around 68°F (20°C). This temperature ensures that the eggs develop at the correct rate: at more than 9°F (5°C) above or below this figure, there is an increased likelihood that the fry will hatch with deformities, because they will develop either too quickly or too slowly.

Physical changes in the fish will also indicate that they are coming into breeding condition. For example, males of the Cyprinidae family display white swellings called tubercles. In fish such as orfe, rudd, goldfish, and koi, the tubercles appear on the gill plates and along the pectoral fins, while

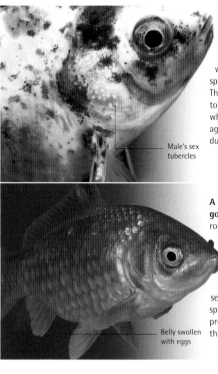

Tubercles are white, pimplelike lumps that male cyprinids, such as Shubunkins, develop when they are in spawning condition. The tubercles may help to arouse the female when they are rubbed against her body during courtship.

Male's sex tubercles

A gravid female goldfish has a more rounded body profile, but she rapidly regains her normal shape after spawning. She may spawn several times during spring and summer, producing several thousand eggs in total.

Belly swollen with eggs

Goldfish become sexually mature before they are fully grown, and it is quite possible for them to spawn when they are barely a year old.

FROGS VERSUS GOLDFISH

In spring, frogs often visit garden ponds to spawn at the same time that goldfish are breeding. On rare occasions, this results in frogs accidentally killing goldfish. If a goldfish swims past a male frog, he may grab the passing fish in the mistaken belief that it is a female of his own species. If the frog grips the fish by its head, closing off its gill covers, the goldfish will suffocate, since the frog's mating embrace lasts a long time, often for hours. There are casualties on both sides, however, since goldfish will sometimes prey on frog tadpoles in their pond. (They avoid toads, which are toxic to them.)

in Red Shiners they run along the top of the head. In some groups, such as sticklebacks, the males become more brightly colored to attract mates and deter rivals. In all coldwater species, the females become fatter-bodied than their male counterparts as they swell with eggs prior to spawning. There is also a significant increase in activity in the pond at this time, as the males chase the females relentlessly, often butting or nuzzling against them.

Breeding strategies

Coldwater pond fish show no long-term pair-bonding; any pairings that do occur are purely temporary. Fertilization is external, with eggs and sperm being released into the water simultaneously. Most pond species, including goldfish and koi, are egg-scatterers. They randomly discharge their sticky eggs, which either sink to the substrate or attach to the stems and leaves of aquatic plants. Only a small proportion of eggs will be fertilized, so the fish compensate by producing large numbers of them—up to 400,000 per spawning in the case of koi.

After spawning, egg-scatterers have no further involvement with either their eggs or offspring, but some pond species take more care to ensure that the maximum number of young will survive. The Fathead Minnow (see p.360), for example, lays its eggs in caves or under rocky overhangs in order to hide them from predators, while male sticklebacks keep a protective watch over both their eggs and the newly hatched fry.

Early life of fry

The reason that coldwater fish spawn in the spring is that this is the time of year when conditions are most favorable for the survival of the young. The algal bloom that grows in spring, and which is often cursed by fishkeepers, is actually crucial to the survival of the fry, since it provides them with their first food. The young fish eat not only the algae but also the microscopic creatures called infusoria (see pp.67–68) that live among them.

Dragonfly larvae, water boatmen, and many other pond invertebrates—not to mention fish (including the fry's own parents)—will readily prey on the young fish. As a result, the fry spend most of their early weeks hiding among aquatic vegetation, rarely straying far from plant cover. It can take between one and seven years for the fish to reach sexual maturity, depending on the species and the temperature of the water in the pond.

The full coloration of goldfish usually takes six months or more to develop, as you can see from these silvery-green juveniles. In a few cases, individuals do not color up at all but remain dark, while others change completely within two months of hatching.

GOLDFISH EGGS AND FRY

The incubation period of goldfish eggs and the growth rate of the fry are both temperature-dependent: generally, the warmer the water, the more rapidly the young develop. The fry, which measure less than $1/4$ in (0.5 mm) long on hatching, are nourished at first by their egg sacs. After a few days, they are free-swimming and actively seeking food.

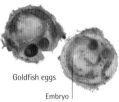

Goldfish eggs

Embryo

Goldfish embryos can be seen curled up inside their eggs in the close-up view above. The eggs usually develop on oxygenating plants (left), held in position by their sticky coating.

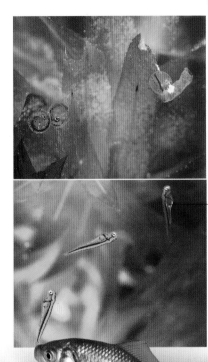

Transparent body

Newly hatched goldfish fry hide among vegetation for about two months. By the time they emerge, they have dark-colored bodies to camouflage them in their murky surroundings.

Breeding

Left to their own devices, a number of coldwater fish will breed readily in a pond environment. However, in a densely planted pond, a successful spawning may go unnoticed until later in the year, when the fry are larger and can be seen feeding alongside the adults. Breeders who like to have more control over the reproductive habits of their fish often spawn them artificially.

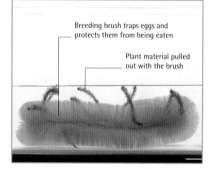

Koi breeding is a bit of a lottery, since there is no guarantee that the coloration and patterning of the fish—even with top-quality specimens—will be transferred to any of the offspring.

When egg-scatterers, such as goldfish and koi, are kept in a group, they may come into breeding condition simultaneously and spawn together. In such cases, having more males in the group than females will ensure that a higher proportion of the eggs are fertilized. If you want to breed particular fish together and be sure of the parentage of the fry, you should keep your chosen breeding stock on their own. It is also a good idea to set up a special spawning pond for them so that you can move the adults back to the main pond after the eggs have been laid and rear the young on their own. After spawning, the adults may be exhausted and float on their sides at the surface. They will soon recover, but make sure they have not sustained any fin damage during spawning, which could lead to infection.

If you choose to hatch and rear the fry in a tank, either outside or indoors, use a mature sponge filter to provide gentle filtration. Feed the fry on infusoria (*see pp.67–68*) at first, or a commercial substitute, and then wean them onto powdered flake. Add them to the pond when they are about 1 in (2.5 cm) long and too large to be eaten. Fish reared indoors should not be introduced to the pond in cold weather—the shock of the cooler pond water may kill them. If necessary, keep them inside until the following spring, when conditions will be warmer.

Hand-stripping

Some breeders of goldfish and koi prefer not to allow their fish to spawn naturally. Instead, a male and female fish are given an injection of pituitary gland extract to bring them into breeding condition. Hand-stripping (massaging the underparts of the fish) is then used to expel eggs from the female and semen from the male. The eggs and sperm are placed together in a mixture of urea and iodine-free table salt called Woynárovich's solution, which aids fertilization by removing the eggs' sticky coating. Finally, the eggs are washed in a tannic-acid solution to protect them against fungus and left in an indoor tank to hatch. With hand-stripping, up to 90 percent of the eggs are fertilized, compared with 50 percent when the fish spawn naturally.

BREEDING GOLDFISH

Goldfish, like many other egg-scatterers, can be persuaded to spawn onto an artificial medium such as a breeding brush or a spawning mop, which is then transferred elsewhere so that the eggs can hatch in safety. The adults should spawn again within a few weeks, particularly if they are kept well fed on nutritious live foods such as bloodworm.

Spawning mops are smaller than breeding brushes and more suitable for aquariums than ponds.

① Place the breeding brush in the pond
Make sure there is enough clearance for the fish to swim over the brush and spawn. Egg-laying usually occurs as the sun starts to warm the water in the early morning.

Breeding brush traps eggs and protects them from being eaten

Plant material pulled out with the brush

② Transfer the brush to an indoor tank
When the fish have spawned, move the brush to the hatching tank without delay. Keep it submerged in dechlorinated water, to ensure that the eggs do not dry out.

COMMERCIAL KOI PRODUCTION

Today, commercial koi breeding practices vary widely from country to country, but in Japan they still follow fairly traditional lines. A female is selected for breeding when she is about five years old. She is kept with two or three males in a large net, on the bottom of which is an artificial medium known as spawning grass. Using multiple males ensures that as many of her eggs are fertilized as possible. This is vital, since only about one in 70,000 fry will be of the highest quality. When the koi have spawned, the eggs are moved to a well-oxygenated hatching pond, which is medicated to minimize the risk of the eggs being attacked by fungus.

The fry hatch after five days and become free-swimming within a further 24 hours. They are then moved to a rearing pond. Successful koi breeding relies on ruthless assessment, and this process begins when the fish are barely a month old. They are carefully caught and examined, and any showing deformities or poor markings are culled, leaving the remaining koi more growing space in the pond. Two further culls are carried out over the summer. In early fall, the koi are assessed for a final time. The best specimens are kept at the farm as breeding stock and taken indoors to overwinter, while the remainder of the group are sold.

The selection process is a daunting task, due to the vast numbers of fish involved. It takes years of experience to identify the best specimens at such an early stage of development.

Koi fry are reared in large, muddy ponds, which are conditioned in advance with chicken manure to encourage the growth of microscopic infusoria for the fry to eat. Plenty of food and spacious surroundings ensure that the young koi develop to their maximum potential.

Large koi are caught in strong, deep nets to minimize the risk of injury

Bringing the breeding stock indoors during winter helps to maintain their appetites and growth rates. Koi must be touched only with wet hands, to protect their delicate mucus coating.

SHOWING KOI

Koi shows are great occasions for breeders, dealers, and hobbyists alike. The fish are exhibited in different size categories. The smallest category typically includes fish up to 8 in (20 cm) long, and the largest includes those in excess of 32 in (80 cm). All the fish are displayed in the same blue vats, to make the task of judging easier, and measured with a floating scale to ensure that they are of the correct size. The judges do not compare the fish directly with each other but instead assess them against what is considered the ideal for that particular koi variety. Color, patterning, body shape, skin quality, and even swimming action are all taken into account. Great care is needed when transporting koi to and from shows, since any blemish, such as a split fin or abnormal reddening of the skin due to stress, can ruin a koi's chances of success.

Judges consider the qualities of two koi at a Japanese show. Koi are bred to be viewed primarily from above, so the fish are not usually caught and examined as part of the judging process but simply assessed in their vats.

DIRECTORY OF
POND FISH

GOLDFISH

Undoubtedly the most widely kept of all fish, goldfish exist in a far wider range of colors than their name implies. Goldfish are suitable for both home aquariums and ponds, although the different color forms vary in terms of their hardiness, and not all are suited to being kept outdoors all year in temperate areas. Goldfish are members of the carp family, but unlike most fish in this group, they lack any barbels around the mouth. This characteristic allows them to be distinguished at a glance from koi.

Common Goldfish (*Carassius auratus*) can become tame in both pond and aquarium surroundings. They may live for more than 40 years—far longer than most other pond and aquarium fish.

Carassius auratus

Common Goldfish

This is not only the most popular goldfish variety but also the hardiest and potentially the largest. It occurs in a range of colors, but solid ("self-colored") fish are usually preferred.

Good specimens display body symmetry, with even curves on the upper and lower body. A short, broad caudal peduncle and a wide, slightly forked caudal fin make common goldfish strong swimmers. These fish can survive in frozen ponds for short periods, provided the water is deep enough for them to avoid becoming trapped in the ice itself.

Red-and-White Common Goldfish The white areas of these variably patterned fish have a silvery sheen.

Pinkish coloration produced by blood vessels

Backward-pointing dorsal fin

Regular pattern of scales

White Common Goldfish This variety, sometimes called the Pearl, is less popular than its colored cousin, but it proves to be equally hardy.

Obvious gill cover

Orange extends to the fins

Even body coloration

Common Goldfish These fish vary from yellow through bright orange to a deep blood-red. In exceptional circumstances, they may reach over 24 in (60 cm) long.

🌐 **ORIGINS** Asia, originally from waterways in southern China.

🔄 **SIZE** Highly variable; often exceeds 12 in (30 cm) in ponds.

🍴 **DIET** Goldfish food (flakes or pellets) and live foods.

💧 **WATER** Temperature 50–77°F (10–25°C) and neutral (pH 7.0).

😊 **TEMPERAMENT** Relatively social, but does not shoal.

Carassius auratus

Comet

This elegant variety originated in the United States during the late 1800s. It is distinguished by its slim, streamlined body and its deeply forked caudal fin, which should be longer than the body when fully extended. Comets are usually variegated in color; the most popular variety is the Sarasa, which is easily recognizable by the deep red-and-white patterning extending over the body and fins. Comets are active by nature and require a spacious aquarium if kept indoors. They will thrive in pond surroundings, although they may prove vulnerable to fin congestion during periods of severe cold weather.

Comet The Comet's caudal and dorsal fins are greatly enlarged. This individual displays some chocolate body patterning.

Variegated dorsal fin

Clear demarcation between colored and white areas

Sarasa Orange may replace the more common red color of these fish. The variegated patterning differs widely between individuals.

PIGMENTATION AND SHEEN

The protective scales on a goldfish form part of the outer layer of the body known as the epidermis. Beneath this is a layer called the dermis, which itself overlies layers of fat and muscle. Distributed among these layers are the pigments that give goldfish their vibrant skin colors. These include reddish-orange and yellow pigments known as lipochromes, and melanin, a black pigment. Lipochromes usually occur in the upper layers, but the location of the black pigment is more variable. If melanin is present just below the scales, the goldfish looks jet black; if located in the lower layers, the fish looks blue (for example, the Blue Pom-Pon, bottom right). When both types of pigment are present in different layers, this creates chocolate or coppery shades. A goldfish that completely lacks pigmentation is silvery in color.

Another factor influencing the appearance of goldfish is the presence in the dermis of cells known as iridocytes. These cells are normally distributed over the entire body, giving goldfish, such as the Blue Pom-Pon, a shiny appearance. However, the upper iridocytes are missing in some goldfish varieties. In such cases, the lower level of cells has a direct effect on the appearance of the goldfish, resulting in a kind of mother-of-pearl sheen. Such individuals are described as nacreous. Shubunkins, for example, are nacreous goldfish *(see p.335)*. When the iridocytes are totally absent, a matt appearance results, as typified by the Black Moor shown above.

Matt coloration extends over the entire body

ORIGINS Asia, originally from waterways in southern China.

SIZE Highly variable; often exceeds 12 in (30 cm) in ponds.

DIET Goldfish food (flakes or pellets) and live foods.

WATER Temperature 50–77°F (10–25°C) and neutral (pH 7.0).

TEMPERAMENT Relatively social, but does not shoal.

ORIGINS AND ANCESTRY

Goldfish are descended from carp that were kept in China about 1,700 years ago. The first records of orange-marked carp date back to 300 CE, but it was only from around 800 CE, during the Sung Dynasty, that people started to breed these colorful cyprinids for ornamental purposes. Goldfish feature prominently in Oriental literature and many other forms of art, including ceramics, and it is possible to track their early development from such sources.

Ancestral lines displaying many of the features seen in today's varieties, including telescope-eyes, were well established by 1600, as were numerous color variants, including some with variegated coloring. The different body shapes and fin types that characterize many of the modern varieties were also beginning to emerge by the early 17th century.

Goldfish were imported to Japan in the 16th century, where still more varieties were bred, but it was to be another 200 years before they became available in the West. They soon became highly sought after, as the pond fish of first choice for the estates of the European aristocracy and were kept in decorative bowls in grand houses. Rather surprisingly, they did not reach North America until 1874. Nevertheless, their popularity grew so rapidly that the first commercial goldfish breeding farm was established in the United States just 15 years later.

Carassius auratus

Shubunkin

This popular variety is very close in appearance to the Common Goldfish. This is especially so in the case of the London Shubunkin, which has an identical body and differs only in terms of the arrangement of its iridocytes. This particular variety was developed by London breeders during the 1920s, by which time enthusiasts in the U.S. had already created the long-tailed American Shubunkin. In due course, the two varieties were crossed by breeders of the Bristol Aquarist Society in western England, creating the Bristol Shubunkin—a very distinctive and different form with large, flowing lobes on its caudal fin, which must not be allowed to droop. Shubunkin coloration is generally very variable, but the orange areas tend to be paler than those of Common Goldfish. They may also display dark speckling, as well as bluish shades that range from pale-whitish through to violet. Darkly marked Shubunkins are highly attractive when seen at close range, but they are less conspicuous in ponds unless the water is particularly clear.

Sloping dorsal fin

Pale orange

London Shubunkin This is the most commonly seen form of the Shubunkin, with a caudal fin resembling that of the Common Goldfish.

Random dark speckling

Bristol Shubunkin This form has rounded lobes on its caudal fin. Enthusiasts strive to breed this and other goldfish varieties to prescribed exhibition standards.

Coloration extends into the fins

Long, flowing caudal fin

American Shubunkin The caudal fin lobes of this variety are much narrower than those of the Bristol Shubunkin; they are tapering rather than rounded in shape.

Carassius auratus

Pearlscale

This ancient Chinese variety can be identified by its rotund body, double caudal fins, and pearl-like markings on the sides of its body. Each scale has a raised whitish center, making it look as if a pearl is embedded in it. The variegated red-and-white form is the most common Pearlscale, but there is also a nacreous variety (see p.333) that resembles the Shubunkin in coloration. Pearlscales are not strong swimmers and are usually kept in aquariums rather than ponds, where their distinctive appearance is easier to appreciate.

Chocolate Pearlscale The depth of chocolate coloration can vary from reddish-brown to a much darker brown.

Symmetrical caudal fin

No pearl markings on the head

Entire body shows pearly markings

Variegated Pearlscale The pearl-like markings are evident on this goldfish, even against the white areas of the body. These goldfish do not grow especially large.

⚙ **ORIGINS** Asia, originally from waterways in southern China.

🗘 **SIZE** Highly variable; often exceeds 12 in (30 cm) in ponds.

🗘 **DIET** Goldfish food (flakes or pellets) and live foods.

🗘 **WATER** Temperature 50–77°F (10–25°C) and neutral (pH 7.0).

🗘 **TEMPERAMENT** Relatively social, but does not shoal.

Carassius auratus

Ryukin

The most obvious feature of this goldfish is the hump between the dorsal fin and the head. The body is relatively short and deep, the dorsal fin is tall, and the elongated caudal fin is divided to form a double tail. Ryukins are generally brightly colored, with a deep-red-and-white coloration being the preferred form. The markings on these goldfish should be symmetrical as far as possible. Chocolate (coppery) individuals are often recognized as a separate form, the Tetsuonaga, especially in Japan. Tetsuonagas have a reputation for both hardiness and the quality of their fin shape, so they are useful in Ryukin breeding. The Ryukin is named after Japan's Ryukyu Islands, where the ancestors of this goldfish were first introduced from China.

Orange-and-White Ryukin Ryukins have either normal eyes, as shown in this largely orange form, or, occasionally, telescope-eyes.

Tail coloration varies

Calico Ryukin Nacreous patterning *(see p.333)* is not common in double-tailed goldfish but is seen in the Ryukin. Calico Ryukins often have bold, contrasting markings.

Carassius auratus

Wakin

This form displays a variegated pattern of orange and white body markings. The vibrantly colored areas, which can vary from yellow through to reddish-orange, should extend around the body so that the white areas do not predominate. Pure-white Wakins, which occasionally occur, are not favored by breeders. Although the reflective metallic form is the most common, a nacreous variety *(see p.333)* also exists. The Wakin has a body shape similar to the Common Goldfish, but it can be instantly distinguished by its double caudal fin. Wakins are lively by nature and grow rapidly; fish reared in ponds can reach 8 in (20 cm) in length by three years of age.

Relatively slender body

Divided caudal fin

Extensive areas of color

Carassius auratus

Jikin

Descended from Japanese Wakin stock, the Jikin is often known in the West as the Peacock Tail. The raised upper lobes of its double caudal fin form an X-shape when viewed from behind.

The Jikin's body should be mainly silvery, with red areas restricted to the fins and around the lips. However, breeding Jikins with this desired arrangement of markings and a well-balanced caudal fin shape always proves difficult, even when the parent fish are both well marked and from a long-established line.

Carassius auratus

Black Moor

The matt-black color of the Black Moor is highly distinctive, as is its corpulent body shape. This goldfish is a telescope-eye variety, with eyes extending out from the sides of the head. The Black Moor is a selective color form of the Veiltail *(see p.339)*. Although developed in the UK, it is now kept worldwide. These fish are not very hardy and are better suited to an aquarium than an outdoor pond, especially through the winter (in temperate areas). Their coloration makes for an attractive contrast with brightly colored goldfish.

Even black coloration over the entire body

Telescope-eyes

Double caudal fin

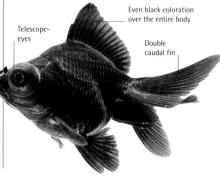

🌐 **ORIGINS** Asia, originally from waterways in southern China.

🔵 **SIZE** Highly variable; often exceeds 12 in (30 cm) in ponds.

🔴 **DIET** Goldfish food (flakes or pellets) and live foods.

🌡 **WATER** Temperature 50–77°F (10–25°C) and neutral (pH 7.0).

🔶 **TEMPERAMENT** Relatively social, but does not shoal.

FEEDING BEHAVIOR OF GOLDFISH

Although many of the foods marketed for goldfish float, and therefore encourage these cyprinids to feed at the surface, this is contrary to their instincts. Like their carp ancestors, goldfish are bottom-feeding fish by nature; they scavenge in the substrate for foods such as worms and other invertebrates. This means that they can be fairly destructive to the decor in the aquarium, since they will instinctively dig around in search of edible items. Larger goldfish varieties can move quite considerable amounts of gravel in this way, sometimes concentrating on a particular area of the tank and even digging right down to the undergravel filter. Plants in an aquarium for goldfish should therefore have their bases well protected by rockwork in order to minimize the likelihood that they will be uprooted. This type of behavior is most likely to arise when the fish are hungry,

and the goldfish may even resort to eating some of the plants in their tank. Substrate digging does not reflect a desire to spawn, as it does in some other fish, notably various cichlids.

Goldfish are unusual in that they lack a stomach at the start of the digestive tract where food can be stored, so they need to eat small quantities of food on a frequent basis. Offering them a large amount of food just once a day will therefore be wasteful and will also lead to a deterioration in water quality as the unwanted food breaks down and pollutes the water. Instead, give them a small amount of food four or five times a day, which will meet their appetite.

There is now a wide range of different goldfish foods available. Some types are designed to be used as growth foods for young goldfish. Others are tailored to suit specific varieties or to enhance particular characteristics,

such as to aid the growth of the distinctive hoods seen in Lionheads and Orandas (see p.338). Goldfish in ponds can benefit from wheat-germ foods, which will prove more digestible at lower water temperatures than most other food types.

Carassius auratus

Lionhead

The absence of a dorsal fin is a key feature of the Lionhead. The result is a smooth back that curves gently to the double caudal fin, the curvature accentuated by the fish's relatively long body. As Lionheads grow older, they develop a distinctive hood that covers the entire head area. This usually starts to become evident at the very top of the head and takes several years to develop to its full extent, when it has a raspberrylike appearance. The hood is more developed in this variety than in any other. Lionheads exist in a wide range of colors, although solid colors such as orange are most commonly seen. They do not thrive at high temperatures, nor are they hardy in temperate areas.

Compact yet broad body shape

Red-and-White Lionhead (above) This young fish has yet to develop its hood. Special diets are available to promote the hood's growth.

Hood is in initial stages of development in this blue individual

Blue Lionhead When fully grown, the hood should cover the entire head, encircling the eyes. The head has a wide appearance when viewed from above.

Carassius auratus

Oranda

The dorsal fin on the back of an Oranda allows it to be distinguished at a glance from other types of hooded goldfish. The Oranda also has a longer body shape and is a more powerful swimmer. The hood, or wen as it is called in Japan, is normally restricted to the top of the head, extending back over the eyes. In mature individuals, the area between the folds of the hood may appear whitish. Although this can look like a sign of disease, it is actually an accumulation of the protective mucus produced by the fish's body.

The coloration of these goldfish is sometimes unstable, just as it can be in other hooded varieties. This is particularly true of black-and-orange individuals, in which the orange areas often become more prominent over time.

Blue Oranda In this increasingly popular color variety, the underparts are usually a lighter shade.

Relatively long, double caudal fin

Red coloration but no swelling

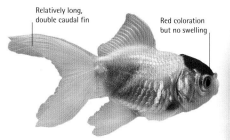

Red-Capped Oranda One of the most common Orandas, this silvery variety is a relatively hardy fish. The red patterning on the head will develop to form the hood.

🌐 **ORIGINS** Asia, originally from waterways in southern China.

📏 **SIZE** Highly variable; often exceeds 12 in (30 cm) in ponds.

🐟 **DIET** Goldfish food (flakes or pellets) and live foods.

💧 **WATER** Temperature 50–77°F (10–25°C) and neutral (pH 7.0).

🌡 **TEMPERAMENT** Relatively social, but does not shoal.

Carassius auratus

Ranchu

Sporting a hood similar to the Lionhead's, the Ranchu is the Japanese counterpart of that ancient Chinese breed. The Ranchu can be differentiated from the Lionhead by its shorter, more steeply curved body. As with Lionheads, not all Ranchus display smooth body curvature from head to tail, and an individual with slight humps along its back is considered to be seriously flawed. The double caudal fin may be only partially divided. In Ranchus of the highest quality, the top edge of the caudal fin should ideally form an angle of 90 degrees with the caudal peduncle. Ranchus, which are also known as Buffaloheads, are the most popular Japanese goldfish. Four principal founding lines are recognized, each of which is named after its creator. The dominant variety is the Ishikawa lineage; the others are Sakuri, Uno, and Takahashi. All these forms display a hood, but some less-common varieties lack this feature. They include the Osaka Ranchu, named after its city of origin, which also has a more rounded body. Another hoodless variety is the Nankin Ranchu, from the Shimane area of Japan, a silvery-white fish with red gill covers, lips, and fins. In addition, there is the rare Nacreous Ranchu, also called the Edonishiki, in which the hood is poorly developed.

Relatively small, symmetrical caudal fin

Black Ranchu This is the darkest variety. Ranchus are not hardy and need to overwinter in aquariums in temperate areas.

Short, broad body gives an impression of strength

Red-and-White Ranchu A mature individual with hood growth on the side of the face is described as *okame* (the name of a Japanese theatrical mask indicating a fat girl).

Red Ranchu All the Ranchu's fins are relatively short; the caudal fin is carried high. The hood has yet to develop in the young specimen shown above.

Carassius auratus

Veiltail

The elegant fins of the Veiltail are easily damaged, so this goldfish should be housed in a spacious aquarium—free from obstructions such as large rocks—rather than in a pond. The long caudal fin of the Veiltail is fully divided so that it hangs down in folds. The dorsal fin is tall, and in a well-proportioned Veiltail it should match the height of the body. The overall body shape of this variety is rounded rather than elongated.

The anal fin is paired and relatively long and tends to flow vertically when the fish is swimming. In addition to individuals with normal eyes, telescope-eye examples of this variety are not uncommon. The breed was developed from Ryukin stock by American breeders around Philadelphia in the late 1800s.

Enlarged dorsal fin

Short caudal peduncle

Bronze Veiltail The Veiltail has been bred in a wide range of colors, including bronze, as seen in this young fish. Even the juveniles display elongated fins.

The dorsal fin should start at the highest point on the back

Red-Capped Veiltail This fish has a variable reddish area on top of its head. Its fins are semitransparent.

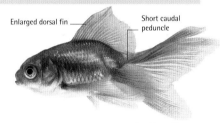

Calico Veiltail The nacreous patterning *(see p.333)* of the Calico Veiltail is highly variable, with darker streaking usually evident in the fins.

⊕ **ORIGINS** Asia, originally from waterways in southern China.

◔ **SIZE** Highly variable; often exceeds 12 in (30 cm) in ponds.

◔ **DIET** Goldfish food (flakes or pellets) and live foods.

◔ **WATER** Temperature 50–77°F (10–25°C) and neutral (pH 7.0).

◔ **TEMPERAMENT** Relatively social, but does not shoal.

Carassius auratus

Fantail

This striking goldfish gets its name from its relatively stiff, double caudal fin, which does not droop. The dorsal fin is also enlarged and is typically about one-third of the fish's body length. Fantails are probably most closely related to Ryukins, but they differ in having a smooth upper profile, with no sign of a hump. The body shape is longer and less rotund than many other double-tailed varieties. Fantails have been developed in a wide range of colors, and telescope-eye forms have also been bred. This is an adaptable variety, suitable for ponds or aquariums.

Red Fantail These fish rank among the most popular goldfish. When their fins have a ragged appearance, it is usually an indication of poor environmental conditions.

Patterning on the fins

Flowing tail

Relatively long, slim body

Nacreous Fantail Although not often available, this attractive color variant has a pale bluish-white background, coppery orange patches, and dark markings.

Carassius auratus

Butterfly

The tail lobes of these goldfish, when extended and viewed from above, resemble the wings of a butterfly. A good specimen should have a strong caudal fin that does not droop down. The fin also needs to be compact and rounded in shape, rather than tapering to a point. There are no embellishments on the head of the Butterfly; however, the telescope-eye characteristic has been bred into this variety.

It is possible that a number of the fry from a single spawning may develop with one eye being normal and the other telescoped, although such individuals are not favored. Butterflies, which exist in a wide range of color forms, have proved to be relatively hardy goldfish, although care should be taken, as with other fancy varieties, not to expose them to very low water temperatures.

Chocolate Butterfly An even depth of color is an important feature of this goldfish, which is sometimes called the Copper or Iron-Colored Butterfly.

Predominantly black and white

Rounded dorsal fin

Telescope-eyes

Well-balanced caudal fin

Panda Butterfly The black-and-white coloration of these goldfish is highly individual, with distinct patches being desirable.

Calico Butterfly (right) This is an attractive, lightly marked nacreous variety of the Butterfly.

ORIGINS Asia, originally from waterways in southern China.

SIZE Highly variable; often exceeds 12 in (30 cm) in ponds.

DIET Goldfish food (flakes or pellets) and live foods.

WATER Temperature 50–77°F (10–25°C) and neutral (pH 7.0).

TEMPERAMENT Relatively social, but does not shoal.

BREEDING GOLDFISH IN PONDS AND AQUARIUMS

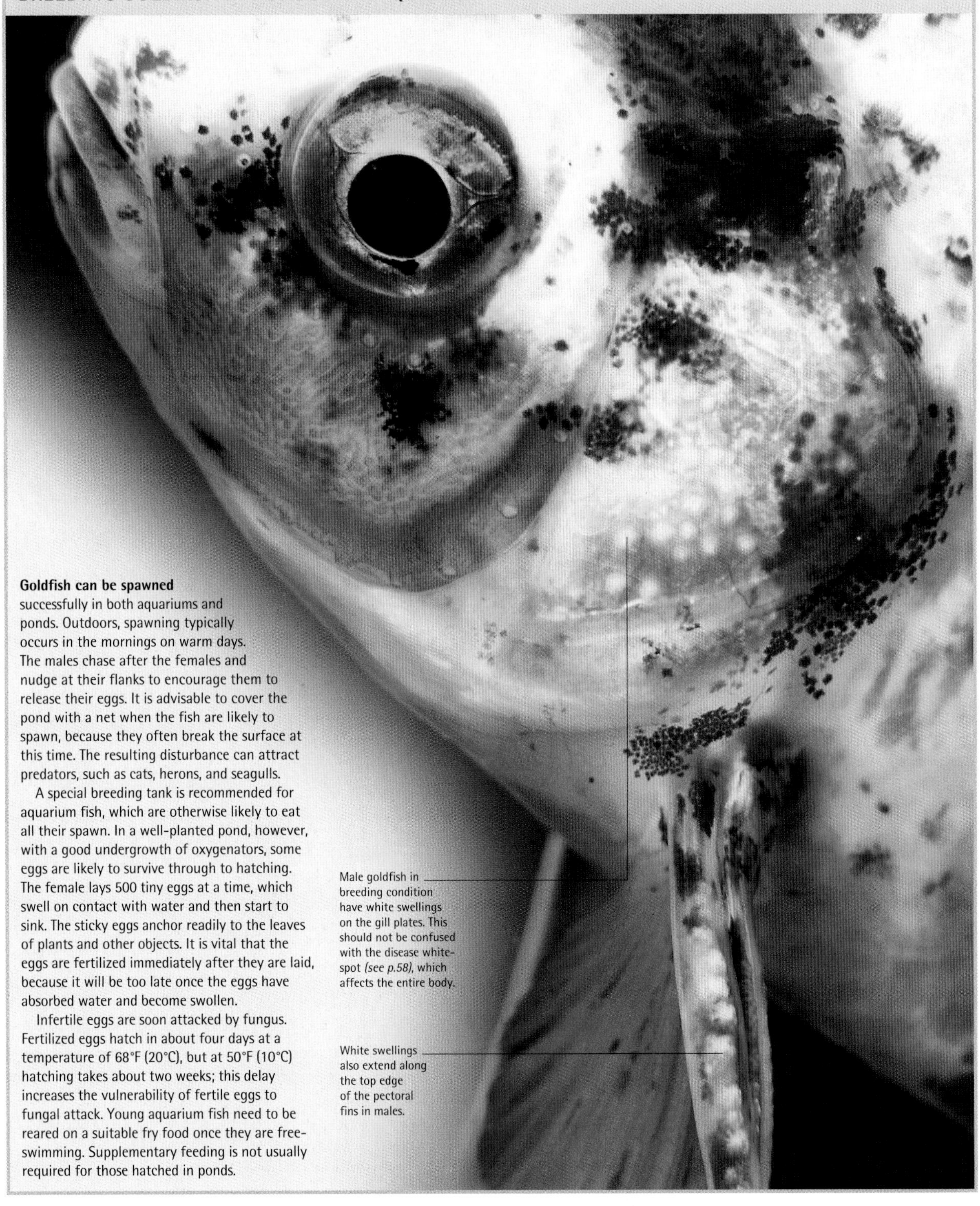

Goldfish can be spawned
successfully in both aquariums and
ponds. Outdoors, spawning typically
occurs in the mornings on warm days.
The males chase after the females and
nudge at their flanks to encourage them to
release their eggs. It is advisable to cover the
pond with a net when the fish are likely to
spawn, because they often break the surface at
this time. The resulting disturbance can attract
predators, such as cats, herons, and seagulls.

A special breeding tank is recommended for
aquarium fish, which are otherwise likely to eat
all their spawn. In a well-planted pond, however,
with a good undergrowth of oxygenators, some
eggs are likely to survive through to hatching.
The female lays 500 tiny eggs at a time, which
swell on contact with water and then start to
sink. The sticky eggs anchor readily to the leaves
of plants and other objects. It is vital that the
eggs are fertilized immediately after they are laid,
because it will be too late once the eggs have
absorbed water and become swollen.

Infertile eggs are soon attacked by fungus.
Fertilized eggs hatch in about four days at a
temperature of 68°F (20°C), but at 50°F (10°C)
hatching takes about two weeks; this delay
increases the vulnerability of fertile eggs to
fungal attack. Young aquarium fish need to be
reared on a suitable fry food once they are free-
swimming. Supplementary feeding is not usually
required for those hatched in ponds.

Male goldfish in
breeding condition
have white swellings
on the gill plates. This
should not be confused
with the disease white-
spot (see p.58), which
affects the entire body.

White swellings
also extend along
the top edge
of the pectoral
fins in males.

Carassius auratus

Celestial

Selective breeding of the goldfish has brought into being numerous variations in eye shape. The Celestial has eyes that protrude very obviously. They are not on the sides of the head, as in most goldfish, but rather in a semihorizontal plane so that they point upward, as if toward the stars (hence the name). The fry hatch with a normal eye arrangement, but the eyes rotate and shift position soon afterward. The bodies of these goldfish are relatively elongated, and they have slightly curved backs, with no dorsal fins. Both metallic and nacreous forms of the Celestial exist.

Orange Celestial Rich orange is common in these fish. The eyes should be equal in size and symmetrically positioned.

Celestial Pom-Pon This unusual form has both the Celestial eye position and the enlarged nasal flaps known as pom-pons *(see opposite)*.

Carassius auratus

Bubble-Eye

This unmistakable variety is characterized by the presence of large, bubblelike sacs under its eyes. As in the case of Celestials, Bubble-Eyes have a long body shape, lack a dorsal fin, and have a double caudal fin. Symmetry is a very important feature of this variety, with the sacs ideally being equal in size and shape. These fluid-filled sacs wobble when the fish swims and become compressed when it searches for food on the floor of the aquarium. In a good specimen, the combined width of the bubbles and head should match that of the body. Bubble-Eyes are suitable only for aquarium surroundings. The tank setup needs to minimize the risk that the fish will damage their bubbles and provide them with plenty of swimming space. Rockwork should not be included, and plants should be restricted to the back and sides of the tank. If a sac is accidentally punctured, it is likely to deflate.

Flowing double caudal fin

Smooth curve to back

Orange Bubble-Eye The coloration of the bubblelike eye sacs can vary; veins in the sacs are sometimes conspicuous as thin red streaks.

Goldfish sees over the top of the sacs

Black Bubble-Eye The color of the bean-shaped sacs under the eyes corresponds to the goldfish's overall coloration.

Calico Bubble-Eye The Calico's under-eye sacs are often almost transparent, although in some individuals they may display orange or bluish markings.

ORIGINS Asia, originally from waterways in southern China.

SIZE Highly variable; often exceeds 12 in (30 cm) in ponds.

DIET Goldfish food (flakes or pellets) and live foods.

WATER Temperature 50–77°F (10–25°C) and neutral (pH 7.0).

TEMPERAMENT Relatively social, but does not shoal.

Carassius auratus

Telescope-Eye Goldfish

True to its name, this fish has exaggeratedly protruding eyes. Ideally, the eyes should be of equal size, and it is not unusual for them to have a different color from the head. A short, round body and a large dorsal fin are other typical features associated with this breed. If a Telescope-Eye Goldfish is allowed to mate with a goldfish that has normal eyes, all their young will have normal eyes. However, some of their offspring will carry the telescope-eye gene, and when two of these mate together, a small percentage of the resulting fry will display the telescope-eye characteristic.

High dorsal fin

Variable coloring around the eyes

Dragon-Eye The telescope-eye characteristic can be combined with any color variety.

Black-and-Gold Telescope-Eye Telescope-Eye Goldfish look as if they are staring intently.

Carassius auratus

Pom-Pon

The nasal flaps, which are inconspicuous in other goldfish, are greatly enlarged in this variety. Known as pom-pons, the flaps may match the surrounding coloration, or they may be entirely different. It is important that the pom-pons are equal in size on a particular specimen, although they may be larger in some individuals than others. The pom-pon characteristic has been introduced to other breeds, notably the Oranda, Lionhead, and Ranchu. It is not uncommon for mature Pom-Pons to develop a small raised area on top of the head.

Red Magpie Pom-Pon In this color variant, the brilliant orange pom-pons are highlighted by the black around the head.

Pom-pons are evenly balanced in size

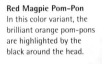

Orange Pom-Pon Pom-Pons can be bred in any color, and the size of the pom-pons can vary considerably.

ORIGINS Asia, originally from waterways in southern China.

SIZE Highly variable; often exceeds 12 in (30 cm) in ponds

DIET Goldfish food (flakes or pellets) and live foods.

WATER Temperature 50–77°F (10–25°C) and neutral (pH 7.0).

TEMPERAMENT Relatively social, but does not shoal.

KOI

Few fish inspire greater devotion than koi. These large ornamental carp have now established a dedicated following in the West, matching that which they enjoy in their Japanese homeland. Koi were developed primarily for the beauty of their colors and patterning, when viewed from above. This led to great interest in breeding and showing koi varieties, and the most desirable individuals now change hands for huge sums of money. To appreciate these attractive pond fish fully, they must be clearly visible through the water; as a result, koi are most often kept in well-filtered water in a pond containing few or no plants.

When assessing the quality of an individual koi, experts focus not only on the color or patterning but also the overall physical appearance or type, including their length and body width.

UNDERSTANDING KOI NOMENCLATURE

Distinguishing between koi varieties can be extremely confusing at first, partly because they are referred to in the West by traditional Japanese descriptions, even if they have been bred in other countries. Knowledge of a few basic terms, such as the words to describe the various colors (right) and the main varieties (below), is a useful introduction to the complex world of koi nomenclature.

KOI VARIETIES

The illustrations below give examples of the main koi varieties, highlighting their individual characteristics. The most popular varieties, known collectively as "Go Sanke," are the Showa, Kohaku, and Sanke. Varieties prefixed with the term "Hikari" are metallic koi, which have an overall reflective luster, while all others are known as nonmetallics.

In addition to the color differences described here, the appearance of a koi is influenced by its pattern of scalation. "Doitsu" koi, for example, may have large mirror scales on either side of the dorsal fin but are scaleless elsewhere, while the scale type known as "Kin Gin Rin," or simply "Gin Rin," is characterized by the sparkling appearance of the scales.

KOI COLORS

Japanese descriptions of color are important in koi nomenclature. Some have multiple names depending on the form in which that color appears.

Color	Name		Color	Name
Red	"Aka" (red background) "Hi" (red markings)		Black	"Karasu" (black background) "Sumi" (black markings)
Blue	"Ai"		Yellow	"Ki" (yellow) "Yamabuki" (pale yellow)
Orange/ Red	"Beni" (orange/red background)		Green	"Midori"
Brown	"Cha"		Gray	"Nezu"/"Nezumi"
Silver (metallic)	"Gin"		Orange	"Orenji"
Gold (metallic)	"Kin"		White	"Shiro"

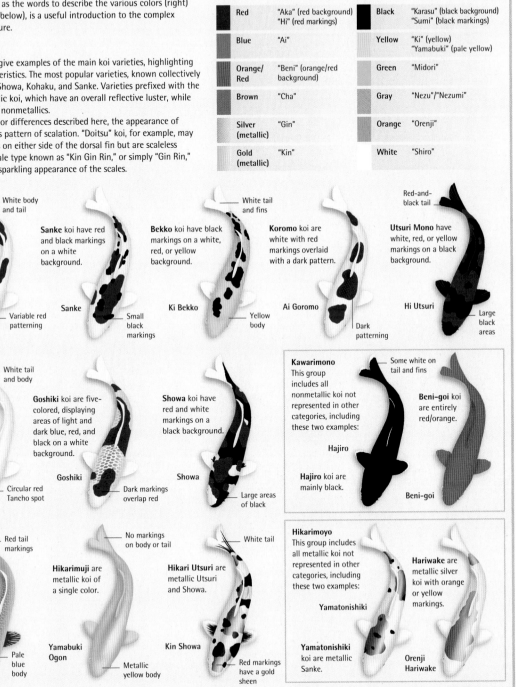

Kohaku koi have red markings on a white background.
— White body and tail
— Variable red patterning
Kohaku

Sanke koi have red and black markings on a white background.
Sanke
— Small black markings

Bekko koi have black markings on a white, red, or yellow background.
Ki Bekko
— Yellow body

Koromo koi are white with red markings overlaid with a dark pattern.
— White tail and fins
Ai Goromo
— Dark patterning

Utsuri Mono have white, red, or yellow markings on a black background.
— Red-and-black tail
Hi Utsuri
— Large black areas

Tancho are white koi with a red mark, ideally circular, on the head but no red on the body.
— White tail and body
Tancho Kohaku
— Circular red Tancho spot

Goshiki koi are five-colored, displaying areas of light and dark blue, red, and black on a white background.
Goshiki
— Dark markings overlap red

Showa koi have red and white markings on a black background.
Showa
— Large areas of black

Kawarimono This group includes all nonmetallic koi not represented in other categories, including these two examples:
— Some white on tail and fins
Hajiro
Hajiro koi are mainly black.

Beni-goi koi are entirely red/orange.
Beni-goi

Asagi koi are pale blue with red and white areas.
— Red tail markings
Asagi
— Pale blue body

Hikarimuji are metallic koi of a single color.
— No markings on body or tail
Yamabuki Ogon
— Metallic yellow body

Hikari Utsuri are metallic Utsuri and Showa.
— White tail
Kin Showa
— Red markings have a gold sheen

Hikarimoyo This group includes all metallic koi not represented in other categories, including these two examples:
Yamatonishiki
Yamatonishiki koi are metallic Sanke.

Hariwake are metallic silver koi with orange or yellow markings.
Orenji Hariwake

Cyprinus carpio

Kohaku

The earliest forerunners of modern koi displayed simple red-and-white markings. Known as Kohaku, these koi rank today as one of the most popular varieties. Kohaku are characterized by their white body color and red (or "hi") patterning. In the highest-quality Kohaku, it is particularly important that the white areas show no trace of yellowing (a fault known as "shimis"), while the red areas should be dense. The border, or "kiwa," at the back of each red patch must be well defined; at the front, however, the white scales overlay the red so the definition is not as sharp. Assessing the potential of young Kohaku can be difficult, because their scales have a translucent nature—a feature described as "kokesuke." All Kohaku stem from six basic breeding lines, which are named after the Japanese breeders who developed them.

Pure white pectoral fins

Distinct "Maruten" head spot

Hiroshima Sakai This Kohaku, of the famous Sakai breeding line, was bred on the Sakai family's farms in Hiroshima.

Kohaku This koi was awarded Kokugyo—best entry for its size out of all varieties—at Japan's prestigious Wakagoi show.

Red lipstick marking

Kuchibeni Hasegawa This koi from the Hasegawa breeding line has a "Kuchibeni" or lipstick marking on the head.

Classic and desirable white area on the caudal peduncle

Good skin quality is important

Marking closest to the caudal fin is known as "ojime"

"Hi" markings in Kohaku are only above the lateral line

Maruten head spot must be distinct from other areas of "hi"

Distribution of markings in five-step Kohaku varies

Maruten Sandan Yagozen The head spot, plus two other colored areas, indicate a Sandan or three-step Kohaku.

Maruten Yondan This koi is a Yondan or four-step Kohaku, because it has a red head marking and three other red patches.

Godan This is the most complex five-step Kohaku. In high-quality examples, the dorsal fin must be unmarked.

Matsunosuke A Kohaku from one of the most notable bloodlines, this koi has an excellent depth of red coloration.

ORIGINS Developed in Japan from carp brought from China.

SIZE Slight varietal differences; typically up to 36 in (90 cm).

DIET Specialized commercial koi foods of various types.

WATER Temperature 37–77°F (3–25°C) and neutral (pH 7.0).

TEMPERAMENT Relatively social but do not shoal.

THE ORIGINS OF KOI

Today's modern koi varieties are descendants of black carp, known as Magoi, which were introduced to Japan from China around 1000 CE. By the 1600s, these plain-looking fish were thriving in the waterways around the paddy fields of Niigata prefecture on Honshu Island, and the local rice farmers caught them for food. Around the early 1800s, individual fish displaying patches of color and patterning on their bodies started to appear, and some of the farmers began to selectively breed for these characteristics. Known as "Nishikigoi," or "brocaded carp," these colorful fish attained public recognition when a group was shown at the 1914 Taisho Exhibition in Tokyo, and a number were then transferred to the moat surrounding the Emperor's Imperial Palace. Their descendants can still be found there today. Koi-keeping and breeding subsequently became extremely popular in Japan, signaling the birth of the lucrative Japanese koi industry of today.

Koi were first introduced to the US in the early 1940s. It took longer for them to gain recognition in Europe; koi were not seen in Great Britain until the 1960s. Since then, they have gained a huge international following and are now bred not only in Japan but in other countries, including the US, Israel, China, Korea, Thailand, and South Africa.

Cyprinus carpio

Sanke

One of the most popular koi varieties, Sanke probably first arose in the late 1800s but only came to prominence in the early 20th century. These tricolored koi display variable black or "sumi" markings superimposed on red-and-white coloration similar to that of Kohaku. The skin color of high-quality Sanke should be snow white, while large areas of red ("hi") free from "sumi" are not considered desirable. In addition, black patches should not be present on the head. Although a symmetry of patterning is not required, the overall effect should be balanced; the "sumi" markings, for example, should be evenly distributed and not simply restricted to one side of the body. It can be difficult to assess the exhibition potential of Sanke until they are at least two years old. Before this, some individuals may resemble Kohaku, while the "sumi" patterning in others, although evident, may not be stable and may even vanish for a period.

Black areas do not extend below lateral line

Extent of black striping on pectoral fins varies

Prominent "sumi" marking

Red areas not broken by "sumi" patterning

Traditional Sanke This koi has "kasane sumi" patterning, in which the black "sumi" markings overlap the red areas.

Hiroshima Sanke Named All-Japan Supreme Champion, this koi displays rare "tsubo sumi" (black-on-white) patterning.

Shintaro Tategoi Koi described as "tategoi" are still developing and should continue improving as they mature.

Small white patches or "windows" in red areas suggest that color may disappear with age

No symmetry in body markings

"Sumi" markings develop slowly in Matsunosuke Sanke

Extensive red coloration

Pure white coloration

Tosai Tategoi The black "sumi" patterning of this one-year-old Sanke has only just started to become visible.

Ogawa Sanke Black markings in the pectoral fins of Sanke koi appear as streaks rather than blotches.

Matsunosuke Sanke One of the most famous bloodlines, these Sanke appear slim when young but broaden with age.

Torazo This koi is from a notable Sanke bloodline named after breeder Torakichi Kawikame's father.

⊕ **ORIGINS** Developed in Japan from carp brought from China.

⬡ **SIZE** Slight varietal differences; typically up to 36 in (90 cm).

⬡ **DIET** Specialized commercial koi foods of various types.

⬡ **WATER** Temperature 37–77°F (3–25°C) and neutral (pH 7.0).

⬡ **TEMPERAMENT** Relatively social but do not shoal.

Cyprinus carpio

Bekko

Bekko are white, red, or yellow koi with black (or "sumi") patterning. Shiro Bekko, which is white with black markings, is the most common form. There are physical variants, too, such as the "Gin Rin," with its shiny appearance, and is often described as partially scaled or matt. Bekko koi are often of Sanke descent. Top-quality examples should not display any "sumi" coloration on the head.

Shiro Bekko This koi is similar in appearance to a Sanke but with no trace of red ("hi") coloration.

Cyprinus carpio

Koromo

The name "Koromo" translates as "robed." This refers to the characteristic dark scale-edging that overlays the colored markings of this variety. This feature should not extend into the white areas and can take years to develop fully. Also known as Goshiki, these koi are classified in the Koromo category in the West but in Japan are still considered part of the Kawarimono group *(see p.356).*

Ai Goromo Dark scale-edging highlights the underlying Kohaku patterning of this koi.

Cyprinus carpio

Utsurimono

Utsurimono are black koi with white markings (Shiro Utsuri), red markings (Hi Utsuri), or yellow markings (Ki Utsuri). They can sometimes be confused with Bekko koi *(see left),* but the black coloration extends below the lateral line and over the head. Shiro Utsuri are the most frequently seen variety, while red-and-black Ki Utsuri are exceptionally rare.

Shiro Utsuri Contrasting black and pure white areas on the head and body characterize this variety.

Cyprinus carpio

Tancho

Named after the Japanese crane *(Grus japonensis),* or Tancho, which has a distinctive red crown, this variety is extremely popular in Japan because its head marking echoes the design of the Japanese national flag. The red (or "hi") marking on the head should be circular in shape and centrally positioned, with no other red areas on the body. Unfortunately, creating the ideal Tancho is exceedingly difficult, and well-marked specimens are highly valued. Slight deviations in patterning have now become acceptable, especially if the "hi" marking is symmetrical. Tancho Kohaku are usually considered to be the most desirable form.

Tancho Kohaku This elegant koi is characterized by its pure snow-white body color and distinctive red head marking.

No "hi" marking on the body

Black markings at the base of the pectoral fins

Black markings do not extend over the Tancho spot

Tancho spot should be centered between the eyes

Tancho Sanke This koi displays obvious black-and-white Bekko patterning, with a prominent red Tancho spot on its head.

Tancho Showa The black Showa coloration extends over the characteristic red Tancho spot in this individual.

ORIGINS Developed in Japan from carp brought from China. | **SIZE** Slight varietal differences; typically up to 36 in (90 cm). | **DIET** Specialized commercial koi foods of various types. | **WATER** Temperature 37–77°F (3–25°C) and neutral (pH 7.0). | **TEMPERAMENT** Relatively social but do not shoal.

FEEDING AND NUTRITION

Koi eat both plant and animal matter, instinctively seeking their food close to the bottom of the pond. The two pairs of barbels on either side of the mouth serve as sensory feelers, helping them to locate edible items, such as worms, hidden in the substrate. Koi are also able to dig quite effectively using their jaws, a behavior that is likely to prove disruptive in a planted pond.

The koi's jaw structure is surprisingly flexible, which allows them to suck fairly large edible items directly into their mouths. At the back of the throat are toothlike structures that grind food before it is swallowed, making it more accessible to digestive enzymes. Koi do not have a stomach where food can be stored and so can digest only small amounts of food at a time. Swallowed food passes directly into the intestinal tract, and nutrients are absorbed as the food passes through before exiting the body. In adult koi, the intestine is two or three times the length of the body, while young koi have much shorter intestines and so require a higher protein content in their food to achieve the same levels of nutrition.

Cyprinus carpio

Hikarimoyo

The Hikarimoyo grouping encompasses all metallic koi of more than one color that are not categorized as Hikari Utsuri (*see p. 352*). This includes metallic forms of Kohaku and Sanke and the popular Hariwake, which are metallic silver koi with orange ("orenji") or yellow ("yamabuki") markings. The metallic appearance of these attractive koi, which first came to prominence in the 1960s, means they are highly visible in the water. The reflective scalation, however, dilutes the depth of their base coloration so that red areas tend to appear more orange, and black coloration appears grayish.

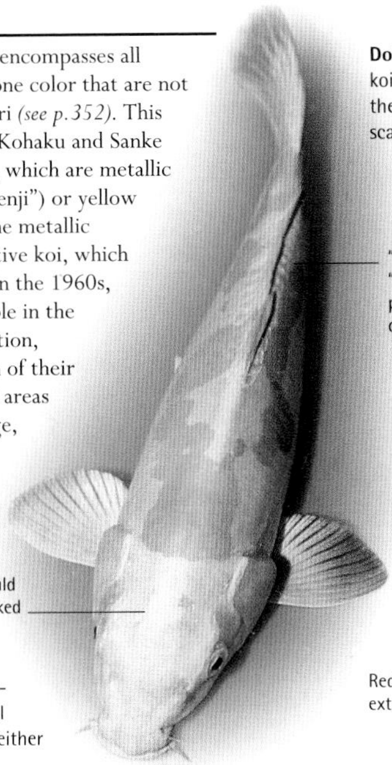

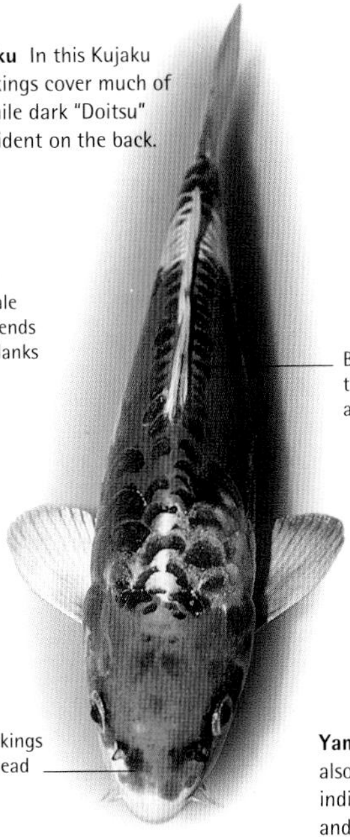

Doitsu Kujaku In this Kujaku koi, red markings cover much of the body, while dark "Doitsu" scales are evident on the back.

"Doitsu" or "mirror" scale pattern extends down the flanks

Metallic luster to body and fins

Black centers of the scales give a pine-cone effect

Head should be unmarked

Doitsu Hariwake This yellow-and-silver koi has symmetrical mirror or "Doitsu" scaling on either side of the dorsal fin.

Red ("hi") markings extend onto head

Yamatonishiki In this variety, also known as Metallic Sanke, individuals with the richest red and black markings are favored.

Cyprinus carpio

Goshiki

The name Goshiki literally means "five-colored," referring to the white, red, light blue, dark blue, and black coloration of these koi. There may even be a sixth color evident, when a blue area is overlaid by black, creating a purple shade. There is considerable variability between the koi of this group. The traditional form is relatively dark in color, but over recent years, some strains have been developed on much more colorful lines. Goshiki are now generally classified with Koromo but were traditionally included in the Kawarimono category (*see p. 356*).

Classic Goshiki This koi has larger areas of red and dark reticulated patterning, with little white coloration on the body.

Bold red ("hi") coloration

Snow-white skin

Characteristic reticulated blue scale pattern on the back

Goshiki This koi is from an Asagi lineage (*see p. 352*) and has bluish coloration with a reticulated pattern on the back.

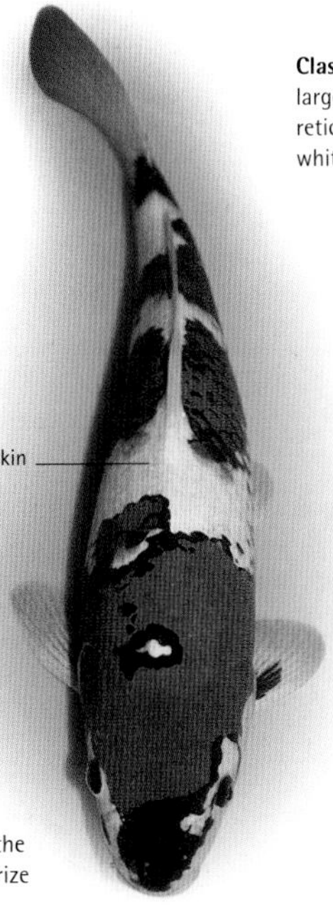

Polo Nippress Goshiki Reticulations on areas of the snow-white skin characterize this notable Goshiki.

Vibrant red ("hi") coloration on the head

ORIGINS Developed in Japan from carp brought from China.

SIZE Slight varietal differences; typically up to 36 in (90 cm).

DIET Specialized commercial koi foods of various types.

WATER Temperature 37–77°F (3–25°C) and neutral (pH 7.0).

TEMPERAMENT Relatively social but do not shoal.

Cyprinus carpio

Hikari Utsuri

This category features metallic koi with Showa *(see p.355)* and Utsuri *(see p.349)* patterning. Hikari Utsuri are often strikingly colored but typically display little refinement in their patterning. Their metallic sheen can negate their depth of coloration; black (or "sumi") markings, for example, are not as vivid in Hikari Utsuri as in their nonmetallic counterparts. This variety was developed when Ogons, which are single-colored metallic koi *(see below)*, were bred with Showa and Utsurimono stock.

"Sumi" coloration more vivid on the fins than on the body

Kin Showa This metallic variety has highly variable patterning. Both red ("hi") and black ("sumi") areas are quite pale.

Kin Ki Utsuri An attractive contrast of gold and black, this koi's "sumi" markings extend around the sides of the body.

Rich "sumi" coloration on body and head

Golden base color

Kikokuryu A fairly new variety, this metallic "Doitsu" koi is sometimes classed as a Hikarimoyo *(see p.351)*.

Parallel lines of dark scales on either side of dorsal fin

Helmet pattern

Cyprinus carpio

Hikarimuji

Members of this group are single-colored, metallic koi. They are all descendants of a single wild, black carp (or Magoi) with a golden stripe along its back that was discovered in Yamakoshi prefecture in 1921. A selective breeding program from this fish, carried out by the Aoki family, produced the first pure-colored metallic koi (or Ogon) 25 years later. Hikarimuji have become immensely popular with koi enthusiasts, because they show well in ponds, are easily tamed, and grow fast.

Head must be free from flecks of orange

Yamabuki Ogon Top-quality Ogons, like this metallic gold koi, must be well muscled, but not fat, and have perfect scaling.

Choguro Purachina This white koi with a lustrous appearance is also known as the Platinum Ogon.

Sparkling (or "furakin") effect created by the metallic nature of skin and scales

---

Cyprinus carpio

Asagi

This unmistakable variety, whose ancestry dates back more than 160 years, is distinguished predominantly by the bluish, scaled pattern over the back, with reddish areas on the fins and on the sides of the head. Symmetry in appearance is highly valued in these koi. Asagi with "Doitsu" scaling *(see p.345)* are known as Shusui, while there is also a colorful red ("hi") form in which blue coloration is overlaid with red.

Pure white, scaleless head contrasts with dark body

⊕ **ORIGINS** Developed in Japan from carp brought from China.

◐ **SIZE** Slight varietal differences; typically up to 36 in (90 cm).

◑ **DIET** Specialized commercial koi foods of various types.

◔ **WATER** Temperature 37–77°F (3–25°C) and neutral (pH 7.0).

◕ **TEMPERAMENT** Relatively social but do not shoal.

CLIMATE ISSUES

Winters can be harsh in places where koi originate, and today's established koi varieties are hardy enough to spend the winter in an outdoor pond in all but the coldest climates. An outdoor koi pond must be sufficiently deep, however, to ensure that the fish will not become trapped in any ice that forms. Pond heaters can help to prevent the surface from freezing over.

As water temperatures drop, koi spend more time at the bottom of the pond and start to eat less. Young fish may be better housed in an aquarium over the winter, since spending time in this torpid state temporarily slows their rate of growth.

Hot weather also brings its hazards. Increasing water temperature can reduce the amount of oxygen in the pond to dangerously low levels. Evaporation increases, and the pond is likely to require regular refilling with dechlorinated water. Fish should be checked more regularly for signs of disease in summer, because infectious agents can multiply more quickly in warm weather. Screening may also be required in very hot weather, to provide shade over the pond and to help prevent pale-colored fish from suffering sunburn *(see p.323)*. Canopies fashioned from bamboo matting on wooden supports are a popular decorative option for this purpose.

GROWTH AND DEVELOPMENT

Few pond fish live longer than koi; indeed, a number of the original eight koi transferred to the moat around Japan's Imperial Palace following the Tokyo exhibition of 1914 *(see p.347)* were still sighted there more than 50 years later. Koi also rank among the largest of all pond fish, with some individuals reputedly reaching up to 6 ft (1.8 m) in length. Overall size is partly dependent on variety; Chagoi *(see p.356)*, for example, naturally grow to a much larger size than most other koi. An individual koi grows to almost half its potential adult size in the first two years of its life and, if kept in optimum conditions, has a growth rate in this time of 1 in (2.5 cm) per month. This rate is largely influenced by the koi's environment—including quality and temperature of the water and the stocking density in the pond—and also by the amount and quality of food provided. After this stage, the rate of growth declines, and a koi will not reach its full adult size until it is 15 years old. The color and patterning of some varieties can change as they grow and develop. This is especially true of Matsukawabake koi, which have unstable black-and-white markings that can alter in response to changes in environmental conditions.

Cyprinus carpio

Showa

One of the most popular koi varieties, Showa were originally developed during the late 1920s. It was not until the 1960s, however, following crosses with Sanke and Kohaku varieties, that the yellowish markings of these early Showa were transformed into the vibrant red that is a feature of the variety today. Showa can be confused with Sanke koi *(see p.348)*, which also have red, black, and white coloration. They can be distinguished by the extent and distribution of black ("sumi") markings on the head and body. "Sumi" patterning is more dominant in Showa than in Sanke, and the black markings extend on to the head and below the lateral line. In contrast, Sanke have only "sumi" on the body and above the lateral line. The patterning of Showa koi can change considerably as they mature, which makes it extremely difficult to assess the potential of young koi of this variety.

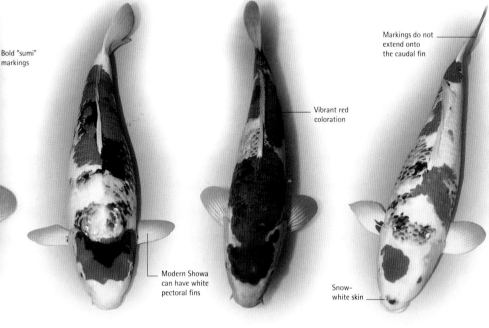

Patterning over the full length of the body to the caudal fin

Semicircular, black markings on the pectoral fins known as "motoguro"

Large black areas in the pectoral fins

"Sumi" markings extend to the head

Kindai Showa Modern ("Kindai") Showa have more extensive white coloration and less black than traditional Showa.

Hi Showa Red ("hi") coloration predominates in this variety, although the white body color can still be seen.

Traditional Showa This popular koi variety has large areas of red ("hi") and jet-black ("sumi") markings.

Bold "sumi" markings

Markings do not extend onto the caudal fin

Vibrant red coloration

Modern Showa can have white pectoral fins

Snow-white skin

Hi Showa This koi has extensive red ("hi") coloration on the head and body, which contrasts with strong "sumi" patterning.

Matsunosuke Kage Showa Shadowy, pale "sumi" markings characterize this Kage Showa with Matsunosuko Sanke ancestry.

Kage Showa This koi has paler "sumi" than a traditional Showa, although these markings often alter as the fish matures.

Kindai Showa This modern Showa is predominantly white with pale, shadowy "sumi" markings on the head and body.

ORIGINS Developed in Japan from carp brought from China.

SIZE Slight varietal differences; typically up to 36 in (90 cm).

DIET Specialized commercial koi foods of various types.

WATER Temperature 37–77°F (3–25°C) and neutral (pH 7.0).

TEMPERAMENT Relatively social but do not shoal.

Cyprinus carpio

Kawarimono

This diverse grouping encompasses all nonmetallic koi ("Kawari") that are not included in other categories, although most are named and recognized varieties in their own right. Among the most popular are single-colored koi of various colors, which can provide a striking contrast to patterned koi in a pond. This category also includes koi that are predominantly black in color; these are considered to be the koi most closely related to the ancestral Magoi *(see p. 347)*. In addition to koi with standard matt coloration, Kawarimono includes partially scaled ("Doitsu") and shiny-scaled ("Gin Rin") varieties. Rarities— unique koi that do not fit into other groups and whose parentage may be unknown—also feature in this group. There is some variation between the koi considered to be Kawarimono in Japan and the standards elsewhere in the world.

"Doitsu" koi have an incomplete covering of body scales

Symmetrical white markings on either side of the dorsal fin

One of the fastest growing varieties

Even, jet-black body color

Paler, saffron-colored individuals are favored

Kumonryu This partially scaled ("Doitsu") koi has variable black-and-white coloration, which can alter with age.

Hageshiro Black coloration predominates on the body of this koi, which has a contrasting white head and pectoral fins.

Chagoi A variable brownish hue is evident on the body of these broad-bodied koi, which readily become tame.

Dark "mirror" scales are a feature of "Doitsu" koi

Silvery (or "Gin Rin") scalation

Red ("hi") coloration predominates on the head

Colored markings extend to the pectoral fins

"Kuchibeni" or lipstick marking

Doitsu Kujaku Nonmetallic forms of the Doitsu Kujaku *(see p. 351)*, such as this individual, are fairly unusual.

Beni Kumonryu This "Doitsu" koi has a black-and-white body and head with orange-red (or "beni") patterning.

Gin Rin Matsukawabake The black-and-white patterning of this variety changes depending on the water temperature.

Gin Rin Ochiba Shigure A newer variety, this striking koi has reflective scales ("Gin Rin") that give it a sparkling appearance.

⚬ **ORIGINS** Developed in Japan from carp brought from China.

⚬ **SIZE** Slight varietal differences; typically up to 36 in (90 cm).

⚬ **DIET** Specialized commercial koi foods of various types.

⚬ **WATER** Temperature 37–77°F (3–25°C) and neutral (pH 7.0).

⚬ **TEMPERAMENT** Relatively social but do not shoal.

KOI AND JAPAN

Most koi of Japanese origin command premium prices in the international marketplace, reflecting not only the individual quality of each koi but also the rigorous selection procedures that breeding stock must meet in their homeland. Koi-breeding in Japan is still largely carried out by a number of well-known families who have koi-breeding lines extending back over centuries; a chosen family name is now often included in the name of an individual koi to indicate that it has this ancestry. A Matsunosuke Sanke, for example, is developed from the famous Matsunosuke line, a breeding line that has come to prominence since the 1960s. There are often subtle but recognizable differences between koi of the same variety but from different breeding lines, not just in the appearance of the adult fish but also in the development of their markings. Kichinai Sanke, for example, have a reputation for having very stable black ("sumi") markings, while the "sumi" patterning on a Sanke with Matsunosuke ancestry is a pale bluish-gray shade at first but subsequently darkens as the koi matures.

OTHER COLDWATER FISH

In addition to goldfish and koi, many other fish from a wide range of families thrive in coldwater ponds, from small, colorful species, such as this Red Shiner *(see also p. 361)*, to prehistoric-looking sturgeon. However, the keeping of coldwater fish has raised environmental concerns, principally that imported exotic species may escape into the wild and endanger populations of native fish. As a result, there are legal restrictions on the sale and movement of some species. Dealers should be familiar with these laws, but you can check with the US Department of Agriculture (or, in Canada, the Department of Fisheries and Oceans) for up-to-date regulations.

The Red Shiner (*Cyprinella lutrensis*) is an attractive and easily maintained cyprinid. Like most coldwater pond fish, it originates from temperate regions.

Etheostoma caeruleum

Orange-Throated Darter

- **ORIGINS** Southeastern Canada and eastern US, near the Great Lakes; also in Louisiana and Mississippi.
- **SIZE** 3 in (8 cm).
- **DIET** Small live foods and the eggs of other fish.
- **WATER** Temperature 39–68°F (4–20°C); hard (100–150 mg/l) and around neutral (pH 7.0).
- **TEMPERAMENT** Breeding males are territorial.

These small, bottom-dwelling fish are difficult to observe in ponds but are more visible in well-filtered coldwater aquariums. In a pond setting, they supplement their diet by feeding on aquatic insect larvae. Male Orange-Throated Darters are more brightly colored than females. A rise in water temperature triggers breeding. The female lays several hundred eggs over the course of two or three days and buries them in mulm on the pond floor.

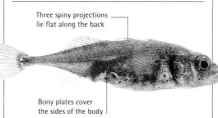

Male

Gasterosteus aculeatus

Three-Spined Stickleback

- **ORIGINS** Occurs widely over much of Europe, northern Asia, and Africa. Also present in North America.
- **SIZE** 5 in (12.5 cm).
- **DIET** Prefers fresh and prepared live foods.
- **WATER** Temperature 39–68°F (4–20°C); soft (50–100 mg/l) and neutral (pH 7.0).
- **TEMPERAMENT** Males become territorial when breeding.

Three spiny projections lie flat along the back

Bony plates cover the sides of the body

These sticklebacks show a distinct difference in coloration between the sexes during the spawning period, when male fish become red and blue. The male builds a nest out of plant matter and lures a succession of females inside so that they can lay their eggs, which he fertilizes. In total, the nest may contain as many as 50 eggs from different females. The male guards the eggs and also watches over the newly hatched fry.

Lepomis humilis

Orange-Spotted Sunfish

- **ORIGINS** North America, where it occurs in rivers and lakes from Texas to North Dakota.
- **SIZE** 4 in (10 cm).
- **DIET** Mainly live foods of different types.
- **WATER** Temperature 39–72°F (4–22°C); hard (100–150 mg/l) and neutral (pH 7.0).
- **TEMPERAMENT** Occasionally aggressive.

The small size of these sunfish, coupled with their attractive appearance, means that they can be kept in coldwater aquariums as well as in ponds. Sexing is quite straightforward, since only males display the distinctive reddish-orange spots, which are brown in females. The white edging around the so-called "ear flap" behind each eye is another point of recognition. Avoid housing them with other sunfish, because they will hybridize readily.

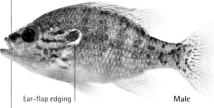

Ear-flap edging

Male

Acipenser gueldenstaedtii

Diamond Sturgeon

- **ORIGINS** Europe and western Asia, inhabiting the Azov, Caspian, and Black Seas; also ventures into rivers.
- **SIZE** 48 in (125 cm) in ponds.
- **DIET** Sturgeon pellets and live foods.
- **WATER** Temperature 50–68°F (10–20°C); hard (150–200 mg/l) and neutral to alkaline (pH 7.0–7.5).
- **TEMPERAMENT** May prey on small fish.

Young Diamond Sturgeon lose their characteristic white patterning as they mature, becoming grayer with age. House them in a large, well-oxygenated pond that is free from blanketweed, among which they can become trapped. Sturgeon are highly susceptible to chemical poisoning, so avoid using algicides, which are toxic to these primitive fish, as are some medications.

Acipenser stellatus

Star Sterlet

- **ORIGINS** Europe and western Asia, inhabiting the Azov, Caspian, and Black Seas; also ventures into rivers.
- **SIZE** 60 in (150 cm).
- **DIET** Pellets high in oil and protein, plus live foods.
- **WATER** Temperature 50–68°F (10–20°C); hard (150–200 mg/l) and neutral to alkaline (pH 7.0–7.5).
- **TEMPERAMENT** Peaceful.

As with other members of the sturgeon family, the Star Sterlet generally grows much smaller in ponds than in the wild. It would naturally spend much of its early life in the sea, heading up rivers to spawn in fresh water when mature. Despite this, it can be kept in an entirely freshwater environment, but it requires a large pond with a volume of at least 1,200 gallons (4,500 liters). Star Sterlets can be kept with koi.

Acipenser baerii

Siberian Sturgeon

- **ORIGINS** Rivers in Siberia, from the Kolyma to the Ob; also in some larger lakes, including Lake Baikal.
- **SIZE** 78 in (200 cm).
- **DIET** Pellets high in oil and protein. Carnivorous.
- **WATER** Temperature 50–68°F (10–20°C); hard (150–200 mg/l) and neutral to alkaline (pH 7.0–7.5).
- **TEMPERAMENT** Peaceful.

This fast-growing fish is gray or black on top, with white underparts. Like other sturgeons, it requires cool surroundings and highly oxygenated water. It digs with its snout for food, which in its natural habitat is mainly invertebrates. Spawning is not an annual event in the wild, but when a female does spawn, she may lay in excess of 400,000 eggs. This species does not normally breed in ponds.

Acipenser ruthenus

Sterlet

- **ORIGINS** Range extends from tributaries of rivers feeding the Azov, Caspian, and Black Seas to parts of Siberia.
- **SIZE** 48 in (120 cm).
- **DIET** Pellets high in oil and protein, plus live foods.
- **WATER** Temperature 50–68°F (10–20°C); hard (150–200 mg/l) and neutral to alkaline (pH 7.0–7.5).
- **TEMPERAMENT** May prey on smaller companions.

Sterlets are the most easily accommodated members of the sturgeon family, typically growing more slowly and reaching a smaller size than their relatives. Their name derives from the star-shaped bony scutes set into the skin. Sterlets are dark in color, although juveniles have white lines along their back and sides and white borders on their pectoral fins. The smaller, so-called albino variant has a pale yellow coloration that shows up well in the clear water that these fish require. In the wild, Sterlets spawn between May and June, with some females producing more than 100,000 eggs, which hatch in about five days. However, these fish rarely breed in ponds. When buying young sturgeon, regardless of the species, avoid individuals with a slightly bent body shape. This is a sign of malnutrition, which may be hard to reverse, even though specialist sturgeon diets are now available.

Pimephales promelas

Fathead Minnow

- ⊕ **ORIGINS** Found through much of North America, from Canada's Great Slave Lake southward to Mexico.
- ⊙ **SIZE** 4 in (10 cm).
- ⊜ **DIET** Flake and pelleted foods.
- ⊜ **WATER** Temperature 50–77°F (10–25°C); hard (100–150 mg/l) and neutral to alkaline (pH 7.0–7.5).
- ⊜ **TEMPERAMENT** Active and social.

A dark stripe extends along the midline

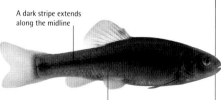

Brown is the natural coloration

Fathead Minnows are naturally brown, but there is also a yellowish strain called the Golden Minnow. These lively shoalers are not entirely hardy, but they can be moved into an indoor aquarium if necessary to protect them from extreme cold. Sexing is easy during spring, when the males develop white spots called tubercles on their gill plates. The female lays her eggs beneath rocks or raised pots, and the male guards them until they hatch about five days later.

Chrosomus erythrogaster

Southern Red-Bellied Dace

- ⊕ **ORIGINS** North America, extending from Minnesota eastward to New York State.
- ⊙ **SIZE** 3½ in (9 cm).
- ⊜ **DIET** Flake, small pellets, and live foods.
- ⊜ **WATER** Temperature 50–77°F (10–25°C); hard (100–150 mg/l) and neutral (pH 7.0).
- ⊜ **TEMPERAMENT** Relatively peaceful.

The small size of these minnows means that their attractive coloration will be difficult to appreciate in a pond setting, and they probably look best in a coldwater aquarium. Good oxygenation and filtration is important, since their natural habitat is fast-flowing streams. Lowering the water temperature over winter and increasing it again in spring should trigger spawning behavior. The female scatters her eggs above the substrate.

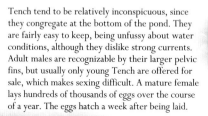

Rhodeus amarus

European Bitterling

- ⊕ **ORIGINS** Europe to the north of the Alps, although it does not naturally occur in Scandinavia or Great Britain.
- ⊙ **SIZE** 4 in (10 cm).
- ⊜ **DIET** Eats a wide variety of foodstuffs, including pellets.
- ⊜ **WATER** Temperature 50–70°F (10–21°C); hard (100–150 mg/l) and around neutral (pH 7.0).
- ⊜ **TEMPERAMENT** Lively and peaceful.

Blue streak in front of the caudal fin

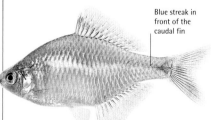

European Bitterling will breed successfully only if their pond houses Swan Mussels (*Anodonta cygnea*). The female lays her eggs inside an open mussel using her prominent egg-laying tube, or ovipositor, which measures about 1½ in (3.75 cm) long. The male then fertilizes the eggs before the mussel closes. The fry hatch and emerge from the mussel's siphon tube about a month later.

Tinca tinca

Tench

- ⊕ **ORIGINS** Occurs naturally throughout much of Europe; absent from the far south and Scandinavia.
- ⊙ **SIZE** 2 ft (60 cm).
- ⊜ **DIET** Eats pellets, which it may take at the surface.
- ⊜ **WATER** Temperature 32–86°F (0–30°C); hard (100–150 mg/l) and around neutral (pH 7.0).
- ⊜ **TEMPERAMENT** Social; needs to be kept in small shoals.

Tench tend to be relatively inconspicuous, since they congregate at the bottom of the pond. They are fairly easy to keep, being unfussy about water conditions, although they dislike strong currents. Adult males are recognizable by their larger pelvic fins, but usually only young Tench are offered for sale, which makes sexing difficult. A mature female lays hundreds of thousands of eggs over the course of a year. The eggs hatch a week after being laid.

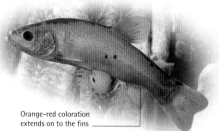

Orange-red coloration extends on to the fins

Red Tench This variety is distinguished by vivid orange-red coloration offset against variable dark markings, typically on the head and along the back. The appearance of Red Tench can be improved by color feeding.

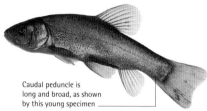

Caudal peduncle is long and broad, as shown by this young specimen

Green Tench This is the natural color form, although its appearance may vary depending on its background. Fish living in water with dense vegetation are a much darker green than those inhabiting sparsely planted ponds.

Red-and-White Tench As in orfe, goldfish, and other carp, this coloration is the result of a natural mutation, which has been enhanced by selective breeding.

Leuciscus idus

Orfe

- 🌐 **ORIGINS** Widely distributed through northern Europe, although it does not occur naturally in Norway.
- **SIZE** 24 in (60 cm).
- **DIET** Pond foods and live foods.
- **WATER** Temperature 32–86°F (0–30°C); hard (100–150 mg/l) and neutral to alkaline (pH 7.0–7.5).
- **TEMPERAMENT** Social and active by nature.

Wild Orfe display the same coloration as the domesticated strain known as the Silver Orfe. All Orfe have a narrow, streamlined body shape and need plenty of space for swimming, especially as they grow larger. Keep these fish in small groups to make them less nervous. On mild summer evenings, Orfe can often be seen patrolling just below the pond's surface in search of gnats. At this time of year, they are especially vulnerable to low oxygen levels in the water. Including a fountain or waterfall in their pond will help to address this problem by improving the level of dissolved oxygen in the water. Orfe are easy to sex in summer, since the females swell with eggs and mature males develop white tubercles on their gill plates and along the edges of the pectoral fins. These cyprinids can breed successfully by the time they are four years old. They lay their eggs among pond plants. Hatching can take nearly three weeks.

Silver Orfe Although this strain corresponds very closely to the wild color form, it is less commonly kept than the Golden Orfe. The life span of Orfe in pond surroundings can be in excess of 15 years.

Blue Orfe The coloration of Blue Orfe looks attractive in isolation but is not especially conspicuous in a pond setting. Unfortunately, however, Orfe generally grow too large to be housed in a coldwater aquarium.

Golden Orfe Black speckles on the upperparts offset the orange-gold coloration, which is much richer in some individuals than others. The depth of orange coloration can be improved by color feeding.

Cyprinella lutrensis

Red Shiner

- 🌐 **ORIGINS** North America, occurring in the Midwest, the Mississippi drainage basin, and northern Mexico.
- **SIZE** 3½ in (9 cm).
- **DIET** Flake, live foods, and small pellets.
- **WATER** Temperature 50–77°F (10–25°C); hard (100–200 mg/l) and neutral to alkaline (pH 7.0–7.5).
- **TEMPERAMENT** Active and social.

A Red Shiner's tank should include lots of swimming space, with planting restricted to the back and sides. The water must be well filtered and oxygenated. Reduce the temperature over winter to mimic the changes that occur in the wild. When you raise the temperature again in spring, males will become more colorful, and females will swell with eggs. Spawning then occurs in the substrate.

Male developing breeding coloration

Scardinius erythrophthalmus

Rudd

- 🌐 **ORIGINS** Widely distributed in northern Europe, but absent from much of Scotland and Scandinavia.
- **SIZE** 18 in (45 cm).
- **DIET** Pond pellets will be eaten readily.
- **WATER** Temperature 32–93°F (0–34°C); hard (100–150 mg/l) and neutral to alkaline (pH 7.0–7.5).
- **TEMPERAMENT** Social, and peaceful with other species.

These cyprinids are active shoalers that should be kept in groups. They are often seen patrolling the upper reaches of the pond. Rudd sometimes nibble aquatic plants, but they prefer to feed on invertebrates at the surface, often darting out from beneath water lilies to snatch insects. The males develop swellings on the head when entering breeding condition. Females can lay more than 100,000 eggs in batches during spring and summer. Hatching may take up to two weeks.

Gold Rudd This is a domesticated variant with a golden hue to its body. This coloration is especially evident on the head and back.

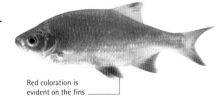

Red coloration is evident on the fins

Silver Rudd This is the natural color form, with a silvery sheen. Rudd can be distinguished from Roach (*Ratilius ratilius*) by the pelvic fins, which are located in front of the dorsal fin, rather than level with it, as in Roach.

DIRECTORY OF
POND PLANTS

MARGINAL PLANTS

These plants are more decorative than functional. However, when planted in containers on the marginal shelf—the ledge around the inside of preformed ponds, about 12 in (30 cm) below the surface—they can provide retreats for young fish. Marginals can also be grown as edging plants, giving the pond a more informal look and creating a barrier that makes it more difficult for predators to reach the fish. Some marginals trail down into the water, which helps to hide the perimeter from view. All the plants featured here are suitable for temperate climates, but some will benefit from protection in winter.

A judicious selection of marginals adds floral interest to the pond throughout much of the year. The **White Skunk Cabbage** *(Lysichiton camtschatcensis)* is shown here.

Rheum palmatum

Chinese Rhubarb

- **ORIGINS** Asia, where its distribution is restricted to Tibet and parts of China.
- **SIZE** Can grow up to 8 ft (2.5 m) tall.
- **WATER** Grows best in moist soil around the perimeter of the pond, rather than in water. Hardy to −20°F (−29°C).
- **PROPAGATION** Easily accomplished by division of the rhizome, although it can also be grown from seed.

This spectacular poolside plant grows well in both partial shade and full sun. In spite of its height, it does not suffer badly from wind damage, so it can be useful in exposed positions. The large leaves are supported on strong stems. The most commonly available cultivars have reddish flowers, while the flower spikes of the wild form are white.

Sagittaria sagittifolia

Arrowhead

- **ORIGINS** North America, where it is widely distributed throughout much of the United States.
- **SIZE** Attains a height of 3 ft (1 m).
- **WATER** Plant in damp soil around the edge of the pond, or on the marginal shelf. Hardy to −10°F (−23°C).
- **PROPAGATION** Easily accomplished by the division of established plants.

The green leaves of this hardy marginal are held vertically and shaped like arrowheads. White flowers on spikes are produced throughout the summer. If you wish to restrict the spread of this fast-growing species, plant it in a container from the outset. It forms small tubers that resemble potatoes, and new plants can be grown from these.

Myosotis scorpioides

Water Forget-Me-Not

- **ORIGINS** Europe and Asia, but it has now become naturalized in parts of North America.
- **SIZE** Typically reaches 6–12 in (15–30 cm) in height.
- **WATER** Fares best in shallow water, rather than being submerged. Hardy to −20°F (−29°C).
- **PROPAGATION** Can be grown from seed or by the division of existing plants.

The Water Forget-Me-Not is easy to grow and can help to create a very informal edging around a pond, growing both around the margins and also in shallow water. A number of different cultivars are now established, of which 'Mermaid' is probably the most free-flowering, while 'Semperflorens' has a more compact growth pattern. The small, pale-blue flowers have tiny yellow eyes at the center, although in 'Alba' the blue coloration is replaced by white.

Oblong leaves, which have a slightly hairy appearance

Carex elata 'Aurea'

Bowles' Golden Sedge

- 🌐 **ORIGINS** The native form is widely distributed throughout temperate regions of the world.
- 🌱 **SIZE** Reaches about 24 in (60 cm) in height.
- 💧 **WATER** Can be grown in shallow water, or alternatively in damp earth. Hardy to –20°F (–29°C).
- 🌼 **PROPAGATION** This can easily be accomplished by the division of existing plants in the spring.

Clumps of this golden-yellow sedge look their best when set alongside contrasting red- or green-leaved marginals. This plant is not particularly invasive, with individual clumps rarely exceeding 18 in (45 cm) in diameter. The brownish flower spikes are hard to see among the arching foliage. At the end of the growing season, cut back old foliage to ensure renewed growth in the spring.

Eriophorum angustifolium

Cotton Grass

- 🌐 **ORIGINS** Widely distributed in temperate regions of the northern hemisphere.
- 🌱 **SIZE** Grows to a height of about 12 in (30 cm).
- 💧 **WATER** Set on the marginal shelf 2 in (5 cm) below the waterline, or in damp soil. Hardy to –30°F (–34°C).
- 🌼 **PROPAGATION** Usually by the division of the rootstock, although it can also be grown from seed.

This member of the sedge family has distinctive white flowers resembling cotton swabs that stand on tall stems above its grasslike leaves. Despite its rather delicate appearance, Cotton Grass grows and spreads rapidly under favorable conditions. Preferring a position in full sun and acidic, peaty soil, this marginal is often found in moorland areas. Cotton Grass spreads underground by its rootstock, so it is likely to overrun the margins of a pond if it is not properly containerized. Hardy and evergreen, Cotton Grass will blend very effectively into an informal planting plan around the perimeter of a pond, especially when it is interspersed with taller, more statuesque plants. A related species, known scientifically as *E. latifolium*, is also occasionally available. It has similar growing needs and can be distinguished from *E. angustifolium* by its wider leaves and the purplish-green coloration of its flowering spikes.

Cyperus involucratus

Umbrella Sedge

- 🌐 **ORIGINS** In eastern parts of Africa, as well as on the island of Madagascar, off Africa's southeast coast.
- 🌱 **SIZE** Reaches a height of up to 24 in (60 cm).
- 💧 **WATER** Typically grows at the perimeter of the pond, or in shallow water. Minimum temperature 40°F (5°C).
- 🌼 **PROPAGATION** By the division of existing plants in fall, although it can also be grown from seed.

Umbrella Sedge is not frost-hardy and requires winter protection in all but the mildest areas. It is best suited to being planted in a container that can be moved indoors before the first autumn frosts. Place the container in a trough of water in well-lit surroundings until late spring, when it will be safe to return the plant to the garden. In milder regions, Umbrella Sedge may survive outside if placed in deeper water than normal, which will prevent the base of the plant from becoming encased in ice. The flowers of Umbrella Sedge form on bracts during late summer. The arching leaves of this plant are an attractive light green when growing but turn brown as fall approaches, with the seed-heads assuming a similar color. For a hardy alternative, Sweet Galingale (*C. longus*) survives to 0°F (–18°C), but it will still need to be set in a container, because its sharp roots may otherwise damage the pond liner.

Zantedeschia aethiopica

African Lily

- ⊛ **ORIGINS** This plant is native to southern and eastern parts of Africa.
- ⊛ **SIZE** May reach up to 3 ft (90 cm) in height.
- ⊛ **WATER** Set in containers on the marginal shelf down to 8 in (20 cm), or in damp ground. Hardy to –10°F (–12°C).
- ⊛ **PROPAGATION** This can be carried out by dividing clumps in the spring and also by taking offsets.

This arum lily's dark green leaves emphasize its large white flowers, with their golden central spadix. The blooms, which may last for a month, are replaced by yellow berries. The plant is most likely to survive outside over winter if it is kept in deep water, which will protect the base from ice. 'Crowborough' is the hardiest cultivar available.

Acorus calamus

Sweet Flag

- ⊛ **ORIGINS** Widely distributed throughout temperate parts of the northern hemisphere.
- ⊛ **SIZE** Can grow to a height of 3 ft (1 m).
- ⊛ **WATER** May be planted in damp ground, or 4 in (10 cm) below the water's surface. Hardy to –30°F (–34°C).
- ⊛ **PROPAGATION** This can be achieved easily by dividing clumps in the spring.

There are two forms of this rush, both of which have swordlike leaves resembling those of an iris. The wild green form grows more vigorously than the smaller variegated form, whose foliage is striped with creamy white. Sweet Flag grows well both in sunny conditions and in partial shade. Plant it in clumps for maximum impact. The flowers are fairly inconspicuous. Regular divisions of mature plants will encourage good growth.

Calla palustris

Bog Arum

- ⊛ **ORIGINS** Temperate regions of northern Europe and Asia, and also North America.
- ⊛ **SIZE** Can grow to a height of 8 in (20 cm).
- ⊛ **WATER** Can be grown on the marginal shelf down to a depth of 4 in (10 cm). Hardy to –30°F (–34°C).
- ⊛ **PROPAGATION** Easily accomplished by the division of the rhizomes in the spring.

Although bearing a superficial resemblance to the African Lily *(see left)*, Bog Arum can be identified by its smaller size and the lighter green coloration of its leaves. The white, flattened flowers may be fertilized by water snails rather than by insects. The reddish berries produced after flowering will maintain the plant's attractive appearance.

Lysichiton camtschatcensis

White Skunk Cabbage

- ⊛ **ORIGINS** Eastern Asia, on Russia's Kamchatka Peninsula, the Kuril Islands and Sakhalin, extending to northern Japan.
- ⊛ **SIZE** Can reach a size of 30 in (75 cm) in height and width.
- ⊛ **WATER** Plant in damp soil at the pond's edge, or in pots 2 in (5 cm) underwater. Fully hardy.
- ⊛ **PROPAGATION** Take offsets from established plants, or grow from seed.

The striking flowers of the White Skunk Cabbage are produced in early spring, thus helping to brighten up the pond at a time when other plants, such as water lilies, are barely stirring into growth. Measuring up to 16 in (40 cm) high, the flowers may sometimes, but not always, have a rather unpleasant odor, as the plant's name suggests. When White Skunk Cabbage is in bloom, the large, deep green leaves, which are heavily veined like those of a cabbage, will emerge and start to unfurl. This plant fares best when set in relatively rich soil at the side of the pond, although it can also be grown in a container on the marginal shelf. It takes time for White Skunk Cabbage to become fully established, but it grows more rapidly from divisions than from seed. When growing from seed, sow from spring to summer, standing the seed pot in a tub of water to keep the seeds damp at all times.

Iris ensata

Japanese Clematis-Flowered Iris

- **ORIGINS** Asia, where it can still be found growing wild in parts of Japan.
- **SIZE** Reaches a height of about 3 ft (1 m).
- **WATER** Grows better around the pond's edges, rather than in the water itself. Hardy to −20°F (−29°C).
- **PROPAGATION** Can only be increased reliably by dividing existing clumps.

This particular iris is one of the most beautiful of all poolside flowers, thanks to its large petals, with their relatively broad, flattened shape. It is now available in a wide range of colors, from white to pink and lavender, to shades of purple or blue. Japanese Clematis-Flowered Irises look best when planted together in groups of the same color. For this reason, you should avoid buying a mixed batch of plants and opt only for named color varieties. These irises dislike being permanently waterlogged and certainly should be removed from the pond before winter if they are set in containers. They can then be planted in the ground but must never be allowed to dry out. Choose a sheltered, sunny position where they will not be affected by the wind.

Iris pseudacorus

Yellow Flag

- **ORIGINS** Extends from parts of Europe southward to North Africa, and eastward into Asia.
- **SIZE** Grows up to 5 ft (1.5 m) in height.
- **WATER** Plant either around the pond or in submerged pots on the marginal shelf. Hardy to −20°F (−29°C).
- **PROPAGATION** Most quickly accomplished by dividing rhizomes. Can also be grown from seed.

This hardy iris has green leaves and buttercup-yellow flowers with reddish markings at the top of the petals. It blooms fairly early in summer, and although the individual flowers last only for about a day, a number are produced in succession up the flower stem. Yellow Flag grows rapidly, and in ponds it is best to set the rhizomes in marginal pots to restrict its spread and protect the liner.

Butomus umbellatus

Flowering Rush

- **ORIGINS** Naturally found in Europe, Asia, and North Africa, but now also occurs in parts of North America.
- **SIZE** Grows up to 5 ft (1.5 m).
- **WATER** Will thrive in boggy ground, or in shallow water to a depth of 5 in (12.5 cm). Hardy to −20°F (−29°C).
- **PROPAGATION** Achieved by dividing established clumps of the plant.

The dark green leaves of this rush are ½ in (1.25 cm) wide and up to 3 ft (90 cm) long, with sharp edges that apparently deter cattle from eating them in the wild. From midsummer onward, the plant produces spikes of reddish-white flowers, which are displayed as clusters on stems above the leaves. Flowering Rush makes an attractive addition to the border of the pond. While it prefers a sunny spot, it tends not to thrive in very hot climates.

Aponogeton distachyos

Water Hawthorn

- **ORIGINS** Originally from southern Africa, but now occurs in Europe, Australia, and South America.
- **SIZE** Leaves may reach 8 in (20 cm) long.
- **WATER** Extremely adaptable, thriving in water up to 2 ft (60 cm) deep. Hardy to 20°F (−7°C).
- **PROPAGATION** Divide the rhizome. Can also be grown from seeds, which it produces readily.

Water Hawthorn will spread across the surface of the pond, readily producing white blooms that have dark centers and a strong smell of vanilla. This marginal grows best in temperate regions, flowering first in the spring and then again in the early fall. Water Hawthorn survives best if set at a depth at which the tubers will not freeze.

Typha latifolia

Common Cattail

- **ORIGINS** Widely distributed throughout much of Europe and North America.
- **SIZE** Can reach 6 ft (2 m) in height.
- **WATER** Thrives in shallow water, down to a depth of about 12 in (30 cm). Hardy to −40°F (−40°C).
- **PROPAGATION** Easily achieved by splitting clumps of established plants.

The cattail's distinctive female flower is soft, dark brown, and measures up to 9 in (22.5 cm) long. The paler male flower is carried above this on the same sturdy stem. The flowers may be cut and dried for use as winter decorations. The cattail can be invasive unless its spread is curtailed by a suitable container. It grows rapidly and can contribute to the silting up of larger ponds, since mud becomes trapped in its dense root network.

Pontederia cordata

Pickerel Weed

- **ORIGINS** This plant occurs naturally in eastern parts of North America.
- **SIZE** May grow to 3 ft (1 m) tall.
- **WATER** Grows well on the marginal shelf, to a depth of 5 in (12.5 cm). Hardy to -40°F (-40°C).
- **PROPAGATION** Can be rooted easily by breaking off pieces from existing rootstock.

This plant has broad, tapering green leaves and upright flowers. Although Pickerel Weed tends to be quite slow growing in spring, it has a long flowering period that lasts well into late summer, when other plants are finished blooming. Pickerel Weed prefers a position in full sun, either planted in a damp spot, such as a bog garden adjacent to a pond, or permanently submerged in the pond itself. In addition to *P. cordata*, some other forms of Pickerel Weed are available. Those found in more southerly parts of the US—such as the White-Flowered Pickerel Weed (*P. alba*), which is naturally restricted to northern Florida—are unlikely to be as hardy as *P. cordata*. The roots of these marginals should not be allowed to become frozen in ice if they are to continue growing again the following year. Pickerel Weed can spread rapidly, and since it looks best in clumps, set the roots in relatively large planting baskets at the outset.

Erythranthe guttatus

Monkey Flower

- **ORIGINS** Occurs naturally in western parts of Canada and the US.
- **SIZE** Can reach up to 3 ft (1 m) in height.
- **WATER** Grows underwater in the winter, and then puts up sprouts during the spring. Hardy to -10°F (-23°C).
- **PROPAGATION** Can be grown quite easily from seed, as well as from cuttings.

This is one of the hardiest of the *Erythranthe* species, which grow wild in the Americas as far south as Chile. Wide hybridization of the Monkey Flower has produced many different varieties. Appreciated for its vivid flowers, it is usually cultivated as an annual, even though some varieties can overwinter. Monkey Flower seeds should be sown in the spring.

Thalia dealbata

Water Canna

- **ORIGINS** Occurs in southern parts of the US and across the border into Mexico.
- **SIZE** Can reach up to 6 ft (2 m) in height.
- **WATER** Needs to be planted quite deep in the pond, down as far as 18 in (45 cm). Hardy to 0°F (-18°C).
- **PROPAGATION** This is easily achieved by the division of the rootstock.

Hardier than its distribution might suggest, this Water Canna's angular leaves unfold off a central stem, which in late summer may be crowned with a purple flower spike. This plant does well in the deeper areas of a pond, especially in temperate zones, since this provides the rootstock with a barrier against ice. For added protection in winter, the stems can be wrapped in burlap.

Lysimachia nummularia

Creeping Jenny

- **ORIGINS** Natural distribution in the wild is restricted to parts of central Europe.
- **SIZE** Can grow to a height of 6 in (15 cm).
- **WATER** Grows best around the edges of ponds, rather than in the water itself. Hardy to -30°F (-34°C).
- **PROPAGATION** Divide established plants or take cuttings. Can also be grown from seed.

This low-growing member of the primula family spreads by shoots, which can reach more than 12 in (30 cm) in length. With yellow, cup-shaped flowers and green leaves, Creeping Jenny provides excellent groundcover around the pond edge, even in shady areas. However, it grows vigorously and may even invade the lawn. The variety *L.n.* 'Aurea' is often a better choice, being less invasive and also more attractive, thanks to its golden leaves.

Geum rivale

Water Avens

- **ORIGINS** Found in temperate regions of Europe, Asia, and North America.
- **SIZE** Can grow to a height of 12 in (30 cm).
- **WATER** Grow in shallow water, or in damp soil around the edges of the pond. Hardy to -40°F (-40°C).
- **PROPAGATION** Take cuttings, which will root easily. Can also be grown from seed.

Water Avens is a good choice for the surroundings of a pond, where low groundcover is needed. Its green leaves are reminiscent of those of strawberries. A number of different cultivars are now established so that plants with flowers of varying colors are available, ranging from the natural purplish-pink to shades of orange and yellow. Water Avens is a hardy perennial that will regrow well in spring after being cut back in fall.

Each stem produces more than one flower bud

Gunnera manicata

Gunnera

- **ORIGINS** South America, where it grows along waterways in Colombia and Brazil.
- **SIZE** Stems can grow up to 6 ft (2 m) tall.
- **WATER** Grow beside the pond, rather than in the water itself. Hardy to −0°F (−18°C).
- **PROPAGATION** Can be carried out by the division of existing plants. May also be grown from seed.

With leaves up to 6 ft (2 m) wide and greenish flower spikes standing 3 ft (1 m) tall, this marginal needs very spacious surroundings. Gunnera, also known as Giant Rhubarb, grows rapidly in a sunny yet sheltered location with rich, damp soil. When it dies back in winter, use the leaves to mulch the crown, which will help to prevent frost damage.

Ranunculus lingua 'Grandiflorus'

Giant Water Buttercup

- **ORIGINS** Found in temperate regions of Europe, Asia, and North America.
- **SIZE** Can grow up to 30 in (75 cm) tall.
- **WATER** Plant in the pond itself, down to 5 in (12.5 cm), or in marshy surroundings. Hardy to −20°F (−29°C).
- **PROPAGATION** Easily propagated by the division of its tuberous rootstock. Can also be grown from seed.

This is the most spectacular of all the buttercups, thanks to its large size and the vivid yellow flowers it produces in spring. Like other members of its family, the Giant Water Buttercup can become rampant if its growth is unchecked, so it should be set in marginal containers from the outset. The leaf shape is variable, being broader and longer on stems that do not form flowering shoots.

Caltha palustris

Marsh Marigold

- **ORIGINS** Widely distributed throughout much of North America, Europe, and Asia.
- **SIZE** Can grow 12 in (30 cm) or more in height.
- **WATER** Plant in damp ground around the edge of the pond, or in shallow water. Hardy to −40°F (−40°C).
- **PROPAGATION** This can be achieved by the division of existing plants, or by growing from seed.

Flowering both readily and early in the year, Marsh Marigolds look best when planted in small groups. Several cultivars now exist, including the double-bloomed 'Flore Plena,' which may flower again in early autumn, and a less vigorous Himalayan white form, *C. p.* var. *alba*. The leaves of Marsh Marigolds may suffer from mildew later in the year.

Asplenium scolopendrium

Hart's-Tongue Fern

- **ORIGINS** This species occurs naturally in temperate parts of Europe.
- **SIZE** Individual fronds can be up to 16 in (40 cm) long.
- **WATER** Grow in moist soil around the water's edge. Never submerge this plant. Hardy to −10°F (−23°C).
- **PROPAGATION** Divide plants or take leaf cuttings. Propagates naturally through the dispersion of spores.

The attractive shape of ferns makes them a popular choice for the surroundings of a pond, and numerous species grow well in such locations. Hart's-Tongue Fern is one of the hardy ferns, ideal for temperate areas. It has an upright growth habit, especially in the case of new fronds, which are pale green but become darker as they mature. The leaves are wavy and up to 2 in (5 cm) wide. Although ferns will not thrive if kept saturated, they do require a relatively high level of humidity. Plant them in a shady spot where they can draw moisture, such as in a crack in the rockwork around the pond, or even adjacent to a waterfall, where they will benefit from the water spray even in dry weather. Kept in favorable surroundings, they will soon start to reproduce. Small offspring may develop in tiny crevices and grow successfully if they are not allowed to dry out. Ferns generally prefer a shady location, out of direct sunlight.

OXYGENATORS

These plants, which grow beneath the water's surface, play a crucial role in creating a healthy environment for pond fish, because they release oxygen into the water as a by-product of photosynthesis. They also help to maintain water clarity by competing for dissolved nutrients with particulate algae (which are responsible for the green hue of pond water). Some species also produce highly attractive flowers. Oxygenators can, however, become rampant, and it may be necessary to remove clumps to ensure that the fish have adequate swimming space.

Bubbles of oxygen stream from the leaves of **Canadian Pondweed** (*Elodea canadensis*). Bubbles produced by oxygenators can be seen rising to the surface of ponds on sunny days when the water is calm.

Ranunculus aquatilis

Water Crowfoot

- **ORIGINS** Occurs naturally in parts of North America and Europe.
- **SIZE** Grows in clumps up to 3 ft (1 m) in diameter.
- **WATER** Grows well in both flowing and still water. Hardy to –20°F (–29°C).
- **PROPAGATION** Take stem cuttings during the growing season, or sow seeds in late summer.

This member of the buttercup family has two leaf forms: finely segmented leaves that grow underwater, and broader ones that float on the surface. Its flowers, which are white with bright buttercup-yellow centers, are often held above the water. As with other oxygenators, cuttings of Water Crowfoot can be rooted in containers set on the pond floor. Start them off on the marginal shelf (*see p. 364*) and then move them to deeper water.

Myriophyllum spicatum

Eurasian Water Milfoil

- **ORIGINS** Grows widely in parts of Europe, Asia, and North Africa; a similar species exists in the US.
- **SIZE** Strands may reach 10 ft (3 m) in length.
- **WATER** Plant up to 3 ft (1 m) deep, in brackish and fresh water. Hardy to –30°F (–34°C).
- **PROPAGATION** Take stem cuttings from established plants in spring or summer.

Eurasian Water Milfoil spreads rapidly, forming dense thickets that look attractive in shallow water. Its delicate whorls are usually green but sometimes have a reddish hue. The plant may produce small yellowish-white flowers during summer. Eurasian Water Milfoil is an invasive plant that can clog rivers and lakes with dense mats of vegetation, and it should never be released into natural waterways.

Elodea canadensis

Canadian Pondweed

- **ORIGINS** Naturally occurs in North America, but now established in Europe.
- **SIZE** Strands can easily grow to 12 in (30 cm) or more.
- **WATER** Thrives in clear water in a sunny position. Hardy to –20°F (–29°C).
- **PROPAGATION** Break off pieces about 6 in (15 cm) from the growing tip. Does not need to be planted.

The relatively small, dark green leaves help to distinguish Canadian Pondweed from similar species. Pondweed grows readily, especially during the warmer months of the year, and is sufficiently hardy to survive the winter outdoors in temperate areas. Pondweed is sold as sprigs that simply need to be attached to a weight so that they sink to the bottom. The sprigs will soon start to grow and provide a valuable refuge for young fry.

Utricularia vulgaris

Common Bladderwort

- **ORIGINS** Widely distributed in temperate parts of Europe, Asia, and North America.
- **SIZE** Stems reach 6 to 18 in (15–45 cm) in length.
- **WATER** Prefers relatively calm water in a sunny position. Hardy to –30°F (–34°C).
- **PROPAGATION** Remove young plantlets from an established plant during the growing season.

This slow-growing, rootless carnivorous plant has bladderlike structures among its foliage; as well as providing buoyancy, they also trap tiny aquatic creatures, including newly hatched fry. In summer, it produces a cluster of yellow flowers held above the water on a strong stem. Bladderwort may become choked by blanketweed *(see p. 319)*.

Ceratophyllum demersum

Hornwort

- **ORIGINS** May have originated in Asia, but now occurs throughout temperate regions of the world.
- **SIZE** Stalks may reach up to 24 in (60 cm) in length.
- **WATER** Not fussy about water chemistry; grows well in both sun and shade. Hardy to –10°F (23°C).
- **PROPAGATION** Break up the stems of established plants during the growing season.

The unusual name of this plant originates from the distinctive broad shape of its growing tip, which is reminiscent of a cow's horn. Hornworts do not root, but if in contact with a substrate, the leaves will start to anchor the plant in place. Over the course of the growing period, hornwort forms long strands. As the leaves start to die back, the budlike tips of the strands drop off (or can be cut off) and sink to the bottom of the pond, and it is from these buds that new plants will develop the following spring. By the end of the growing season, Hornwort becomes very straggly, so it is best to pull out the plants at this stage. Hornwort is strictly aquatic and dies back if exposed to the air for any length of time. It is also fragile, and breaks easily when handled.

Fontinalis antipyretica

Willow Moss

- **ORIGINS** Widely distributed in parts of Europe, Asia, North Africa, and North America.
- **SIZE** Stems can grow to a length of 20 in (50 cm).
- **WATER** Prefers clear water, but tolerates either sun or shade. Hardy to –30°F (–34°C).
- **PROPAGATION** Break off branches from established plants and attach them to submerged objects.

Although this hardy moss does not flower, it has an attractive appearance. It fares best in ponds free of filamentous algae and is particularly suited to areas around waterfalls, since it naturally occurs in fast-flowing streams. Willow Moss will attach itself by its roots to submerged objects, such as planting containers and rocks. Hold pieces in place with a rubber band until the roots get a firm grip.

Leaves vary in color from green to olive-brown

Hottonia palustris

Water Violet

- **ORIGINS** Found naturally in the wild throughout much of Europe.
- **SIZE** Can grow to a height of more than 3 ft (1 m).
- **WATER** Thrives best under acidic water conditions with a pH of 6.0–6.5. Hardy to –20°F (–29°C).
- **PROPAGATION** Divide clumps during the growing season, or take cuttings.

Despite its name, this plant is not related to the violet but actually belongs to the primrose family. The large surface area of Water Violet's fine foliage makes it a valuable oxygenator. During summer, plants develop flower spikes that stand more than 12 in (30 cm) above the water's surface. The leaves on the flower spikes are more compact than the fine, feathery foliage that Water Violet displays on its submerged parts. The flower color itself can be quite variable, ranging from white through pinkish-lilac to blue. As the flowers fade, the flower stems falls back into the water, and the seedheads develop. Water Violet dies back naturally in the fall, when the plants form so-called winter buds, or turions, from which new plants will grow again the following spring. Although Water Violet is hardy, it tends to thrive only in clear water, and it will be adversely affected by any buildup of filamentous algae in the pond.

FLOATING PLANTS

These plants are not renowned for their flowers, but they give the pond a more natural feel and are important for the well-being of the fish. They spread to form a dense mass, protecting the fish from predators, especially birds, and also from sunburn, in addition to curbing algal growth. Furthermore, they provide spawning sites and food for some species of fish. Floating plants are easy to establish—simply let them drift on the surface until they find a suitable position—and they develop much faster in a new pond than other types of plants, such as marginals and water lilies. Some popular varieties originate from warm climates and are not hardy in temperate areas. They should be brought inside to protect them from winter frosts.

Water Chestnut
(Trapa natans) now occurs far beyond its native habitat and may create environmental problems in these new localities.

Azolla cristata

Carolina Fairy Moss

- ⚙ **ORIGINS** Occurs naturally from the US to South America, but now naturalized in parts of Europe.
- 🌿 **SIZE** Leaves each measure about ¹/₂ in (1.5 cm) long.
- 💧 **WATER** This plant needs clear water in order to thrive. Hardy to 0°F (−18°C).
- 🌱 **PROPAGATION** Reproduces asexually, so simply divide up a clump, preferably in spring.

This floating green fern spreads rapidly over the pond's surface, so its growth may have to be kept in check. Fairy Moss becomes more reddish during summer. It dies back in fall and sinks to the bottom but resurfaces again in spring. In temperate areas, overwinter some of the fern indoors in an aquarium, or in a plastic container of water on a windowsill; otherwise, the entire stock may be destroyed by very cold weather.

Trapa natans

Water Chestnut

- ⚙ **ORIGINS** A native European species, it is now naturalized in parts of the US and Australia.
- 🌿 **SIZE** Rosettes can reach up to 30 in (75 cm) in diameter.
- 💧 **WATER** Prefers still or slow-moving water. Min. temp. 32°F (0°C). This annual plant dies off before winter.
- 🌱 **PROPAGATION** Grown easily from its chestnutlike seeds, which can be set in pots of aquatic soil.

This annual plant has serrated edges on its green, purple-centered leaves, which grow in the form of a rosette. The inconspicuous white flowers are followed by large black fruits, which can be left to overwinter in the pond. Otherwise, store them indoors in damp sphagnum moss. They must not dry out if they are to germinate in the spring.

Eichhornia crassipes

Common Water Hyacinth

- ⚙ **ORIGINS** South America, but now naturalized in many other areas, including Africa. Banned in some places.
- 🌿 **SIZE** Leaves 4 in (10 cm) long; flower spikes 6 in (15 cm).
- 💧 **WATER** Prefers calm water, so keep away from fountains. Hardy to 32°F (0°C).
- 🌱 **PROPAGATION** Split small plantlets off the sides of existing clumps.

This plant gets its name from its mauve blooms, which resemble hyacinth flowers. Air trapped in the leaf bases provides buoyancy and enables the plants to float. Hidden beneath the surface are long, trailing roots that provide spawning sites for goldfish and protection for fry. Common Water Hyacinth spreads rapidly in warm climates and should never be released into natural waterways, where it can cause serious environmental and economic damage.

Lemna species

Duckweed

- **ORIGINS** Found throughout the world, in both temperate and tropical regions outside polar areas.
- **SIZE** Tiny leaves measure about ⅓ in (0.8 cm) across.
- **WATER** Grows well in any depth of water, but prefers relatively little movement. Hardy to –30°F (–34°C).
- **PROPAGATION** Simply split off a few pieces from a mat, and these will soon start to replicate.

Duckweed is often accidentally introduced into a pond with other plants. Once present, it spreads rapidly, providing cover for fish and even food for some species. Scoop it out with a net if it threatens to form a suffocating mat over the entire pond. You can restrict its spread by using a fountain to create surface movement. Duckweed does not compete well with other floating plants or lilies.

Stratiotes aloides

Water Soldier

- **ORIGINS** Occurs naturally in parts of Europe and as far east as northwest Asia.
- **SIZE** Can reach up to 20 in (50 cm) in height.
- **WATER** Prefers hard water with little or no movement, and a sunny location. Hardy to –20°F (29°C).
- **PROPAGATION** Detach winter buds, or remove young plantlets in spring.

This plant is seen only on the surface in summer, when it produces white flowers on short stems. Distinct male and female plants do occur, but Water Soldier usually reproduces by division, rather than by seeding. The calcium carbonate it absorbs from its hard-water surroundings causes it to sink to the bottom in winter, where it throws out side shoots called turions. These produce new plants in spring.

Pistia stratiotes

Water Lettuce

- **ORIGINS** Originally from Florida and the Gulf Coast of the US; now present in warmer areas worldwide.
- **SIZE** Can reach 6–12 in (15–30 cm) in height.
- **WATER** Prefers still water and a sunny location. Minimum temperature 50°F (10°C).
- **PROPAGATION** Separate young plantlets from large plants. Water lettuce may occasionally set seed.

The green, velvety leaves of Water Lettuce grow in the form of a rosette above the water's surface, while its long, feathery roots, measuring up to 18 in (45 cm) long, provide valuable breeding sites and cover for fish. The small, whitish flowers are easily overlooked. Water Lettuce is sensitive to cold temperatures and so must be brought inside if it is to survive the winter in temperate areas.

Orontium aquaticum

Golden Club

- **ORIGINS** Occurs naturally in eastern parts of Canada and the US.
- **SIZE** Spread is 24–30 in (60–75 cm).
- **WATER** Start off in a shallow, sunny position; can later be moved to deeper water. Hardy to –10°F (–23°C).
- **PROPAGATION** Divide mature plants in spring or fall, or sow seeds during summer.

The versatile Golden Club can be cultivated either as a floating plant or as a marginal in shallow water around the edge of the pond, where it will look particularly attractive against waterside irises and primulas in early summer. The appearance of Golden Club varies accordingly, with the lance-like leaves measuring about 18 in (45 cm) in the shallows but rarely exceeding 12 in (30 cm) when floating in deeper water. The leaves are mid-green above and often purplish beneath. Golden Club blooms from late spring to midsummer, producing unusual blooms that are white at the base and yellow toward their tips. Golden Club is hardy in temperate areas and capable of forming large clumps. Plant the rhizomes of Golden Club in deep containers, since this species develops a large root system. Do not allow plants to root by themselves, because it is difficult to move clumps successfully once they have become established.

WATER LILIES AND LOTUSES

Water lilies are among the most attractive of all pond plants and relatively easy to keep. They help to maintain the water quality in the pond, because their roots absorb nitrates produced by the decomposition of fish feces. A mat of lily leaves on the pond's surface will reduce the amount of sunlight entering the water, protecting the fish from sunburn and restricting algal growth. It will also enable the fish to dart out of sight when a predator's shadow darkens the pond. Do not plant tropical lilies in water that is below 75°F (24°C); otherwise, they may not grow but simply remain dormant or even rot.

Like many of today's water lilies, **Red Laydeker** (*Nymphaea* x *laydekeri* 'Fulgens') is hardy, since its ancestry includes *Nymphaea alba* var. *rubra*, which grows wild in Sweden. Tropical varieties are suitable only for indoor ponds in cold climates.

Nymphaea 'Pearl of the Pool'

Pearl of the Pool

- **ORIGINS** Bred from *N.* 'Pink Opal' crossed with *N. marliacea* 'Rosea'.
- **SIZE** Leaves may spread to 5 ft (1.5 m).
- **WATER** Not fussy about water chemistry. Hardy to −30°F (−34°C).
- **PROPAGATION** Divide the rhizome. Plants can be placed down to a depth of 30 in (75 cm).

This water lily, which was created in the US, became the first hardy cultivar to be patented in 1946. It remains popular today, since it flowers quite freely once established, especially when planted in a large container. The pink blooms are cup-shaped when they open but become stellate (starlike) as they mature. They are also fragrant, adding to the plant's appeal. The maximum leaf diameter is about 10 in (25 cm).

Nymphaea pygmaea 'Helvola'

Helvola

- **ORIGINS** Thought to be the result of crossings involving *N. tetragona* and *N. mexicana*.
- **SIZE** Leaves may spread to about 3 ft (1 m).
- **WATER** Not fussy about water chemistry. Hardy to −30°F (−34°C).
- **PROPAGATION** Divide the rhizome. May also set seed on occasions.

This water lily's dainty yellow flowers, which measure no more than 2 in (5 cm) in diameter, are produced in profusion. 'Helvola' tends to flower later in the season than its larger relatives. In addition, the blooms open and close later in the day than those of other water lilies. The small olive leaves are another attractive feature, being heavily blotched with purple on their upper surface.

Nymphaea 'Aurora'

Aurora

- **ORIGINS** Believed to have been created by crossings of *N. alba* var. *rubra* with *N. mexicana*.
- **SIZE** Leaves may spread to about 3 ft (1 m).
- **WATER** Not fussy about water chemistry. Hardy to −30°F (−34°C).
- **PROPAGATION** Divide the rhizome at the start of the growing period.

One of the so-called Marliac hybrids, the free-flowering 'Aurora' was created at the end of the 19th century in France by Joseph Latour-Marliac. The blooms, which typically last for three days, undergo a dramatic change in color. They are yellowish-apricot at first, darkening to orange-red on the second day, and finally appearing burgundy-red before dying off. The leaves are green on top, with purple undersides.

Nymphaea 'Black Princess'

Black Princess

- **ORIGINS** Resulted from cross-breeding between a red hardy water lily and a blue tropical variety.
- **SIZE** Leaves spread to about 4 ft (1.2 m).
- **WATER** Not fussy about water chemistry. Hardy to –30°F (–34°C).
- **PROPAGATION** Divide the rhizome at the start of the growing period.

With its highly distinctive deep-red double flowers, which centrally appear blackish and measure about 5.5 in (14 cm) in diameter, this free-flowering water lily has become very popular, since being created in the US by Perry Slocum in about 1998. It is the darkest water lily currently available, with its nearest rival being *Nymphaea* 'Almost Black,' and has proved to be very hardy and vigorous, in spite of its part-tropical ancestry.

Nymphaea 'Gonnère'

Gonnère

- **ORIGINS** Bred from *N. tuberosa* 'Richardsonii' crossed with another water lily of unknown origins.
- **SIZE** Leaves may spread to about 4 ft (1.2 m).
- **WATER** Not fussy about water chemistry. Hardy to –30°F (–34°C).
- **PROPAGATION** Divide the rhizome at the start of the growing period.

This variety is often available under the name of 'Snowball.' It has ball-shaped white flowers with upward-curving petals and dark green leaves that measure up to 10 in (25 cm) across. Although it grows vigorously, it has a short flowering season. Like all water lilies, it may take a year or two to become established.

Nymphaea 'Froebeli'

Froebeli

- **ORIGINS** This cultivar was developed from a seedling of *N. alba* var. *rubra*.
- **SIZE** Leaves may spread to about 3 ft (1 m).
- **WATER** Not fussy about water chemistry. Hardy to –30°F (–34°C).
- **PROPAGATION** Divide the rhizome at the start of the growing period.

This late-flowering variety is an example of the dedication that can be involved in breeding water lilies. It was created in the 19th century over a period of 40 years by its creator, Otto Froebeli of Zurich, who started from a single seedling. Its petals should be burgundy-red, offset against orange-red stamens with yellow anthers. The squarish blooms are about 4 in (10 cm) across. The leaves are bronze when they first unfurl and then turn green. They are relatively small, rarely exceeding 6 in (15 cm) in width. 'Froebeli' flowers freely, creating an impressive display, and it is especially suitable for growing in colder areas, although it prefers a sunny site. This cultivar can even be grown successfully in a patio tub, where fish may be housed temporarily over the summer.

Nymphaea 'René Gérard'

René Gérard

- **ORIGINS** Ancestry is unclear; this variety was produced by Joseph Latour-Marliac's breeding program.
- **SIZE** Leaves may spread to about 5 ft (1.5 m).
- **WATER** Not fussy about water chemistry. Hardy to –30°F (–34°C).
- **PROPAGATION** Divide the rhizome at the start of the growing period.

Created at the Latour-Marliac nursery in France in 1914, 'René Gérard' has endeared itself to pond enthusiasts because of the ease with which it can be grown. Another factor underlying its popularity is its free-flowering nature. The star-shaped blooms are relatively large compared to the leaves, measuring up to 9 in (23 cm) in diameter. The lightly fragranced flowers are predominantly rose pink, with darker flecking especially evident on the outer petals. The central area of the flower is a crimson shade, while the stamens are yellow. Variable darker streaking extends into the pale pink areas, although this inconsistency in coloring does not have universal appeal. At the back of the green, almost circular leaves there is a deep, V-shaped indentation. Mature leaves typically reach up to about 11 in (28 cm) in diameter; new leaves display an attractive bronzy tone on their upper surface.

Nymphaea odorata var. *minor*

Odorata Minor

- **ORIGINS** Found in Newfoundland through eastern North America to the Caribbean.
- **SIZE** Leaves may spread to about 4 ft (1.2 m).
- **WATER** Not fussy about water chemistry. Hardy to –30°F (–34°C).
- **PROPAGATION** Divide the rhizome at the start of the growing period. May self-seed.

N. odorata var. minor is a stable cultivar whose fragrant, pure white petals contrast with the bright yellow stamens and green leaves. Flower size can be increased by planting it in special aquatic soil. In the wild, its flowers vary greatly in appearance across its extensive range, with some forms having much broader petals than others.

Nymphaea odorata 'Sulphurea Grandiflora'

Sunrise

- **ORIGINS** Created in France in 1888, probably from *N. odorata* var. *gigantea* x *N. mexicana*.
- **SIZE** Leaves may spread to about 5 ft (1.5 m).
- **WATER** Not fussy about water chemistry. Hardy to –30°F (–34°C).
- **PROPAGATION** Divide the rhizome at the start of the growing period.

The eye-catching yellow flowers of 'Sunrise' rank among the largest of all the hardy water lilies, up to 10 in (25 cm) in diameter. It grows well only in reasonably warm localities, however, and produces twisted leaves in the spring if the weather is cold. This plant was given the alternative name *Nymphaea* 'Sunrise' by a California supplier around 1930.

Nymphaea 'Rose Airey'

Rose Airey

- **ORIGINS** Created in the US in 1913, probably from *N. odorata* stock, but precise origins are unknown.
- **SIZE** Leaves may spread to about 5 ft (1.5 m).
- **WATER** Not fussy about water chemistry. Hardy to –30°F (–34°C).
- **PROPAGATION** Divide the rhizome at the start of the growing period. May also self-seed.

This strain was created by breeder Helen Fowler at Kenilworth Gardens, Washington, D.C., and named after her cousin. It requires a large basket, about 24 x 24 x 12 in (60 x 60 x 30 cm), to allow the rhizomes to multiply. It grows slowly, but the pink flowers are fragrant and beautifully proportioned. The green leaves are purple when they first unfurl.

Nymphaea 'Lucida'

Lucida

- **ORIGINS** From the breeding program of Joseph Latour-Marliac; its ancestry is unknown.
- **SIZE** Leaves may spread to about 5 ft (1.5 m).
- **WATER** Not fussy about water chemistry. Hardy to –30°F (–34°C).
- **PROPAGATION** Divide the rhizome at the start of the growing period.

As is the case with a number of other Marliac cultivars of uncertain origin, it is thought that 'Lucida' may have arisen simply from bee pollination during the course of Joseph Latour-Marliac's breeding program, rather than from deliberate crossings between plants. 'Lucida' is essentially a red variety, although the outer petals are more pinkish in color, with the stamens being deep yellow. The color of the individual flowers becomes more intense with age. The leaves, too, are attractive—their upper surfaces are green with prominent purple mottling. 'Lucida' flowers freely, but care needs to be taken because it is more susceptible than many other water lilies to the disease called crown rot, which is caused by *Phytophthora* fungus. (Make sure plants are not affected before buying.) Removing an affected water lily from the pond and treating it separately with a fungicide may resolve the problem.

Caroliniana Nivea

Nymphaea caroliniana 'Nivea'

- **ORIGINS** Created by Joseph Latour-Marliac in 1893, this cultivar includes *N. odorata* in its parentage.
- **SIZE** Leaves may spread to about 5 ft (1.5 m).
- **WATER** Not fussy about water chemistry. Hardy to −30°F (−34°C).
- **PROPAGATION** Divide the rhizome at the start of the growing period.

The popularity of this white cultivar has faded over recent years in the face of competition from other, more free-flowering varieties. Nevertheless, it produces large, very fragrant blooms, typically up to 6 in (15 cm) across, with the leaves being entirely green on both surfaces. Plant 'Caroliniana Nivea' so that the rhizomes have space to spread.

Madame Wilfron Gonnère

Nymphaea 'Madame Wilfron Gonnère'

- **ORIGINS** Created soon after 1912, but this cultivar's ancestry is unclear.
- **SIZE** Leaves may spread to about 4 ft (1.2 m).
- **WATER** Not fussy about water chemistry. Hardy to −30°F (−34°C).
- **PROPAGATION** Divide the rhizome at the start of the growing period.

This beautiful pink water lily is easy to grow and will flower for many years once established, producing double blooms resembling those of a peony. It should not be confused with the cultivar known as 'Gonnère' *(see p. 377)*, although it does have a similar flower shape. 'Madame Wilfron Gonnère' is yet another cultivar created at Joseph Latour-Marliac's nursery in Temple-sur-Lot, near Bordeaux, France, although it was not developed until after his death in 1911. The flowers of 'Madame Wilfron Gonnère' are rather ball-shaped when in bud, but they open rapidly and stay open until late in the afternoon. There is a slight color change as the flowers mature, with the petals becoming a darker shade of pink. Fully open, they measure about 5 in (12.5 cm) in diameter. The leaves, which can be double the flower size, are green, with early leaves often displaying a yellowish stripe in the spring.

Vesuve

Nymphaea 'Vesuve'

- **ORIGINS** Created by M. Latour-Marliac in 1906, but this cultivar's ancestry is unknown.
- **SIZE** Leaves may spread to about 4 ft (1.2 m).
- **WATER** Not fussy about water chemistry. Hardy to −30°F (−34°C).
- **PROPAGATION** Divide the rhizome at the start of the growing period.

'Vesuve' blooms over a longer period than most hardy water lilies. The stellate flowers, with their dark orange stamens, are predominantly red, becoming a deeper, fiery shade as they age. The concave petals are quite distinctive and often appear to be folded along their length. Because of its glowing color, this water lily was named after Mount Vesuvius, the Italian volcano.

Fulgens

Nymphaea x *laydekeri* 'Fulgens'

- **ORIGINS** Created by Joseph Latour-Marliac in 1895, but this cultivar's ancestry is unknown.
- **SIZE** Leaves may spread to about 5 ft (1.5 m).
- **WATER** Not fussy about water chemistry. Hardy to −30°F (−34°C).
- **PROPAGATION** Divide the rhizome at the start of the growing period.

The flowering period of 'Fulgens' begins in early spring and can continue into fall, with the cup-shaped, deep red blooms darkening as they mature. The stamens are fiery red, while the outer sepals around the bud are streaked with rose pink. The leaves are purplish-green at first, turning fully green as they age. This fast-growing hybrid establishes itself quickly, flowering profusely yet not choking the pond with its leaves. It can be included in a small pond but is ideal for large expanses of water, where a number planted close together create a spectacular display. Choose a sunny site, both to encourage early growth in the spring and to maximize the flowering period. The scientific name commemorates Joseph Latour-Marliac's son-in-law, Maurice Laydeker, who took over the running of the nursery after his father-in-law's death.

Nymphaea 'James Brydon'

James Brydon

- ✿ **ORIGINS** Ancestral species probably included *N. alba* var. rubra, *N. candida*, and a *N.* x *laydekeri* hybrid.
- ✿ **SIZE** Leaves may spread to about 4 ft (1.2 m).
- ✿ **WATER** Not fussy about water chemistry. Hardy to −30°F (−34°C).
- ✿ **PROPAGATION** Divide the rhizome at the start of the growing period.

Created by Dreer Nurseries in Philadelphia during the late 1890s, 'James Brydon' soon built up an international following, which it maintains to this day. This colorful cultivar is very resistant to disease, especially to fungal crown rot. It will flower even when sited in partial shade and can grow well in shallow water. The raised goblet shape of the flowers, which reach up to 6 in (15 cm) in diameter, is particularly striking. The flowers are naturally two-toned, with paler outer petals and a more reddish center, offset against yellow stamens. The fragrance of the flowers is also unusual, being said to resemble that of ripe apples. The leaves are an attractive purplish-brown when they first unfurl but gradually change to green. The leaf shape is decidedly rounded, with just a slit rather than a V-shaped area at the rear edge.

Nymphaea x *marliacea* 'Chromatella'

Chromatella

- ✿ **ORIGINS** This cultivar was bred in France from *N. alba* crossed with *N. mexicana*.
- ✿ **SIZE** Leaves may spread to about 3 ft (1 m).
- ✿ **WATER** Not fussy about water chemistry. Hardy to −30°F (−34°C).
- ✿ **PROPAGATION** Divide the rhizome at the start of the growing period.

'Chromatella' (also known as Golden Cup) is a proven variety of long standing. Its yellow blooms have golden centers and measure up to 6 in (15 cm) across. The green leaves with reddish-brown blotches are attractive in their own right. Chromatella grows well even in shady places and flowers throughout the season. Check the rootstock of this vigorous plant regularly, and divide it as required. The rhizomes bear an unmistakable resemblance to pineapples.

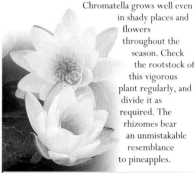

Nymphaea 'Pink Sensation'

Pink Sensation

- ✿ **ORIGINS** This cultivar was developed in the US as a variant of *N.* 'Lustrous'.
- ✿ **SIZE** Leaves may spread to about 4 ft (1.2 m).
- ✿ **WATER** Not fussy about water chemistry. Hardy to −30°F (−34°C).
- ✿ **PROPAGATION** Divide the rhizome at the start of the growing period.

The flowers of this mid-20th-century American introduction are up to 8 in (20 cm) across and have silver-tipped petals. They are soft pink and cup-shaped at first, but the pink grows stronger with age and the shape becomes more stellate. There is a slight scent to the flowers, which remain open late into the day. The leaves are dark green, with reddish undersides.

Nymphaea 'Blue Beauty'

Blue Beauty

- ✿ **ORIGINS** Crossings of *N. caerulea* and *N. capensis* var. *zanzibariensis* at the University of Pennsylvania in 1897.
- ✿ **SIZE** Leaves may spread to about 6 ft (2 m).
- ✿ **WATER** Not fussy about water chemistry. Minimum temperature 30°F (−1°C).
- ✿ **PROPAGATION** Divide the rhizome at the start of the growing period.

Although the blue coloration is not as vivid as that of its *N. capensis* var. parent, 'Blue Beauty' flowers readily and will grow in slightly shaded places. The large, fragrant daytime blooms, up to 12 in (30 cm) across, rise above the water surface on stems. The leaves, which may be twice as wide as the flowers, are green and brown on top.

Nymphaea 'William C. Uber'

William C. Uber

- ✿ **ORIGINS** Nothing has been documented regarding the origins of this water lily.
- ✿ **SIZE** Leaves may spread to about 6 ft (2 m).
- ✿ **WATER** Not fussy about water chemistry. Minimum temperature 30°F (−1°C).
- ✿ **PROPAGATION** Divide the rhizome at the start of the growing period.

Introduced in 1970 by Van Ness Water Gardens of California, this tropical day-blooming cultivar has grown in popularity, thanks to its striking fuschia-pink coloration and attractive scent. The leaves are green on both surfaces and measure up to 12 in (30 cm) in width, while the mature flowers can reach 9 in (22.5 cm) in diameter.

Nymphaea capensis

Cape Blue Water Lily

- ⊕ **ORIGINS** This popular species occurs naturally in the Cape of Good Hope area of South Africa.
- ⊕ **SIZE** Leaves may spread to about 8 ft (2.5 m).
- ⊕ **WATER** Not fussy about water chemistry. Minimum temperature 30°F (−1°C).
- ⊕ **PROPAGATION** Divide the rhizome at the start of the growing period. Can also be grown from seed.

This day-blooming species fares well even in relatively small conservatory ponds, producing a succession of stellate, light blue blooms raised on stems. The green leaves are round, with serrated edges. The Cape Blue's seeds can be germinated in small containers in a brightly lit aquarium with a water temperature of about 75°F (24°C).

Nymphaea 'White Delight'

White Delight

- ⊕ **ORIGINS** Introduced as recently as 1984, the origins of this day-blooming water lily are unknown.
- ⊕ **SIZE** Leaves may spread to about 6 ft (2 m).
- ⊕ **WATER** Not fussy about water chemistry. Minimum temperature 30°F (−1°C).
- ⊕ **PROPAGATION** Divide the rhizome at the start of the growing period.

Despite being called 'White Delight,' this water lily's dahlialike blooms are light yellow with a darker center. The stamens, which are also yellow, blend in with the inner petals. The flowers open during the day (some tropical species are night-flowering) and measure up to 12 in (30 cm) in diameter. They have a strong scent, which is best appreciated in a conservatory pond setting. New leaves are purple-mottled on top but gradually become all green. The leaves are relatively large and tend to be raised slightly above the water. 'White Delight' reproduces by rhizomes, but some other tropical species produce small plantlets at the center of their leaves, which split off to form new water lilies.

Nelumbo 'Mrs. Perry D. Slocum'

Mrs. Perry D. Slocum

- ⊕ **ORIGINS** Bred from a crossing carried out by Perry Slocum involving *N. lutea* and *N.* 'Rosea Plena.'
- ⊕ **SIZE** Grows to a height of about 5 ft (1.5 m).
- ⊕ **WATER** Not fussy about water chemistry. Min. temp. 30°F (−1°C); set 6 in (15 cm) deep to protect from frost.
- ⊕ **PROPAGATION** Divide the rhizome at the start of the growing period. Can also be grown from seed.

Lotuses will thrive only outdoors in areas where summer temperatures are 75–84°F (24–29°C). This particular cultivar is named after the wife of one of the most famous US breeders of water lilies and lotuses, Perry D. Slocum. Its aniseed-scented flowers measure up to 12 in (30 cm) across. On opening, they are pinkish with a yellow flush, but the following day they develop a more even pink-and-yellow coloration. On maturity, the flowers are again transformed, this time to cream with a slight pink suffusion. A well-established lotus may thus display several different-colored blooms at once. After the flowers die back, the seed capsules form, being yellow at first, then turning green. As well as being suitable for inclusion in ponds, lotuses are often grown on patios in half whiskey barrels. If fish are going to be included, put a pond liner in the barrel in case there are any residues that may be toxic. Plastic containers are a safer option.

Glossary

Acidic A reading below 7.0 on the pH scale.

Activated carbon A manufactured form of carbon that is highly porous.

Adipose fin A small fin between the dorsal and caudal fins, most notable in tetras and other characoids.

Adsorption The way that dissolved waste binds onto the surface of a filter medium, such as activated carbon, during chemical filtration.

Aeration Adding extra air into water to improve its oxygen content.

Aerobic bacteria Bacteria that require oxygen to survive.

Airstone A porous device that splits up the airflow from an air pump into small bubbles to improve water movement and oxygenation.

Algae Primitive aquatic plants that photosynthesize.

Alkaline A reading that is above 7.0 on the pH scale.

Anal fin An unpaired fin on the underside of the body, near the vent.

Annual A plant that grows, flowers, sets seed, and dies within a year.

Aquarium peat Peat that can be added to the aquarium filter to acidify the water. Unlike garden peat, it is free from additives.

Barbels Sensory growths around the mouths of various bottom-dwelling fish, including catfish and koi.

Biological filtration The use of aerobic bacteria to break down waste matter in aquariums and ponds.

Blackwater Water that has been acidified and darkened by tannin produced by decaying vegetation. Blackwater is used to encourage spawning in fish such as tetras.

Brackish water A mix of fresh and salt water found in estuaries.

Breeding brush An artificial spawning medium that is typically used in ponds. Fish spawn over the brush and their eggs attach to its bristles. The brush is then removed from the pond so that the eggs can be hatched elsewhere in safety.

Breeding trap A device used to separate a gravid female livebearer from other tank occupants. It also prevents her from eating her own fry.

Brood The offspring produced by a pair of fish; typically a group of fry that is guarded by one or both adults.

Brood pouch A pouch on the body of male sea horses and pipefish, in which the eggs are incubated.

Bubble nest A nest made by many anabantoids and some catfish to protect eggs and fry. It consists of air bubbles trapped in mucus and is usually anchored to vegetation.

Cartilage A tough, flexible body tissue. In some fish, the skeleton is made entirely of cartilage.

Caudal fin The tail fin, which is often divided into lobes.

Caudal peduncle The muscular shaft that links the body of a fish to the tail (caudal) fin.

Chemical filtration The use of chemicals to remove dissolved waste by adsorption.

Chromosome A gene-carrying structure found in the nucleus of every living cell. Genes determine the characteristics of all organisms.

Classification A method of grouping living things to show the relationships between them.

Community tank An aquarium that houses a number of different yet compatible species.

Compatibility The degree to which fish interact favorably with one another; also, the bonding of a pair of fish prior to spawning.

Conditioning Managing the fish and their surroundings to encourage breeding. Conditioning also refers to the way in which water is treated to make it safe for the tank occupants.

Coral sand Crushed coral, with a particle size similar to that of sand. Coral sand can be added to the filter to make the water more alkaline.

Crown The central area of a plant, from which new growth develops.

Cultivar A cultivated example of a plant that does not occur in the wild. The common name is often suffixed by "cv," meaning "cultivated variety."

Cutting Part of a plant that is removed and used for propagation.

Dechlorinator A chemical preparation that removes chlorine-based compounds from tap water.

Dorsal fin The unpaired fin that runs along the center of the back. Some fish have a divided dorsal fin.

Ectothermic Describes an animal whose internal body temperature varies according to its surroundings.

Egg-layer Any fish in which the eggs are fertilized and hatched outside the body.

Egg spots Egg-shaped markings on the anal fin of some male cichlids. As mouth-brooding females nibble at the egg spots, they take in sperm released by the male, ensuring that the eggs in their mouth become fertilized.

Environment A living thing's surroundings. For a fish, this includes the water, the substrate, and the plants and animals with which it interacts.

Evolution The origin of species by development from earlier forms.

Family A group of related genera.

Fancy Describes fish that have been bred to emphasize ornamental qualities, such as coloration or fin shape, that are not seen in wild forms.

Filter bed A layer of substrate through which water passes during the filtration process.

Filter feeder An invertebrate that feeds by sifting tiny food particles from water.

Filter medium Any material used in a filtration system to remove waste, or which can be colonized by beneficial aerobic bacteria.

Filtration The removal of waste from water in aquariums and ponds.

Fins The projections on a fish's body that enable it to move through water. They are also often used for display and even mating purposes.

Flake Manufactured fish food in the form of thin, waferlike fragments that float at the surface. Suitable for both freshwater and marine fish.

Flexible liner A sheet of butyl rubber or PVC that is used to form a watertight lining for a pond.

Flow adjuster A valve that regulates the movement of water through a pump.

Frost-hardy Describes a plant that is able to survive frost unprotected, but not extreme cold.

Fry Newly hatched or newborn fish.

Genetics The branch of science that deals with the way characteristics pass from one generation to the next.

Genital pore An opening on the underside of the body, marking the entrance to the genital tract.

Genus (plural genera) A group of closely related species.

Gills The main respiratory organs of a fish, located on each side of the head behind and below the eyes. Gills extract dissolved oxygen from water.

Gonopodium The modified anal fin of male livebearers, which is used for mating purposes.

Gravid Describes a female whose body is swollen with eggs or developing young.

Habitat The place where a plant or animal naturally lives.

Half-hardy Describes a plant that is likely to be killed off by frost if left unprotected.

Hand-stripping The manual removal of eggs from a female fish and sperm from a male.

Hard water Fresh water containing a high level of dissolved calcium and magnesium mineral salts, typically in excess of 150 milligrams per liter.

Hardy Describes a plant that can withstand regular exposure to freezing winter temperatures.

Hermaphrodite An animal with both male and female sexual organs.

Hospital tank A small, simply equipped tank that allows sick fish to be treated and recover in isolation from those in the main aquarium.

Hybridization The cross-breeding of different species together.

Hydrometer A device used to measure the specific gravity—and hence the salinity—of water.

Invertebrates Animals without a backbone (vertebral column).

Ion-exchange column A water-softening device that uses resins to remove mineral salts from tap water.

Isolation tank A tank in which new fish are quarantined before being added to the main aquarium, to make sure that they are free from disease.

Labyrinth organs Auxiliary respiratory organs, located close to the gills, that enable anabantoids to breathe air at the water's surface.

Larva (plural larvae) The post-hatching stage in an invertebrate's life cycle. Marine fish fry are often called larvae, since they are poorly developed when they hatch.

Lateral line A fluid-filled canal that runs horizontally along each side of a fish's body. It contains sensory pores that detect pressure changes in the water, aiding navigation and giving early warning of impending danger.

Length A standard dimension for describing fish, taken from the snout to the end of the caudal peduncle, but not including the caudal fin.

Livebearer A fish whose eggs are fertilized and develop inside the body.

Live foods Invertebrates used as fish food. They can be bred at home and fed to the fish alive or purchased in frozen or freeze-dried form.

Live rock Marine rock that harbors many different invertebrate and plant organisms, some of which will not be apparent to the naked eye.

Marginals Plants that can be grown in shallow water around the pond edge (includes so-called bog plants).

Mechanical filtration A method of filtration that relies on straining particulate matter out of the water.

Metabolism The biochemical processes that occur in living things.

Midline The central horizontal axis of a fish's body.

Morph A naturally occurring color variant of a species.

Mouth-brooder An egg-laying fish that incubates and hatches its eggs in the mouth. Some cichlids and anabantoids are mouth-brooders.

Mulm Waste matter, such as fish feces, unwanted food, and decaying plant debris, that builds up on the floor of the aquarium or pond.

Mutation An unexpected change in the genetic makeup of an organism.

Nauplius (plural nauplii) The larval stage in the life cycle of the brine shrimp, which is an important rearing food for fry.

New tank syndrome The sudden illness and/or death of fish that occurs when inefficient filtration allows toxic chemicals, such as nitrite and ammonia, to build up in the water.

Nitrogen cycle The natural process by which nitrogenous waste is recycled in aquatic environments. Ammonia excreted by fish is broken down by bacteria, first into nitrite and subsequently into nitrate. Plants absorb the nitrate and use it for growth. Fish then eat the plants, completing the cycle.

Nuchal hump The swelling on the forehead of mature male cichlids.

Offset A new plant that develops when a runner sets down roots and detaches from the parent plant.

Operculum (plural opercula) The covering over the gills, sometimes called the gill flap.

Ovipositor An extendable tube used by some egg-layers to deposit their eggs at breeding times.

Oxygenators Plants that mostly grow below the surface and give off streams of oxygen bubbles in sunlight.

Parasite An organism that lives on or in the body of a host animal or plant and feeds off it.

Pectoral fins Paired fins, one on each side of the body behind the gills.

Pelvic fins Paired fins on the underside of the body in front of the anal fin.

pH An expression of the hydrogen ion content of water. The pH scale runs from 1 to 14, with a pH of 7 being neutral. Low numbers, indicate acidity, and higher numbers alkalinity.

Pharyngeal teeth Projections in the throat of cyprinids and some other fish, which help to break down food.

Photosynthesis The process by which plants use light energy to make food from carbon dioxide and water, releasing oxygen as a waste product.

Plankton Microscopic plant and animal life found in the sea. Plankton provides a rich source of nourishment for many marine creatures.

Power filter A self-contained filtration system that incorporates a motorized pump to move water through the filter.

Powerhead A small pump that creates surface water movement to improve aeration.

Prepared foods Foods that have been specifically formulated to meet the nutritional needs of fish.

Protein skimmer A filtration device that creates and collects an electrically charged foam, which draws waste products from the water.

Rays The bony supports found in the fins of many different types of fish.

Reverse osmosis (RO) A process used by some water-softening units, in which water is forced through a membrane to remove mineral salts.

Rhizome A swollen plant stem that spreads underground and produces shoots along its length, which then develop above ground.

Rotifer A component of plankton, used by fishkeepers as a first food for rearing marine fish.

Runner A creeping horizontal stem that grows above the ground and on which new plants (offsets) develop.

Salinity The concentration of dissolved mineral salts, especially sodium chloride, in water.

Scales Small protective platelets that cover the bodies of most fish.

School A group of fish that associate together, usually (but not always) of the same species.

Scute A scale modified into a bony plate, found especially in some catfish.

Sessile Attached to a surface.

Sexual dimorphism A difference in the appearance of the sexes. In fish, this may include differences in size, color, patterning, or fin shape.

Shoal A group of fish of the same species that swims together.

Siphon A tube for removing water from an aquarium; also, a tube through which water enters and leaves the body of mussels and other invertebrates.

Soft water Fresh water containing less than 100 milligrams per liter of dissolved calcium and magnesium mineral salts.

Spawning The process of egg-laying and fertilization.

Spawning mop Strands of synthetic yarn attached to a float, on which fish can be persuaded to spawn.

Spawning pit An area of substrate excavated by some species of fish, in which they spawn or guard their fry.

Species A group of animals or plants with similar characteristics that can breed together in the wild to produce fertile offspring.

Specific gravity (SG) The density of a liquid containing dissolved minerals compared to that of pure water, which has an SG of 1.00. SG can be used to measure salinity.

Strain A selectively bred form that has distinctive characteristics.

Substrate The material on the floor of an aquarium, pond, river, or lake.

Subspecies A distinct population within a species.

Swim bladder A fish's buoyancy organ. It may also produce sounds.

Symbiosis A beneficial relationship between two different species.

Taxonomy The study of the naming and classification of living things.

Territorial The readiness of an animal to defend a particular area.

Trace element Minerals, such as iron, that an organism needs in small amounts to ensure its well-being.

Tube feet Projections on the underside of starfish and some other marine invertebrates, which propel the animal over the sea floor.

Tuber The swollen storage organ of a plant, which normally grows at least partially underground.

Tubercles White swellings seen on male cyprinids prior to spawning.

Undergravel filter A filtration system in which a perforated plate sits under the gravel substrate. This allows oxygenated water to flow through the gravel, promoting the growth there of aerobic bacteria, which break down waste matter.

Variety Another word for a strain.

Vent The anogenital opening, which is located close to the anal fin in fish.

Ventral fins An alternative name for the pelvic fins.

Vertebrate An animal with a backbone (vertebral column).

Water conditioner A preparation that makes tap water habitable for fish. It combines a dechlorinator with other ingredients, such as aloe vera.

Yolk sac The part of an egg that nourishes a young fish before and immediately after hatching.

Zeolite A clay-based compound that removes ammonia from water.

Zooxanthellae Single-celled algae that live inside animals such as corals in a symbiotic relationship. The algae grow in the relative safety of the coral's body and provide food for the coral when they photosynthesize.

Useful websites

Most aquatic organizations have websites that give up-to-date contact telephone numbers and mailing addresses. The following is a selection of currently operating sites:

American Cichlid Association
www.cichlid.org
Features news, advice, contacts, and a gallery and chat room for fans of these popular aquarium fish.

American Killifish Association
www.aka.org
Helps to promote the keeping and breeding of these attractive and interesting egg-layers.

American Livebearer Association
www.livebearers.org
Livebearer photos and links.

Aquatic Gardeners Association (AGA)
www.aquatic-gardeners.org/links.html
A list of resources relating to aquatic plants, both for the home aquarium and ponds, with links also to local aquarium clubs in the US and further afield.

Australia New Guinea Fishes Association (ANGFA)
www.angfa.org.au
A site devoted to the freshwater fish of Australia and New Guinea, especially rainbowfish. ANGFA links both hobbyists and scientists.

Betta Source.com
www.bettasource.com
Information and advice on keeping, breeding, and showing bettas.

Canadian Association of Aquarium Clubs (CAOAC)
www.caoac.ca/clubs.html
Details about the wide range of both local and specialist aquatic clubs to be found in Canada.

FishBase
www.fishbase.org/home.htm
An invaluable, comprehensive online taxonomic resource, which also provides a great deal of information about individual species, many of which are illustrated.

Goldfish Society of America
www.goldfishsociety.org
An essential point of reference for fans of these coldwater fish.

International Fancy Guppy Association
www.ifga.org
A well-established organization dedicated to these attractive livebearers, with regional North American clubs and associations listed here.

International Federation of Online Clubs and Aquatic Societies (IFOCAS)
www.ifocas.org
Umbrella body linking organizations in more than 50 countries.

Koivet
www.koivet.com
Site specializing in the health of koi and goldfish; covers koi herpes.

Loaches Online
www.loaches.com
Includes a species database, details of disease treatment, and much more.

Marine Aquarium Societies of North America (MASNA)
www.masna.org
An extensive online resource dealing with all aspects of marine fishkeeping.

North American Native Fishes Association
www.nanfa.org
Covers native species occurring in Canada, the US, and Mexico.

Pet Industry Joint Advisory Council (PIJAC)
www.pijac.org
US-based trade organisation involved with regulatory issues. PIJAC is also represented in Canada.

Planet Catfish
www.planetcatfish.com
In-depth website, with information on L-numbers and breeding.

Index of common and scientific names

General index

Acknowledgments

Author's acknowledgments

Producing a book of this type relies on the skills of many people, and I have been fortunate to be working with two very talented teams at cobalt id and Dorling Kindersley. Many thanks to everyone involved, for your enthusiasm, commitment, and hospitality, which helped to make it such an enjoyable experience. I'd also like to thank the illustrators, as well as all the photographers, especially Max Gibbs and his team at PhotoMax, and Nigel Caddock of Nishikigoi International for their essential input, and the unsung enthusiasts who so readily provided various fish for photographic purposes. They included the late Derek Lambert, who did much to promote the study of livebearers and will be sadly missed. I'd also like to express my gratitude to Marshall Meyers and his colleagues; Michael Kokoscha, editor of the German aquarium magazine *Datz*; my agent, Sheila Watson; and my daughters Isabel and Lucinda for their help.

Publisher's acknowledgments

Cobalt id would like to thank the following for their assistance with this book: Aadithyan Mohan and Vatsal Verma for editorial assistance; Hilary Bird for indexing; Kate Humby for proofreading; Christine Heilman for Americanization; Max Gibbs and Craig Wardrop for their wonderful pictures and invaluable help and advice; Barry Allday at the Goldfish Bowl; the late Derek Lambert for providing hard-to-find fish species to be photographed; Bill Zarnick at Animal Graphics; Nigel Caddock at Nishikigoi International; and Chris Clarke and Kevin Webb of the Anabantoids Association of Great Britain. Thanks also to Neal Cross at Aquadesign; Nicholas Stantifort at Aquarium Technology; Richard Goldberg at Aquarium Design, NY (www.aquariumdesign.com); Tim Gallantree at BiOrb (www.bi-orb.com); Paul Trott and Clair Fitton at Rolf C. Hagen UK Ltd (www.hagen.com); David Zoltowski and Sabine Schulz at Eheim (www.eheim.com); Kylie Arthur at Pet-Mate (www.pet-mate.com); Maryja Johnson at Dransfield Owens de Silva, for kindly providing a picture of their indoor pond (Project architect: Richard Truscott; Interior designer: Marjorie Abéla; Photographer: Rupert Truman); the Press Office, London Aquarium (www.londonaquarium.co.uk); and Chermayeff, Sollogub and Poole, Inc.

Commissioned photography:
PhotoMax/Max Gibbs and Craig Wardrop

Commissioned illustration:
Debbie Maizels, John Plumer

Picture credits

The publisher would like to thank the following for their kind permission to reproduce their photographs:

Abbreviations key: a=above; b=below/bottom; c=center ; f=far; l=left; r=right; t=top. Pictures within boxes contain the word *box* followed by a number indicating position. Pictures in columns are numbered top to bottom; pictures in rows are numbered left to right.

Alamy Stock Photo: Andy Cutler-Davies 2–3; Bruno Cavignaux / Biosphoto 44–45; Malie Rich-Griffith 8–9; Mark Collinson 313bl; Paul Springett 10; pintailpictures 344.

Animal Graphics: Scott W. Michael 16box3, 105cc, 115tr, 120tc, 218box1, 287tr, 289tl, 290bl, 291tl, 291tr, 291bl, 292tl, 292tr, 292br, 295bl, 296br, 297tl; William J. Zarnick 21box6, 86tc, 86b, 105tr, 106tr, 106cr, 108bl, 109tl, 109tr, 110tr, 158tr, 192br, 193tc, 193bl, 193br, 194bl, 196tl, 196tc, 196tr, 197tl, 197bc, 197br, 198bl, 198bc, 199tr, 199br, 213box1, 213box3, 213box6, 297tr, 316bl, 333box1, 338br, 339tr1, 339bl, 340bl, 342bl, 342cr, 373tl.

Aquarium Design/Richard Goldberg: Howard Barash 23tl.

Ardea: Ken Lucas 22tr; Liz Bomford 67tl; Liz & Tony Bomford 64b; Paulo Di Oliviera 156bl.

Bruce Coleman: Hans Reinhard 17tl; Jane Burton 67b; Jim Watt 232tr; Kim Taylor 185br; Luiz Claudio Marigo 14b3; Photobank Yokohama 15br.

Corbis: Brandon D. Cole 20box2; Jeff Albertson 326bl; Kevin R. Morris/Bohemian Nomad Picturemakers 311bl; Richard T. Nowitz 301tl; Vince Streano 323br.

David Alderton: 374br, 375cl, 376tl, 377tr.

Dreamstime.com: Coolkengzz 143tc; Dwight Smith / Dwights 372br; Jesue92 143ca; Nawaj Panichphol 143tl; Sombra12 143cra; Thiradech 142tc.

Frank Lane Picture Agency: Dos Winkel/Foto Natura 14b1; Linda Lewis 143br; Lode Greven / Foto Natura 14b2; Michel Gunther/Foto Natura 23tr; Wil Meinderts/Foto Natura 60tr.

Garden Picture Library: Howard Rice 320tl; J. S. Sira, 368tc; John Baker 368tr; John Glover 373br; Steven Wooster 304tr.

Getty Images: Bryan Mullennix 325; Georgette Douwma 6–7; James F. Housel 314bl; Jeffrey Sylvester 315; Jonelle Weaver 303; Lionel Isy-Schwart 4–5.

iStockphoto.com: Imagenavi 366br; Kororokerokero 364tr.

K. A. Webb: 21box5, 105cr, 105tc, 106tl, 106br, 107bl, 108bc, 109tc, 115tc.

Masterfile: Brad Wrobleski 321; Carl Valiquet 298–299.

NASA (photographersdirect.com): 14tr.

National Geographic Image Collection: Nick Caloyianis 210cc; Wolcott Henry 200–201.

Natural Visions: Heather Angel 47, 318bl, 318br, 319tr, 322br, 326br, 370bc, 371tl, 371bl, 372tr; Ian Took 15bl; Jeff Collett 207b; Norman T. Nicoll 20box3; Soames Summerhays 289tr.

NHPA: B. Jones & M. Shimlock 15bc2; Eric Soder 183tr; Ernie Janes 318tr; Gerard Lacz 19box2; Image Quest 3-D 370tr; Linda Pitkin 231bl; LUTRA 359tr; Pete Atkinson 231tr; Stephen Dalton 370br.

Nigel Caddock: 310tr, 328tr, 329cl, 329cc, 329cr, 329br, 346tl, 346tc, 346tr, 346bl, 346bc1, 346bc2, 346br, 348tl, 348tc, 348tr, 348bl, 348bc1, 348bc2, 348br, 349tl, 349tc, 349tr, 349bl, 349bc, 349br, 350, 351tl, 351tc, 351tr, 351bl, 351bc, 351br, 352tl, 352tc, 352tr, 352bl, 352bc, 352br, 353, 354, 355tl, 355tc, 355tr, 355bl, 355bc1, 355bc2, 355br, 356tl, 356tc, 356tr, 356bl, 356bc1, 356bc2, 356br, 357.

Oxford Scientific Films: Alan Root/SAL 187tr; Larry Crowhurst 371br; Mark Deeble & Victoria Stone 347; OSF 206box2; Paul Kay 19b; Paulo De Oliveira 63tr, 67cl; Tony Bomford 69br.

PhotoMax: 1b, 12–13, 15tr, 15box3, 17tr, 17tc, 17box3, 17br, 18bl, 19box1, 19box3, 19box4, 19box5, 20b, 20tr, 20b, 21box2, 21box4, 21bl, 22bl, 24–25, 26tl, 26tr, 27br, 28bl, 28br, 28tc, 28tr, 29, 30bl, 33br, 35b, 36bl, 37br, 38tr, 40tr, 41br, 44tl, 45tl, 48b, 49tl, 50tr, 50b, 51tl2, 51tr, 52bl, 53, 54tl1, 54tl2, 54tr, 55tr, 55b, 59, 60bl, 61tl1, 61tl2, 61tr, 61br, 62tl, 62br, 63tl, 63bl, 63br, 64tr, 65br1, 65br2, 66tl, 68tr, 69tl1, 69tl2, 69tl3, 69tr, 70bl,

PhotoMax continued: 770bc, 70br, 71tr, 71cr, 71bl, 71bc, 71br, 72–73, 4tr, 74bl, 74br, 75bl, 75br, 75tr, 76tl, 76tr1, 76tr3, 76bc, 76br, 77tl, 77tc, 77b, 78tl, 78tr, 78bc, 79tc, 79br, 80tc, 80bl, 80br, 81t, 81br, 82t, 82tr, 82bc, 83tl, 83tr, 83b, 84tr, 84tl, 84bc, 85tc, 85bc, 87tl, 87tr, 87bc, 88tr, 88bl, 89tr, 89tc, 89br, 90tl, 90tc, 90bl, 91tr, 91bl, 91bc, 92tr, 92bc, 92br, 93tl, 93tc, 93bl, 93br, 94tr, 94bc, 94br, 95tl, 95tc, 95bl, 95br, 96t, 96bl, 97tl, 97tr, 97bl, 98tl, 98bc, 99tl, 99tr, 99br, 100tr, 100bl, 100bc, 100br, 101tc; 101br, 102tc, 102tr, 102bl, 103tl, 103bl, 104tr, 104bc, 104br, 105bl, 105br, 106bl, 107tr, 107bc, 107br, 108tr, 108c, 108br, 109br, 110tl, 110tc, 110bl, 110br, 111tr1, 111tr2, 111br, 112tr, 112bl, 112br, 113tr1, 113bl, 113br, 114tr, 114bc, 115bl, 115br, 116tr, 116bc, 116br, 117tl, 117tc, 117bl, 118tl, 118tr, 118bl, 118br, 119t, 119br, 120bl, 120br, 121tr, 121bc, 121br, 122tr, 122bl, 122br, 123tl, 123tr, 123br, 124tl, 124tr, 124bl, 124br, 125tr, 125bl, 125bc, 125br, 126tl, 126tr, 126bc, 126br, 127tl, 127br, 127tc, 127br, 128tl, 128tc, 128tr, 128bl, 128br, 129t, 129bc, 129br, 130tl, 130tc, 130tr, 130br, 131tl, 131tr, 131bl, 131bc, 131br, 132tl, 132tc, 132tr, 132br, 133tl, 133tr, 133bl, 133br, 134tl, 134tr, 134bl, 134br, 135tl, 135tr, 135b, 136tr, 136bl, 136br, 137tl, 137tr, 137br, 138tl, 138tr, 138b, 139tl, 139tr1, 139tr2, 139bl, 139br, 140tc, 140tr, 140b, 141tl, 141tr, 141cr, 141br, 142tr, 142b, 143tr, 143bl, 144tl, 144tr, 144bl, 145tr, 145bl, 145br1, 145br2, 146tc, 146tr, 146br, 147tl, 147tr, 147bl, 147bc, 147br, 148tl, 148tr, 148bl, 148bc, 148br, 149tl, 149tc, 149tr, 149br, 150tl, 150tc, 150tr, 150br, 151tl, 151tc, 151tr, 151bl, 151br, 152tl, 152tc, 152tr, 152bl, 152br, 153tr, 153bl, 153bc, 153br, 154tr, 154bl, 154br, 155bl, 156tr, 156bc, 157tr, 157bl, 157br, 158tl, 158bl, 158br, 159tl, 159tr, 159bc, 160tr, 160bl, 161tl, 161tr, 161bl, 162tr, 162cc, 162b, 163tr, 163b, 164tr, 164cl, 164br, 165tr, 165tc, 165bl, 165br, 166tr, 166bl, 166br1, 166br2, 167tl, 167tr, 167c, 167bl, 167bc, 167br, 168tl, 168tr, 168b, 169tr,

169bl, 170tr, 170bc, 171tc, 171bl, 171br, 172tl, 172tr, 172bl, 172bc, 173tl, 173tc, 173tr, 173b, 174tl, 174bl, 174br, 175tl, 175tc, 175br, 176tr, 176bc, 176br, 177tl, 177tc, 177tr, 177bl, 177br, 178tr, 178bc, 179tc, 179bl, 179br, 180tr, 180bl, 180br, 181tc, 181bl, 181br, 182tr, 182bl, 182br, 183tl, 183bl, 184tr, 184bl, 184br, 185tc, 185bl, 185br, 186tr, 186bc, 187tl, 187tc, 187b, 188tr, 188bc, 189tl, 189tc, 189tr, 189bl, 189br, 190–191, 192tr, 194tr, 194bc, 194br, 195tr, 195b, 196br, 197tl, 198tr, 199bl, 202cr, 202bl, 203cl, 203b, 204cl, 204br, 205, 206bl1, 207tl1, 207tl2, 212tr, 213box2, 213box4, 213box5, 213b, 214tr, 217, 218b, 219bl, 220tr, 223, 224tr, 224tl, 224box1, 224box2, 225bl, 229, 230tr, 230bl, 232l1, 232l2, 232l3, 232l4, 233br, 234–235, 236tr, 236bl, 236bc, 236br, 237tc, 237tr, 237bl, 237br, 238tl, 238tr, 238bl, 238br, 239tl, 239tc, 239tr, 239br, 240tr, 240bc, 241tc, 241tr, 241bl, 241br, 242tr, 242bl, 242br, 243tl, 243tc, 243tr, 243bl, 243br, 244tr, 244bl, 244bc, 244br, 245tl, 245tr, 245bl, 245br, 246tr, 246bl, 246bc, 246br, 247tl, 247tr, 247bl, 247br, 248tr, 248bl, 248bc, 248br, 249tl, 249tr, 249bl, 249br, 250tr, 250bl, 250bc, 250br, 251tc, 251tr, 251bl, 251br, 252tr, 252bc, 253tc, 253bl, 253br, 254tl, 254tr, 254bl, 254bc, 254br, 255tl, 255tc, 255tr, 255br, 256tr, 256bl, 256bc, 256br, 257tc, 257tr, 257bl, 257br, 258tl, 258tc, 258tr, 258br, 259tr, 259bl, 259bc, 259br, 260tr, 260bl, 260bc, 260br, 261tl, 261tr, 261tc, 261bl, 261br, 262tr, 262bl, 262bc, 262br, 263tl, 263tc, 263bl, 263br, 264tr, 264bl, 265tl, 265tr, 265bl, 265bc, 265br, 266tr, 266bl, 266bc, 266br, 267bl, 267tc, 267tr, 267br, 268tr, 268bc, 268br, 269tl, 269tc, 269tr, 269bl, 269br, 270tl, 270tr, 270bc, 270br, 271tr, 271bl, 271br, 272tl, 272tr, 272tc, 272bl, 272br, 273tr, 273bl, 273bc, 273br, 274tl, 274tr, 274bl, 274br, 275tr, 275bl, 275bc, 276tl, 276tr, 276bl, 276br, 277tl, 277tr, 277bl, 277br, 278tl, 278tr, 278bl, 278br, 279tl, 279bl, 279br, 280tr, 280bl, 280bc, 280br, 281tl, 281tc, 281tr, 281bl, 281br, 282tl, 282tc, 282tr, 282bl, 282br, 283tr, 283bl,

283bc, 283br, 286tr, 286bl, 286bc, 286br, 287tl, 287bl, 287bc, 287br, 288tr, 288bl, 288bc, 288br, 289bl, 289br, 290tr, 290bc, 290br, 291tc, 291br, 292bl, 293tr, 293bl, 293br, 294tr, 294bl, 294br, 295tl, 295tc, 295tr, 295br, 296bl, 296tr, 296bc, 297tc, 297br, 308b3, 312tr, 312bl, 313tr, 316tr, 317tl, 317br, 322bl, 322br1, 322br2, 323tl, 326tr1, 326tr2, 327br, 327tr2, 327tr3, 330–331, 332tr, 332cr, 332bl, 333tr, 333br2, 334, 335tr1, 335tr2, 335tr3, 335br, 336tl, 336tr, 336bc, 337, 338tl, 339tr2, 339br2, 340tr, 340br1, 340br2, 341, 342tl, 342tr, 342br2, 343tl, 343tr, 343bl, 343br, 358tr, 358bl, 359tl, 359tc, 359bl, 360tc, 360bl, 360br1, 361tr1, 361tr2, 361tr3, 361bl, 361br1, 362–363, 371tr, 372bl, 374tr.

Red Cover: Chris Tubbs 208tr; Ken Hayden 300tr.

Reuters Picture Agency: Simon Kwong 70tr.

Science Photo Library: John Walsh 233bl; Peter Scoones 20box11, 155tl; Tom McHugh 66br.

seneye.com: 46cr.

The Bridgeman Art Library: Kobayashi Eitaku 22br.

Uwe Werner: 112 tc.

Every effort has been made to trace the copyright holders. The publisher apologizes for any unintentional omissions and would be pleased, in such cases, to place an acknowledgment in future editions of this book.

All other images © Dorling Kindersley
For further information, see: www.dkimages.com

AUTHOR

David Alderton is a widely respected expert when it comes to the care and breeding of aquarium and pond fish, and he shares his knowledge in this field widely, both as a writer and a broadcaster. Based in the UK, he has worked with various organizations in this field, including the Pet Industry Joint Advisory Council (PIJAC) headquartered in Washington, DC. His interest in fish and their care began in childhood with a pet goldfish, and he has retained a lifelong fascination with members of this group and koi, even traveling to parts of China and Japan to view them there. David's broad experience also extends to tropical fish—he has kept and bred many species—and in the area of aquatic plants, he is particularly fascinated by water lilies.

PHOTOGRAPHER

Max Gibbs, the principal photographer for this book, first kept fish at the age of 11. The pastime became a business when he started a retail aquarium in the mid-1950s, which grew to become The Goldfish Bowl of Oxford. He took up photography in 1990 and has built one of the world's largest collections of high-quality marine and freshwater fish photographs, the Photomax Picture Library.

CONSULTING EDITORS

Marshall Meyers is the Senior Advisor and former CEO of PIJAC in Washington, DC. He has very extensive knowledge of the aquatic industry, both personally and through his board members, who represent a number of the most significant companies operating in the supply of fish of all types, as well as of fishkeeping equipment.

Timothy Hovanec, PhD, is the president of DrTim's Aquatics, a company that provides natural products for aquariums and ponds. He was the Chief Science Officer for Marineland Laboratories, Moorpark, California. A keen fishkeeper himself, his expertise also extends to the fields of public aquariums and aquaculture, and he is the author of numerous scientific and popular articles on microbial ecology and fishkeeping topics.

Bob Weintraub was actively involved in fishkeeping for more than 55 years, initially as a hobbyist breeding rare species. He was a founding partner of Aquarium Pharmaceuticals, an international manufacturer of fish medicines and aquarium and pond treatments. Toward the end of his career, he was engaged both in the field of pond and aquarium design, and as a koi consultant in the US and Japan